Fasting Superfoods

Anil Kumar • Jitendra Kumar Tiwari
Dinesh Chandra Joshi • Rajeev K. Varshney
Editors

Fasting Superfoods

Cultivation, Nutrition & Market Potential

Springer

Editors
Anil Kumar
Plant Molecular Biology and Biotechnology
Rani Lakshmi Bai Central Agricultural University
Jhansi, Uttar Pradesh, India

Jitendra Kumar Tiwari
Genetics and Plant Breeding
Rani Lakshmi Bai Central Agricultural University
Jhansi, Uttar Pradesh, India

Dinesh Chandra Joshi
Crop Improvement Division
Vivekananda Parvatiya Krishi Anusandhan Sansthan
Almora, Uttarakhand, India

Rajeev K. Varshney
Centre for Crop and Food Innovation
Murdoch University
Murdoch, WA, Australia

ISBN 978-981-95-2761-8 ISBN 978-981-95-2762-5 (eBook)
https://doi.org/10.1007/978-981-95-2762-5

This Springer imprint is published by the registered company Springer Nature Singapore Pte Ltd.
The registered company address is: 152 Beach Road, #21-01/04 Gateway East, Singapore 189721, Singapore

Foreword

In an era defined by rapid urbanization, poor dietary habits, and mounting health challenges, the revival of traditional food systems offers a powerful pathway toward nutritional security, cultural preservation, and environmental resilience. Among these, the practice of fasting—deeply rooted in India's spiritual and cultural heritage—stands out not only for its ritualistic significance but also for its resonance with modern scientific insights into health and well-being.

This timely and insightful volume, *Fasting Superfoods: Cultivation, Nutrition, and Market Potential*, addresses a vital yet often overlooked aspect of sustainable food systems: the underutilized, nutrient-rich crops that form the foundation of traditional fasting diets. These include pseudocereals, millets, root crops, minor fruits, oilseeds, and medicinal plant species—crops deeply embedded in India's agrarian and cultural landscape, yet many of which remain underexploited in mainstream agriculture and food markets.

Bringing together diverse perspectives from agronomy, plant genetics, nutrition, ethnobotany, and market analysis, the editors and contributors present a comprehensive examination of 19 such crops. Each chapter opens a window into the unique value of these species, illuminating their ecological adaptability, genetic potential, traditional significance, and commercial promise. In doing so, the book serves as an indispensable reference for researchers, policymakers, agri-entrepreneurs, and all who are interested in shaping sustainable, health-conscious food systems.

What makes this work especially commendable is its seamless integration of ancient wisdom with cutting-edge science. By interweaving Ayurveda, indigenous knowledge, and modern disciplines such as plant breeding and nutritional science, the book offers a holistic framework that is both academically rigorous and socially relevant. At a time when the global community seeks climate-resilient agricultural solutions and functional nutrition strategies, this book provides both inspiration and a practical roadmap.

I extend my heartfelt congratulations to the editors and contributors for their dedication to reviving and revaluing these fasting superfoods. This work is not only a scholarly contribution to food and agricultural sciences but also a meaningful step towards restoring harmony in the way we grow, consume, and relate to food.

Vice Chancellor
Rani Lakshmi Bai Central Agricultural University
Jhansi, Uttar Pradesh, India

Ashok Kumar Singh

Preface

Nearly 7,000 plant species are known to be edible to humans, yet today more than 75% of the world's food supply is derived from just 12 major crops. Wheat, rice, and maize alone account for nearly half of global dietary intake. This narrowing of agricultural biodiversity heightens our vulnerability to climate risks and nutritional deficiencies, while eroding traditional food systems that have evolved over millennia through ecological and cultural adaptation. To secure long-term food and nutritional security, it is essential to diversify both diets and agricultural practices by promoting resilient, underutilized, and nutritionally rich crops.

India holds a rich tradition of linking food with spiritual and health practices, with fasting as a central pillar of this harmony. Rooted in Ayurveda and Hindu philosophy, fasting (*vrata*) is believed to balance the body's energies (*Vata*, *Pitta*, and *Kapha*), cleanse the digestive system, and promote longevity. Aligned with lunar and solar cycles, these traditions have given rise to dietary systems that emphasize *satvik* foods—light, nutrient-rich, and easily digestible ingredients that nourish both body and mind.

Modern science is now affirming the value of these ancient practices. The Nobel Prize-winning research of Dr. Yoshinori Ohsumi on autophagy has shown how fasting triggers cellular repair, slows aging, and protects against various diseases. In this light, the age-old wisdom of fasting gains renewed relevance in global health discussions. At the heart of these fasting traditions lies a diverse array of crops that are not only culturally significant but also nutritionally superior and environmentally resilient—collectively referred to in this book as "fasting superfoods."

Fasting Superfoods: Cultivation, Nutrition, and Market Potential is a pioneering effort to document, analyze, and promote these crops. It explores 19 plant species that form an integral part of traditional fasting diets in India. From pseudocereals and millets to root crops, minor fruits, herbs, and functional foods, the book offers an interdisciplinary perspective on their historical roles, cultivation methods, genetic potential, nutritional attributes, and emerging commercial prospects. These crops are increasingly valued for their adaptability to marginal environments, rich nutraceutical profiles, and compatibility with organic and low-input farming systems.

The book is organized into five thematic parts:

- *Part I* covers oilseeds such as groundnut (*Arachis hypogaea*) and linseed—protein-rich crops often consumed during fasting.

- *Part II* focuses on pseudocereals and minor millets, including amaranth, buckwheat, water chestnut, and makhana, which are gaining popularity as gluten-free, nutrient-dense alternatives, alongside barnyard millet and proso millet, noted for their fast growth, climate resilience, and health benefits.
- *Part III* discusses minor vegetables, tubers, and roots, such as *Kalingda* (*Citrullus colocynthis*), a desert melon with medicinal value, as well as sweet potato, tapioca, and *tikhur*, which serve as key energy sources during fasting.
- *Part IV* is dedicated to minor fruits and trees, including mahua and amla, celebrated for their therapeutic properties.
- *Part V* concludes with herbs and functional crops—such as poppy seeds, saffron, tulsi, chia seeds, and halim seeds—that bridge ancient uses with modern nutritional science.

Each of the 19 chapters has been authored by experts in agriculture, nutrition, food science, and traditional knowledge systems. They present a comprehensive account of each crop's origin, traditional usage, agronomic traits, genetic resources, nutritional profile, and potential for value addition and commercialization. While every effort has been made to ensure coherence and avoid redundancy, some overlap has been retained where necessary to preserve thematic and contextual continuity.

We extend our sincere gratitude to all contributing authors for their scholarly inputs, timely submissions, and constructive revisions. We also acknowledge with deep appreciation the patience and support of our families throughout the editorial process.

We hope this book will serve as a valuable resource for students, researchers, policymakers, agri-entrepreneurs, and health-conscious consumers. By promoting awareness, innovation, and investment in these underutilized fasting superfoods, we aim to contribute to a future where tradition, nutrition, sustainability, and resilience are harmoniously aligned.

Jhansi, Uttar Pradesh, India — Anil Kumar
Jhansi, Uttar Pradesh, India — Jitendra Kumar Tiwari
Almora, Uttarakhand, India — Dinesh Chandra Joshi
Murdoch, WA, Australia — Rajeev K. Varshney

Contents

Part I Oily Ferus Crops

1 **Groundnut (*Arachis hypogaea*): The Protein-Rich Oilseed of Adaptability and Nutrition** ... 3
Jitendra Kumar Tiwari, Yogendra Kumar Mishra, Bhavna Tiwari, Anjali Singh, Dinesh Chandra Joshi, and Anil Kumar

2 ***ALSI*: Omega-3-Rich Source for Health and Vitality** ... 27
Sonal Tiwari, Ashish Kumar Tiwari, and Anil Kumar

Part II Pseudocereals and Minor Millets

3 **Grain Amaranth (*Amaranthus* spp.): The Sacred Seed of Ancient Civilizations** ... 45
Navjot Kaur, Shubham Verma, and Satbeer Singh

4 **Buckwheat (*Fagopyrum esculentum*): The Resilient Crop with Nutritional Riches** ... 77
Rintu Jha, Bandana Devi, Akanksha Singh, and Prashant Singh

5 **Water Chestnut (*Trapa natans* L.): The Aquatic Treasure of Traditional Diets** ... 103
Rima Kumari, Pankaj Kumar, and Anil Kumar Singh

6 **Makhana (*Euryale ferox*): The Super Snack from Water** ... 115
Pankaj Kumar, Rima Kumari, and Anil Kumar Singh

7 **Barnyard Millet (*Echinochloa* spp.): The Fast-Growing Nutritious Staple** ... 125
Shivam Yadav, Rumana Khan, Mitali Tiwari, Narendra Kumar Singh, Silpa Sahoo, Sushrut Kumar Samanta, and Jitendra Kumar Tiwari

8 **Proso Millet (*Panicum miliaceum*): The Prodigious Millet for Diverse Uses** ... 155
C. Nandini, Dinesh Chandra Joshi, Vetriventhan Mani, B. Nandini, B. Niranjan Kumar, and J. Meenakshi

Part III Minor Vegetable, Tuber and Root Crops

9 ***Kalingda (Citrullus colocynthis)*: The Desert Melon with Medicinal Power** . . . 187
Devendra Upadhyay, Ashish Kumar Tiwari, Sonal Tiwari, Anjali Sharma, and Jitendra Kumar Tiwari

10 **Sweet Potato (*Ipomoea batatas* L.): The Nutrient-Rich Root of Resilience** . . . 221
Devendra Upadhyay, Padmakshi Thakur, Oshin Pali, and Ashish Kumar Tiwari

11 **Sago (Tapioca) *Manihot esculenta*: The Starchy Sustainer of Energy** . . . 261
Sasmita Priyadarsini Dash, Prachi Pattnaik, Nitish Kumar Jena, Padmakshi Thakur, and Jitendra Kumar Tiwari

12 **Tikhur (East Indian Arrowroot) (*Curcuma angustifolia* Roxb.): The Starchy Root of Versatile Nutrition** . . . 285
Padmakshi Thakur, Sasmita Priyadarsini Dash, Omesh Thakur, Vikas Ramteke, Upendra Naik, Devendra Upadhyay, Rohit Maurya, and Jitendra Kumar Tiwari

Part IV Minor Fruits and Trees

13 **Aonla (Indian Gooseberry) (*Emblica officinalis* Gaertn.): The Vitamin C Powerhouse for Immunity** . . . 303
Neetu Saroj, Manoj Kundu, Sukanya Misra, and Ranjit Pal

14 **MAHUA (*Madhuca indica*): The Multifunctional Tree of Tradition and Sustenance** . . . 329
Shalini Purwar, Vinita Bisht, and Ena Gupta

Part V Herbs and Functional Crops

15 **Poppy Seeds (Khus Khus) (*Papaver somniferum*): The Tiny Seed of Flavor and Nutrient Density** . . . 361
Sivendra Joshi, K. T. Venkatesha, Anil Kumar Gupta, Rajendra Chandra Padalia, Rakesh Kumar Upadhyay, Dipender Kumar, Amit Chauhan, Amit Kumar Tiwari, and Sonveer Singh

16 **Saffron (*Crocus sativus* L.): Gold of the Spices** . . . 381
Mudasir Hafiz Khan, Niyaz Ahmad Dar, Sher Ahmad Dar, Ghulam Hassan Mir, Firdous Ahmad Nehvi, Bashir Ahmad Alie, Mehraj Din Sofi, Uzma Fayaz, Ulfat Jan, and Tawkeer Nabi

17 Tulsi (Holy Basil) (*Ocimum tenuiflorum*): The Sacred Herb of Wellness 415
Channayya Hiremath and B. S. Chandana

18 Chia Seeds (*Salvia hispanica*): The Tiny Powerhouse of Omega-3 and Fiber 435
Ashish Kumar Tiwari, Sonal Tiwari, Jitendra Kumar Tiwari, and Dinesh Chandra Joshi

19 Halim Seeds (Garden Cress): The Ancient Seed of Iron and Immunity 453
Sarita Agrawal, Bhavana Tiwari, and Aprajita Tiwari

Editors and Contributors

About the Editors

Anil Kumar is the Director of Education at Rani Lakshmi Bai Central Agricultural University, Jhansi, India. With 30 years of experience in research, teaching, and administration, he has held key academic positions in molecular biology and genetic engineering. His expertise includes plant molecular and cell biology, functional genomics, and immunodiagnostics in agriculture. He has authored over 250 publications, including books and research papers, and received multiple national awards for his contributions to biotechnology education and agricultural research.

Jitendra Kumar Tiwari is Associate Professor and Head of Genetics and Plant Breeding at Rani Lakshmi Bai Central Agricultural University, Jhansi, India. With a PhD from IGKV, Raipur, he specializes in crop improvement, varietal development, and germplasm research. He has developed 12 crop varieties, registered two germplasm lines, and published 70 research articles and a book. His work has been recognized with awards like the Young Scientist Award and BARC Young Scientist Research Award.

Dinesh Chandra Joshi is a Senior Scientist in the Crop Improvement Division at ICAR-VPKAS, Almora, India. Leading the Minor Millets and Pseudocereals Improvement Programme, he has developed eight cultivars and registered seven genetic stocks. His expertise includes genomics, trait mapping, and genome sequencing. He has published over 40 peer-reviewed articles and edited a book on millets. He has received awards such as the IACBG Fellowship and the Best Performing Centre Award for millet research.

Rajeev K. Varshney is a Professor at the Food Futures Institute, Murdoch University, Australia. With over 20 years of experience in genomics and molecular breeding, he has worked across multiple countries, contributing to global food security. He has developed genomic resources for tropical crops, improving drought and pest tolerance. With over 600 publications, he has been recognized as a Clarivate Highly Cited Researcher (2014–2023) and has received numerous international awards for his pioneering research.

Contributors

Sarita Agrawal Indira Gandhi Agricultural University, Raipur, Chhattisgarh, India

Bashir Ahmad Alie Advanced Research Station for Saffron and Temperate Seed Spices, SKUAST-Kashmir, Srinagar, India

Vinita Bisht Department of Silviculture and Agroforestry, Banda University of Agriculture and Technology, Banda, Uttar Pradesh, India

Amit Chauhan Division of Plant Breeding and Genetic Resources Conservation, CSIR-Central Institute of Medicinal and Aromatic Plants (CIMAP), Research Centre, Pantnagar, Uttarakhand, India

B. S. Chandana Divisional Unit, Plant Breeding and Genetic Resource Conservation, CSIR-Central Institute of Medicinal and Aromatic Plants, Research Centre, Bangalore, India

Niyaz Ahmad Dar Advanced Research Station for Saffron and Temperate Seed Spices, SKUAST-Kashmir, Srinagar, India

Sher Ahmad Dar Advanced Research Station for Saffron and Temperate Seed Spices, SKUAST-Kashmir, Srinagar, India

Sasmita Priyadarsini Dash IAS, Siksha 'O' Anusandhan Deemed to be University, Bhubaneswar, Odisha, India

Bandana Devi Department of Botany, Institute of Science, Banaras Hindu University, Varanasi, Uttar Pradesh, India

Channayya Hiremath Divisional Unit, Plant Breeding and Genetic Resource Conservation, CSIR-Central Institute of Medicinal and Aromatic Plants, Research Centre, Bangalore, India

Uzma Fayaz Advanced Research Station for Saffron and Temperate Seed Spices, SKUAST-Kashmir, Srinagar, India

Anil Kumar Gupta Division of Plant Breeding and Genetic Resources Conservation, CSIR-Central Institute of Medicinal and Aromatic Plants (CIMAP), Research Centre, Pantnagar, Uttarakhand, India

Division of Plant Breeding and Genetic Resources Conservation, CSIR-Central Institute of Medicinal and Aromatic Plants (CIMAP), Near Kukrail Picnic Spot, Lucknow, India

Ena Gupta Department of Community Science, University of Allahabad, Prayagraj, Uttar Pradesh, India

Ulfat Jan Advanced Research Station for Saffron and Temperate Seed Spices, SKUAST-Kashmir, Srinagar, India

Nitish Kumar Jena IAS, Siksha 'O' Anusandhan deemed to be University, Bhubaneswar, Odisha, India

Rintu Jha State Key Laboratory of Crop Gene Resources and Breeding, Institute of Crop Sciences, Chinese Academy of Agricultural Sciences, Beijing, China

Dinesh Chandra Joshi Crop Improvement Division, Vivekananda Parvatiya Krishi Anusandhan Sansthan, Almora, Uttarakhand, India

Sivendra Joshi Division of Plant Breeding and Genetic Resources Conservation, CSIR-Central Institute of Medicinal and Aromatic Plants (CIMAP), Research Centre, Pantnagar, Uttarakhand, India

Navjot Kaur Academy of Scientific and Innovative Research, Ghaziabad, Uttar Pradesh, India

Agrotechnology Division, Council of Scientific and Industrial Research- Institute of Himalayan Bioresource Technology, Palampur, Himachal Pradesh, India

Mudasir Hafiz Khan Division of Genetics and Plant Breeding, Faculty of Agriculture, SKUAST-Kashmir, Wadura, Sopore, J&K, India

Rumana Khan Rani Lakshmi Bai Central Agricultural University, Jhansi, UP, India

Anil Kumar Rani Lakshmi Bai Central Agricultural University, Jhansi, Uttar Pradesh, India

Dipender Kumar Division of Plant Breeding and Genetic Resources Conservation, CSIR-Central Institute of Medicinal and Aromatic Plants (CIMAP), Research Centre, Pantnagar, Uttarakhand, India

Rima Kumari Department of Molecular Biology and Genetic Engineering, Bihar Agricultural University, Sabour, Bihar, India

Division of Plant Biotechnology, College of Agricultural Biotechnology, Bihar Agricultural University, Sabour, Bihar, India

Pankaj Kumar Department of Molecular Biology and Genetic Engineering, Bihar Agricultural University, Sabour, Bihar, India

Division of Plant Biotechnology, College of Agricultural Biotechnology, Bihar Agricultural University, Sabour, Bihar, India

Manoj Kundu Department of Fruit Science, College of Horticulture & Forestry, Rani Lakshmi Bai Central Agricultural University, Jhansi, Uttar Pradesh, India

Vetriventhan Mani International Crops Research Institute for the Semi-Arid Tropics (ICRISAT), Hyderabad, India

Rohit Maurya KVK Turki, DRPCAU, Muzaffarpur, Bihar, India

J. Meenakshi AICRP on Small Millets, University of Agricultural Sciences, GKVK, Bengaluru, Karnataka, India

Ghulam Hassan Mir Advanced Research Station for Saffron and Temperate Seed Spices, SKUAST-Kashmir, Srinagar, India

Yogendra Kumar Mishra Rani Lakshmi Bai Central Agricultural University, Jhansi, Uttar Pradesh, India

Sukanya Misra Department of Fruit Science, College of Horticulture & Forestry, Rani Lakshmi Bai Central Agricultural University, Jhansi, Uttar Pradesh, India

Tawkeer Nabi Advanced Research Station for Saffron and Temperate Seed Spices, SKUAST-Kashmir, Srinagar, India

Upendra Naik S.G. College of Agriculture and Research Station, IGKVV, Jagdalpur, Chhattisgarh, India

B. Nandini College of Horticulture, Kolar University of Horticultural Sciences, Bagalkot, Karnataka, India

C. Nandini Zonal Agricultural and Horticultural Research Station, Babbur Farm, Hiriyur, KSNUAHS, Shivamogga, Karnataka, India

Firdous Ahmad Nehvi Division of Genetics and Plant Breeding, Faculty of Agriculture, SKUAST-Kashmir, Wadura, Sopore, J&K, India

B. Niranjan Kumar Agricultural and Horticultural Research Station, Kattalagere, KSNUAHS, Shivamogga, Karnataka, India

Rajendra Chandra Padalia Division of Plant Breeding and Genetic Resources Conservation, CSIR-Central Institute of Medicinal and Aromatic Plants (CIMAP), Research Centre, Pantnagar, Uttarakhand, India

Ranjit Pal Department of Fruit Science, College of Horticulture & Forestry, Rani Lakshmi Bai Central Agricultural University, Jhansi, Uttar Pradesh, India

Oshin Pali Dau Kalyan Singh College of Agriculture and Research Station, Indira Gandhi Krishi Vishwavidyalaya, Bhatapara, Chhattisgarh, India

Prachi Pattnaik IAS, Siksha 'O' Anusandhan Deemed to be University, Bhubaneswar, Odisha, India

Shalini Purwar Department of Basic and Social Science, Banda University of Agriculture and Technology, Banda, Uttar Pradesh, India

Vikas Ramteke S.G. College of Agriculture and Research Station, IGKVV, Jagdalpur, Chhattisgarh, India

Silpa Sahoo Rani Lakshmi Bai Central Agricultural University, Jhansi, UP, India

Sushrut Kumar Samanta Rani Lakshmi Bai Central Agricultural University, Jhansi, UP, India

Neetu Saroj Department of Fruit Science, College of Horticulture & Forestry, Rani Lakshmi Bai Central Agricultural University, Jhansi, Uttar Pradesh, India

Akanksha Singh Institute of Biosciences and Biotechnology, Chhatrapati Shahu Ji Maharaj University Campus, Kanpur, Uttar Pradesh, India

Anil Kumar Singh Directorate of Research, Bihar Agricultural University, Sabour, Bihar, India

Narendra Kumar Singh Govind Ballabh Pant University of Agriculture and Technology, Uttarakhand, India

Anjali Sharma Rani Lakshami Bai Central Agricultural University, Jhansi, Uttar Pradesh, India

Anjali Singh Rani Lakshmi Bai Central Agricultural University, Jhansi, Uttar Pradesh, India

Prashant Singh Department of Botany, Institute of Science, Banaras Hindu University, Varanasi, Uttar Pradesh, India

Satbeer Singh Agrotechnology Division, Council of Scientific and Industrial Research- Institute of Himalayan Bioresource Technology, Palampur, Himachal Pradesh, India

Sonveer Singh Division of Plant Breeding and Genetic Resources Conservation, CSIR-Central Institute of Medicinal and Aromatic Plants (CIMAP), Research Centre, Pantnagar, Uttarakhand, India

Mehraj Din Sofi Division of Genetics and Plant Breeding, Faculty of Agriculture, SKUAST-Kashmir, Wadura, Sopore, J&K, India

Omesh Thakur College of Agriculture and Research Station, IGKV, Gariyaband, CG, India

Padmakshi Thakur S.G. College of Agriculture and Research Station, IGKV, Jagdalpur, Chhattisgarh, India

Amit Kumar Tiwari Division of Plant Breeding and Genetic Resources Conservation, CSIR-Central Institute of Medicinal and Aromatic Plants (CIMAP), Research Centre, Pantnagar, Uttarakhand, India

Aprajita Tiwari Ayurvedic Medical Practitioner, Jhansi, Uttar Pradesh, India

Ashish Kumar Tiwari Indira Gandhi Krishi Vishwavidyalaya, Raipur, Chhattisgarh, India

Sant Vinova Bhave College of Agriculture and Research Station, IGKV, Marra, Patan, Chhattisgarh, India

Bhavna Tiwari Rani Lakshmi Bai Central Agricultural University, Jhansi, Uttar Pradesh, India

Jitendra Kumar Tiwari Genetics and Plant Breeding, Rani Lakshmi Bai Central Agricultural University, Jhansi, Uttar Pradesh, India

Mitali Tiwari Govind Ballabh Pant University of Agriculture and Technology, Uttarakhand, India

Sonal Tiwari Mahatma Gandhi University of Horticulture and Forestry, Durg, Chhattisgarh, India

Mahatma Gandhi Udyaniki Evam Vaniki Vishwavidyalaya, Durg, Chhattisgarh, India

Department of Genetics and Plant breeding, Mahatma Gandhi Udyaniki Evam Vanki Vishwavidyalaya, Sankra-Patan, Chhattisgarh, India

Devendra Upadhyay Dau Kalyan Singh College of Agriculture and Research Station, Indira Gandhi Krishi Vishwavidyalaya, Bhatapara, Chhattisgarh, India

Rakesh Kumar Upadhyay Division of Plant Breeding and Genetic Resources Conservation, CSIR-Central Institute of Medicinal and Aromatic Plants (CIMAP), Research Centre, Pantnagar, Uttarakhand, India

K. T. Venkatesha Division of Plant Breeding and Genetic Resources Conservation, CSIR-Central Institute of Medicinal and Aromatic Plants (CIMAP), Research Centre, Pantnagar, Uttarakhand, India

Shubham Verma Academy of Scientific and Innovative Research, Ghaziabad, Uttar Pradesh, India

Agrotechnology Division, Council of Scientific and Industrial Research- Institute of Himalayan Bioresource Technology, Palampur, Himachal Pradesh, India

Shivam Yadav Govind Ballabh Pant University of Agriculture and Technology, Uttarakhand, India

Part I

Oily Ferus Crops

1 Groundnut (*Arachis hypogaea*): The Protein-Rich Oilseed of Adaptability and Nutrition

Jitendra Kumar Tiwari, Yogendra Kumar Mishra, Bhavna Tiwari, Anjali Singh, Dinesh Chandra Joshi, and Anil Kumar

Abstract

Groundnut (*Arachis hypogaea* L.), a widely cultivated legume and oilseed, is a potent source of plant-based protein, healthy fats, essential vitamins, and bioactive compounds, making it an ideal candidate for inclusion in fasting diets and functional foods. Known for its adaptability to diverse agroclimatic conditions and its ecological contribution through nitrogen fixation, groundnut plays a dual role in supporting both human health and sustainable agriculture. This chapter explores the nutritional and therapeutic attributes of groundnut relevant to fasting regimens, including its low glycemic index, satiety-inducing properties, and cardioprotective benefits. The crop's rich composition of monounsaturated fatty acids (MUFA), dietary fiber, antioxidants (resveratrol, tocopherols), and essential amino acids contributes to improved metabolic health, sustained energy release, and reduced oxidative stresskey considerations during intermittent and prolonged fasting. Groundnut-based products such as roasted kernels, nut butter, and flour serve as convenient and nutrient-dense options during fasting, especially in vegetarian and vegan diets. Additionally, the chapter highlights the cultural, botanical, and varietal diversity of groundnut, alongside recent advancements in postharvest handling, value addition, and its growing role in global health food markets. Emphasis is placed on its utility in dryland farming systems, where groundnut supports food, nutrition, and economic security.

J. K. Tiwari (✉) · Y. K. Mishra · B. Tiwari · A. Singh · A. Kumar
Rani Lakshmi Bai Central Agricultural University, Jhansi, Uttar Pradesh, India
e-mail: jitendratiwari.gpb@rlbcau.ac.in

D. C. Joshi
Crop Improvement Division, Vivekananda Parvatiya Krishi Anusandhan Sansthan, Almora, Uttarakhand, India

A. Kumar et al. (eds.), *Fasting Superfoods*,
https://doi.org/10.1007/978-981-95-2762-5_1

Keywords

Groundnut · Fasting nutrition · Monounsaturated fatty acids · Functional foods and dryland crop

1.1 Introduction

Arachis hypogaea L., commonly referred to as peanut, groundnut, or *moongphali* in Hindi, belongs to the Leguminosae/Fabaceae family (Krapovickas and Gregory 1994). *Arachis hypogaea* Linn, the scientific name for groundnut, is a combination of two Greek words: *Arachis*, which means legume, and *hypogaea*, which means below ground and refers to the development of pods in the soil. It is an annual plant that can be either upright or prostrate. It is thought to have started in the South American region that stretches from southern Bolivia to northern Argentina. With only two allotetraploids, the majority of the 80 species in the genus *Arachis* are diploid (2n = 2x = 20). The sole species in the genus that is grown is *Arachis hypogaea* L. It is allotetraploid (AABB, 2n = 4x = 40) and is thought to be the product of chromosome doubling and hybridization between two wild species: *Arachis duranensis* (AA-genome, 2n = 2x = 20), known as the "A-genome ancestor," and *Arachis ipaensis* (BB-genome, 2n = 2x = 20), known as the "B-genome ancestor." The two subspecies of the cultivated plant, *A. hypogaea* subsp. *hypogaea* and *A. hypogaea* subsp. *fastigiata*, are distinguished by the patterns of reproductive and vegetative branching as well as by the morphology of the pods. Botanical variations are another division of the subspecies. While the subsp. *fastigiata* is separated into fastigiata (Valencia), vulgaris (Spanish), peruviana, and aequatoriana (Krapovickas and Gregory 1994), the subsp. *hypogaea* is separated into hypogaea (Virginia) and hirsuta (Fig. 1.1). The Virginia runner, Virginia bunch, Valencia, and Spanish bunch

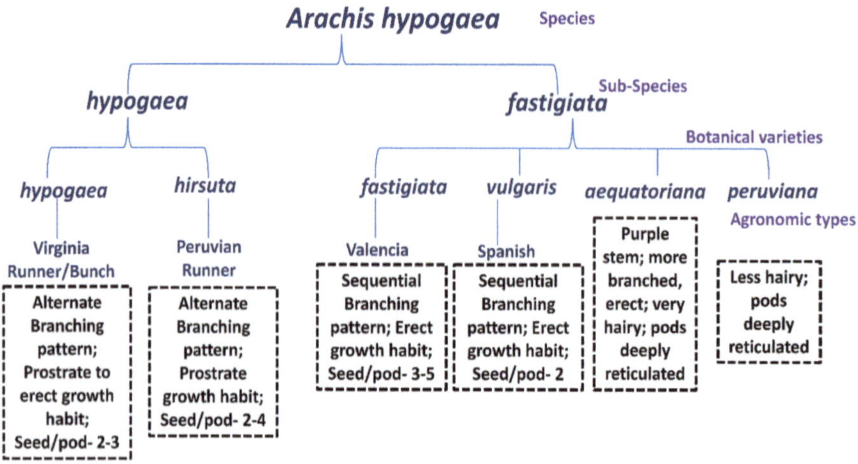

Fig. 1.1 A tree of subspecies and sub-subspecies of groundnut

are the four botanical varieties of groundnuts; their chemical makeup and oil quality vary. Spanish bunch kinds have lower oil and chemical concentrations than Virginia bunch seeds, whereas Virginia runner seeds have more soluble carbohydrates, Valencia runner seeds have more protein and better oil. The variety of testa determine its color, which can range from red, brown, and purple to white. Normally, the kernel and germ are white. Groundnut also plays a crucial role in sustainable agriculture. Through nitrogen 50 fixation, groundnut improves soil fertility; it promotes biodiversity by fostering a balanced ecosystem.

1.2 Health Benefits of Groundnut

Over the world, because of its taste, affordability, and nutritional value, groundnuts are a popular legume consumed worldwide. It is a priceless source of calories, protein, fiber, vitamins, minerals, and important fatty acids. Groundnut is often considered as "superfood" or "nutritional powerhouse" or "functional food" because of its balanced composition of micro- and macronutrients. It is eaten in different forms such as raw kernels, peanut snacks (salted, roasted), oil, confectionary products, soups and desserts (sweet bars, *chikki*, and *gajak*), and peanut butter. Groundnut consumption can cater the requirement of majority part of recommended daily allowance (RDA) of several micro- and macronutrients essential for the human body (Fig. 1.2). Availability of various cultivar groups provides diverse preference for particular use as cultivars differ in flavor, oil content, size, and shape. The root enriches the soil by fixing nitrogen and adding organic matter; vegetative plant part or haulm is a good source of fodder for cattle and biodiesel; kernels are source of edible oil and useful in numerous ways, whereas shells are used for fuel, soil conditioner, filler in fertilizers and feeds or is processed as substitute for cork or hardboard or composting with the aid of lignin decomposing bacteria. Groundnut cakes are used in the animal feed industry or as fertilizer.

Groundnut kernels contain 40–60% oil and are composed of 80% unsaturated fatty acids (UFA) and 20% saturated fatty acids (SFA). Unsaturated fatty acids

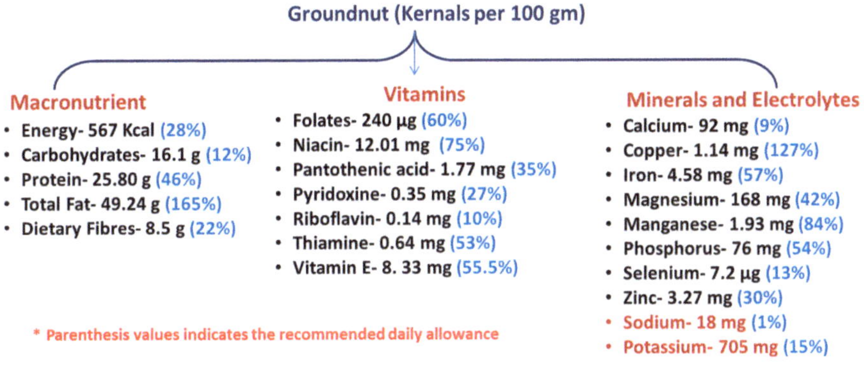

Fig. 1.2 Nutrient quality of groundnut kernels

consisted of 42% monounsaturated fatty acids (MUFA) while 37% poly unsaturated fatty acid (PUFA) (Wang et al. 2015). Among fatty acid proportion of groundnut oil, oleic acid (36–52%), linoleic (24–43%), and palmitic acid (8–11%) share more than 90% and remaining 10% is comprised of stearic, arachidic, behenic, lignoceric, and gadoleic acid. The development of groundnut cultivars with higher oleic/linoleic acid is prime objective for longer shelf life as well as for reducing the risk of cardiovascular diseases as it reduces low-density lipoproteins (LDL) levels in blood vessels. Excellent quality groundnut oil is used for human consumption and poor quality oil is used in preparation of soaps, fuel, cosmetics, leather dressings, lubricants, and so on. Peanuts contain essential functional nutrients such as fibers, antioxidants, polyphenols, and vitamins and minerals. Bioactive substances are abundant in groundnuts such as phytosterols [β-sitosterol (>80% of total phytosterols), campesterol (about 10%), and stigmasterol (<5%)], phenolic compounds (Resveratrol-73 mcg/g without skin, *p*-hydroxybenzoic acid, chlorogenic acid, *p*-coumaric acid), stilbenes, lignans, and isoflavonoids, and these substances successfully protect chronic illnesses such as diabetes, cancer, and cardiovascular disease. They are a valuable nutritional component due to their accessibility and cost, particularly in areas where protein-energy deficits or malnutrition are prevalent.

Beside oil extraction, groundnut is valued for as a high protein source (24–36%) and on an equal weight basis (kg for kg), it contains more protein than meat and about two times higher than eggs. Groundnut protein contains all the 20 amino acids in variable proportions, and is the richest source of "arginine" amino acid with relatively higher biological value and net protein utilization without flatulence factor and beany flavor. Groundnut protein is deficient in sulfur-containing amino acids such as methionine and cysteine. The protein content of defatted oil cakes can exceed 50%, and they are employed as both an industrial raw material and a protein supplement in animal feed regimens. Groundnut haulms are a good source of protein and can be used to make silage or hay or as fresh or dry animal feed. Haulm has a high digestibility of about 53% and contains 8–15% protein, 1–3% fats, 9–17% minerals, and 38–45% carbohydrates. The aforementioned information provides a solid foundation for the use of groundnuts in therapeutic foods, and the United Nations World Health Organization encourages the use of groundnut-based "ready-to-use therapeutic foods" (RUTF) in an effort to eradicate severe malnutrition.

1.3 History

The cultivated groundnut (*Arachis hypogaea* L.) is an ancient legume crop that traces its origin to the New World, specifically the region encompassing southern Bolivia and northwestern Argentina, where archaeological evidence suggests its cultivation as early as 1000 B.C. Traditionally valued by indigenous South American civilizations for its nutritional and oil-rich seeds, groundnut was integrated into both daily diets and ceremonial practices. Its global dissemination began during the sixteenth and seventeenth centuries, facilitated by the transoceanic voyages of Spanish, Portuguese, British, and Dutch explorers (Krapovickas 1969, 1973; Gregory et al.

1980; Hammons 1982; Isleib et al. 1994). As it spread to Africa, Asia, Europe, and the Pacific Islands, groundnut was quickly adopted due to its adaptability and multipurpose use. In many cultures, it became embedded in traditional cuisines, folk remedies, and socioreligious rituals. In India, for instance, groundnut is not only consumed in various forms but also holds cultural significance during festivals and local customs, symbolizing abundance and nourishment. Today, groundnut is cultivated across tropical and subtropical regions between 40°S and 40°N latitudes, thriving in areas with 500–1200 mm of rainfall and mean daily temperatures above 20 °C, continuing its legacy as a crop of both historical and cultural importance.

1.4 Utilization Pattern

The utilization pattern of groundnut varies significantly across regions, affected by cultural preferences, industrial demands, and economic factors. Understanding these patterns is crucial for optimizing production strategies, enhancing value addition, and ensuring sustainable utilization of groundnut resources worldwide. Groundnut is eaten in various forms, including roasted nuts, boiled nuts, peanut butter, peanut flour oil, snacks, and confectionery products. Its versatility makes it a popular ingredient in cuisines worldwide, used in savory dishes, desserts, and sauces. Groundnut oil, extracted from the seeds, serves as a cooking medium and ingredient in processed foods, while groundnut-based snacks are favored for their taste and nutritional value.

Major groundnut oil consuming countries in Asia include India, China, Myanmar, and Vietnam. In North America, groundnut is used primarily as a food staple, with 74% of production allocated for direct consumption. Additionally, 16% of groundnut is crushed for oil extraction, while the remaining 10% is utilized for various industrial purposes. In South America, groundnut serves both culinary and industrial purposes, with 39% allocated for food consumption and a significant 53% crushed for oil extraction. The remaining 8% is utilized in other industrial applications, showcasing a balanced utilization pattern. Across different subregions of Africa, groundnut is predominantly utilized for food consumption. East Africa dedicates 49% for food use and crushes 46% for oil, while Southern Africa and West Africa allocate 71% and 55%, respectively, for food consumption. The proportion crushed for oil extraction varies, with Southern Africa at 17% and West Africa at 33%. In Europe, particularly Western Europe, groundnut is predominantly utilized as a food staple, with 95% allocated for direct consumption. A minimal portion of 4% is crushed for oil extraction, while 1% is utilized in other industrial applications. Eastern Europe solely utilizes groundnut for food consumption (Table 1.1).

Asia demonstrates diverse utilization patterns. In East Asia, 40% of groundnut is used for food consumption, while a significant 53% is crushed for oil extraction. In Southeast Asia, 69% groundnut is utilized for food and 23% for oil extraction. However, Southwest Asia primarily focuses on oil extraction, with only 9% allocated for food consumption and a substantial 78% crushed for oil.

Table 1.1 Utilization pattern of groundnut kernels in different regions of the world

Region	Subregion	Food use (%)	Crushed for oil (%)	Other uses (%)
Europe	Eastern Europe	100	–	–
	Western Europe	95	4	1
America	North America	74	16	10
Africa	Southern Africa	71	17	12
Asia	Southeast Asia	69	23	8
Africa	West Africa	55	33	12
	East Africa	49	46	5
Asia	East Asia	40	53	7
America	South America	39	53	8
Asia	Southwest Asia	9	78	13
World	–	41	49	10

Source: Nigam (2014)

Table 1.2 Per capita consumption of groundnut oil in Indian rural and urban population

Year	per capita consumption (kg/year)	
	Rural	Urban
1993–94	1.44	2.88
1999–2000	1.44	2.76
2004–2005	0.84	1.92
2009–2010	0.65	1.51
2022–23	0.25	0.57

Source: NITI Aayog report (2024)

In India, groundnuts are primarily used as a feed crop, oilseed, and food. Either raw, edible groundnuts or "prepared" groundnuts (roasted, salted, flavored, etc.) or groundnut butter/paste are made from shelled groundnuts. Additionally, the seeds can be crushed to extract oil and groundnut meal, which is used as animal feed. With a neutral flavor and odor and a high smoke point (4400F), groundnut oil is used as high-quality cooking oil.

Globally, groundnut serves as a staple food, with 41% of production allocated for direct consumption. A significant portion of 49% is crushed for oil extraction, highlighting its importance in the oil industry. The remaining 10% is utilized for various industrial purposes.

Groundnut oil is clear, edible, light yellow oil that tastes nice and has a pleasant scent. Peanut oil has somewhat acceptable fatty acid content, making it easier for humans to absorb and digest. In India, groundnut oil consumption has steadily decreased over time among both rural and urban populations. In a span of three decades, it declined from 1.44 kg to 0.25 kg in the rural population and from 2.88 kg to 0.57 kg in the urban population (Table 1.2).

Products made from groundnuts satisfy a wide range of consumer tastes, from gourmet recipe seekers to health-conscious people looking for wholesome snacks. Traders play a vital role in facilitating the supply chain, sourcing groundnut from farmers, processing facilities, and international markets, ensuring a steady flow of quality products to consumers.

In most of the developing countries, kernels are used for oil extraction, food, and as an ingredient in confectionery products. After extraction, the leftover cake is processed mostly for animal feed; however, it can also be consumed by humans.

1.5 Production

1.5.1 Package of Practice for Cultivation

Suitable package of practice is available for scientific cultivation of groundnut in Bundelkhand, India, for increasing productivity and income of farmers (Table 1.3).

1.6 Protection: It Involves Major Pests, Diseases, and Management Strategies

1.6.1 Insect and Pests

The major groundnut insects in Bundelkhand include defoliators, sucking pests and soil insects, which lead to reduced yields and economic losses. Some of the major groundnut insects include the following.

Defoliators
Four major defoliators cause significant damage to groundnut crops:

- *Bihar hairy caterpillar* (*Spilosoma obliqua* Walker).
- *Red hairy caterpillar* (*Amsacta moori* Butler, *A. albistriga* Walker).
- *Tobacco caterpillar* (*Spodoptera litura* Fabr., *S. exigua* Hub.)
- *Fall armyworm* (*Spodoptera frugiperda*).

The larvae mainly feed on leaves, resulting in defoliation. Early instar larvae are gregarious feeders and scrape green tissue, giving a mesh-like appearance to damaged leaves that can be spotted easily from a distance. Older larvae can cause complete defoliation.

Management Practices

- Plant castor or sunflower as trap crops for *Spodoptera* spp.
- Install 10–12 light traps/ha to collect and destroy adult moths.
- Practice deep summer ploughing to expose hibernating stages.
- Collect and destroy egg masses and early larval stages.
- Dig trenches around fields to trap migrating hairy caterpillars.
- Use pheromone traps: 4–5/ha for monitoring or 10–12/ha for mass trapping of *Spodoptera*.
- Release egg parasitoid *Telenomus remus* @ 50,000/ha, four times at 7–10 day intervals.

Table 1.3 Package of practice for groundnut cultivation in Bundelkhand region

Practice	Details
Climate and soil requirements	Groundnut is a rainfed Kharif crop, suitable for semiarid climates with temperatures between 25 and 35 °C. Preferred soils are light sandy loams with pH 6.0–7.5. Well-drained soils with good aeration are ideal
Land preparation	1. Deep ploughing (15–20 cm depth) to loosen the soil 2. Harrowing twice for fine tilth 3. Application of FYM (farmyard manure) at 5–10 tons/ha during land preparation for improved soil fertility
Seed selection	1. Use high-yielding and drought-tolerant varieties that are recommended to different zones, region, or country 2. Seed rate: 100–125 kg/ha for spreading types and 80–100 kg/ha for semi-spreading types
Seed treatment	1. Treat seeds with *Trichoderma viride* @ 5 g/kg or Carbendazim @ 2 g/kg to prevent soil-borne diseases 2. Use Rhizobium inoculant @ 200 g/10 kg seed to enhance nitrogen fixation
Sowing time	Kharif season: Mid-June to early July (before the monsoon begins) Rabi season: October to early November (irrigated conditions)
Sowing method	1. Line sowing with a spacing of 30 cm × 10 cm (plant to plant and row to row) 2. Depth of sowing: 4–6 cm for optimal germination 3. Ensure proper seed-to-soil contact for better germination rates
Nutrient management	1. Basal dose: Apply 20–25 kg/ha of nitrogen (N), 40–50 kg/ha of phosphorus (P_2O_5), and 40–50 kg/ha of potassium (K_2O) 2. Micronutrients: Application of gypsum @ 400 kg/ha at flowering stage enhances pod formation 3. Zinc @ 5 kg/ha may be required in zinc-deficient soils
Irrigation	1. Rainfed crop: Sowing should be timed with the onset of monsoon. Supplemental irrigation may be required during dry spells 2. Irrigated crop: Requires 4–5 irrigations at critical stages (flowering, pod formation, and maturity) 3. Avoid waterlogging to prevent fungal diseases
Weed management	1. Preemergence herbicide: Apply Pendimethalin @ 1.0 kg/ha within 2 days of sowing 2. Manual weeding: 2–3 hand weeding at 20–25 DAS (days after sowing) and 40–45 DAS 3. Mulching with organic materials helps reduce weed growth and conserve soil moisture
Harvesting	1. Harvesting time: 90–120 DAS when leaves turn yellow and pods mature 2. Method: Uproot plants manually or use a digger 3. Postharvest: Dry pods for 4–5 days to reduce moisture content to 8–10% for safe storage
Yield	Average yield: 2.0–3.0 tons/ha depending on variety and management practices
Storage	Store groundnuts in dry, ventilated conditions to prevent aflatoxin contamination. Use jute bags and stack them on pallets to avoid direct contact with the floor

- Apply biologicals: NPV @ 250 LE/ha, *Bacillus thuringiensis* @ 1–1.5 kg/ha, or neem seed kernel extract @ 5%.
- In case of severe infestation, spray Chlorantraniliprole 18.5% SC (0.3–0.5 mL/L) or Emamectin benzoate 5% SG (1–1.5 g/L), repeating 2–3 times at 7–10 day intervals.

Leaf Miner (Aproaerema modicella Zell)

The larvae mine tender leaflets, feeding on mesophyll and creating small brown blotches. At later stages, larvae web leaflets together and feed on green matter, reducing photosynthesis and seed production. Heavily infested fields appear "burnt" from a distance.

Management Practices

- Plant cowpea or soybean as trap crops.
- Rotate with nonleguminous crops.
- Release egg parasitoid *Trichogramma chilonis* @ 50,000/ha, four times at 7–10 day intervals.
- For severe infestations, spray Chlorpyrifos 20 EC (2–4 mL/L) or Fipronil 5% SC (2–3 mL/L), repeating 2–3 times at 7–10 day intervals.

Sucking Insects

1. *Aphid (Aphis craccivora Koch):* Nymphs and adults suck sap from leaves, shoots, and flowers, causing stunting and deformation. Honeydew secretion encourages sooty mold. Aphids also transmit rosette virus.
2. *Jassid (Empoasca kerri Pruthi):* Both nymphs and adults suck sap, causing whitening of veins and V-shaped chlorotic patches at leaflet tips. Infested crops turn yellow and appear scorched ("hopper burn").
3. *Thrips (Scirtothrips dorsalis Hood and Thrips palmi Karny):* Nymphs and adults lacerate leaflets and suck oozing sap, producing white patches and distortion of young leaves. Severe infestation leads to stunted plants. *T. palmi* transmits peanut bud necrosis disease.

Management Practices

- Apply balanced nitrogenous fertilizers.
- Conserve natural enemies such as ladybird beetles, praying mantis, green lacewings, dragonflies, and spiders.
- Install yellow sticky traps (20–25/ha).
- Spray neem seed kernel extract @ 5%.
- In case of severe infestation, spray Imidacloprid 17.8 SL (0.5–1 mL/L) or Thiamethoxam 25 WG (0.2 g/L).

Soil Insect-Pests

1. Termite (*Odontotermes obesus* Rambur)

Colonies of worker termites damage plants by hollowing out taproots and stems, boring into pods, and feeding on seeds. They also weaken pod shells, facilitating *Aspergillus* infection and aflatoxin contamination.

Management Practices

- Destroy termite mounds and queens.
- Apply well-rotted FYM to reduce termite incidence.
- Irrigate fields to prevent termite infestation.
- Treat seed with Chlorpyrifos 20 EC @ 6.5 mL/kg or apply Chlorpyrifos dust @ 30–40 kg/ha.
- In severe cases, spray Imidacloprid 17.8% SL (0.5–1 mL/L) or Chlorpyrifos 20 EC (1.5–2 mL/L).

2. White Grub (*Holotrichia consanguinea* Blanch. and *H. insularis* Bren)

Both adults and grubs damage plants. Mature grubs feed on roots and pods, while young grubs attack rootlets. Infested plants wilt and die, giving fields a patchy appearance. Adult beetles feed on foliage at night, sometimes causing total defoliation.

Management Practices

- Deep plough in summer to expose hibernating stages.
- Apply well-decomposed FYM (attracts beetles).
- Collect and destroy adult beetles from host trees (jujube, guava, ber, khejri, neem) after early rains.
- Install 1–2 light traps/ha to collect beetles.
- Erect 10–12 bird perches/ha.
- Use aggregating pheromone (Methoxy benzene trap @ 4–5/ha).
- In severe infestations, apply Fipronil 0.6% GR or Chlorantraniliprole 4% GR @ 4 kg/acre mixed with sand or manure.

Diseases of Groundnut

Major diseases include fungal, bacterial, and viral infections that reduce yield and cause economic losses.

- *Early leaf spot (Tikka disease): Cercospora arachidicola* (yield loss up to 50%).
- *Late leaf spot (Tikka disease): Phaeoisariopsis personata* (yield loss up to 50%).
- *Rust: Puccinia arachidis* (up to 60% yield loss if uncontrolled).
- *Aflatoxin contamination: Aspergillus flavus* and *A. parasiticus* infect pods/seeds pre- and postharvest, leading to rejections and market losses.
- *Stem and pod rot: Sclerotium rolfsii* (significant yield and quality loss).
- *Groundnut bud necrosis virus (GBNV):* transmitted by thrips, causing necrosis and stunted growth.

- *Collar rot: Aspergillus niger*, leading to seedling mortality.
- *Alternaria leaf spot: Alternaria alternata*, causing severe defoliation and reduced yields.

Disease Management Strategies

- Seed treatment with fungicides (Thiram, Captan) to prevent soil- and seed-borne infections.
- Use of resistant varieties suited to Bundelkhand conditions.
- Integrated Pest and Disease Management (IPDM) combining resistant varieties, biocontrol agents, and cultural practices.
- Cultural practices: timely planting, crop rotation with nonhost crops, and strict field sanitation.
- Weed management to reduce competition and ease interculture/harvesting.
- Soil health management: apply organic matter and balanced fertilizers to improve plant resistance.

1.7 Groundnut Production Scenario

Groundnut cultivation spans across continents, with major producing regions located in West Africa, Asia, North and South America. It is grown in more than 100 countries and global groundnut production in 2024 was 50.387 million tons as compared to 2023 (54.4 million tons) and 2022 (49.4 million tons). China is leader in groundnut production and accounts for nearly 37% of global production followed by India with approx. 13.84% and Nigeria with 8.34% of global production. India has shown significant increase in yield, with a rise from 1.0 t/ha in 1999–2001 to 1.8 t/ha in 2020–2022, and an impressive growth rate of 80%. However, the United States of America (USA) maintains the global lead with a yield of 1.8 t/ha and a 50% growth rate. The top ten groundnut producing nations include China, India, Nigeria, the USA, Sudan, Burma, Senegal, Argentina, Tanzania, and Guinea and collectively account for approximately 82.3% of global production (Table 1.4).

Data on productivity indicates that whereas India has made significant improvement (1.0 t/ha in 1999–2001 to 1.8 in 2020–2022), still we are far behind the USA and China (Fig. 1.3). China has nearly 3.84 t/ha yield, whereas the Indian average is around 1.77 t/ha with Gujarat showing maximum productivity of 2.2 t/ha. A few newly bred varieties have shown potential of 3.5–4.0 t/ha in yield trials. Certified quality seed replacement rate (SRR) of groundnut seed has remained stagnant to around 25% in the last decade.

1.7.1 Indian Production Scenario

In India the total production of groundnut was 10.18 million ton from 4.71 million ha area with a productivity of 2163 kg/ha during 2023–24 (Table 1.5). The

Table 1.4 Top ten groundnut producing countries in 2024

Rank	Country	Percent of total production	Production ('000 tons)
1	China	38.0	19,000
2	India	14.0	7100
3	Nigeria	9.0	4300
4	The USA	6.0	2925
5	Burma	3.0	1750
6	Senegal	3.0	1700
7	Argentina	3.0	1400
8	Tanzania	2.0	1100
9	Guinea	2.0	1000
10	Sudan	2.0	1000
Share of top ten producer		*82.0*	*41,275*
Other countries		18.0	9112
Total		*100.0*	*50,387*

Source: https://ipad.fas.usda.gov/-https://ipad.fas.usda.gov/cropexplorer/cropview/commodityView.aspx?startrow=1&cropid=2221000&sel_year=2024&rankby=Production

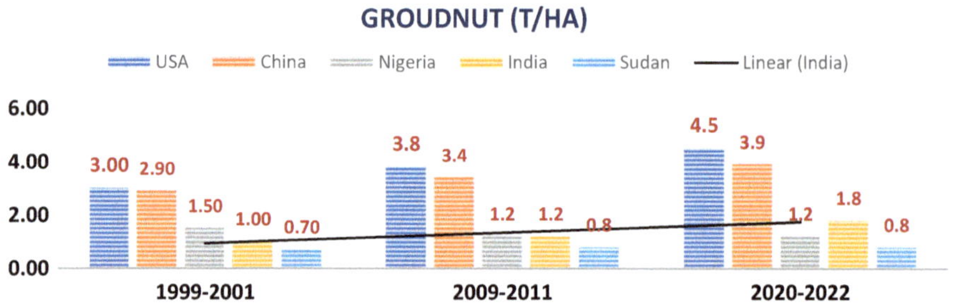

Fig. 1.3 Yield comparison across major producer countries (1999–2022). Source: NITI Aayog report (2024)

corresponding figure in 2022–23 was 10.30 million tons production from 4.96 million ha area with a productivity of 2075 kg/ha (Source: https://oilseeds.dac.gov.in/StatisticsCropWise.aspx). Groundnut acreage in the country has shown the declining trend since 2021–22 but production remains at constant level due to a jump in productivity from 1703 to 2163 kg/ha.

In 2023–24, Gujarat occupied a maximum share of 36.01% in groundnut acreage area and accounted for 45.061% of total production with a productivity of 2739 kg/ha (Fig. 1.4). It was followed by Rajasthan with 18.92% area and 19.90% production share with a productivity of 2276 kg/ha. Gujarat recorded an 18.4% rise in productivity from 1.9 to 2.3 t/ha over the last decade.

In 2023–24 in Uttar Pradesh, the total area under groundnut cultivation was 0.126 million hectares, yielding a production of 1.47 lakh tons with an average yield of 1172 kg/ha (Table 1.5). Meanwhile, in Madhya Pradesh, the total area was higher at 0.479 million hectares, with a production of 9.83 lakh tons with an average yield

Table 1.5 Groundnut area production and yield during 2022–23 and 2023–24 in major producing states along with coverage under

	2022–23					2023–24				
State	Area (million ha)	Production (million tons)	Yield (kg/ha)	% Share in total production	% Share in total area	Area (million ha)	Production (million tons)	Yield (kg/ha)	% Share in total production	% Share in total area
Gujarat	1.763	4.531	2570	44.00	35.54	1.695	4.644	2739	45.62	36.01
Rajasthan	0.805	1.932	2400	18.76	16.23	0.891	2.027	2276	19.91	18.93
Madhya Pradesh	0.389	0.744	1912	7.23	7.841	0.479	0.983	2051	9.66	10.18
Tamil Nadu	0.351	0.912	2598	8.86	7.075	0.3.26	0.878	2694	8.62	6.92
Andhra Pradesh	0.594	0.601	1011	5.84	11.97	0.3.12	0.323	1038	3.17	6.62
Maharashtra	0.209	0.267	1278	2.59	4.217	0.2.26	0.285	1257	2.80	4.80
West Bengal	0.078	0.199	2561	1.93	1.572	0.085	0.259	3060	2.54	1.80
Karnataka	0.465	0.581	1248	5.64	9.373	0.373	0.242	648	2.38	7.94
Telangana	0.103	0.252	2448	2.45	2.076	0.092	0.225	2450	2.21	1.95
Uttar Pradesh	0.098	0.115	1168	1.12	1.975	0.126	0.148	1172	1.45	2.67
All India	*4.961*	*10.297*	*2075*			*4.707*	*10.18*	*2163*		

Source: https://data.desagri.gov.in/website/crops-apy-report-web

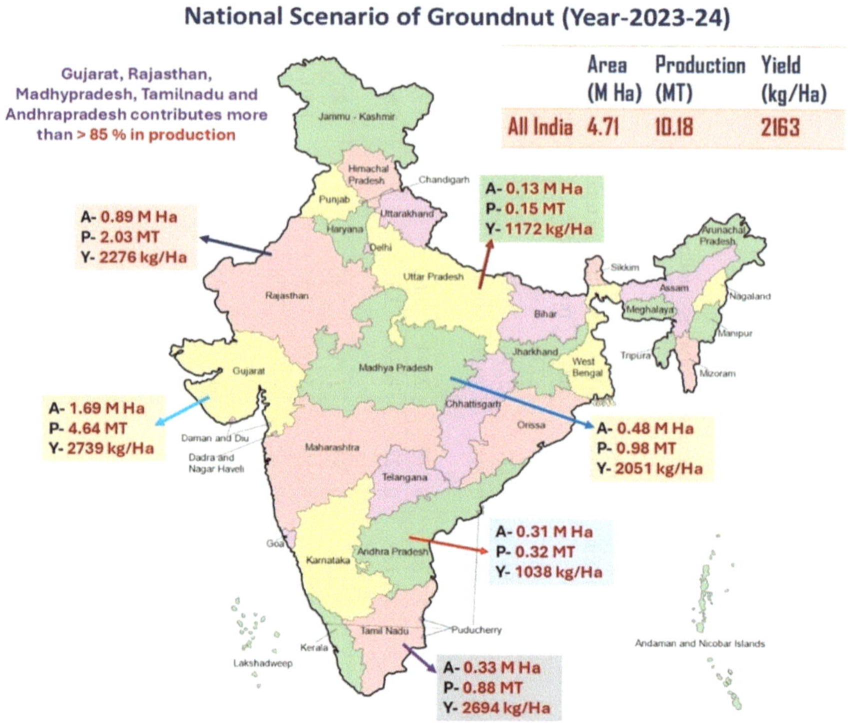

Fig. 1.4 National scenario of groundnut production in 2023–24

of 2051 kg/ha. Significant jump of 464 kg/ha has been recorded by Madhya Pradesh during the last 5 years (2019–20 to 2023–24).

Groundnut acreage in the country has fluctuated over the years and remains almost equal as 1950–51, but analysis of seven decades data indicates that there is near stagnant productivity during 1950–51 to 1980–81. However, in the last 24 years, groundnut production has seen a positive trajectory, with yield increasing from 977 kg/ha in 2000–01 to 2163 kg/ha in 2023–2024, reflecting an impressive 121% growth rate. The production during the period has gone up from 3.48 million tons to 10.18 million tons, showing an impressive increase of 192.5%. There was a significant jump during 1990–91 when the production, productivity, and area have increased significantly probably due to the technology mission of oilseed during the period (Table 1.6).

Over the past 5 years, production under kharif has essentially remained unchanged, but production throughout the summer has increased. The overall output over the past 5 years has essentially remained unchanged, with very little variations in annual production numbers. Over 80% of the entire output is produced during the Kharif season (Table 1.5).

Groundnut area in the country has shown inconsistency over the years, and area has declined as farmers are shifting to other remunerative crops. Groundnut area is

Table 1.6 Groundnut cultivation and productivity trend of last seven decades

Year	Area (million ha)	Production (million tons)	Yield (kg/ha)	Area under irrigation (%)
1950–51	4.49	3.48	775	–
1960–61	6.46	4.81	745	3.02
1970–71	7.33	6.11	834	7.55
1980–81	6.80	5.01	736	13.30
1990–91	8.31	7.51	904	19.63
2000–01	6.56	6.41	977	16.89
2010–11	5.86	8.26	1411	23.86
2020–21	6.01	10.24	1703	41.06
2021–22	5.70	10.13	1777	37.75
2022–23	4.96	10.30	2075	–
2023–24	4.71	10.18	2163	–

Source: Agricultural Statistics at a Glance 2023, Economics, Statistics and Evaluation Division, Department of Agriculture and Farmers Welfare, Government of India

Table 1.7 Trend growth rate of area, production, and productivity (1980–2022)

	1980–2022	1980–1990	1991–2000	2001–2010	2011–2022
Area	−1.27	1.64	−2.3	−0.96	−0.25
Production	0.5	3.75	—1.25	0.77	2.76
Yield	1.8	2.07	0.78	2.76	4.24

significantly shrinking continuously while productivity has been increasing in the last 42 years (1980–2022), primarily due to increase in productivity mainly in major groundnut producing states such as Rajasthan and Gujarat (Table 1.7).

1.8 Groundnut Improvement

In groundnut breeding programs, key target traits include enhanced pod yield, elevated oil content, improved shelling percentage, early maturity, fresh seed dormancy, high oleic acid concentration, and development of confectionery and dual-purpose types. Additionally, resistance to both biotic (e.g., foliar diseases, aflatoxin, rosette) and abiotic (e.g., drought, heat) stresses is a critical breeding objective.

In developed nations such as the USA, where high-input agricultural systems are prevalent, breeding efforts have traditionally prioritized yield maximization. However, in recent years, the emphasis has shifted toward enhancing quality attributes such as oil composition and flavor, along with improving resistance to drought and key diseases. In contrast, breeding programs in Asia and Africa primarily focus on increasing pod yield and oil content while enhancing tolerance to biotic and abiotic constraints, aligning with the resource-limited production systems in these regions. Conventional breeding techniques, including introduction, selection, hybridization (pedigree, backcross, and single-seed descent), and induced mutagenesis, continue to play a vital role in varietal development. While mutation breeding was widely utilized in the USA from the 1950s to the early 1970s, its current application is minimal (Holbrook and Stalker 2003). In contrast, India continues to leverage mutation breeding, notably at institutions such as the Bhabha Atomic Research Centre (BARC) (Mondal et al. 2007), for groundnut improvement.

The utilization of conserved genetic resources from gene banks has been instrumental in broadening the genetic base of breeding programs. These germplasm collections serve as valuable sources of novel alleles for various target traits. Recent advancements in genomics and the application of marker-assisted selection (MAS) have enabled more precise and efficient breeding, facilitating accelerated genetic gains.

Multiple breeding lines and elite germplasm have been developed with desirable traits such as drought tolerance, resistance to foliar diseases (e.g., late leaf spot, rust), rosette virus resistance, aflatoxin mitigation, and improved oil quality. However, genotype × environment interactions significantly influence pod yield and other complex traits (Bucheyeki et al. 2008; Jogloy et al. 2009; Makinde et al. 2013; Janila et al. 2016), necessitating extensive multi-location and multi-season trials for reliable varietal performance assessment.

To ensure varietal adoption and relevance, farmer participatory varietal selection (FPVS) has emerged as an effective approach, particularly in South Asia and Sub-Saharan Africa (Ntare et al. 2007). This participatory method integrates farmers' preferences and facilitates better alignment of breeding outputs with ecological conditions and market demands. Between 2000 and 2016, several improved varieties were released in these regions, characterized by traits such as early maturity, high yield, drought tolerance, and resistance to foliar diseases and rosette virus (Desmae et al. 2017).

1.9 Processing

Efficient postharvest handling and primary processing are crucial for preserving the quality and marketability of groundnuts. These practices reduce postharvest losses, maintain nutritional value, and ensure safe storage for consumption and further processing (Fig. 1.5). The primary processing of groundnut includes several key steps:

Harvesting:

Groundnuts are harvested when the lower leaves turn yellow and pods are mature with firm shells. Delayed harvesting can lead to yield loss due to shattering or aflatoxin contamination, especially under humid conditions.

Drying:

Immediately after harvest, pods are sun-dried to reduce moisture content from 35–40% to 8–10%. Proper drying is essential to prevent mold growth and aflatoxin production. Groundnuts are dried on clean surfaces or raised platforms to avoid contamination from soil or pests.

Cleaning:

The dried pods are cleaned to remove soil, plant debris, stones, and immature or damaged pods. This is typically done through mechanical sieves, gravity separators, or air blowers. Cleaned produce ensures higher grade and market value.

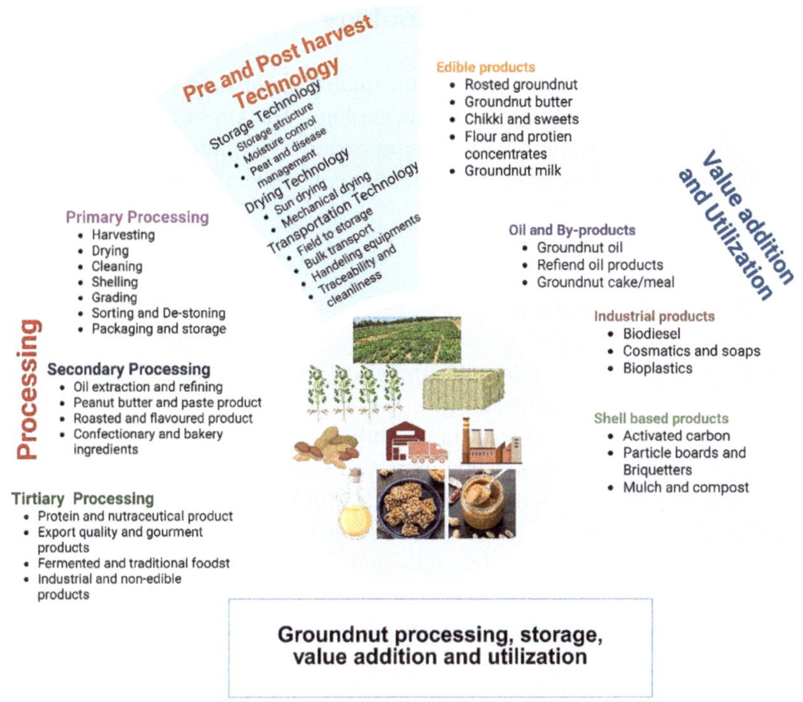

Fig. 1.5 Groundnut processing and value chain development a holistic approach

Shelling:

Shelling involves removing the kernels from the pods. Manual or mechanical shellers are used, depending on the scale of operation. Mechanical shellers improve efficiency but must be adjusted to avoid kernel breakage.

Grading:

After Post shelling, the kernels are graded based on size, weight, and appearance using sieves and gravity separators. Grading ensures uniformity in processing and helps fetch premium prices in the market.

Sorting and de-stoning:

Kernels are manually or mechanically sorted to eliminate defective, discolored, shriveled, or infested nuts. De-stoning machines are employed to remove any remaining stones or heavy impurities.

Roasting (optional for edible use):

For edible use, kernels may undergo dry or oil roasting to enhance flavor, aroma, and shelf-life. Roasting also reduces anti-nutritional factors and microbial load.

Packaging and storage:

Processed kernels are packed in moisture-proof bags or containers to prevent reabsorption of moisture. Ideal storage conditions include low humidity (below 65%) and cool temperatures (15–20 °C) to minimize insect infestation and aflatoxin risk.

1.10 Pre- and Postharvest Technology

Reducing losses and preserving groundnut quality from field to market depend heavily on efficient pre- and postharvest technology. The key components, namely storage, drying, and transportation require attention to hygiene, moisture control, and handling practices to prevent spoilage, aflatoxin contamination, and pest infestation as depicted in Fig. 1.5.

1.10.1 Storage Technology

Proper storage of groundnut pods and kernels is essential to preserve seed viability, reduce spoilage, and ensure food safety.

Storage structures: Traditional structures (kothis, bamboo bins, earthen pots) are still used in many rural areas, while improved structures (metal bins, Pusa bins, gunny/polypropylene bags) are recommended for better protection.

Moisture control: Groundnuts must be stored at moisture content below 8% to inhibit mold growth and aflatoxin production.

Pest and fungal management: Use of botanical repellents (neem leaves), inert dusts (ash, lime), or safe chemical treatments (aluminum phosphide) help manage storage pests.

Aeration: Ventilated storage areas reduce heat and humidity build-up, preventing quality deterioration.

Seed storage: For sowing purposes, seeds are stored in airtight containers with desiccants (e.g., silica gel) to preserve viability.

1.10.2 Drying Technology

In order to guarantee safe storage and lower biological activity in harvested groundnuts, drying is an essential step.

Sun drying: This is the most commonly practiced method where groundnut pods are spread thinly on clean floors, mats, or raised platforms. Pods are dried until they reach 8–10% moisture.

Mechanical dryers: In large-scale operations or during adverse weather, forced-air or solar dryers are employed for uniform and faster drying.

Drying duration: Typically takes 3–7 days depending on weather. Overdrying may cause kernel shrinkage and underdrying increases risk of fungal infestation.

1.10.3 Transportation Technology

Transportation plays a vital role in maintaining quality during movement from field to market or processing centers.

Field to storage: After harvesting, groundnut should be transported in clean, dry, and ventilated containers to prevent contamination and heating.

Bulk transport: For large quantities, jute or woven polypropylene bags are used. Proper stacking and covering during transit prevent mechanical damage and moisture exposure.

Handling equipment: Use of conveyors, trolleys, and loaders is preferred over manual carrying to reduce pod breakage and contamination.

Traceability and cleanliness: Transport vehicles should be cleaned and disinfected before loading to avoid cross-contamination, especially for edible or export-grade groundnut.

1.11 Value Addition and Utilization

Groundnuts have high oil, protein, vitamin, and bioactive chemical content, because of this qualities groundnuts have a lot of potential for value addition (Fig. 1.5). Transforming raw groundnuts into diversified, high-value products not only increases shelf life and profitability but also opens up new market opportunities both domestic and international.

1.11.1 Edible Products

Groundnut can be processed into a wide range of nutritionally enriched and consumer-friendly products, such as:

Roasted groundnuts: Popular snack item with long shelf-life. Variants include salted, spiced, and coated nuts;

Groundnut butter: A protein-rich spread gaining popularity as a plant-based alternative to dairy butter;

Chikki and sweets: Traditional products such as groundnut *chikki* (brittle) and *laddus* have strong market appeal;

Flour and protein concentrates: Used in bakery items, infant foods, and protein supplements;

Groundnut milk: A lactose-free, plant-based milk alternative gaining demand among health-conscious consumers.

1.11.2 Oil and By-products

Groundnut oil is a major cooking medium in India and abroad. By-products from oil extraction also hold economic significance:

Groundnut oil: Premium edible oil known for flavor, stability, and health benefits;

Refined oil products: Blends for salad oils, margarine, or shortening; and.

Groundnut cake/meal: High-protein animal feed or base material for protein isolates and organic fertilizers.

1.11.3 Industrial Products

Groundnut and its derivatives are used in various non-food applications, including:
Biodiesel: Groundnut oil serves as a potential feedstock for eco-friendly fuel;
Cosmetics and soaps: Oil and its extracts are used in moisturizers, lotions, and soap formulations;
Bioplastics: Groundnut shells and proteins are being explored in biodegradable plastic development.

1.11.4 Shell-Based Products

Groundnut shells, often discarded, can be converted into value-added products, contributing to waste utilization:
Activated carbon: Used in water purification, filters, and medical applications;
Particle boards and briquettes: For construction and fuel applications;
Organic mulch and compost: Useful in sustainable agriculture and horticulture.

1.12 Marketing

One of the most significant food legume and oilseed crops traded worldwide is groundnut. Its versatility—used for industrial applications, edible items, and oil extraction—drives consistent demand both domestically and abroad. For farmers, processors, and exporters to increase revenue, effective marketing techniques are crucial

1.12.1 Global Demand

- *Food and nutritional demand*: Groundnut consumption is rising due to its health benefits, including plant-based protein, healthy fats, and micronutrients. Demand is especially strong for snack products, peanut butter, and roasted nuts.
- *Oil demand*: Groundnut oil remains a staple in South Asia, Southeast Asia, and parts of Africa, with growing interest in cold-pressed and organic oils in Western countries.
- *Health-conscious consumers*: Increased awareness of plant-based diets, veganism, and heart-healthy foods has fueled global demand for groundnut-derived products.

1.12.2 Global Trade

India holds a significant position in the global groundnut market. In 2020–21, India exported 6.38 lakh tons of groundnuts, valued at Rs 5381 crore, primarily to nations

Table 1.8 Total export of groundnut vis-a-vis oilseed (2019–20 to 2021–22)

	2019–20		2020–21		2021–22	
	Quantity ('000 tons)	Value (billion rupees)	Quantity ('000 tons)	Value (billion rupees)	Quantity ('000 tons)	Value (billion rupees)
Groundnut	664.4	51.0	638.3	53.8	514.2	47.0
Total oilseed	1080.9	95.0	1072.8	94.3	860.7	85.5

Source: DGCI&S

in the United Arab Emirates, China, Russia, Ukraine, Indonesia, Vietnam, the Philippines, Malaysia, Thailand, and Nepal. India exported 38,200, 24,100, 20,100 tons of groundnut oil in 2019–20, 2020–21, 2021–22, respectively. Similarly, the quantity of groundnut oil meal exported was 4100, 16,000, 3600 tons in 2019–20, 2020–21, 2021–22, respectively (Table 1.8).

India exports groundnuts to over 50 countries. The major export destination comprises of Southeast Asian countries such as Indonesia, Malaysia, Vietnam, and Philippines, mostly in the form of raw or shelled nuts; Middle East countries such as the UAE, Saudi Arabia, and other Gulf nations often as snacks and processed food; European Union in various forms; China especially for oil extraction; African countries (such as Sudan and Egypt); and North America in value-added forms such as peanut butter or roasted peanuts.

India has certain advantages in the growing international groundnut market due to competitive pricing, varietal diversity suited to different consumer preferences, organic groundnuts, and so on. The Agricultural and Processed Food Products Export Development Authority (APEDA) has taken various initiatives to promote and streamline the ground export through websitepeanut.net, buyer registration, batch processing, application for certificate of export, aflatoxin analysis, and issue of stuffing certificate by laboratories, issuance of certificate of exports by the APEDA, and so on. These initiatives have promoted the export for the benefit of producers as well as entrepreneurs.

1.12.3 Challenges in Groundnut Exports

- *Aflatoxin contamination*: One of the most significant challenges for Indian groundnut exports is aflatoxin contamination, a naturally occurring toxin that affects groundnuts in humid conditions. According to Indian Council of Medical Research (ICMR)—Lucknow, 21% of groundnut in India is unfit for human consumption due to aflatoxin. Another study conducted by the International Crops Research Institute for the Semi-Arid Tropics (ICRISAT) reveals that the level of aflatoxin in some samples of Indian groundnut is 40 times more than permissible limits. The permitted level of aflatoxin in groundnut for human consumption, according to international standards, is 4 parts per billion (ppb) in the European Union and 20 ppb in the USA, and 30 ppb in India. The European Union, in particular, has very strict aflatoxin limits, which has caused rejections of Indian shipments in the past. Improving storage, drying, and quality control processes is critical for reducing contamination.

- *Quality and grading standards*: Indian groundnuts need to meet stringent international quality and grading standards. Ensuring uniformity in size, color, and quality, along with proper packaging, is essential for tapping high-end markets.
- *Logistics and infrastructure*: Inadequate postharvest infrastructure such as storage, drying facilities, and transportation can lead to quality degradation, especially in perishable products like groundnuts.
- *Tariff and nontariff barriers*: Although India's groundnut exports are often competitive, market access may be impacted by high tariffs and nontariff barriers in certain nations. For instance, Indian exporters are required to abide by strict sanitary and phytosanitary requirements enforced by nations such as the USA and China.

1.12.4 Market Trends

- *Processed products*: There is a shift from raw exports to value-added products such as roasted peanuts, groundnut butter, snacks, and flavored coated nuts.
- *Organic and non-GMO labels*: International markets, especially in the EU and the USA, prefer certified organic, aflatoxin-free, and non-GMO groundnuts.
- *E-commerce and direct-to-consumer channels*: Digital platforms are increasingly used for marketing groundnut-based products, especially gourmet and health foods.
- *Sustainability and traceability*: Consumers and buyers demand eco-friendly production, fair trade certification, and traceable supply chains.

1.12.5 Market Linkages and Value Addition

- *Cooperatives/Farmer Producer Organizations (FPOs)*: The formation of FPOs or cooperatives will help to strengthen collective bargaining and marketing, reduce input costs, and improve access to markets.
- *Agri-business promotion*: Support small-scale processing units for groundnut oil, roasted peanuts, and peanut butter to promote value addition.
- *Market information systems*: By using local agricultural offices and mobile apps to provide market price information, you can improve the connections between farmers and markets. Farmers can use this to better schedule their sales and increase revenues.
- *Promotion of export markets*: Explore export opportunities for Bundelkhand's groundnuts, especially in international markets where Indian groundnuts are in demand.

- *Enhancing marketing infrastructure*: To improve the realizable income of farmers, ensuring procurement at the "minimum support price" (MSP) is essential. Facilitating direct marketing of output with NAFED/state-owned oilseed federations can act as a catalyst for promoting groundnut cultivation.
- *Setting up testing laboratories in the mandis*: Indian markets need to depend more on technologies as traders rely mostly on visual inspection. This subjectivity leads to inconsistencies in pricing and disadvantages producers of higher-quality oilseeds. Establishing standardized quality parameters and implementing mandatory testing procedures at mandis (agricultural markets) is vital to address this gap. Testing laboratories may be set up in public–private partnership (PPP) mode utilizing the expertise of agricultural universities and ICAR institutions.
- *Strategic partnerships for sustainable production*: Using private sector experience in technology, marketing, seed production, and region promotion, public–private partnerships are essential for increasing groundnut output. Governmental organizations, grower cooperatives, regional nongovernmental organizations, and the corporate sector must work together.
- *Value chain integration*: Strong collaboration along the groundnut value chain by promoting vertical integration among producers, processors, distributors, and retailers needs to be developed. Contract farming arrangements, forward linkages, and backward linkages should be encouraged to ensure a steady demand-and-supply chain of quality produce at suitable price.

1.13 Conclusion

Groundnut (*Arachis hypogaea* L.) stands out as a versatile and nutritionally dense legume that fulfils the dietary, agricultural, and economic requirements of both traditional and modern food systems. Its unique combination of high-quality plant protein, heart-healthy fats, essential micronutrients, and functional bioactive compounds makes it an ideal component of fasting diets, particularly in vegetarian and culturally diverse populations. From a health perspective, groundnut offers sustained energy, satiety, and metabolic support during fasting, while its low glycemic response and antioxidant-rich profile contribute to chronic disease prevention and overall well-being. The diversity in botanical types and wide range of value-added products from roasted nuts and nut butter to flour and oil further enhances its appeal as a functional food for diverse culinary and health applications. Agronomically, groundnut's adaptability to marginal lands, drought resilience, and soil-enriching nitrogen fixation make it a vital crop for sustainable agriculture, especially in dryland regions such as Bundelkhand. Significant prospects for revenue generation, rural development, and nutritional security are presented by postharvest technologies, processing methods, and integration into regional and international markets.

As consumer interest grows in plant-based, clean-label, and fasting-compatible foods, groundnut is well-positioned to emerge not only as a traditional staple but also as a *future-forward fasting superfood* with broad-reaching benefits across health, agriculture, and economy.

References

Bucheyeki TL, Shenkalwa EM, Mapunda TX, Matata LW (2008) The groundnut pyramidal seeds system: an important delivery pathway for improved groundnut cultivars to farmers in Tanzania. Afr J Agric Res 3(11):809–822

Desmae H, Janila P, Sriswathi M, Vishnuvardhan K, Kumar PL, Nigam SN, Varshney RK (2017) Genetic improvement of peanut (Arachis hypogaea L.) for productivity and resistance to diseases: a review. Int J Mol Sci 18(11):2522

Gregory WC, Krapovickas A, Gregory MP (1980) Structure, variation, evolution, and classification in Arachis. In: Wilson CT, Demen MW (eds) Peanut Science and Technology. American Peanut Research and Education Society, pp 47–74

Hammons RO (1982) Origin and early history of the peanut. In: Wilson CT, Demen MW (eds) Peanut Science and Technology. American Peanut Research and Education Society, pp 1–20

Holbrook CC, Stalker HT (2003) Peanut breeding and genetic resources. Plant Breed Rev 22:297–356

Isleib TG, Wynne JC, Beute MK (1994) Groundnut breeding. In: *Plant breeding reviews*, vol 12, pp 489–526

Janila P, Manohar SS, Patne N, Variath MT, Nigam SN (2016) Genotype × environment interaction effects for pod yield and its components in groundnut in India. Crop Pasture Sci 67(8):755–765

Jogloy S, Vorasoot N, Akkasaeng C, Kesmala T, Patanothai A (2009) Breeding for drought resistance and water use efficiency in peanut. SABRAO J Breed Genet 41(2):113–127

Krapovickas A (1969) Origen, domesticaci6n y distribución de Arachis hypogaea L. *Actas y Memorias del 4° Congreso Nacional de Botánica 1*:49–54

Krapovickas A (1973) El género Arachis L. en la Argentina. Rev Investig Agropecu 10(1):35–117

Krapovickas A, Gregory WC (1994) Taxonomy of the genus Arachis (Leguminosae). Bonplandia 8:1–186

Makinde SCO, Ariyo OJ, Akinbowale JT (2013) Genotype × environment interaction for pod yield and other quantitative traits in groundnut (Arachis hypogaea L.). Afr J Biotechnol 12(11):1231–1238

Mondal S, Badigannavar AM, Kale DM, Murty GSS (2007) Induction of genetic variability in a disease-resistant groundnut breeding line. BARC Newsletter 185:237–246

Nigam SN (2014) Groundnut at a glance. 121 pp

NITI Aayog (2024) Annual Report 2023–24. Government of India

Ntare BR, Diallo AT, Ndjeunga J, Waliyar F (2007) Farmer participatory evaluation and dissemination of improved groundnut varieties in Sub-Saharan Africa. International Crops Research Institute for the Semi-Arid Tropics (ICRISAT), Patancheru

Wang ML, Chen CY, Tonnis B, Barkley NA, Pinnow DL, Pittman RN, Pederson GA (2015) Oil content and fatty acid composition variability in wild peanut species. Plant Gen Res 13(1):36–42

ALSI: Omega-3-Rich Source for Health and Vitality

2

Sonal Tiwari, Ashish Kumar Tiwari, and Anil Kumar

Abstract

Linseed has long serve as a dual-purpose crop, yielding mutually high-quality fiber and oil-rich seeds. Flax fiber is sturdy, flexible, and highly absorbent, making it valuable in textiles, ropes, specialty papers, and biodegradable composites. The seeds are loaded in alpha-linolenic acid (ALA), an essential omega-3 fatty acid, along with dietary fiber, minerals, lignans, vitamins, and protein. ALA has been linked to many health benefits, including reduced risks of cardiovascular disease, hypertension, stroke, and inflammatory disorders. It serves as a plant-based precursor to EPA and DHA, significant for immune and cellular functions. With increasing interest in functional foods, flaxseed is being more and more recognized for its therapeutic potential. It is now included into a variety of food products and dietary interventions aimed at improving health. Although it has nutritional advantages, the awareness of flaxseed's health benefits is limited. Further research and public education are needed to increase its acceptance and utilization. From an industrial standpoint, flaxseed oil contains over 50% linolenic acid, appropriate for industrial applications but less advantageous for cooking. Low linolenic acid varieties such as LINOLA and SOLIN have been developed to address this, with similar breeding efforts underway in India. Future breeding should focus on reducing anti-nutritional factors, enhancing omega-3 bioavailability, and improving tolerance to drought and disease using both conventional and molecular tools. Thus, linseed offers huge potential as a nutrition-

S. Tiwari (✉)
Mahatma Gandhi Udyaniki Evam Vaniki Vishwavidyalaya, Durg, Chhattisgarh, India

A. K. Tiwari
Indira Gandhi Krishi Vishwavidyalaya, Raipur, Chhattisgarh, India

A. Kumar
Rani Lakhsmi Bai Central Agricultural University, Jhansi, Uttar Pradesh, India

A. Kumar et al. (eds.), *Fasting Superfoods*,
https://doi.org/10.1007/978-981-95-2762-5_2

ally rich, multipurpose crop. Advancing its agronomic traits and promoting its health benefits through research and innovation will improve its role in sustainable agriculture, nutrition, and industry.

Keywords

Flaxseed · Omaga-3 fatty acids · Alpha-linolenic acids · Nutritional health benefits · Cultivation

2.1 Introduction

Linseed (*Linum usitatissimum* L.), also known as flaxseed, is an old crop basically known to originate in the European center. The species *usitatissimum* means having numerous uses. It has been used for fiber as well as oil content for a long time. It is a self-pollinate crop with chromosome number $2n = 30$ and belong to the genus *Linum* of the family Linaceae (Gill 1987). India ranks third in area under linseed followed by Canada, China, whereas in production it stands fifth (Annual report Linseed, 2024). It is grown up for oil in tropical and subtropical countries and in Asian parts of the USSR, whereas in European countries, namely Canada, Argentina, Poland, European parts of the USSR, and China, it is cultivated for both oil and fiber.

The root system of linseed is branched taproot growing more than a meter deep. The stem is slender and upright with branches beginning from the base of stem (Diederichsen and Richards 2003). Leaf arrangement is alternate and is sessile. The dimensions of leaf vary from 1.5 to 5.5 cm in length and 0.3 to 1.3 cm in width. Flowering occur at the terminal position of branches and the flowers are generally regular, hypogynous, and tetracyclic. The flowers are pentamerous with a radially symmetrical arrangement (Hayward 1938; Schewe et al. 2011). The petals of flowers are found in different forms, i.e., funnel, disk-shaped, star-shaped, and tubular (Dillman and Brinsmade 1938; Diederichsen and Richards 2003). Color of the petals varies from light blue, blue, dark blue, violet, to red-violet, pink, white, and further variations may also possible (Hayward 1938; Diederichsen and Richards 2003). Linseed crop at full bloom showing complete flowering is presented in Fig. 2.1. The flowers open during mornings and the petals drop off by noon (Diederichsen and Richards 2003). The anthers are introrse, i.e., when the flowers close during noon, the filaments bend down and shed the pollen inward inside the flower thus causing self-pollination. The ovary is pentalocular with two locules per chamber, which are separated by a false septum. Each chamber has an ovule inside (Hayward 1938).

Linseed is loaded with polyunsaturated fatty acids and other nutritional components, making it a potential food for health and nutraceutical industry. The nutritional profile comprises mainly of omega-3 fatty acid along with dietary fiber, minerals, lignans, vitamins, and protein, which are further discussed in the chapter. Further, the chapter also covers the production technology and ways to incorporate linseed in our diet.

Fig. 2.1 Linseed crop at full bloom flowering

2.2 Nutritional Profile of Linseed (Flaxseed)

Linseed or flaxseed (*Linum usitatissimum*) is considered as a functional food due to its very rich composition of essential fatty acids, dietary fiber, lignans, vitamins, high-quality protein, and minerals. This nutrient dense seed provide a wide range of health promoting compounds, making it valuable not only in nutrition but also in precautionary healthcare.

2.2.1 Lipids (Fats)

Flaxseed is exceptionally rich in fats, especially polyunsaturated fatty acids (PUFAs), on dry matter basis; the content of lipid in flaxseed is around 45%. These lipids are favorable to humans as they are in a nice ratio of polyunsaturated and monounsaturated, which appeals to consumes in terms of nutrition (Daun et al. 2003). Flaxseed has a high concentration of polyunsaturated fatty acids such as alpha-linolenic acid (ALA) (about 55%) and linoleic acid (about 14%) and a medium concentration of monounsaturated fatty acids, primarily oleic acid (18%). Some amount of sterols such as tocopherol and tocotrienol are found in linseed. Flaxseeds are the finest source of omega-3 fatty acids for those who do not consume seafood. Flaxseed products have been suggested as nutritious supplements in the creation of many food items, including meat extenders, bread, muffins, fiber bars, salad toppings, ready-to-eat cereals, baked cereal goods, and spaghetti (Singh 2001).

2.2.1.1 Omega-3 Fatty Acids (Alpha-Linolenic Acid, ALA)

The major form of fatty acid found in linseed is alpha-linolenic acid (ALA), an essential omega-3 fatty acid, which makes flax the richest source of vegetarian plant-based omega-3. Omega-3 and omega-6 fatty acids are both essential polyunsaturated fats, which means that the body needs to take them through diet and are not possible for the body to produce by itself. Naturally, fishes are found rich in omega-3 but vegetarian sources of omega-3 are less (Fig. 2.2). Therefore, the impact of linseed in nutraceuticals is very high. There are two main types of omega-3, the first is eicosapentaenoic acid (EPA) and the second form is docosahexaenoic acid (DHA). These are vital for hormone production, immune function, blood clotting, cell growth, and maintaining healthy cell membranes. Plant-based sources such as flaxseed and walnuts contain ALA, which is the precursor to EPA and DHA.

Studies have shown that ALA exhibits anti-inflammatory, anti-thrombotic, and anti-arrhythmic properties (Simopoulos 2002). Nutritionists worldwide emphasize the inclusion of omega-3 fatty acid sources in the diet, making flaxseed an ideal choice for non-fish eaters as it serves as an excellent omega-3 fatty acid source.

More than half of the fat content in flaxseeds is made up of alpha-linolenic acid (ALA), an essential omega-3 fatty acid. Omega-3 fatty acids are beneficial for healthy infant growth and development. Omega-3 fatty acids are essential fats that offer a wide range of health benefits. They support heart health by reducing triglyceride levels, lowering blood pressure, and preventing the buildup of arterial plaque. Omega-3s also have powerful anti-inflammatory properties, helping to reduce chronic inflammation and alleviate symptoms of conditions like arthritis. These

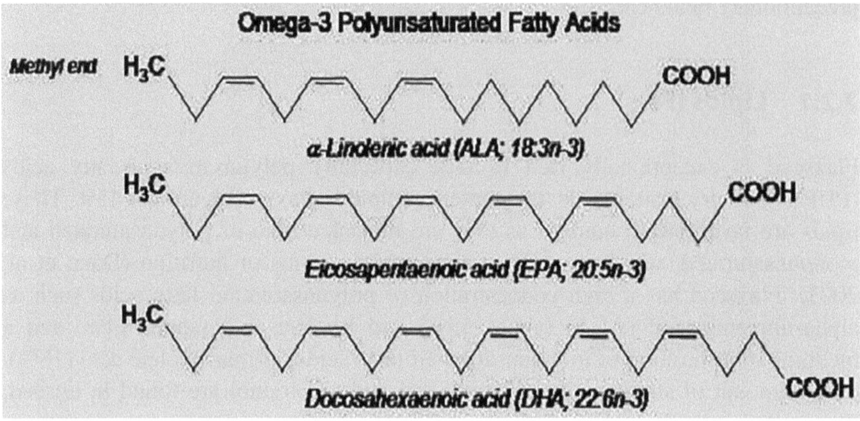

Fig. 2.2 Chemical structure of omega-3 polyunsaturated fatty acids found in linseed, which impart different health benefits

healthy fats are crucial for brain function, improving memory, mood, and even reducing the risk of cognitive decline or mental health issues such as depression. Additionally, omega-3s promote eye health, enhance skin hydration, and contribute to healthy joints and bones. With their ability to support weight management, improve metabolic health, and even reduce the risk of certain cancers, omega-3 fatty acids are a cornerstone of overall health and well-being. Nutritionists from all over the world are now emphasizing the incorporation of omega-3 fatty acid sources in the diet. This makes linseed an ideal choice for vegans because it is an excellent omega-3 fatty acid source in plant kingdom.

Flaxseed is available in various edible forms, including whole flaxseed, ground meal, and extracted oil or mucilage. These products are being proposed as nutritional additives in the preparation of a wide array of dietary items such as baked cereal products, ready-to-eat cereals, fiber bars, salad toppings, meat extenders, bread, muffins, and spaghetti (Singh 2001).

Despite the numerous clinical evidences supporting the health benefits of flaxseed, many people remain unaware of its nutritional and therapeutic advantages. Raising awareness about flaxseeds as a functional food could lead to better health outcomes and an increased interest in incorporating this nutrient-rich ingredient into daily diets.

- ALA constitute around 50–55% of the total fat content.
- ALA is a plant-based essential omega-3 fatty acid that the human body cannot synthesize and must be obtained through diet.
- It is a precursor to EPA and DHA, which are very important for brain health, preventing cardiovascular diseases and reducing inflammation.
- Habitual ALA consumption has been linked to lower risks of heart disease, stroke, hypertension, and rheumatoid arthritis.

2.2.1.2 Omega-6 Fatty Acids (Linoleic Acid)

Linseed has slighter amounts of omega-6 fatty acids (linoleic acid) around 15–18% of total fat. Omega-6 is also essential but should be balanced with omega-3 intake to prevent pro-inflammatory effects.

2.2.1.3 Saturated and Monounsaturated Fats

Linseed or flaxseed contains low levels of saturated fats (around 3.5–4%) and moderate monounsaturated fats (~7.5%). As a result, flaxseed oil has been shown to be very healthy and useful, and it is better for heart health compared to other oilseeds.

2.2.2 Protein

Linseed is a very good source of plant-based protein (18–20%), and they contain all nine essential amino acids, though some (like lysine) are in lower amounts. Linseed protein is globulin- and albumin-rich, easily digestible, and suitable for vegetarian and vegan diets. It contributes to muscle repairs, immune function, and enzyme synthesis.

2.2.3 Carbohydrates

2.2.3.1 Dietary Fiber

Flaxseed is extraordinarily rich in fiber, with just about 27 g per 100 g, mostly insoluble fiber, but also some soluble mucilage. Insoluble fiber helps to improve bowel movements and prevents constipation and soluble fiber (mucilage gum) forms a gel in the gut, slowing digestion and absorption of nutrients. High fiber content supports gut health, promotes satiety, weight management, help to regulate blood sugar and cholesterol levels.

2.2.4 Lignans: 300–800 mg/100 g

Flaxseed is the richest dietary source of lignans, predominantly secoisolariciresinol diglucoside (SDG). Lignans are phytoestrogens plant-based compounds with estrogen-like activity and antioxidants. They have been shown to reduce the risk of hormone-related cancers (breast, prostate), improve cardiovascular health, and help in hormonal balance, especially in postmenopausal women.

2.2.5 Vitamins and Minerals

Flaxseeds are a good source of different vitamins and minerals, which is shown in the following table:

Vitamin	Function
Vitamin B1 (thiamine)	Supports nervous system and carbohydrate metabolism
Vitamin B6	Involved in amino acid metabolism and neurotransmitter synthesis
Folate (B9)	Essential for DNA synthesis, especially significant during pregnancy
Vitamin E (alpha-tocopherol)	Antioxidant that protects cell membranes, though in lower amounts in flax

Mineral	Function	Content per 100 g
Magnesium	Muscle and nerve function, enzyme activity, bone development	~392 mg (98% DV)
Phosphorus	Energy production (ATP) and bone health	~642 mg (92% DV)
Iron	Hemoglobin formation and oxygen transport	~5.7 mg (32% DV)
Zinc	DNA synthesis, immune function, and wound healing	~4.3 mg (39% DV)

Mineral	Function	Content per 100 g
Calcium	Neuromuscular function and bone health	~255 mg (25% DV)
Manganese	Antioxidant enzyme function and metabolism of carbohydrates and amino acids	~2.5 mg (108% DV)
Selenium	Thyroid health, antioxidant, and immunity	~25 mcg (46% DV)
Copper	Helps form red blood cells and maintains nerve cells and immune system	~1.2 mg (60% DV)

2.2.6 Ant Nutritional Factors

Anti-nutritional factors are those compounds that may interfere with nutrient absorption of a particular food item or cause toxicity if consumed in large amounts. Linseed contains compounds such as phytic acid that binds to minerals such as iron and zinc, reducing their bioavailability. Other compounds include cyanogenic glycosides, which may release cyanide under certain conditions, and Trypsin inhibitors, which may interfere with protein digestion. Both of the above-mentioned compounds can be reduced by heat, i.e., by either roasting or boiling of linseed (Plaha et al. 2023).

2.2.7 Water Content

Linseed has low moisture content (~6–8%), which gives it a long shelf life if stored properly in airtight, cool, dark conditions.

2.3 Methods to Incorporate Linseed in Diet

Linseed is gaining importance in the global food market in the form of a functional food, i.e., the food ingredients offering health benefits and help in preventing and/or curing diseases. The increasing interest of people in health-conscious diets has emerged a new fascination with flaxseed because of its unique and outstanding nutritional profile and potential health benefits.

Before incorporating linseed, it is important to understand that the way of consuming linseed makes a huge difference in the absorption of nutrients in the body. There is a difference between whole and ground forms. Consuming whole seeds make it inappropriate as the seeds mostly pass through the digestive tract undigested, meanwhile if ground linseed is consumed, it becomes more available, i.e., the body can absorb its nutrients better. Therefore, consuming ground linseed is recommended for maximum health benefits. Flaxseed oil alone contains neither the fiber nor the phytochemicals of whole flaxseed meal. Since the flaxseed contains highly unsaturated oils, it is likely to get rancid soon. Hence, it needs to be consumed immediately once it is used for any dishes. People with inflammatory bowel conditions, pregnant and lactating mothers, women with fibroids endometriosis and polycystic ovary should not consume flaxseed.

We can include linseed in our daily diet by utilizing it as smoothies, porridge, in breads, salad dressings, salad sprinklers, crackers, energy bowls, and so on. Linseed is a popular vegan egg replacement in baking. Flaxseed oil can be used in ice creams at a rate of 2% (w/w) without changing the ice cream quality and enhance nutritional quality. A study on utilization of linseed in diet has shown that microencapsulation of linseed powder in yoghurt provides a good delivery of omega-3 fatty acid (Goyal et al. 2016). Breakfast cereals or snacks with extruded flaxseed may have health benefits for consumers (Wu et al. 2007). Pasta enriched with flaxseed. The health promoting properties of the linseed are ensured by its content, which is a good source of protein. Protein and omega-3 fatty acid of wheat chip samples are enhanced with the inclusion rates of 10% flaxseed (Yuksel et al. 2014).

The beneficial lipid profile of flaxseeds alters apparent omega-3 polyunsaturated fatty acid contents and profiles among snack foods. Foods such as *chapatti*, *khakhra*, vegetable *chilla*, and Manchurian are prepared by incorporation of flaxseed flour in varying ratios. Flaxseed can be used to increase the quality of some traditional animal origin foods, such as beef patties.

2.4 Production Technology

2.4.1 Land Requirements

The best performance is seen on black or clay loams soils. It survives in regions with high rainfall. The best soil is well drained with soil pH ranging from 5.0 to 7.0.

Ploughing of the land must be done for three to four times followed by two to three harrowing to achieve a fine tilth. To conserve soil moisture, creating a surface mulch using a hoe is recommended.

2.4.2 Sowing Time

The sowing time of linseed is from mid-October to mid-November. This time varies as per the agro-climatic zones in various states. If rainfed cultivation is practiced, early sowing should be encouraged because it helps in better utilization of residual soil moisture. Early sowing also helps in avoidance of pest and diseases such as powdery mildew and rust.

2.4.3 Spacing, Seed Rate, and Seed Treatment

Spacing must be maintained, 23 cm row spacing and 7–10 cm plant to plant. The depth of sowing plays an important role as it helps the plant to better utilize the nutrients from the soil. In linseed, 4–5 cm deep sowing should be practiced. Generally, linseed is a small seeded oilseed and requires a seed rate of 15–18 kg/acre. Seeds are treated with Bavistin or Thiram at a rate of 2 g/kg of seed prior to sowing for prevention of soil borne infections and diseases.

2.4.4 Nutritional Requirements

Twenty-five kilograms of nitrogen, which can be procured using 55 kg urea/acre, is needed as nitrogen dose for crop growth. Sixteen kilograms of phosphorus, which is equivalent to 100 kg of superphosphate, is needed. In case of enough irrigation facilities, half of nitrogen along with complete dose of phosphorus is applied as basal dose. The remaining nitrogen should be applied during the first irrigation, approximately 35 days after sowing.

2.4.5 Irrigation

There are two major critical periods for irrigation: first, after sowing and the second, when flowers begin to develop into seeds.

2.4.6 *Utera* Farming (Relay Cultivation) in Linseed: Another Major Faming System

Utera cultivation is also known as *paira* or relay cropping, which includes broadcasting of linseed in the field of standing paddy crop few days prior to its harvest. In this method of linseed cultivation, residual soil moisture and other nutrients left in the field after previous crops are used by linseed for its initial growth. It helps in escaping the initial land preparation and irrigation, resulting in significant decrease of cost of cultivation. The farmers having low input resources can therefore opt for this method of sowing. Although the yield in case of *utera* cultivation is low as compared to irrigated or rainfed conditions, but is still favorable because of the overall benefit–cost ratio owing to reduced input requirements. Therefore, *utera* proves to be a cost-effective and resource-efficient option, especially in areas where the farmers do not harvest a second crop due to a lack of irrigation and land preparation inputs.

Fig. 2.3 Linseed crop at compete maturity ready for harvest

2.4.7 Plant Protection Techniques

Bud fly is one of the major insect pests of linseed. Fortnightly spray of Spinosad 45 SC (0.015%) for two times reduces up to 78% bud fly infestation. This practice can enhance the yield by up to 63%. Considering the use of biopesticides, two applications of neem-based commercial formulation containing Azadirachtin 300 ppm help to reduce 63% bud infestation, resulting in yield enhancement by 40%.

2.4.8 Harvesting and Yield

The crop is ready for harvest when the seed capsules (bolls) turn completely brown, usually by April, and the yield of linseed estimated in rainfed condition 8–10 qt/ha and irrigated condition 16–20 qt/ha. Figure 2.3 represents linseed crop at compete maturity ready for harvest.

2.5 Threshing

After harvesting, the linseed plants are bundled and laid out on a threshing area to dry for 4–5 days. Traditionally, linseed threshing is done by minimally beating the dried plants with sticks or under the feet of bullocks. A more resource-efficient method is a two-step process where the plants are firstly thrashed on the ground for loosening out of seeds and then stems are pulled out to complete the process.

2.6 Linseed Improvement Strategies

The major breeding objective is largely determined by its fatty acid composition, especially the levels of linoleic and α-linolenic acids. Wild-type linseed oil typically contains 45–65% α-linolenic acid, a polyunsaturated fatty acid that, while beneficial for industrial applications, is highly prone to oxidation. This makes it unstable for edible use, as autoxidation during storage can lead to undesirable off-flavors (Green and Marshall 1984). To address this, two mutant flax genotypes with reduced linolenic acid content were developed (Green and Marshall 1984; Rowland 1991). These efforts culminated in the release of the low-linolenic acid cultivar "LINOLA™-947," which produces oil with improved oxidative stability, making it more suitable for the food industry (Dribnenki and Green 1995).

2.6.1 Molecular Approaches in Flax Breeding

The development of multiplex PCR-based marker systems such as AFLP (Vos et al. 1995) has enabled the construction of high-density genetic linkage maps in many crops, including potato, barley, rice, maize, and soybean (Castiglioni et al. 1999; Lin et al. 1996). These tools have proven valuable for studying genetic traits and accelerating crop improvement. For such approaches to be effective in flax, marker-trait associations must be validated across diverse cultivated germplasm, relying on the presence of linkage disequilibrium.

2.6.2 Marker-Assisted Selection (MAS)

Molecular markers are involved in identifying the genetic basis of both monogenic and polygenic traits. Marker-assisted selection (MAS) allows breeders to select for complex traits, especially those with low heritability, such as disease resistance more accurately and cost-effectively than traditional phenotypic methods (Knapp 1998; Young 2000). MAS is particularly useful in perennial crops and in developing Backcross Introgression Libraries (BILs), which serve both breeding and QTL discovery purposes (Eshed et al. 1992; Jeuken and Lindhout 2004). However, successful MAS depends on strong linkage between markers and target genes, supported by dense genetic maps and high-quality genotyping.

In flax, molecular marker techniques have been used to analyze genetic diversity within and between species, often through cluster analysis and diversity indices. Optimization of AFLP fingerprinting and linkage map development is crucial for accurate QTL mapping. By integrating molecular marker data with trait values, researchers can explore the genetic architecture of key traits and expand the understanding of the flax gene pool. The goal is to discover new alleles and enable marker-assisted breeding (MAB), particularly for disease resistance and quality traits.

2.7 Value Addition in Linseed and Processing Strategies

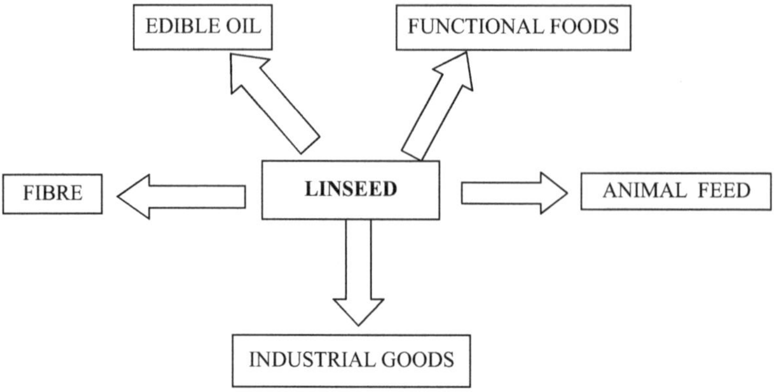

Flax is a multipurpose crop that has high-value uses in industrial oil, paper/pulp, textiles, composites, and nutrition. The two main products are linseed oil, which comes from seeds, and fiber, which comes from stems. Both flaxseed and flax fiber have significant growth potential, despite making up only 1% of the world's oilseed production and 3% of its natural fiber production, respectively.

2.7.1 Value-Added Food Items from Linseed

There are many types of products that can be prepared from linseed to boost its use as functional food in daily diet.

Omega-3 Capsules

Since the vegetarian source of omega-3 are very limited, nowadays nutraceutical companies are offering vegan source of omega-3. They are produced by creating gelatin-coated capsules of linseed oil. The optimum level of consumption of these capsules offers great benefits for cardiovascular health and gut health.

Linseed Chutney Powder

To prepare this delicacy, roasted linseed is grounded along with curry leaves, rock salt, and chilies as per choice to prepare the powdered chutney, which can be consumed daily and also along with curd during fasting.

Salted Linseed

A simple add-on after meals, it can be prepared using roasted linseed along with regular salt or rock salt for fasting purposes.

Flaxseed Laddoo
Roasted and ground linseed is mixed with small jaggery pieces. *Laddoos* are prepared using clarified butter. This *laddoo* is rich in omega-3 and iron content and is nutritious.

Linseed Burfi
Roasted linseed, cashews, and walnuts are ground and assed to sugar syrup. The mixture is then cooled and cut into pieces.

Other preparations include linseed tortillas, holige, biscuits, and so on, which can be used in addition to fasting as well.

2.8 Challenges in Linseed Production

Despite of having tremendous use in various sectors and extensive breeding and agronomic efforts, the adaptability and productivity of linseed is not up to the mark. The following points may contribute to the above statement.

2.8.1 Agronomic Limitations

The growth pattern of linseed is indeterminate, the size of seed is small and seed retention is also less, hampering efficient production.

2.8.2 Market Competition

The market sufficiently contains more stable and low-cost alternatives to linseed oil. The competition among the products in oil market like linseed with edible oils, e.g., rapeseed in the food industry, is high. This stiff competition reduces linseed's global market value.

2.8.3 Processing Difficulties

Flax cultivation and processing is labor-intensive and seed and fiber maturity are generally nonsynchronous.

2.8.4 Issues with Oil Stability

Since linseed oil is highly unsaturated because of the presence of high concentration of omega-3, it is therefore prone to oxidation. It limits the shelf life of linseed oil, affecting its market acceptance.

Despite these problems, linseed remains valuable because of its unique oil composition and increasing recognition of its health and industrial benefits.

2.9 Conclusion

Linseed is unique for its fatty acid profile, specifically omega-3 content, making it a highly valuable and demanding crop in food industry. Its consumption is highly beneficial and offers great health benefits. However, due to certain constraints the market value of this crop is affected and can be tackled by developing value-added markets for its oil and fiber, advancing genomic research, and adopting molecular breeding tools. The demand for health products, animal feed, and sustainable fibers is rising, propelling linseed to emerge as a vital industrial and agricultural crop.

References

Castiglioni P, Ajmone-Marsan P, van Wijk R, Motto M (1999) AFLP markers in a molecular Linkage map of maize: codominant scoring and linkage group distribution. Theor Appl Genet 99:425–431

Daun JK, Barthet VJ, Chornick TL, Duguid S (2003) Structure, composition, and variety development of flaxseed. Flaxseed in Human Nutrition, Champaign, Illinois: AOCS Press 03:1–40

Diederichsen A, Richards K (2003) Cultivated flax and the genus Linum L., Taxonomy and germplasm conservation. In: Muir AD, Westcott ND (eds) Flax: The genus Linum. Taylor and Francis, London

Dillman AC, Brinsmade JC (1938) Effect of spacing on the development of the flax plant. Agron J 30:267–278

Dribnenki JCP, Green AG (1995) LinolaTM'947' low linolenic acid flax. Can J Plant Sci 75:201–202

Eshed Y, Abu Abied M, Saranga Y, Zamir D (1992) Lycopersicon esculentum lines containing small overlapping introgressions from L. pennellii. Theor Appl Genet 83:1027–1034

Gill KS (1987) Linseed. Indian Council of Agricultural Research, New Delhi, p 386

Green AG, Marshall DR (1984) Isolation of induced mutants in linseed (Linum usitatissimum) having reduced linolenic acid content. Euphytica 33:321–328

Hayward HE (1938) Linum usitatissimum. In: Hayward HE (ed) The structure of economic plants. Macmillan, NewYork, pp 371–410

Jeuken MJW, Lindhout P (2004) The development of lettuce backcross inbred lines (BILs) for exploitation of the Lactuca saligna (wild lettuce) germplasm. Theor Appl Genet 109:394–401

Knapp SJ (1998) Marker-assisted selection as a strategy for increasing the probability of selecting superior genotypes. Crop Sci 38:1164–1174

Lin JJ, Kuo J, Ma J, Saunders JA, Beard HS, MacDonald MH, Kenworthy W, Ude GN, Matthews BF (1996) Identification of molecular markers in soybean comparing RFLP, RAPD and AFLP DNA mapping techniques. Plant Mol Biol Report 14:156–169

Plaha NS, Kaushik N, Awasthi S, Singh M, Kaur V, Langyan S, Kumar A, Kalia S (2023) Comparison of nutritional quality of fourteen wild Linum species based on fatty acid composition, lipid health indices, and chemometric approaches unravelling their nutraceutical potential. Heliyon 9(11). https://doi.org/10.1016/j.heliyon.2023.e21192

Rowland GG (1991) Growing flax: Production, management and diagnostic guide. Flax Council of Canada and Saskatchewan Flax Development Commission. Can J Pl Sci 71:393–396

Schewe LC, Sawhney VK, Davis AR (2011) Ontogeny of floral organs in flax (Linum usitatissimum; Linaceae). Am J Bot 98:1077–1085

Simopoulos AP (2002) Omega-3 fatty acids in inflammation and autoimmune diseases. J Am Coll Nutr 21:495–505

Singh DN (2001) Heritability and genetic advance in linseed (Linum usitatissimum)

Vos P, Hogers R, Bleeker M, Reijans M, van de Lee T, Hornes M, Frijters A, Pot J, Peleman J, Kuiper M, Zabeau M (1995) AFLP: a new technique for DNA fingerprinting. Nucleic Acids Res 23:4407–4414

Wu ZY, Wang JX, Tang YC (2007) Huma should be Linum usitatissimum, not Sesamum indicum—with special reference to the source of confusion of names for traditional Chinese medicine. Acta Phytotax Sin 45:458–47

Young ND (2000) A cautiously optimistic vision for marker-assisted breeding. Molecul Breed 5:505–510

Yuksel F, Karaman S, Kayacier A (2014) Enrichment of wheat chips with omega-3 fatty acid by flaxseed addition: Textural and some physicochemical properties. Food Chem 145:910–917

Part II

Pseudocereals and Minor Millets

Grain Amaranth (*Amaranthus* spp.): The Sacred Seed of Ancient Civilizations

3

Navjot Kaur, Shubham Verma, and Satbeer Singh

Abstract

Amaranth is a multispecies crop gaining attention for its remarkable potential to enhance global food security, particularly in developing countries, due to its exceptional nutritional value, strong resilience to abiotic stresses, and adaptability to marginal environments. It serves as a highly nutritious, rich in protein, essential amino acids (like lysine), fiber, and essential minerals. Owing to its easy digestibility over these benefits, amaranth is often referred to as a fasting superfood. Amaranth holds a significant position in the food sector due to the rising demand for gluten-free products and the awareness of its health advantages. Its oil is the richest plant-based source of squalene, which has been used in various industrial and pharmaceutical applications. Additionally, amaranth's pharmacological characteristics, such as its antioxidant and anticancer activities, point to its potential as a source of medicinal compounds. It is a cost-effective crop for farmers due to its capacity for multiple harvests within a single growing season. Despite being a traditionally underutilized crop, recent breeding advancements have begun to unlock its potential. Grain amaranth species have high genetic variation emphasized as a vital source for breeding initiatives aimed at enhancing crop productivity and resilience. This chapter highlights the necessity of identifying diverse parental lines in germplasm collections through comprehensive characterization, as this can aid in the creation of superior hybrid variet-

N. Kaur · S. Singh (✉)
Academy of Scientific and Innovative Research, Ghaziabad, Uttar Pradesh, India

Agrotechnology Division, Council of Scientific and Industrial Research- Institute of Himalayan Bioresource Technology, Palampur, Himachal Pradesh, India
e-mail: satbeer@ihbt.res.in

S. Verma
Agrotechnology Division, Council of Scientific and Industrial Research- Institute of Himalayan Bioresource Technology, Palampur, Himachal Pradesh, India

A. Kumar et al. (eds.), *Fasting Superfoods*,
https://doi.org/10.1007/978-981-95-2762-5_3

ies using current breeding methods. The study also addresses the significance of efficient postharvest management techniques in preserving the viability, quality, and nutritional integrity of amaranth grains, which are essential for maximizing the benefits of amaranth in food systems. Thus, amaranth serve as a climate-resilient crop capable of addressing future food and nutritional security challenges.

Keywords

Geographic distribution · Botanical description · Cultivation practices · Genetic diversity

3.1 Introduction

3.1.1 Origin

The Amaranthaceae family is known as the family of Amaranth, which includes approximately 70–80 species (Han et al. 2025). The word *Amaranthus* came from the Greek word Anthos, meaning eternal or unwilting (Allen 1961). Currently, it is referred to as a third-millennium crop plant. Taxonomically, the family is separated into two parts *Blitopsis dumort* and *Amaranthus saucer*, with almost an equal number of species (Ascherson). The species are classified according to how they are used: grain, vegetable, weedy, and ornamental amaranth (Sauer 1967). The grain amaranth comprises three species: *Amaranthus cruentus*, *Amaranthus hypochondriacus*, and *Amaranthus caudatus*. *Amaranthus cruentus* (Fig. 3.1) is a dual-purpose crop that is mostly farmed for grain seed, but some people also harvest tender leaves for flavor. However, in the United States, it is initially classified as the grain species. It can grow to a height of 2 m and produces a single dominant center

Amaranthus caudatus *Amaranthus cruentus* *Amaranthus hypochondriacus*

Fig. 3.1 Field view of grain amaranth species during flowering

taproot with a red-colored and ribbed stem (Aderibigbe et al. 2022). *Amaranthus hypochondriacus* is the most domesticated, and Mexico produces a considerable amount of it. *Amaranthus hypochondriacus* is a strong, erect plant that can grow up to 40–200 cm tall (Achigan-Dako et al. 2014). *Amaranthus caudatus* name comes from its blood-red small petal-less blossoms (Fig. 3.1). It takes 45–75 days to mature, making it the season with the shortest duration (Berkelaar and Alemu 2006). Cultivars with green-yellow blooms are available along with weedy species: *Amaranthus hybridus*, *Amaranthus powellii*, and *Amaranthus quitensis*. *Amaranthus tricolor* species is known for its vegetable and ornamental purposes.

Amaranth is a crop that grows quickly and is frequently referred to as the poor man's vegetable because of its low cost of production, making it one of the most affordable crop in tropical markets. In contrast to other green vegetables, it is grown in the summer when there will not be any other green veggies in the market. It is also used in the pharmaceutical business as a lubricant and for natural dyes. People with gluten allergies can benefit from amaranth flour because it is gluten-free. Amaranth leaves are particularly beneficial for cancer patients because they have anticancer properties that stop the development of cancer cells in the breast, colon, and liver (Li et al. 2015). Furthermore, numerous well-known goods on the market are derived from amaranth grains (Mlakar et al. 2009). Amaranth is one example of a dicotyledonous plant that uses the C4 cycle for photosynthesis, which may serve as a model plant for research on the genetic regulation of photosynthesis (Joshi et al. 2018). The utilization of amaranth as a grain and vegetable crop that provides nutrition to poor people in developing nations.

Food security is one of the biggest concerns for future generations. Amaranth has readily available plant sources of micronutrients that may be grown in the environmental conditions of the majority of countries, hence reducing micronutrient malnutrition. Compared to main crops, amaranth has experienced less domestication pressure. It is renowned for the great diversity of its germplasm. Their genetic improvement is necessary, as is investigation into the breeding behavior of these crops, the potential for population variation, and the use of molecular markers to examine variety.

3.1.2 Geographic Distribution

The genus *Amaranthus* is commonly distributed across subtropical, temperate, and tropical regions of the world (Sauer 1967). While majority of amaranth species are prevalent in the Americas, few (15) are indigenous to Asia, Africa, and Europe. Grain amaranth was found in a Mexican cave, where the seeds of *A. cruentus* and *A. hypochondriacus* estimated to be up to 6000 years old. However, the distribution implies that *A. hypochondriacus* genotypes both wild and domesticated were most likely imported to Latin America from Asia in the early 1700s via sea and land routes. Over the past 50 years, numerous theories regarding the three species of grain amaranth have been put forward. Sauer (1967) proposed hypotheses, including single or several domestication events from distinct ancestors that might give rise to grain species and the possibility that all three species originated from *Amaranthus hybridus*. Studies based on DNA polymorphism, meiotic behavior of hybrid chromosomes, and allozyme analysis suggest that *A. hybridus* is their common ancestor.

In ancient Greek times, the amaranth was known as Elichous and Chrusanthemon and considered sacred by Ephesian Artemis. They believed that amaranth had healing properties and used it as an immorality symbol in the adornment of tombs and gods. The Aztecs and prior American tribes relied significantly on amaranth as a staple meal. The Aztecs thought it has mystical properties that gave it powerful. Because of this notion, amaranth was employed as a grain in religious rites, comparable to corn (Assad et al. 2017).

The amaranth was wiped out in the United States after Mexico was taken over by Spanish Conquistadors in the early 1500s. It was reintroduced in the 1970s by US researchers (Hancock 2022). Due to scaled-up US research, amaranth seed production was significant through the late 1970s, despite the annual output being confined to a few acres. Currently, only a minimal amount of grain is raised in the Aztec. According to evidence, the grain amaranth with pale seeds was grown in German in the sixteenth century. By the 1700s, it was widely distributed throughout Europe and employed as an ornamental and herb. The cultivation of amaranth spread over the world and was widely used for food in countries such as Africa, India, and Nepal whether in the form of leaves or seeds. Farmers in Russia, Italy, China, Australia, Greece, and South America are also increasingly cultivating it (Stallknecht and Schulz-Schaeffer 1993). Amaranth remerged as a cultivated crop in Mexico more than two decades ago. The *A. caudatus* is also grown for its grain in South America and the southwestern United States (Orona-Tamayo and Paredes-López 2024). Surprisingly, amaranth tops the list of well-known classic greens from China and Southeast Asia, followed by Chinese spinach. Amaranth is gaining popularity in East Africa. Grain amaranth is rarely accepted and commercialized throughout Africa, except in Kenya. This may be the reason why the FAO production statistics do not include grain amaranth production linked to other grains or cereals (Rosentrater and Evers 2018).

3.1.3 Significance

The grain amaranth is one of the most significant crop among the pseudocereals. The outstanding nutritional profile of the grain amaranth is among its most notable thing that are deprived in most of the cereals. It is a great addition to vegetarian and plant-based diets since it is particularly high in lysine, an amino acid that is scarce in the majority of other grains. Amaranth served as a useful diet for preventing malnutrition, especially in susceptible groups including children, expectant mothers, and the elderly people (Maurya and Arya 2018). Amaranth is often referred to as a fasting superfood due to its high nutritional content and its widespread use during religious fasts in many cultures.

Grain amaranth is a crop that is hardy and adaptable from an agricultural perspective. It grows well in a variety of agroclimatic locations because of its resistance to drought, it is especially well adapted to dry and semiarid areas. The plant grows well in poor soils and does not require a lot of fertilizer or pesticides, thus it requires very little agricultural input (Nazeer et al. 2022). This feature makes it the perfect crop for farmers with minimal resources in areas where rainfall patterns are unpredictable or land degradation is a problem. Its C4 photosynthetic pathway increases its carbon fixation and water use efficiency, making it a good fit for agriculture.

3.2 History: Traditional Uses and Cultural Significance

Grain amaranths are rich in cultural and historic importance that crosses centuries and continents, which makes them more than an agricultural crop. Many cultures have grown and worshipped these plants, especially in the Americas, where they have been essential to indigenous people spiritual and nutritional rituals (Sauer 1950). Since grain amaranths have been grown for ancient times, they have formed a close historical bond with human communities that had a lasting impact on food and agricultural practices. The way these grains have been incorporated into various cultures everyday routines and ceremonies demonstrates their recognition. Moreover, amaranth seeds were employed by the ancient Mexicans to manufacture zoale, a dough used to build idols (Andrews 2000). In the late 1800s, Edward Palmer saw that people in a town close to Guadalajara were used grain amaranths to make rosaries out of "suale," or amaranth dough. During religious events, especially on the Feast of Cat, these were wrapped in maize husks. The main usage of grain amaranths nowadays is for alegrias, which are little cakes produced from popped seeds that have been bonded with syrup (Hernandez-Rodriguez 2021). Despite being commercial confectioneries, these are particularly well-liked on various saint's days, showing that they still have a place in community meetings and cultural festivals. The development of grain amaranths names also reflects their historical significance. In various places of Mexico, the common term "huauhtli" has steadily been substituted by synonyms (Baker 2006). This change in language demonstrates how the interaction between humans and these plants has evolved. Its historical significance in the agricultural history is highlighted by the fact that early authors such as Sahaguin recognized huauhtli as "bledo," connecting it to the popular Spanish word for amaranths. Grain amaranths are grown throughout Asia, especially in the Himalayas, and are known by a variety of regional names, including "ramdana," which means grain devoted to the god Rama. The spiritual importance of amaranths in various cultures is highlighted by this identification with divinity, which connects them to religious rites and beliefs (Sauer 1950).

3.3 Utilization Pattern: Traditional and Modern Uses

Amaranth is known for the multifunctional purpose as it can be used as food, industrial, pharmaceutical, and fodder crop. The utilization of amaranth in various prospects has been mentioned in Fig. 3.2.

3.3.1 Food

Amaranth is an excellent source of nutrition. It has a higher quantity of minerals and proteins than cereals such as rice, wheat, sorghum, and millet. Additionally, it contains an adequate amount of unsaturated fatty acids as well as essential amino acids, especially lysine, lacking in many other grains (Mustafa et al. 2011). The protein has comparatively high sulfur-rich amino acids, which are typically low in pulse or cereals crops. Due to the grain amaranth balanced amino acid composition, it has been said that its protein content is extremely close to the FAO/WHO-recommended standards (O'Brien and Price 2008). Over the past, amaranth has been a staple crop in many regions, especially in the areas where it was first cultivated. It is a versatile crop that has been used as a grain and leafy vegetable, making it a rich resource for the traditional cuisines across the world. One of the most prominent traditional uses of grain amaranth in Mexico was the creation of a confection produced by popping the amaranth seeds and combining them with unrefined sugar or honey to create bars. Originally, this was used in ceremonial offerings that originated in the Aztec civilization and later became a common snack. In addition, amaranth flour was used to produce *atole*, a thick beverage made by boiling it with water and sugar. The main purpose of *atole* in traditional Mexican diets was reflected in its ability to provide warmth and energy when consumed during festivals (Kauffman 1992). Moreover, in Peru, Bollos were made from the popped seeds along with syrup. Amaranth, commonly referred as *rajgira* in India, is very important during periods of fasting. It is

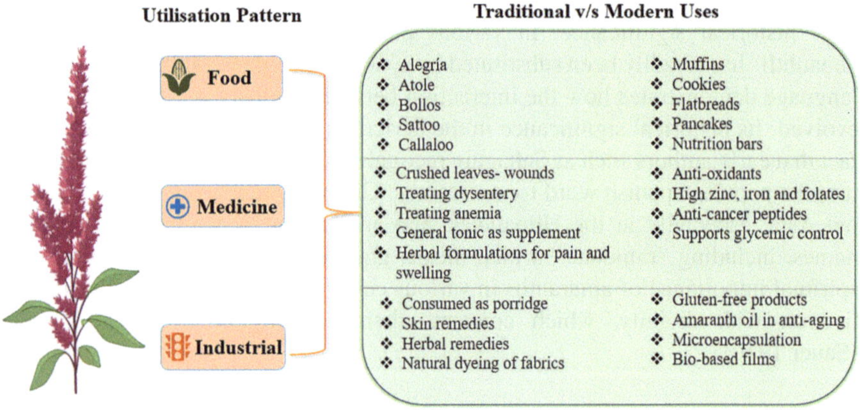

Fig. 3.2 Traditional and modern uses of grain amaranth in food, medicine, and industry

a component of many dishes, including *chikki* (a jaggery-based brittle), *laddoo* (sweet balls), and *rajgira puri*, a deep-fried bread energy dish that is especially appreciated for its nutritional value and efficiency of preparation (Solarov et al. 2022). In Nepal's Himalayan region, amaranth is used to make *sattoo*, a roasted flour that is combined with water, sugar, and occasionally spices to make a porridge or drink (Dwivedi et al. 2024). In alpine tribes, this energy dish is especially appreciated for its nutritional value and efficiency of preparation. Likewise, in East Africa, especially in Kenya and Uganda, amaranth is frequently boiled into a thick porridge that is used as a baby food or as a staple meal (Marimo and Otieno 2025). Its superior protein and micronutrient composition make it a conventional method of preventing malnutrition. In Trinidad and Jamaica, the amaranth plant is utilized for its leaves in recipes like callaloo, while being less popular as a grain throughout the Caribbean and sub-Saharan Africa (Grivetti and Ogle 2000). Nowadays, gluten-free baking is one of its most prominent applications. Amaranth flour is frequently used in baked goods for people with celiac disease or gluten intolerance, such as muffins, cookies, flatbreads, and pancakes (Alvarez-Jubete et al. 2010a). It is also used in making protein-enriched foods. Moreover, amaranth is used in snack meals and multigrain breakfast cereals because of the texture and nutritious qualities, especially when it is puffed or flaked. It is frequently found in nutrition bars, oats, and popped snack crisps. According to scientific research, amaranth improves the fiber content and mineral bioavailability (Olagunju et al. 2022).

3.3.2 Medicine

Traditional medicine uses natural crude extracts from plants to cure numerous ailments. *Amaranthus* spp. is among them; however, its comprehensive therapeutic uses are still unknown. Amaranth was regarded as a sacred and therapeutic grain. It served by the Aztecs in healing practices. Edible figures made from popped amaranth and honey were employed to encourage healing, spiritual harmony, and physical power. It was also thought to help with blood loss and exhaustion, particularly after fights or prolonged travel. For instance, the potash has been made by burning of *Amaranthus* crop in Benin. For instance, in Senegal, the roots were boiled with honey and served to infants. In Ghana, people used macerated amaranth plant to treat joint pain. Moreover, *Amaranthus cruentus* was used as a tapeworm repellent in Ethiopia (Adriana 2013). In Gabon and Sudan, *Amaranthus* stem ash has been used to treat wounds. It helps in promoting the count of red blood cells. Young children, lactating women, and people suffering from constipation, fever, bleeding, anemia, or renal problems can all benefit from the medicinal qualities of *Amaranthus* leaves (Rastogi and Shukla 2013). Phytochemical examination of aerial sections of several *Amaranthus* species confirmed the existence of active elements—flavonoids, alkaloids, phenolic acids, glycosides, saponins, vitamins, terpenoids, catechuic, betaine, carotenoids, tannins, and amino acids. These all can be employed in the formulations of medicines. Amaranth has served as an astringent property. The presence of saponins, protoalkaloids, and betacyans are most likely to be responsible for this impact. The PDR for Herbal Medicines states that amaranth has been used to treat diarrhea, ulcers, and pharyngitis (Capasso et al. 2003). The plant has also been used to treat heavy menstruation, skin conditions such as acne and dermatitis, and as a mouthwash for sore mouth (Esiyok et al. 2004). The scientific literature contains studies of amaranth's therapeutic impact on the cardiovascular (Girija and Lakshman 2011) and hypoglycemic effect (Tiengo et al. 2011), antibacterial activity (Maiyo et al. 2010), and antioxidant properties. Amaranth is frequently employed in the pharmaceutical business to make medical medicines against atherosclerosis (Montoya-Rodríguez et al. 2014), stomach ulcers, anticancer (Maldonado-Cervantes et al. 2010), as well as antipyretic (*viridis* Linn, *caudatus* Linn, & *spinosus* Linn) and hepatoprotective activity (Escudero et al. 2011).

3.3.3 Industry

Ancient cultures have long used *Amaranthus* spp. for a variety of purposes, including as a staple diet and in rudimentary industrial processes. Amaranth seeds were used extensively in religious and ceremonial preparations of food in ancient Mesoamerican civilizations like the Aztecs and Mayans. They were frequently popped and tied with honey to create ceremonial or festival snacks. It is also is utilized to make energy-dense dishes such as *rotis* and *laddoos* during Hindu fasting rituals in India due to its high protein content (Bressani 2018). The grain's nutritional and medicinal qualities overlapped with small-scale processing methods, such as grinding into flour or boiling into gruels. Beyond its traditional culinary purposes, grain amaranth has grown progressively recognized in the modern industrial applications for its many practical benefits. Using its seeds, the food industry creates gluten-free baked goods, breakfast cereals, snack bars, and nutritional supplements for consumers with celiac disease and other dietary limitations as well as those who are health conscious (Ballabio et al. 2011). The cold-pressed oil made from amaranth seeds is ideal for use in pharmaceutical and cosmetic compositions because it contains squalene, with antioxidant and skin-rejuvenating qualities (Ott et al. 2015). Additionally, betalain pigments, which are found in amaranth, are being researched as natural, nontoxic food coloring agents and substitutes for artificial dyes (Cai et al. 2005). Furthermore, its starch content has created interest for the creation of biodegradable films and environment friendly packaging materials. All of these developments show how grain amaranth has evolved from an ancient crop to a modern, multipurpose industrial crop that may help bio-based businesses and sustainable food systems (Chandla et al. 2017).

3.4 Production

3.4.1 Cultivation Practices

Amaranth is a pseudocereal with tiny seeds. It requires fine tilth for better germination. This does not prevent the cultivation of it using conventional seed machines. It is important for farmers to maintain planting depths of 3–5 mm. Crop emergence will be affected if the planting depth is more than 5 mm. It is an extremely versatile species that can live anywhere from 0 to 2400 m above sea level in a variety of temperatures and habitats. Although it can adapt to adverse situations, it produces considerable yields when grown on suitable soil. Farmers may successfully plant amaranth in both mixed and single-production systems. Maize is considered as the best crop for mixed agriculture (Das 2016). Seeds are planted in the field by spreading or placing them in hills. Broadcasting allows you to disperse the seeds across the plot or in rows with a planter or by hand. This method is not cost-effective. When seedbanks are abundant, direct planting is an option. It is also known as sowing in situ, which means that the seeds will not be moved. If the seedling emergence is favorable, this method enables amaranth seedlings to take off early and have a good crop stand. The optimum seed rate is 1.5–3 kg per hectare (Palada 2003). Mixing seeds with sand ensures that they are evenly distributed. The spacing of rows in grain amaranth ranges from 45 to 75 cm, while the plant spacing ranges from 30 to 45 cm. If the rows are wide, it will produce higher yields and allow for the use of row cultivators for weed management.

3.4.2 Climatic and Soil Requirements

Amaranth thrives in tropical, subtropical as well as in temperate regions (Rastogi and Shukla 2013). It grows well in temperatures ranging from 18 to 35 °C, with a minimum temperature of 18 °C when planting (Wolosik and Markowska 2019). The germination rate depends upon the warm and sunny weather. Even with only 500 mm of rain per year, amaranth has sufficient yield. During the dry season, it is important to adjust irrigation. Medium–heavy, loose, and friable soils are ideal for amaranth growth. The pH range of soil is 5.0–8.0 (Rastogi and Shukla 2013). Soils with low clay content will assist decrease crusting problems, which can lead to poor seedling emergence. Amaranth is a resilient plant that survive harsh situations such as full sun, insufficient hydration, and high temperatures (Emire and Arega 2012).

3.4.3 Global Production Statistics

Global data on the area and production of amaranth are limited, but available figures provide useful insights. China is the leading global producer of grain amaranth, followed by Russia (Tyrus and Lykhochvor 2022). In Europe, particularly in Slovakia, Hungary, and Italy, about 1000 ha are under cultivation (Wrigley et al. 2015). Meanwhile, in South and Central America, seed yields can reach up to 7 metric tons per hectare under optimal conditions (CropTrust 2024). Yield variability is considerable, with typical grain yields ranging from 1500 to 7200 kg/ha, depending on species, environmental factors, and crop management practices (Mukuwapasi et al. 2024). Although comprehensive global production statistics are not available, these regional insights highlight the potential for amaranth as a high-yielding crop in suitable conditions. Continued efforts in breeding, agronomy, and market development could help realize this potential further, particularly in regions facing food and nutritional security challenges.

3.5 Protection

Amaranth cultivation faces significant challenges due to various pests and diseases that adversely affect both yield and quality. Common insect pests include leaf webbers, leaf rollers, leaf miners, cutworms, flea beetles, aphids, and mites, which damage the foliage and reduce the photosynthetic capacity of the plant (Seni 2018). While major diseases include Leaf Spot, Damping Off, Downy Mildew, and White Rust, which can cause leaf lesions, stunted growth, and, in severe cases, plant death (Ahmmed et al. 2024). Effective management of these pests and diseases is crucial for ensuring good yields and marketable produce. Sustainable pest control is best achieved through Integrated Pest Management (IPM) approaches that combine cultural practices, biological control methods, and the careful, limited use of chemical treatments. Regular monitoring, timely interventions, and use of resistant varieties, where available, further contribute to successful management of amaranth pests and diseases (Aparna 2015).

3.5.1 Major Pests and Management Strategies

3.5.1.1 Leaf Webbers (*Psara basalis* and *Hymenia recurvalis*)

Symptoms of Damage

Leaf webbers are highly destructive pests of amaranth found throughout tropical and subtropical regions. They are most active in warm weather, during rainy seasons, and in early winter. These insects inflict severe foliar damage by webbing silken threads to knit leaves together and feeding from within. Their larvae rasp away the epidermal and palisade layers of the leaves, resulting in characteristic webbed clusters. As a result, the affected leaves lose their chlorophyll content, dry up, and ultimately reduce the plant's photosynthetic capacity and overall vigor (Jambagi and Augustine 2022).

Management

Chemical insecticides should be minimized whenever possible; neem-based formulations provide an effective and preferred means of managing common pests (Hsu and Srinivasan 2012). However, when chemical control becomes necessary, nonpersistent chemicals such as 0.1% malathion or malathion 50 EC (500 mL/ha) or dichlorvos (375 mL/ha) in 500 L may be used.

It is important to adopt integrated management strategies, such as removing and destroying wild amaranth plants growing near the crop, as they can serve as hosts for pests. Additionally, affected plant parts, including those with caterpillars or adult insects, should be removed and destroyed to reduce pest populations. Installing light traps (one per hectare) can also help in attracting and killing adult pests. For small-scale control, malathion 50 EC can be sprayed at a concentration of 1 mL per liter of water.

3.5.1.2 *Amaranthus weevil* (*Hypolixus truncatulus*)

Symptoms of Damage

The *Amaranthus* weevil is a notable pest of amaranth, causing characteristic damage that includes irregular, zig-zag tunnels in the pith region of the stem, often filled with frass or excreta. The feeding activity of the weevil larvae leads to gall-like swellings and longitudinal splitting of the stem, which weakens the plant structure (Seni 2018). Infected plants show a suppression of both root and leaf development, resulting in stunted growth, reduced vigor, and ultimately, lower yields.

Management

To control *Amaranthus* weevil effectively, eliminate and destroy wild amaranth growing around the crop, since it serves as an alternate host. Additionally, remove and destroy affected plant parts along with grubs and adult insects to reduce pest populations. After harvesting the leaves, spray insecticides such as malathion 50 EC at 2 mL per liter of water to control any remaining pests and prevent further infestations (Ezeh et al. 2015).

3.5.1.3 Aphids (*Aphis craccivora*)

Symptoms of Damage

Aphids frequently infest amaranth, feeding by extracting sap from young leaves and stems. This leads to leaf curling, yellowing, and stunted growth of the plants. In addition to direct damage, aphids excrete a sticky substance known as honeydew, which provides a favorable environment for the development of sooty mold (Nampeera 2022). This black fungal growth further reduces the plant's ability to photosynthesize and compromises the overall health and yield of the crop.

Management

For effective control of aphid infestations in amaranth, spray Dimethoate 30 EC at 2 mL per liter of water or Imidacloprid 17.8 SL at 0.5 mL per liter of water. Encouraging natural predators such as ladybird beetles and lacewings in the field can also aid in controlling pest populations naturally. Additionally, installing yellow sticky traps in the field is useful for monitoring and reducing pest numbers, helping to manage infestations more effectively. These strategies, when used together, provide a comprehensive approach to pest management in amaranth crops (Nampeera 2022).

3.5.2 Major Diseases and Management Strategies

3.5.2.1 Leaf Spot Disease (*Cercospora* spp. and *Alternaria* spp.)

Symptoms

The disease starts with the appearance of multiple white spots on the leaves, which eventually spread and cause complete damage to the leaf. This problem is more severe in red amaranth varieties than in green ones (Yashaswini et al. 2021).

Management

To manage leaf blight in amaranth effectively, a combination of eco-friendly practices is recommended. Foliar spraying with a 2% solution of *Pseudomonas fluorescens* along with soil application of *Trichoderma viride* at 2% can help suppress the disease (Uppala et al. 2010). Additionally, a mixture of turmeric powder and baking soda in a 5:1 ratio per liter of water can be applied as a foliar spray and for soil drenching. Another useful method involves foliar spraying and soil drenching with a combination of 2% cow dung slurry supernatant and 2% *Pseudomonas fluorescens*. These integrated approaches can enhance plant health and minimize leaf blight incidence in amaranth crops.

3.5.2.2 Damping Off (*Pythium* spp. and *Rhizoctonia* spp.)

Symptoms

The disease typically causes seedlings to suddenly collapse at the soil line; their stems become soft, waterlogged, and eventually rot, and the affected seedlings often wilt (Thanyasiriwat et al. 2025).

Management

For prevention, seeds must be treated with *Trichoderma harzianum* or Captan at a rate of 2–3 g per kilogram before sowing. Proper field management is equally important, ensure good drainage and avoid overwatering, as excess moisture promotes disease development (Parveen and Khatun 2022). If infection is observed, soil drenching with fungicides such as Metalaxyl or Copper oxychloride is recommended to manage the disease and protect young seedlings from further damage. These practices help in reducing disease pressure and promoting healthy crop establishment.

3.5.2.3 Downy Mildew (*Peronospora farinosa*)

Symptoms

Early symptoms include yellow patches on the upper leaf surfaces. As the disease advances, a grayish or purplish fungal growth appears on the underside of the leaves.

Management

To manage downy mildew in amaranth, it is recommended to spray a combination of Metalaxyl and Mancozeb at the rate of 2 g per liter of water (Hedayetullah and Zaman 2018). Additionally, following crop rotation practices and maintaining right spacing between plants to avoid overcrowding can help reduce disease incidence by improving air circulation and reducing humidity levels around the crop.

3.5.2.4 White Rust (*Albugo bliti*)

Symptoms

White Rust causes leaf distortion, reduced growth, and overall stunting of the plant, which can lead to significant yield losses.

Management

To manage this disease, it is recommended to spray Mancozeb 75 WP at 2 g per liter of water or a combination of Metalaxyl and Mancozeb at the same dosage. Additionally, it is important to regularly monitor the crop, remove and destroy infected plants to prevent the spread of the disease, and maintain good field hygiene for effective control (Ahmmed et al. 2024).

3.6 Improvement (Genetics and Genomics)

3.6.1 Breeding Efforts

Breeding efforts in amaranth have evolved significantly, early programs primarily focused on selecting superior lines based on phenotypic performance. Grain amaranth is predominantly self-pollinated crop with a low rate of cross-pollination. Various emasculation and controlled pollination methods were tested and developed. These techniques have made it possible to exploit hybrid vigor more effectively, by crossing genetically diverse parents from distinct heterotic pools. Such crosses result in F1 hybrids that are not only genetically uniform but also exhibit significantly improved traits compared to both parental lines. This uniformity is beneficial for modern agricultural practices, including mechanical harvesting, which requires consistency in plant height and maturity. Furthermore, despite amaranth traditionally being bred as line varieties with limited genetic improvement, notable heterotic effects have been observed even in interspecific crosses, highlighting the potential for further genetic enhancement (Lehmann et al. 1991).

Genetic data can be used to identify diverse groups in amaranth from germplasm collections, as demonstrated in various cereals. Characterizing germplasm collections yields valuable insights into genetic distances among populations, which can be used to establish heterotic pools (Wu and Blair 2017). Among three grain species, *A. cruentus* and *A. hypochondriacus* are related, while *A. caudatus* is distant from them. This will offer a similarity range for hybrid pool detection. Heterotic groups must be confirmed through the test cross among various pools. In hybrid production, homozygous lines were maintained for six to eight generations through repeated self-pollination and selection. This process is laborious and time-consuming. Moreover, the double haploid technique enables the production of entirely homozygous lines in one generation, which is widely used in many crops (Forster and Thomas 2005). Although this approach has not been tried for grain amaranth, it might be a useful way to boost genetic gain through hybrid breeding.

Mutation breeding is one of the promising fields used efficiently in amaranth to establish cultivars and generate new diversity. So far, mutation breeding has produced four amaranth cultivars: Centenario in Peru for enhanced grain yield; New Asutake in Japan for early maturity; Sterk in Russia for improved heat-stress and moisture tolerance; and Pribina in Slovakia (Das 2016). Gamma irradiation in *A. cruentus* considerably increased the 1000 seed weight, and promising mutant lines C26 and C82 with greater 1000 seed weight were established (Hricova et al.

2016). Moreover, in other study, mutant lines containing 1000 seeds or more from a single panicle were isolated, providing excellent breeding material. Radiation mutagenesis in amaranth can produce mutants with valuable economic features, thereby expanding the genetic base and breeding potential of the crop (Hudym et al. 2021). Another method for mutation breeding is induced polyploidy that has been shown to increase yield and improve quality attributes in grain amaranth. It has been widely proven that induced polyploidy has a significant impact on protein, lysine content, and grain size while not affect fertility or grain quality (Pal et al. 1982).

These efforts collectively signify a promising shift toward more structured and efficient breeding programs in amaranth, paving the way for the development of high-yielding, uniform, and agronomically superior varieties suited for modern cultivation practices.

3.6.2 Genetic Diversity

Global crop productivity has declined in recent decades due to rising temperatures and more frequent floods, particularly in the mid-latitudes (Kundzewicz et al. 2007). Utilizing existing plant genetic resources has given plant breeders a great chance to create better cultivars with more desirable traits (Govindaraj et al. 2015). In recent years, significant developments in molecular genetics have provided experts with innovative tools and techniques for preserving the resources of plants and effectively assessing diversity among plant species (Mondini et al. 2009).

Among these advancements, mutation breeding has emerged as a valuable approach, offering opportunities to broaden genetic variability and complement existing germplasm for crop improvement. Mutants produced by the gamma irradiation, i.e., K-433 hybrid and Ficha cultivar of amaranth, were analyzed using ISSR markers, which included six microsatellite- PCR markers. The ISSR method was initially used to analyze gamma-irradiated amaranth mutant lines. The study identified primers that can differentiate mutant lines from control plants. The genetic dissimilarity of the mutant Ficha cultivar lines was 0.16. Their ISSR profiles were more similar to the Ficha than those of K-433 hybrid lines. The K-433 mutant lines (D54, D279, D282) have a 0.22 genetic dissimilarity to the control plant but are more unique than the Ficha mutant lines. Mutant lines have different ISSR fingerprints compared to Ficha cultivars and K-433 hybrids. This could be due to the complicated reaction of the intergenic region to gamma radiation (Ziarovska et al. 2013).

The genetic diversity within and between species was studied in 21 accessions of *A. hypochondriacus*, 19 of *A. cruentus*, and 16 of *A. caudatus* were examined using 11 ISSR primers. Two genotypes of *A. caudatus* belonged to India, grouped with *A. hypochondriacus*, and 1 genotype of *A. hypochondriacus* belonged to Nepal grouped with *Amaranthus cruentus*. All 55 genotypes that were examined had an average similarity index of 0.522, with a range of 0154–1000. Compared to *A. hypochondriacus* accessions to either *A. cruentus* or *A. caudatus* accessions, the greatest intraspecific variability is noted for these accessions (Štefúnová et al. 2014).

Moreover, seven *Amaranthus* cultivars, collected from various agro-ecological regions in India, produced 313 bands at 14 anchored loci of ISSR markers. Eighty of the 84 amplicons that were obtained were polymorphic, whereas the remaining four were monomorphic. Breeding programs can benefit from the studied material in selecting *Amaranthus* cultivars and genotypes specific to a given region (Rathod et al. 2021). In another study, a set of 11 polymorphic ISSR primers generated 114 amplicons, 98 were polymorphic in nature. Each primer produced 8.91 polymorphic amplicons, with average polymorphism percent and PIC value of 87.15% and 0.853, respectively. Furthermore, distinct amplicon marker-producing primers that will be used for identification of genotype were found (Gelotar et al. 2019). Nine AFLP and 14 ISSR primers were investigated to check the diversity of grain amaranth. In all, 92 ISSR loci were discovered, with 86 found to be polymorphic. ISSR and AFLP genetic similarity had a 0.77 correlation, explaining 59% of the association (Oduwaye et al. 2014).

Moreover, 17 accessions among grain amaranth species were studied using RAPD to explore their genetic diversity. *A. cruentus* (69.2%) has the highest polymorphism rate, followed by *Amaranthus caudatus* (38.5%) and *Amaranthus hypochondriacus* (15.4%). The study found significant genetic similarities between *A. hypochondriacus* and *A. caudatus* (Dehmer 2003). By the usage of marker data for phylogenetic analysis, accessions of *A. hybridus* were assigned to two of the three grain amaranth clades, implying that these three farmed species evolved from separate *A. hybridus* progenitors (Mallory et al. 2008). The similar study using RAPD markers for ten species from the genus *Amaranthus* was examined for genetic diversity and evolutionary relationships. It has been found that the closest grain amaranth to *A. hybridus*, according to the overall mean similarity index, was *A. hypochondriacus*, followed by *A. caudatus* and *A. cruentus* (Sammour et al. 2020). The genetic variation of 58 grain amaranth accessions, including 15 parental genotypes and 43 F1 hybrids, was assessed using RAPD analysis. Out of 40 primers examined, four effectively differentiated all genotypes. The intraspecific percentage polymorphism was 84.85 (*A. tricolor*), 100 (*A. cruentus*), 75.76 (*A. edulis*), 100 (*A. caudatus*), and 71.43 (*A. hypochondriacus*) with hybrids showing varying degrees of polymorphism ranging from 42.86 to 71.43% (Patel et al. 2014). Thirty amaranth accessions were evaluated for variation using ten RAPD primers and 15 phenotypic features. The large proportion of accessions had green-pigmented leaves, and statistically significant phenotypic diversity of accessions occurred throughout this phase. This enables the design of effective hybrid breeding programs by strategically integrating genetic diversity with information on phenotypic variation and the traits that contribute to it (Oduwaye et al. 2019). A similar kind of study has revealed that RAPDs genetic relationships among amaranth species that align with their morphological features (Anjali et al. 2013). *A. cruentus*, the most valued cultivar in Benin, is prone to saline stress. To determine the most reliable sample for mutant lines to additional salinity tolerance, and to evaluate their molecular diversity between 19 γ-irradiated *A. cruentus* local cultivar and its related mutant lines were investigated. Six different genotypes were found to be indicative of the 19 mutant lines produced by *A. cruentus* that were the subject of the study.

This discovery offers a diversity of material to take into account for salinity or other breeding purposes (Kpocheme et al. 2022). The intraspecific and interspecific variability of 21 *A. hypochondriacus*, 18 *A. cruentus*, and 16 *A. caudatus* accessions were analyzed using RAPD and ISSR markers. Three microsatellite and random primers amplified a total of 1013 ISSR and 1126 RAPD scorable fragments were used to assess the potential of both approaches. The percentage of polymorphisms ranged from 90 to 100% in ISSR or from 70 to 100% in RAPD based on the intraspecific variability. Except for three accessions, cluster analysis found that one genotype of *A. hypochondriacus* originated in Nepal and was associated with *A. cruentus* accessions, while two genotypes of *A. caudatus* originated in India (Štefúnová et al. 2015). These findings showed that these markers have a wide range of potential applications in studying inter- and intra-specific genetic diversity, also the evolutionary links between farmed and wild amaranths (Lee et al. 2008). In another study by using this approach, 203 loci were detected, 173 of them appear to be polymorphic, 30 are monomorphic (found in all genotypes examined), and 13 are unique (found in only one genotype). Particularly for breed certification, which is crucial for plant material identification and plant genetic variability monitoring, unique and monomorphic DNA loci can be employed as specialized genetic markers (Lymanska et al. 2017).

To attain the overall objective of sustainable agriculture, it is essential to identify superior genotypes for various breeding applications by careful examination of the population structure and comprehensive dissection of genetic diversity within germplasm collections. As a result, the study was carried out to analyze the informativeness of the genomic SSR markers. In order to develop markers, 13,051 genome-wide microsatellite motifs were found using the genome of *A. hypochondriacus* (JPXE01.1). Forty-five microsatellite motifs were also used to investigate the evolutionary relatedness and cross-species transferability of 10 *Amaranthus* species. These species were found to have a total of 97 alleles. Heterozygosity, PIC, gene diversity, and major allele frequency had respective averages of 0.733, 0.347, 0.06, and 0.291, respectively. Between *A. tricolor* and *A. hybridus* were closely related with 0.0625, whereas between *A. lividus* and *A. viridis*, showed a high genetic divergence 0.7918 (Tiwari et al. 2021). Similarly, 94 different *A. hypochondriacus* accessions, including 92 native and 4 exotic lines from the United States and Russia, were evaluated to investigate the population structure and molecular diversity. There was considerable genetic variability in the germplasm indicated high levels of heterozygosity and allelic frequency, which can be further utilized in next breeding initiatives (Vats et al. 2023). Furthermore, population structure and genetic diversity of 70 amaranth accessions received from Southeast Asia and South were assessed by using 14 SSR markers. They discovered that the genetic exchange between Southeast Asian and South amaranth species was comparatively low (Wang and Park 2013).

Crop yield will be impacted by global climate change, which requires crop variety adaptation. Plants have successfully colonized new environments because of the recent global spread. One of these is the incorporation of Amaranth into India. The study used Whole Genome Sequencing to study its successful introduction. The finding indicated that imported crops can conserve genetic variety, increasing adaptability and resilience to future environmental changes (Singh and Stetter 2024). The study aimed to analyze the genetic diversity of 178 Peruvian amaranth genotypes utilizing 96 biallelic markers of SNPs. KASPar™ assays on a 96.96 Fluidigm EP1 array system were used to create 96 codominant, biallelic markers. PCA, PCO, and cluster analysis findings separated the accessions into six distinct groups. Clusters showed no clear organization across accessions, indicating that diversity remain intact across wide area of Peruvian Andes (Jimenez Rondan 2011). The similar technique was used to document the first amaranth single nucleotide polymorphism (SNP) analysis. They also present the species' first complete genetic linkage map. All 411 markers were categorized by 16 linkage groups, corresponding to 16 haploid chromosomes found in amaranth. The resulting genetic map covers a total length of 1288 (cM), with an average marker spacing of 3.1 cM (Maughan et al. 2011). The examination of marker × trait relationships require consideration of population structure, as demonstrated by the consistency of genotyping-by-sequencing (GBS) classification between seed color and geographic origin. In contrast to the outgroup that consists of two wild species *A. retroflexus* and *A. powellii*, the matrix of distances based on common alleles provides insights into close genetic relationships between the wild and domesticated South American species, as well as two domesticated Central American species (Wu and Blair 2017). One particularly common method of subsampling plant gene banks is through core collections. A gene bank germplasm curator often creates a core using accession DNA analysis or phenotypic trait characterization (Schafleitner et al. 2022). The study established a 2899 gene bank accessions core, of which 1090 of them have up-to-date georeferencing for coordinates of latitude and longitude, using collecting site data of farm and town. They used 90 SNP markers to assess the 260 *Amaranthus* genotypes that made up the core. In addition to being more effective than physical features at rapidly differentiating between Andean and Mesoamerican cultivars, molecular markers are also helpful in tracking interspecies crosses and introgression (Blair et al. 2023). Moreover, the KASP markers among wide panel of amaranth genotypes, which included *A. hypochondriacus* and *A. cruentus* have been used to evaluate the varied collection of numerous accessions of three farmed grain and their closest wild relatives, amaranth species. They chose the markers with the greatest PIC values of the earlier SNP data. Despite being regarded as a weedy relative, *Amaranthus quitensis* was found to be closely clustered with the South American species *Amaranthus caudatus*, which was distant from the two Mesoamerican species *Amaranthus hypochondriacus* and *Amaranthus cruentus*, according to phylogeny. While the latter both were most likely to live in a hybrid condition (wild and cultivated), the first pair of species most likely interbred (Thapa et al. 2021). A valuable technique for identifying significant candidate genes linked to significant plant features is GBS data from a variety of genotypes. A genetically diverse set of 192 amaranth accessions was characterized in the current study using the GBS technique. As a result, thousands of single-nucleotide polymorphisms (SNPs) were segregated inside each accession, and an estimated 41,931 SNPs were segregated throughout the collection (Pandey et al. 2023).

Next-generation sequencing (NGS) breakthroughs have altered comparative genomics, improving our knowledge of genome and gene architecture, functions, and dynamics. The comprehensive study of major genomic features of five amaranth species, including functional and structural annotation of genes for protein-coding, analysis of repetitive element, and phylogenetic arrangements based on single-copy orthologs among species of Amaranthaceae. According to a phylogenetic analysis using single-copy of orthologous genes (406), *Amaranthus hypochondriacus* was distantly related to *A. cruentus* and most closely related to *A. hybridus* (Singh et al. 2024).

3.6.3 Recent Advancements

In recent years with advancements in molecular biology and genomics, significant progress has been made in the knowledge of amaranth genome and in the development and application of genetic marker systems (Clouse et al. 2016). These resources have the potential to significantly improve amaranth breeding methodologies. For genome-enabled breeding, one of the most crucial steps is having access to a reference genome resource. Prior to this, resources such as draft genomes, chloroplast genomes (Chaney et al. 2016), transcriptome, and genetic map (Maughan et al. 2011) were already present for this genus. The diversity of amaranth has been studied using a number of genome-wide marker systems in contrast to these genomes. These include whole-genome resequencing and PCR-based SNP markers. Genetic marker systems can be integrated into breeding programs at various stages. This will be especially valuable for assessing the genetic diversity of ex situ collections and to identify QTLs through genome-wide association studies (GWAS) (Lightfoot et al. 2017). There have been no reports of similar studies in amaranth, but they will considerably advance our understanding of trait genetic architecture and identify targets for MAS. Many of the disadvantages of MAS can be overcome via novel genomic selection (GS) technique. In contrast to MAS, GS predicts the genomic estimated breeding values (GEBVs) and uses a high number of genome-wide distributed DNA markers (Meuwissen et al. 2001). This method enables the evaluation of individuals based on genetic information, eliminating the need for extensive field testing. Unlike traditional phenotype-based selection, which involves both phenotyping and genotyping a large number of individuals in the field, the genome-wide model is trained on a subset of individuals called as training population (Desta and Ortiz 2014). Depending upon trait's heritability, GS can help to select the best individuals to pass on to the next generation or eliminate very low-performing individuals. Establishing a strong training population made up of advanced breeding lines is necessary to fully realize the potential of GS in breeding of amaranth. In order to better understand trait variation and enhance the selection of desired traits in breeding programs, low-cost, high-throughput phenotyping tools are being created with advancements in crop genomics. The ability to phenotype complex quantitative features with high resolution, accuracy, and precision has been made possible by the recent development of next-generation phenotyping, which is equipped by 3D imaging, software, and remote sensing (Cobb et al. 2013). Likewise, automatic optic seed sorting assays could greatly boost quality seed production and accelerate genetic improvement efforts (Pearson 2010). Large seed batches might be rapidly and precisely purified using automatic optic seed sorting equipment as compared to conventional assays based on surface texture, color, size, and shape (Tanabata et al. 2012). Automated seed phenotyping systems would be especially useful in small seeded crops like amaranth, where conventional approaches are labor-consuming and frequently unable to discriminate seed size. The genetic makeup of breeding targets in amaranth can also be better understood by integrating high-density genotypic data with high-throughput phenotypic records.

Apart from that, genome editing technique have emerged as a powerful tool for revolutionizing basic research and the improvement of crop plant traits by enabling rapid, precise, and predictable genome modifications (Bortesi and Fischer 2015). Desired alleles or gene complexes can be directly translated into an elite genetic background using genome editing, even if the normal population does not have these allelic combinations. Plant genome editing has been done using a variety of technologies. However, due to their complex structures, transcription activator-like effector nucleases (TALENs) and zinc-finger nucleases (ZFNs) have essentially been replaced by the most recent CRISPR/Cas9 system. One of the most critical prerequisites for CRISPR/Cas9-based gene editing is the presence of an effective transformation process, as the successful introduction of the T-DNA construct containing the guide RNA and Cas9 into the plant is crucial. Although established genome editing methods are currently lacking in amaranth, developments in genetic transformation raise the prospect of this technique to improve a number of traits in the near future (Ma et al. 2016).

3.7 Primary Processing

Efficient postharvest handling is critical for the long-term viability of any crop's supply chain. Comparable to other cereals, amaranth needs postharvest strategies and practices to maintain its nutrition value, stay viable, and acquire a high price in the market. Grain and seed postharvest management is similar, while before harvest, operations for seed and grain are different. After harvesting, management of grain includes drying the panicles, threshing and cleaning of seeds, followed by processing and storage. If grain losses are to be reduced, these few processes must be practiced with the detailed care they merit. Crop harvesting is mostly done by hand, with sickles, knives, and cutters being the most common instruments. However, combined harvesters are also preferable. The time of harvesting and manner are critical factors in determining seed loss during harvest (Manikandan and Srimathi 2014). The grains have high moisture content during early harvesting, which raises the cost of drying, making it more responsive to the growth of molds. If not thoroughly dried, it will result in a yield loss. Many of the cultivars grow quickly and can be harvested within 6–8 weeks of sowing. Following drying of panicles, the seeds are threshed and cleaned by hand or machines (Ronoh et al. 2010). The threshing process detaches the grain from the panicles, whereas cleaning separates them from other foreign objects such as sand, chaff, and stones or broken grains. After cleaning and drying the grain, keep it in heavy-duty bags made of paper or wooden storage boxes until ready for use.

Moreover, the processing techniques affect the digestibility of proteins, antinutritional factors, antioxidant activity, and minerals in flour or grains of pseudocereals and cereals. There are number of processing methods employed in amaranth grains to create various products based on how people use or consume them. Amaranth grains are tiny, hence special equipment is needed to milling them. Although planetary ball and roller mills can also be utilized, especially for making starch-rich flours. Abrasive mills, which are frequently used for rice or sorghum, are also efficient (Roa et al. 2014). Ball milling can be used to create modified amaranth flour because it increases the solubility, decreases the size of the particles, and increases the stability of liquid foods. The bran, endosperm, and germ of amaranth grains are successfully separated by differential milling, producing flours that are higher in fiber, protein, and carbohydrates, because it produces no effluents and does not require dissolving or filtration, this dry milling process is environmentally beneficial. This procedure produces three different granulometric fractions when paired with pneumatic classification: high-fiber fraction (6.5–7.1%), high-starch flour (46.2–47.6%), and high-protein flour (16.4–16.7%) including oil. Pneumatic classification can further concentrate the fiber content, resulting in flour that contains up to 64% insoluble fiber and 7% soluble fiber. This strategy makes it possible to produce fiber-rich flours with only physical processes, which have little negative effects on the environment (Tosi et al. 2001). As a result of moisture loss during milling, nutrients such as proteins and lipids concentrate in the bran or germ, affecting flour yield and particle size. Most cost-effective ways to improve amaranth nutritional and functional qualities include germination and sprouting (Hübner and Arendt 2013). Amaranth sprouts are in trend due to their health advantages, although being less prevalent. However, germination decreases antinutrients such as phytates and enhances the digestibility of proteins and carbohydrates as well as amino acid profiles. Soaking, steaming, and kilning are further methods that increase phenolics and antioxidants (Alvarez-Jubete et al. 2010b). Pretreatments such as boiling and steaming preserve essential minerals but deplete certain vitamins, such as folate. Although further studies are required to determine how germination and heat treatments affect functional qualities. Cereals have traditionally been processed via fermentation. It has been discovered that fermenting Ethiopian amaranth increased protein digestibility (5–7.5%) and preserved the majority of free amino acids better than popping, while it resulted in a 20% reduction in methionine, lysine, and cysteine (Amare et al. 2015). Using amaranth sourdough, Sterr et al. discovered *Lactobacillus plantarum* and *Pediococcus pentosaceus*, which successfully reduced pH and maintained fermentation between 25 and 35 °C. Because of the lactic acid bacteria (LAB) activity, sourdough produced with these strains exhibited elasticity similar to wheat (Sterr et al. 2009). A heat-based processing method called roasting improves the grain flavor, consistency, and digestion. It has been found that roasting *A. cruentus* and *A. hypochondriacus* at 200 °C for 8 min produced gruels with a higher viscosity because of the breakdown of starch, which makes roasted flour a better thickening agent than popped flour. While the total phenolic content was relatively constant, antioxidant activity increases, because of rise in flavonoid concentration (Muyonga et al. 2014).

3.8 Pre- and Postharvest Technology

3.8.1 Storage

Storage is the critical step in the food chain, and many studies have found that maximum losses frequently occur during this period. Harvested grains are usually stored for either short-term or long-term periods used as seed stock for the next planting season and as food reserves because most crops are grown seasonally (Costa 2014). Mostly storage is preferred in conventional containers such as metal bins, bags, and earthen pots in the house for personal consumption and the following plantation period (Abedin et al. 2012). To avoid infection, seeds should be carefully dried and stored in a tight container. After drying, amaranth seed should not have a moisture content of more than 12%. Any higher amount could encourage the growth of mold and compromise its purity (Aderibigbe et al. 2022). Additionally, keeping the grain in containers on high pallets that permit heat transfer is preferable (Tibagonzeka et al. 2018).

3.8.2 Drying

Drying is an important process after harvesting to preserve storage losses, lower transportation costs, and preserve crop quality (Abass et al. 2014). Drying can be done mechanically and naturally. The natural drying is referred to as sun-drying, and it is the traditional method. A moisture content of less than 12% is preferable for long-term storage (Kitinoja 2013). Grain that is sun-dried in the open is susceptible to bird and bug consumption, and it becomes polluted with animal excreta, dust, other foreign things, and stones. Farmers use mats and plastic sheets to disperse the grains, reducing contamination and making grain gathering easier. However, mechanical drying eliminates some of the issues associated with open drying in the sun. Convection solar tent dryers are an alternative approach for open drying under the sun. In the study, freshly harvested *Amaranthus cruentus* grains have 64% moisture content. After dried, it reached up to 7% steady rate.

3.8.3 Transport

Transportation has an important role in the value chain of grain by enabling the flow of commodities from the field to facilities for processing or storage, and ultimately to markets. Food product damage from spills and bruises is frequently caused by inadequate transportation facility. The well-maintained road networks with proper loading and unloading systems at both field and processing sites guarantee effective handling with little to no damage, saving developed countries from suffering significant transportation costs. However, open trollies or bullock carts are frequently used for grain transportation at the field level in South Asia (Baloch and Mejia 1999). Typically, open trucks, bicycles, bullock carts, or small motor vehicles are used to transport grains in bags for domestic consumption. Significant spills and a higher risk of contamination of grains result from the use of inappropriate and badly maintained transportation techniques in conjunction with inadequate road infrastructure. Significant transportation losses are also a result of handling the crop repeatedly and transferring it several times while in transit (Kumar and Kalita 2017).

3.9 Value Addition and Products

Amaranth is acknowledged as a super-grain with an optimistic outlook and significant potential to develop products commercially (Table 3.1), especially in the baking sector of the food industry. In contrast to wheat, it serves as a raw ingredient for the manufacturing of gluten-free goods. Many procedures for processing may be used to enhance the value of amaranth grain so that it can be used as food and offer the necessary nourishment to consumers. Proper production and storage of amaranth grain can enhance the security of food and nutrition, as well as provide raw materials for industries to increase production and employment. Traditionally, grain amaranth can be cooked and consumed as porridge (Mugalavai 2013). It is known that amaranth can be roasted or popped to create flour, which can then be used on its own or in combination with other foods to make paste, crackers, bread, biscuits, pancakes, spaghetti, muffins, and breakfast cereals (Emire and Arega 2012). Ogi is a fermented culinary product obtained from corn, millet grains, or sorghum, also made from the amaranth grains (Akingbala et al. 1994). It can be used as a starchy material in spirit production, as well as a source of protein concentrates and flours (Corke et al. 1997). Amaranth grain (raw and processed) or green sections have been evaluated as feed for many animals, including rabbits (Bamikole et al. 2000), lambs, rats, pigs, ruminants, and broiler chickens (Ravindran et al. 1996). The effects of antinutritional factors from different species and genotypes of amaranth, as well as differential responses due to age and natural variation, on forage intake, conversion, and live weight gains in animals were found to be significant (Zraly et al. 2023). Nutraceutical foods are viewed not merely as a source of nutrients but also as having the potential to improve consumer health. However, the efficacy of nutraceutical supplements in disease prevention is dependent on the active ingredient bioavailability. In this context, nano- and microencapsulation in the food business has grown due to the numerous benefits that these technologies can provide to the encapsulated material. These include increasing stability under circumstances encountered in food preparation (oxygen, temperature, light) or in the gastrointestinal system (enzymes and pH). Electrospun

Table 3.1 Utilization of grain amaranth in value-addition

Amaranthus species	Applications	Sources	References
Amaranthus spp.	Thickening and stabilizer agent in food	Starch	Corke et al. (1997)
Amaranthus hypochondriacus	Energy supplement in broiler diets	Refined grain	Ravindran et al. (1996)
Amaranthus cruentus	Ogi manufacturing	Grains	Akingbala et al. (1994)
Amaranthus spp.	Concentrating rabbit diet	Unprocessed grain	Bamikole et al. (2000)
Amaranthus cruentus	Blending agent wheat bread up to 15%	Grain flour	Ayo (2001)
Amaranthus cruentus	Dietary fiber	Air-dried grain	Tosi et al. (2001)
Amaranthus cruentus	Used in bakeries	Popped grain	Bodroža-Solarov et al. (2008)
Amaranthus cruentus	Oil and squalene	Grain	He et al. (2002)
Amaranthus spp.	A fat substitute in newborn milk formula	Oil	Pina-Rodriguez and Akoh (2009)
Amaranthus caudatus	Value-added products	Grain	Emire and Arega (2012)
Amaranthus hypochondriacus	Food-grade electrospun fibers	Protein concentrate	Aceituno-Medina et al. (2013)
Amaranthus hypochondriacus	Functional beverage	Grain	Milán-Carrillo et al. (2012)

nanofibers have limited use in food and agricultural systems. This is most likely due to the primary composition of synthetic polymers. However, advances in the manufacturing of nanofibers from food biopolymers has accelerated significantly. In terms of food encapsulation, this approach was just recently used to encapsulate antioxidants (Aceituno-Medina et al. 2013).

Moreover, it also be malted to produce beer. The significance of foods and beverages in illness prevention has drawn more attention in recent years. Functional meals and drinks offer physiological advantages beyond their nutritional content. They are the most popular functional food category because of handy, meet consumer preferences for appearance and packaging, are easily distributed, and remain stable when chilled. Furthermore, they are an effective way to distribute nutrients and bioactive chemicals to a variety of groups. The global sales of functional beverages were predicted to be US$ 105.5 billion (Corbo et al. 2014). Extrusion of grains in amaranth were result in an instant functional beverage (Milán-Carrillo et al. 2012). Apart from beverages, amaranth oil contains high amounts of oleic acid, palmitic acid, and linoleic acid. Chemical or enzymatic interesterification can be used to create structured lipids (SLs) with distinct functional and nutritional properties (Pina-Rodriguez and Akoh 2009). Moreover, it is also rich in squalene content among plant sources. The squalene has high antioxidant activity and protects the oxidative damage created by the free radicals (He et al. 2002).

3.10 Marketing

3.10.1 Global Demand

The demand for pseudo-cereals is growing annually in Europe, making it a niche sector. However, information regarding production levels and the market relationships between importing and producing nations is lacking. The CBI Factsheet of year 2014 estimates that 6000 tons of particular grains, mostly amaranth seeds, were brought from developing nations. Germany is the primary amaranth seed consumer market, according to the survey (Aderibigbe et al. 2022). However, in other industrialized nations such as Sweden, France, the United Kingdom, Belgium, and the Netherlands, amaranth has grown well-known. Either popped, raw, or milled, it is usually marketed in small containers weighing up to 500 g and is mainly consumed as a health food. Due to a lack of set standards, low product supply, and insufficient demand, amaranth crop markets in Africa are still developing and constrained. Because of a lack of strong ties to major customers, smallholder farmers' ability to sell grain amaranth is limited. African leaders and important industry players might interest to create strong market links with European nations and guarantee that easily accessible information on areas with high demand is provided in order to address this. Such initiatives could aid in the creation of income and help reduce poverty (Aderibigbe et al. 2022). Notable nations that grow amaranth crop and control a portion of the market include China, India, Mexico, Peru, Russia, and Kenya (Ainebyona et al. 2012). The expansion of the food processing sector and the rising use of organic ingredients in cosmetic and pharmaceutical products are expected to boost the grain amaranth market.

3.10.2 Trade

India and Mexico are two of the world's major suppliers of *Amaranthus*, which is shipped in 100 of metric tons annually. The price of amaranth, which India sold to Canada, the Netherlands, Germany, Australia, and the United States in 2015, ranged from 0.86 to 2.08 US$ per kilogram (Gordon 2015). The primary market for these exports became the United States. According to the USDA (2015), seeds make up the majority of the traded commodity and provide almost 30% more protein per gram than similar grains such as rye, sorghum, and rice. Amaranth leaves, like its seeds, are high in vitamin C, vitamin A, and folate when cooked. As a supplemental source of vitamins, including niacin, thiamine, and riboflavin, as well as other dietary minerals, they also offer extra nutrition (Santra et al. 2024).

3.10.3 Market Trends

The global amaranth market is divided into four segments: seed, oil, leaf, and flour. The seed segment of amaranth had the greatest share in market of about 52.45% in 2023, and it is predicted to rise at compound annual growth rate (CAGR) of about

11.6% across worldwide amaranth market over the forecast period (MaximizeMarketResearch 2004). Amaranth seed contains various minerals, including calcium, iron, and phosphorus. Such minerals may drive the expansion of the amaranth section during the foreseeable term. Also, the demand for amaranth as a weed is likely to increase. In 2020 the personal care segment maintained a substantial share of 62.05%. Because of greater customer acceptance, the industry has more established enterprises and is predicted to grow rapidly, especially in emerging markets. However, the food and beverage category are predicted to inclined at CAGR of about 10.81% in the worldwide market of amaranth over the upcoming period. The rising necessity for the commodity in flour form or as granola, noodles, flour in bread cereals, pancakes and cookies is predicted to help drive the growth of products. Amaranth has traditionally been eaten during fasts and is generally recommended for recovery from diseases. Gluten-free consumers will find the products of amaranth as an ideal option (MaximizeMarketResearch 2004).

3.11 Conclusion

This chapter emphasizes amaranth's enormous potential as a nutritive and adaptable crop that may combat malnutrition and improve food security, especially in poor nations. It is gluten-free and contains high proteins, vitamins, and minerals content. The amaranth is a great substitute for people who are allergic to gluten, which increases its commercial appeal. The genetic variation present in grain *Amaranthus* species is an essential resource for breeding efforts aimed at enhancing crop resilience and productivity. The study emphasizes the importance of characterizing germplasm collections to find heterotic groups, which can facilitate the development of superior hybrids and increase genetic gains through hybrid breeding. By overcoming the traditional constraints, modern breeding techniques such as doubled haploidy and CRISPR/CAS9 approach offer promising opportunities to speed up the generation of superior amaranth cultivars. Maintaining the quality and viability of amaranth grains is essential for preserving the nutritional advantages for customers, which is made possible by efficient postharvest management techniques. With its adaptability and suitability for a variety of food applications, amaranth holds a strong position in the food industry, driven by the rise in demand for gluten-free products and awareness of its health promoting properties. Furthermore, amaranth's pharmacological qualities, such as its antioxidant and anticancer effect, emphasize its worth beyond just as a food source and raise the possibility of developing novel medicines. In conclusion, amaranth shows promise as a crop capable of addressing security of food, nutrition deficiency, and promoting sustainable agriculture practices. Future research should be conducted on utilizing its genetic diversity and enhancing cultivation methods to optimize its potential within world food systems.

Acknowledgments The authors appreciatively acknowledge the Director of the IHBT (Council of Scientific and Industrial Research), Palampur, India, for continuous support and encouragement for this work.

References

Abass AB, Ndunguru G, Mamiro P, Alenkhe B, Mlingi N, Bekunda M (2014) Post-harvest food losses in a maize-based farming system of semi-arid savannah area of Tanzania. J Stored Prod Res 57:49–57

Abedin M, Rahman M, Mia M, Rahman K (2012) In-store losses of rice and ways of reducing such losses at farmers' level: an assessment in selected regions of Bangladesh. J Bangladesh Agric Univ 10(1):133–144

Aceituno-Medina M, Mendoza S, Lagaron JM, López-Rubio A (2013) Development and characterization of food-grade electrospun fibers from amaranth protein and pullulan blends. Food Res Int 54(1):667–674

Achigan-Dako EG, Sogbohossou OE, Maundu P (2014) Current knowledge on Amaranthus spp.: research avenues for improved nutritional value and yield in leafy amaranths in sub-Saharan Africa. Euphytica 197:303–317

Aderibigbe O, Ezekiel O, Owolade S, Korese J, Sturm B, Hensel O (2022) Exploring the potentials of underutilized grain amaranth (Amaranthus spp.) along the value chain for food and nutrition security: a review. Crit Rev Food Sci Nutr 62(3):656–669

Adriana A (2013) Nutritious underutilized species – Amaranth. Biodiversity International, Rome

Ahmmed ANF, Islam M, Ferdous R (2024) Effective management of white rust disease in red amaranth: a field study in Dhaka, Bangladesh. arXiv preprint arXiv:2409.07579

Ainebyona R, Mugisha J, Kwikiriza N, Nakimbugwe D, Masinde D, Nyankanga RO (2012) Economic evaluation of grain amaranth production in Kamuli District, Uganda. J Agric Sci Technol A 2:178–190

Akingbala J, Adeyemi I, Sangodoyin S, Oke O (1994) Evaluation of amaranth grains for ogi manufacture. Plant Foods Hum Nutr 46:19–26

Allen P (1961) Die Amaranthaceen Mittleleuropas Carl Verlag. Munchen

Alvarez-Jubete L, Arendt EK, Gallagher E (2010a) Nutritive value of pseudocereals and their increasing use as functional gluten-free ingredients. Trends Food Sci Technol 21(2):106–113

Alvarez-Jubete L, Auty M, Arendt EK, Gallagher E (2010b) Baking properties and microstructure of pseudocereal flours in gluten-free bread formulations. Eur Food Res Technol 230:437–445

Amare E, Mouquet-Rivier C, Servent A, Morel G, Adish A, Desse Haki G (2015) Protein quality of amaranth grains cultivated in Ethiopia as affected by popping and fermentation

Andrews T (2000) Nectar and ambrosia: an encyclopedia of food in world mythology. Bloomsbury Publishing USA

Anjali K, Joshi A, Maloo S, Sharma R (2013) Assessment of the morphological and molecular diversity in Amaranthus spp. Afr J Agric Res 8(19):2307–2311

Aparna V (2015) Characterization of rhizosphere and endophytic microflora from organically grown amaranth for management of leaf spot diseases. College of Horticulture, Vellanikkara

Ascherson P. uP Graebner: Synopsis der mitteleuropäischen Flora. 2. Band. Leipzig (1898–1902)

Assad R, Reshi ZA, Jan S, Rashid I (2017) Biology of amaranths. Bot Rev 83:382–436

Ayo JA (2001) The effect of amaranth grain flour on the quality of bread. Int J Food Prop 4(2):341–351

Baker JM (2006) "It's good for many things": Wixárika (Huichol) ethnoecology of amaranth

Ballabio C, Uberti F, Di Lorenzo C, Brandolini A, Penas E, Restani P (2011) Biochemical and immunochemical characterization of different varieties of amaranth (Amaranthus L. ssp.) as a safe ingredient for gluten-free products. J Agric Food Chem 59(24):12969–12974

Baloch UK, Mejia D (1999) Wheat: post-harvest operations

Bamikole M, Ezenwa I, Adewumi M, Omojola A, Adetimirin V, Arigbede O, Orisadeyi S (2000) Alternative feed resources for formulating concentrate diets of rabbits. 1. Unthreshed grain amaranth seedhead. World Rabbit Sci 8(3)

Berkelaar D, Alemu J (2006) Grain amaranth. ECHO Dev Notes 91:1–5

Blair MW, Londoño JM, Buitrago-Bitar MA, Wu X, Brenner DM (2023) Differentiation of Andean and Mesoamerican accessions in a proposed core collection of grain amaranths. Front Plant Sci 14:1144681

Bodroža-Solarov M, Filipčev B, Kevrešan Ž, Mandić A, Šimurina O (2008) Quality of bread supplemented with popped Amaranthus cruentus grain. J Food Process Eng 31(5):602–618
Bortesi L, Fischer R (2015) The CRISPR/Cas9 system for plant genome editing and beyond. Biotechnol Adv 33(1):41–52
Bressani R (2018) Composition and nutritional properties of Amaranth. In: Biology, chemistry, and technology. CRC Press, pp 185–205
Cai Y-Z, Sun M, Corke H (2005) Characterization and application of betalain pigments from plants of the Amaranthaceae. Trends Food Sci Technol 16(9):370–376
Capasso F, Gaginella TS, Grandolini G, Izzo AA, Capasso F, Gaginella TS, Izzo AA (2003) Herbal product regulations. In: Phytotherapy: a quick reference to herbal medicine, pp 73–85
Chandla NK, Saxena DC, Singh S (2017) Amaranth (Amaranthus spp.) starch isolation, characterization, and utilization in development of clear edible films. J Food Process Preserv 41(6):e13217
Chaney L, Mangelson R, Ramaraj T, Jellen EN, Maughan PJ (2016) The complete chloroplast genome sequences for four Amaranthus species (Amaranthaceae). Appl Plant Sci 4(9):1600063
Clouse J, Adhikary D, Page J, Ramaraj T, Deyholos M, Udall J, Maughan P (2016) The amaranth genome: genome, transcriptome, and physical map assembly. Plant Genome 9(1):plantgenome2015.2007.0062
Cobb JN, DeClerck G, Greenberg A, Clark R, McCouch S (2013) Next-generation phenotyping: requirements and strategies for enhancing our understanding of genotype–phenotype relationships and its relevance to crop improvement. Theor Appl Genet 126:867–887
Corbo MR, Bevilacqua A, Petruzzi L, Casanova FP, Sinigaglia M (2014) Functional beverages: the emerging side of functional foods: commercial trends, research, and health implications. Compr Rev Food Sci Food Saf 13(6):1192–1206
Corke H, Wu H, Yue S, Sun H (1997) Developing specialty starches from new crops: a case study using grain amaranth. In: Cereals: novel uses and processes, pp 91–102
Costa SJ (2014) Reducing food losses in sub-Saharan Africa. An 'Action research' evaluation trial from Uganda and Burkina Faso
CropTrust (2024) Crops, countries, and genebanks
Das S (2016) Amaranthus: a promising crop of future. Springer
Dehmer K (2003) Molecular diversity in the genus Amaranthus. Schriften zu Genetischen Ressourcen 22:208–215
Desta ZA, Ortiz R (2014) Genomic selection: genome-wide prediction in plant improvement. Trends Plant Sci 19(9):592–601
Dwivedi S, Singh V, Mahra K, Sharma K, Baunthiyal M, Shin J-H (2024) Functional foods in the northwestern Himalayan Region of India and their significance: a healthy dietary tradition of Uttarakhand and Himachal Pradesh. J Ethn Foods 11(1):1–11
Emire SA, Arega M (2012) Value added product development and quality characterization of amaranth (Amaranthus caudatus L.) grown in East Africa. Afr J Food Sci Technol 3(6):129–141
Escudero NL, Albarracin GJ, Lucero Lopez RV, Giménez MS (2011) Antioxidant activity and phenolic content of flour and protein concentrate of Amaranthus cruentus seeds. J Food Biochem 35(4):1327–1341
Esiyok D, Otles S, Akcicek E (2004) Herbs as a food source in Turkey. Asian Pac J Cancer Prev 5(3):334–339
Ezeh A, Ogedegbe A, Ogedegbe S (2015) Insect pest occurrence on cultivated Amaranthus spp in Benin City, Edo State, Nigeria. J Appl Sci Environ Manag 19(2):335–339
Forster BP, Thomas WT (2005) Doubled haploids in genetics and plant breeding. Plant Breed Rev 25:57–88
Gelotar MJ, Dharajiya DT, Solanki S, Prajapati N, Tiwari KK (2019) Genetic diversity analysis and molecular characterization of grain amaranth genotypes using inter simple sequence repeat (ISSR) markers. Bull Natl Res Cent 43:1–10
Girija K, Lakshman (2011) Anti-hyperlipidemic activity of methanol extracts of three plants of Amaranthus in triton-WR 1339 induced hyperlipidemic rats. Asian Pac J Trop Biomed 1(1):S62–S65. https://doi.org/10.1016/S2221-1691(11)60125-1

Gordon A (2015) Exporting traditional fruits and vegetables to the united states: trade, food science, and sanitary and phytosanitary/technical barriers to trade considerations. In: Food safety and quality systems in developing countries. Elsevier, pp 1–15

Govindaraj M, Vetriventhan M, Srinivasan M (2015) Importance of genetic diversity assessment in crop plants and its recent advances: an overview of its analytical perspectives. Genet Res Int 2015(1):431487

Grivetti LE, Ogle BM (2000) Value of traditional foods in meeting macro-and micronutrient needs: the wild plant connection. Nutr Res Rev 13(1):31–46

Han J, Lin C, Zhu T, Liu Y, Yan J, Qi Z, Yan X (2025) Comprehensive chloroplast genomic insights into Amaranthus: resolving the phylogenetic and taxonomic status of A. powellii and A. bouchonii. Plants 14(5):649

Hancock JF (2022) Agriculture in the Americas before the European Conquest. In: World agriculture before and after 1492: legacy of the columbian exchange. Springer, pp 31–50

He H-P, Cai Y, Sun M, Corke H (2002) Extraction and purification of squalene from Amaranthus grain. J Agric Food Chem 50(2):368–372

Hedayetullah M, Zaman P (2018) Disease management of nonleguminous and nongraminaceous forages. In: Forage crops of the world, vol II: Minor forage crops. Apple Academic Press, pp 349–366

Hernandez-Rodriguez R (2021) Food cultures of Mexico

Hricova A, Fejer J, Libiakova G, Szabova M, Gazo J, Gajdosova A (2016) Characterization of phenotypic and nutritional properties of valuable Amaranthus cruentus L. mutants. Turk J Agric For 40(5):761–771

Hsu Y-C, Srinivasan R (2012) Desert horse purslane weed as an alternative host for amaranth leaf webber, Hymenia recurvalis in Taiwan. Formosan Entomol 32:297–302

Hübner F, Arendt EK (2013) Germination of cereal grains as a way to improve the nutritional value: a review. Crit Rev Food Sci Nutr 53(8):853–861

Hudym O, Lymanska S, Goptsiy T, Turchynova N, Mykhailenko V, Kryvoruchenko R, Stankevych S (2021) Amaranth variability at different doses of gamma radiation. Ukr J Ecol 11(8):146–151

Jambagi SR, Augustine N (2022) Leaf Webber, Diaphania. Res Today 4(7):533–535

Jimenez Rondan FR (2011) Assessment of genetic diversity among Peruvian Amaranth (Amaranthus caudatus L.). Germplasm Using SNP Markers

Joshi DC, Sood S, Hosahatti R, Kant L, Pattanayak A, Kumar A, Stetter MG (2018) From zero to hero: the past, present and future of grain amaranth breeding. Theor Appl Genet 131:1807–1823

Kauffman CS (1992) Realizing the potential of grain amaranth. Food Rev Int 8(1):5–21

Kitinoja L (2013) Innovative small-scale postharvest technologies for reducing losses in horticultural crops. Ethiop J Appl Sci Technol 1:9–15

Kpocheme AOEK, Missihoun AA, Gnanvi BN, Atou R, Wouyou A, Montcho D, Agbangla C (2022) Molecular genetic characterization of Amaranthus cruentus L. mutant lines derived from local and preferred Amaranthus cultivar. J Appl Biosci 175(1):18124–18134

Kumar D, Kalita P (2017) Reducing postharvest losses during storage of grain crops to strengthen food security in developing countries. Foods 6(1):8

Kundzewicz ZW, Mata LJ, Arnell N, Doll P, Kabat P, Jimenez B, Shiklomanov I (2007) Freshwater resources and their management

Lee J-R, Hong G-Y, Dixit A, Chung J-W, Ma K-H, Lee J-H, Park Y-J (2008) Characterization of microsatellite loci developed for Amaranthus hypochondriacus and their cross-amplifications in wild species. Conserv Genet 9:243–246

Lehmann J, Clark R, Frey K (1991) Biomass heterosis and combining ability in interspecific and intraspecific matings of grain amaranths. Crop Sci 31(5):1111–1116

Li H, Deng Z, Liu R, Zhu H, Draves J, Marcone M, Tsao R (2015) Characterization of phenolics, betacyanins and antioxidant activities of the seed, leaf, sprout, flower and stalk extracts of three Amaranthus species. J Food Compos Anal 37:75–81

Lightfoot D, Jarvis DE, Ramaraj T, Lee R, Jellen E, Maughan P (2017) Single-molecule sequencing and Hi-C-based proximity-guided assembly of amaranth (Amaranthus hypochondriacus) chromosomes provide insights into genome evolution. BMC Biol 15:1–15

Lymanska S, Miroshnichenko L, Goptsiy T, Korneeva O (2017) Polymorphism of RAPD and ISSR markers in grain amaranth species. Vavilov J Genet Breed 21(2):189–197

Ma X, Zhu Q, Chen Y, Liu Y-G (2016) CRISPR/Cas9 platforms for genome editing in plants: developments and applications. Mol Plant 9(7):961–974

Maiyo ZC, Ngure RM, Matasyoh JC, Chepkorir R (2010) Phytochemical constituents and antimicrobial activity of leaf extracts of three Amaranthus plant species. Afr J Biotechnol 9(21):3178–3182

Maldonado-Cervantes E, Jeong HJ, León-Galván F, Barrera-Pacheco A, De León-Rodríguez A, De Mejia EG, De La Rosa APB (2010) Amaranth lunasin-like peptide internalizes into the cell nucleus and inhibits chemical carcinogen-induced transformation of NIH-3T3 cells. Peptides 31(9):1635–1642

Mallory MA, Hall RV, McNabb AR, Pratt DB, Jellen EN, Maughan PJ (2008) Development and characterization of microsatellite markers for the grain amaranths. Crop Sci 48(3):1098–1106

Manikandan S, Srimathi P (2014) Studies on post-harvest seed handling techniques on grain amaranth (Amaranthus hypochondriacus L.) cv. Suvarna

Marimo P, Otieno G (2025) A review of select Neglected and Underutilized Species (NUS) in Uganda

Maughan PJ, Smith SM, Fairbanks DJ, Jellen EN (2011) Development, characterization, and linkage mapping of single nucleotide polymorphisms in the grain amaranths (Amaranthus sp.). Plant Genome 4(1)

Maurya NK, Arya P (2018) Amaranthus grain nutritional benefits: a review. J Pharmacogn Phytochem 7(2):2258–2262

MaximizeMarketResearch (2004) Amaranth Market: Global Industry Analysis and Forecast (2024–2030). Retrieved from https://www.maximizemarketresearch.com/market-report/global-amaranth-market/81299/

Meuwissen TH, Hayes BJ, Goddard M (2001) Prediction of total genetic value using genome-wide dense marker maps. Genetics 157(4):1819–1829

Milán-Carrillo J, Montoya-Rodríguez A, Gutiérrez-Dorado R, Perales-Sánchez X, Reyes-Moreno C (2012) Optimization of extrusion process for producing high antioxidant instant amaranth (Amaranthus hypochondriacus L.) flour using response surface methodology

Mlakar SG, Turinek M, Jakop M, Bavec M, Bavec F (2009) Nutrition value and use of grain amaranth: potential future application in bread making. Agricultura 6(4):43–53

Mondini L, Noorani A, Pagnotta MA (2009) Assessing plant genetic diversity by molecular tools. Diversity 1(1):19–35

Montoya-Rodríguez A, Milán-Carrillo J, Dia VP, Reyes-Moreno C, González de Mejía E (2014) Pepsin-pancreatin protein hydrolysates from extruded amaranth inhibit markers of atherosclerosis in LPS-induced THP-1 macrophages-like human cells by reducing expression of proteins in LOX-1 signaling pathway. Proteome Sci 12:1–13

Mugalavai VK (2013) Effect of amaranth maize flour ratio on the quality and acceptability of ugali and porridge (Kenyan cereal staples). ARPN J Agric Biol Sci 8:1–12

Mukuwapasi B, Mavengahama S, Gerrano AS (2024) Grain amaranth: a versatile untapped climate-smart crop for enhancing food and nutritional security. Discov Agric 2(1):44

Mustafa AF, Seguin P, Gélinas B (2011) Chemical composition, dietary fibre, tannins and minerals of grain amaranth genotypes. Int J Food Sci Nutr 62(7):750–754

Muyonga JH, Andabati B, Ssepuuya G (2014) Effect of heat processing on selected grain amaranth physicochemical properties. Food Sci Nutr 2(1):9–16

Nampeera EL (2022) Management of green peach aphid, Myzus persicae in amaranth using host plant resistance and seed treatment. JKUAT-CoANRE

Nazeer S, Zubair Akram M, Ali M (2022) Amaranth as nutrition-rich and climatic resilient crop: a review. Agric Conspec Sci 87(4):281–293

O'Brien G, Price M (2008) Amaranth grain and vegetable types. Echo technical note. Revised by Larry Yarger

Oduwaye OA, Ojo DK, Popoola AR, Daniel IO, Baránek M, Čechová J (2014) Genetic diversity assessment in amaranth germplasm using AFLP and ISSR markers. J Crop Improv 28(4):518–529
Oduwaye O, Ayo-Vaughan M, Porbeni J, Oyelakin O (2019) Genetic diversity in Amaranth (Amaranthus spp.) based on phenotypic and RAPD markers. Niger J Biotechnol 36(1):62–68
Olagunju AI, Arigbede TI, Makanjuola SA, Oyebode ET (2022) Nutritional compositions, bioactive properties, and in-vivo glycemic indices of amaranth-based optimized multigrain snack bar products. Measurement: Food 7:100039
Orona-Tamayo D, Paredes-López O (2024) Chia – the new golden seed for the 21st century: Nutraceutical properties and technological uses. In: Sustainable protein sources. Elsevier, pp 443–470
Ott C, Lacatusu I, Badea G, Grafu IA, Istrati D, Babeanu N, Meghea A (2015) Exploitation of amaranth oil fractions enriched in squalene for dual delivery of hydrophilic and lipophilic actives. Ind Crop Prod 77:342–352
Pal M, Pandey R, Khoshoo T (1982) Evolution and improvement of cultivated amaranths: IX. Cytogenetic relationship between the two basic chromosome numbers. J Hered 73(5):353–356
Palada M (2003) Suggested cultural practices for vegetable amaranth. International Cooperators Guide
Pandey D, Chauhan R, Prabhakaran S, Tiwari A, Joshi D, Chandora R, Jaiswal J (2023) Genetic diversity and genome wide population structure analysis of a grain amaranth (Amaranthus hypochondriacus) diversity panel using Genotyping by Sequencing
Parveen M, Khatun S (2022) Integrated disease management of root rot and stem decay disease in Amaranthus hybridus caused by Fusarium solani. J Mycopathol Res 60(3):363–375
Patel K, Parihar A, Chaurasia P, Patel P, Pachchigar K (2014) Molecular characterization in grain amaranthus (Amaranthus spp.) using RAPD marker. Res J Biotechnol 9:29–35
Pearson T (2010) High-speed sorting of grains by color and surface texture. Appl Eng Agric 26(3):499–505
Pina-Rodriguez AM, Akoh CC (2009) Synthesis and characterization of a structured lipid from amaranth oil as a partial fat substitute in milk-based infant formula. J Agric Food Chem 57(15):6748–6756
Rastogi A, Shukla S (2013) Amaranth: a new millennium crop of nutraceutical values. Crit Rev Food Sci Nutr 53(2):109–125
Rathod KJ, Jadeja BA, Nakar RN (2021) Morphological, biochemical and molecular characterization of Amaranthus varieties in India. Proc Natl Acad Sci India Sect B Biol Sci 91(3):681–694
Ravindran V, Hood R, Gill R, Kneale C, Bryden W (1996) Nutritional evaluation of grain amaranth (Amaranthus hypochondriacus) in broiler diets. Anim Feed Sci Technol 63(1-4):323–331
Roa DF, Santagapita PR, Buera MP, Tolaba MP (2014) Amaranth milling strategies and fraction characterization by FT-IR. Food Bioprocess Technol 7:711–718
Ronoh E, Kanali C, Mailutha J, Shitanda D (2010) Thin layer drying kinetics of amaranth (Amaranthus cruentus) grains in a natural convection solar tent dryer. Afr J Food Agric Nutr Dev 10(3)
Rosentrater KA, Evers A (2018) Introduction to cereals and pseudocereals and their production. Kent's Technology of Cereals:1–76
Sammour RH, Mira M, Radwan S, Fahmey S (2020) Genetic diversity and phylogenetic relationships among and within Amaranthus spp. using RAPD markers. Revista mexicana de biodiversidad 91
Santra D, Schoenlechner R, Brenner D (2024) Sustainability, processing, applications, and international trade of amaranth grain. In: Sustainable Protein Sources. Elsevier, pp 399–412
Sauer JD (1950) The grain amaranths: a survey of their history and classification. Ann Mo Bot Gard 37(4):561–632
Sauer JD (1967) The grain amaranths and their relatives: a revised taxonomic and geographic survey. Ann Mo Bot Gard 54(2):103–137

Schafleitner R, Lin YP, Dinssa FF, N'Danikou S, Finkers R, Minja R, Wu TH (2022) The world vegetable center Amaranthus germplasm collection: core collection development and evaluation of agronomic and nutritional traits. Crop Sci 62(3):1173–1187

Seni A (2018) Insect pests of amaranthus and their management. Int J Environ Agric Biotechnol 3(3):1100–1103

Singh A, Stetter MG (2024) Going new places: successful adaptation and genomic integrity of grain Amaranth in India. BioRxiv:2024.2009.2004.610786

Singh A, Maurya A, Rajkumar S, Singh AK, Bhardwaj R, Kaushik SK, Singh R (2024) Genome-wide comparative analysis of five Amaranthaceae species reveals a large amount of repeat content. Plants 13(6):824

Solarov MB, Šimurina O, Kojić J, Krulj J, Filipović J, Cvetković B, Ilić N (2022) Utilization of Amaranthus spp. grains in food. Food Feed Res 49(1):37–52

Stallknecht GF, Schulz-Schaeffer J (1993) Amaranth rediscovered. In: New crops. Wiley, New York, pp 211–218

Štefúnová V, Bežo M, Labajová M, Senková S (2014) Genetic analysis of three Amaranth species using ISSR markers. Emir J Food Agric (EJFA) 26(1)

Štefúnová V, Bežo M, Žiarovská J, Ražná K (2015) Detection of the genetic variability of Amaranthus by RAPD and ISSR markers. Pak J Bot 47(4):1293–1301

Sterr Y, Weiss A, Schmidt H (2009) Evaluation of lactic acid bacteria for sourdough fermentation of amaranth. Int J Food Microbiol 136(1):75–82

Tanabata T, Shibaya T, Hori K, Ebana K, Yano M (2012) SmartGrain: high-throughput phenotyping software for measuring seed shape through image analysis. Plant Physiol 160(4):1871–1880

Thanyasiriwat T, Kawicha P, Macdonald S, Sangdee A, Tongyoo P, Denby K (2025) Pathogenicity and genome assembly of a Pythium aphanidermatum isolate causing damping-off in amaranth in controlled environment agriculture. BioRxiv:2025.2004.2010.648180

Thapa R, Edwards M, Blair MW (2021) Relationship of cultivated grain amaranth species and wild relative accessions. Genes 12(12):1849

Tibagonzeka JE, Akumu G, Kiyimba F, Atukwase A, Wambete J, Bbemba J, Muyonga JH (2018) Post-harvest handling practices and losses for legumes and starchy staples in Uganda. Agric Sci 9(1):141–156

Tiengo A, Motta EMP, Netto FM (2011) Chemical composition and bile acid binding activity of products obtained from amaranth (Amaranthus cruentus) seeds. Plant Foods Hum Nutr 66:370–375

Tiwari KK, Thakkar NJ, Dharajiya DT, Bhilocha HL, Barvaliya PP, Galvadiya BP, Solanki S (2021) Genome-wide microsatellites in amaranth: development, characterization, and cross-species transferability. 3 Biotech 11:1–12

Tosi E, Ré E, Lucero H, Masciarelli R (2001) Dietary fiber obtained from amaranth (Amaranthus cruentus) grain by differential milling. Food Chem 73(4):441–443

Tyrus M, Lykhochvor V (2022) Yielding capacity of amaranth grain (Amaranthus hypochondriacus) depending on fertilizers. J Cent Eur Agric 23(4):800–806

Uppala S, Beena S, Chapala M, Bowen KL (2010) Bioefficacy of endophytes in the management of leaf blight disease of amaranth. In: Plant growth promotion by Rhizobacteria for sustainable agriculture. Scientific Publishers, Jodhpur, pp 524–530

Vats G, Das D, Gupta R, Singh A, Maurya A, Rajkumar S, Kaushik SK (2023) Validation of genome-wide SSR markers developed for genetic diversity and population structure study in grain amaranth (Amaranthus hypochondriacus). Agriculture 13(2):431

Wang XQ, Park YJ (2013) Comparison of genetic diversity among amaranth accessions from South and Southeast Asia using SSR markers. Korean J Med Crop Sci 21(3):220–228

Wolosik K, Markowska A (2019) Amaranthus Cruentus taxonomy, botanical description, and review of its seed chemical composition. Nat Prod Commun 14(5):1934578X19844141

Wrigley CW, Corke H, Seetharaman K, Faubion J (2015) Encyclopedia of food grains. Academic

Wu X, Blair MW (2017) Diversity in grain amaranths and relatives distinguished by genotyping by sequencing (GBS). Front Plant Sci 8:1960

Yashaswini M, Nysanth N, Anith K (2021) Endospore-forming bacterial endophytes from Amaranthus spp. improve plant growth and suppress leaf blight (Rhizoctonia solani Kühn) disease of Amaranthus tricolor L. Rhizosphere 19:100387

Ziarovska J, Razna K, Labajová M (2013) Using of inter microsatellite polymorphism to evaluate gamma-irradiated Amaranth mutants. Emir J Food Agric 25(9):673

Zraly Z, Kracmar S, Trckova M, Herzig I (2023) Nutritional value of amaranth (genus Amaranthus L.) grain in diets for broiler chickens. Czech J Anim Sci UZPI (Czech Republic) 50(12)

Buckwheat (*Fagopyrum esculentum*): The Resilient Crop with Nutritional Riches

4

Rintu Jha, Bandana Devi, Akanksha Singh, and Prashant Singh

Abstract

Buckwheat possesses remarkable nutritional and medicinal properties, making it a valuable crop for enhancing the food industry through nutrient-dense bakery, non-bakery, and fermented products. Its stress resilience, high flavonoid content, and potent antioxidant activity position it as a promising "smart food." This study provides a comprehensive review of buckwheat's breeding and genetic advancements, bioactive compounds, biological functions, and potential applications in food and pharmaceuticals. Identifying superior cultivars is crucial for developing gluten-free foods, beverages (such as tea, beer, and wine), and nutraceuticals. Rich in bioactive compounds, buckwheat exhibits antioxidant, antimicrobial, anti-glycemic, and anticancer properties, highlighting its therapeutic potential. This chapter summarizes current research on buckwheat's role in human nutrition and health, along with future prospects for its utilization.

Keywords

Buckwheat · Nutrition · Bioactive compounds · Food processing · Buckwheat-enriched food stuffs

R. Jha (✉)
State Key Laboratory of Crop Gene Resources and Breeding, Institute of Crop Sciences, Chinese Academy of Agricultural Sciences, Beijing, China

B. Devi · P. Singh
Department of Botany, Institute of Science, Banaras Hindu University, Varanasi, Uttar Pradesh, India

A. Singh
Institute of Biosciences and Biotechnology, Chhatrapati Shahu Ji Maharaj University Campus, Kanpur, Uttar Pradesh, India

A. Kumar et al. (eds.), *Fasting Superfoods*,
https://doi.org/10.1007/978-981-95-2762-5_4

4.1 Introduction: Origin, Geographical Distribution, and Significance

Buckwheat belongs to the *Fagopyrum* genus within the Polygonaceae family, comprising a total of 21 recognized species (Zhang et al. 2021a, b). Among the species, common buckwheat (*Fagopyrum esculentum*) and Tartary buckwheat (*Fagopyrum tataricum*) are the most extensively cultivated species and for decades, this gluten-free crop is used as a staple diet in many high-altitude regions, including Bhutan, Nepal, and southwestern China (Zhang et al. 2021a, b). Buckwheat has been grown for almost 4000 years, having been domesticated in the southwest of China (He and Zhou 2022). Additionally, morphology and geographic distribution point to southwestern China as the origin point for buckwheat production (Tang et al. 2016). However, buckwheat may have originated in northern China (Boivin et al. 2012), according to palynological and archaeological data, and was likely brought to central and western Europe 1500 years ago via Siberia (Boivin et al. 2012). The origin, domestication, and dissemination history of buckwheat are still unknown (He et al. 2024a, b). Along with having a balanced nutritional composition, buckwheat is abundant in a number of vital bioactive chemicals (Huda et al. 2021; Joshi et al. 2020). Of the species, *F. cymosum* (*F. dibotrys*, golden buckwheat) is used in traditional Chinese medicine, while common and Tartary buckwheat are used in human nutrition (Zhu 2016; Giménez-Bastida et al. 2015; Jha et al. 2024).

With the increasing population, there is still a challenge to global food security in terms of both quantity and quality (Jha et al. 2022). Non-staple crops must be given greater consideration than staple ones in order to provide food security and nutrition in order to fulfill the world's growing food demand (Kumari and Chaudhary 2020). After being processed into flour and the husk removed, buckwheat seeds are usually eaten as cereal. Typically, buckwheat flour is utilized as a raw ingredient to create buckwheat-enriched foods including bread, pasta, noodles, cakes, biscuits, and certain valued fermented goods such as vinegar, wine, and beer (Giménez-Bastida et al. 2015; Suzuki et al. 2020; Jha et al. 2024). Buckwheat consumption can help to recover from a number of chronic malfunctions, including obesity, cardiovascular disease, high blood pressure, and gallstones (Wieslander 2020). Buckwheat has been shown to have anti-inflammatory, anti-tumor, antidiabetic, antioxidant, and hepatoprotective properties in both in vitro and in vivo pharmacological tests (Huda et al. 2021; Zhu 2016). Buckwheat's flavonoids and phenolics are commonly utilized to treat a variety of illnesses, including cancer, microbial diseases, diabetes, hypertension, and glycemic disease (Huda et al. 2021). Buckwheat-derived food products are gaining global attention due to their high nutritional and medicinal value (Jha et al. 2024; He and Zhou 2022). The well-known traditional Chinese herbal remedy, known as golden buckwheat (*Fagopyrum dibotrys* or *Fagopyrum cymosum*), has edible seeds and leaves that are high in flavonoids, which have antidiabetic and antioxidant properties (Jing et al. 2016; Li et al. 2021a, 2021b). An historical treatise on

Chinese herbal medicine called Bencaoshiyi claims that the rhizome of golden buckwheat was used as a folk remedy to cure rheumatism, bacterial dysentery, lung conditions, and stomach aches. It has anti-inflammatory, analgesic, antipyretic, antioxidant, and antiproliferation properties, according to recent research (Chen and Li 2016). A nontoxic Chinese herbal remedy called WeiMaiNing (WMN), made from dried golden buckwheat rhizomes, has been utilized to treat several cancers (Chan 2003; Ke et al. 2021). According to Xu et al. (2019), quercetin is a flavanol with potent antioxidant properties that is widely used to treat a number of illnesses, including diabetes, cancer, depression, inflammation, and cardiovascular conditions. For the treatment of various illnesses, the extraction of these chemicals from golden buckwheat seems like a promising substitute. Therefore, the immense of its nutritional and medicinal value leads buckwheat crop as a substantial source for nutritional and health security and play an functional role in the food industries.

4.2 History: Traditional Uses and Cultural Significance

Self-pollinated Tartary buckwheat and self-incompatible common buckwheat are currently the two most commonly grown buckwheat species (Zhang et al. 2021a, b). Pharmacological classics like the "Compendium of Materia Medica," "Qian Jin Yao Fang," and the "Dictionary of Traditional Chinese Medicine" state that Tartary buckwheat has mental-calming, heart-strengthening, anti-inflammatory, and weight-loss-promoting properties. Nonetheless, a common set of characteristics known as the domestication syndrome, which includes decreased seed dormancy, increased seed size, and loss of seed breaking, is present in domesticated Tartary buckwheat bears as compared to wild accessions (Hunt et al. 2018). Furthermore, the history of buckwheat's domestication and spread is still hotly debated. Buckwheat and southern China are closely related, as linguistic evidence indicates that the Chinese term for the grain was taken from eastern Tibeto-Burman speakers to the southwest of the Han Chinese (Bradley 2011). There is also evidence of a direct connection between European buckwheat and Mongolia, as the English name Tartary is derived from Tatars, the name of the Mongols according to "Marco Polo's Travels," "Dell'Historia della China," and "Matteo Ricci's Reading Notes about China." Additionally, morphology and geographic distribution point to southwestern China as the starting point for buckwheat production (Weisskopf and Fuller 2014; Tang et al. 2016). However, buckwheat may have originated in northern China (Boivin), according to palynological and archaeological data, and was likely brought to central and western Europe 1500 years ago via Siberia (Boivin et al. 2012; Hughes and Hensen 1934). The origin, domestication, and dissemination history of buckwheat are still unknown since it is difficult to locate old buckwheat seeds and it is impossible to differentiate between the pollen fossils of wild and farmed buckwheat (Yao et al. 2023).

4.3 Production: Cultivation Practices, Climatic and Soil Requirements, and Global Production Statistics

Buckwheat is widely cultivated in the Northern Hemisphere, from Eastern Europe to Japan, for local consumption or as a commodity crop. It can also be found on the American continent with a smaller sowing area. Successful buckwheat growing depends on several agroclimatic factors. Buckwheat can grow and thrive well even in harsh environment due the wider adaptability (Zhang et al. 2021a, b). However, according to Campbell et al. (1997), buckwheat's life cycle normally lasts 70–90 days, but in unfavorable conditions, it can last as long as 130–150 days (Michiyama and Hayashi 1998). Even though buckwheat may grow in a variety of environmental situations, it thrives in damp, cool climates. One crucial factor for buckwheat productivity, the sowing time, is determined by the combination of existing environmental factors, mainly temperature, and should be suitable for germination and survival (Farooq et al. 2016). Since buckwheat is a frost-sensitive plant (Clark 2007), and its sensitivity is greatest at the growing stage of one to two leaves, the date of late spring frosts is also a determining factor in the start of sowing. It should be also considered that the optimal soil temperature for germination should be more than 10 °C. The ideal temperature range for plant growth is between 18 and 23 °C. Low temperatures (below 10 °C) typically delay or prevent flowering (Bjorkman 2000), while high temperatures (over 30 °C) have been linked to ovule abortion and deformation of the embryo sac (Gang and Yu 1998). According to Olson (2001), the ideal drying conditions are 45 °C and 16% moisture content. Due to the protracted ripening, it is heterogeneous and in the temperate zone the first frosts in autumn often cause drying and death of the plants. Buckwheat has high water demand and due to its shallow rooting, it reacts sensitively to water deficits in the topsoil. On hot summer days, the foliage can show the lack of water, mainly on loose soils, but turgor pressure can recover without damage at night or in the morning. Lack of water can also cause flower abortion and fertility problems, but at the same time, on soils with higher nutrient supply a vegetative predominance can develop in rainy conditions, which has negative effect on seed production. Although it thrives in drained wetlands, marginal land, and acidic soils (pH 5) (Hore and Rathic 2002), it prefers well-drained sandy soils. It gives good results on most types of soil, except for extremely loose sand and hard, cold, and strongly calcareous soils. The buckwheat is generally indifferent to day length, but in the case of some Asian genotypes are facultative short-day plants. Some of the pictorial view of buckwheat plant has been demonstrated in Fig. 4.1.

Its economic importance is mainly due to its exceptionally favorable nutritional and physiological characteristics. Its versatile, beneficial effect on the human body makes it suitable for use both as a food plant and as a medicinal plant. Different nutritional composition have been illustrated in Fig. 4.2 and uses of buckwheat-derived food stuffs described in Table 4.1.

In recent years, buckwheat production has fluctuated globally according to regional demand and meteorological factors. The Food and Agriculture Organization (FAO) (2023) estimates that 2.3 million metric tons of buckwheat were produced

Fig. 4.1 Field demonstration and physical appearance of buckwheat plant and its inflorescence

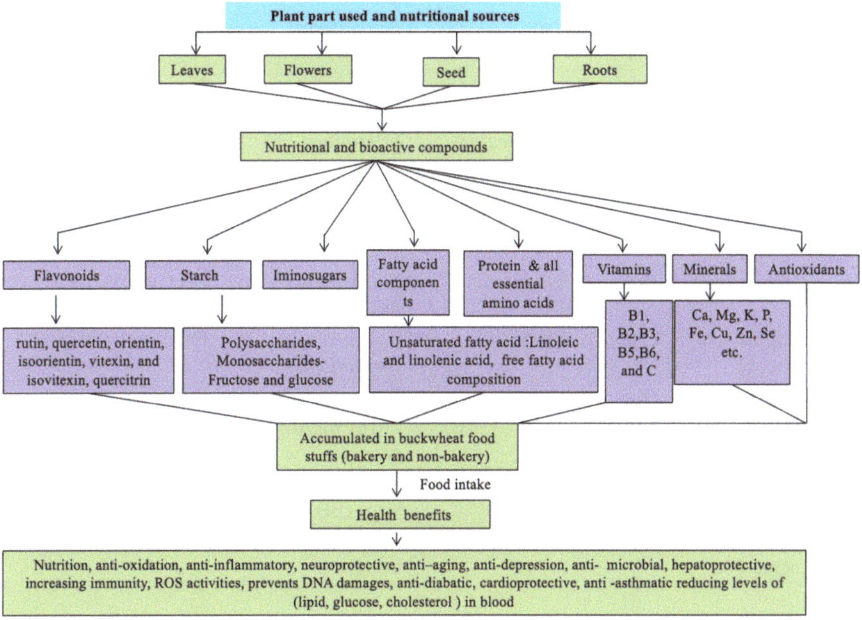

Fig. 4.2 Demonstrates different nutritional and bioactive compounds of buckwheat and their beneficial role in health

worldwide in 2022, with France, China, and Russia producing the most. Nearly 40% of the world's output came from Russia alone, with China coming in second at about 30%. Europe has also produced a lot, especially in Poland and Ukraine, but yields have been affected by geopolitical considerations. Because of its short growing season and minimal input requirements, buckwheat cultivation is still essential for sustainable farming and gluten-free food markets (FAO 2023). In order to satisfy the growing demand from health-conscious consumers, efforts are being made to increase production in North America and Asia.

Table 4.1 Demonstrated diseases and their causal organism occurred in buckwheat crop

S. no.	Diseases	Species	Pathogens	References
1.	Leaf spot	*Fagopyrum esculentum*	*Bipolaris zeae*	Chen et al. (2021)
2.	Leaf spot	*Fagopyrum tataricum*	*Nigrospora osmanthi*	Shen et al. (2021)
3.	Leaf spot	*Fagopyrum esculentum*	*Alternaria alternata*	Li et al. (2021a, b)
4.	Leaf spot	*Fagopyrum esculentum*	*Alternaria alternata*	Maurya et al. (2021)
5.	Powdery mildew	*Fagopyrum esculentum* and *F. tataricum*	*Erysiphe aff. betae*	Hsiao et al. (2022) and Subedi et al. (2020)
6.	Downy mildew	*Fagopyrum esculentum*	*Perenospora* sp.	Gopi et al. (2015)
7.	Downy mildew	*Fagopyrum esculentum*	*Peronospora* cf. *ducometii*	Econopouly et al. (2017)
8.	Downy mildew	*Fagopyrum* sp.	*Peronospora fagopyri*	Subedi et al. (2022)
9.	Rust	*Fagopyrum* sp.	*Piccinia fagopyri*	Manandhar et al. (2016)
10.	Root rot	*Fagopyrum esculentum* Moench and *F. tataricum* Gaert	*Sclerotinia sclerotiorum*	Mondal et al. (2002)
11.	Leaf blight	*Fagopyrum* sp.	*Botrytis cinera*	Manandhar et al. (2016)
12.	Stem canker	*Fagopyrum tataricum*	*Rhizoctonia solani* AG-4 HGIII	Li et al. (2021a, b)
13.	Stipple spot disease	*Fagopyrum* spp.	*Bipolaris sorokiniana*	Babu et al. (2018)
14.	Stem rot	*Fagopyrum* spp.	*Botrytis cinerea*	Babu et al. (2018)
15.	Blight	*Fagopyrum* spp.	*Phytopthora parasilka*	Babu et al. (2018)
16.	Viral diseases	*Fagopyrum* spp.	Buckwheat burn virus (BBV)	Sindarovska et al. (2014)
17.	Root-knot nematode	*Fagopyrum esculentum*	*Meloidogyne incognita*	Ferreira et al. (2023)

4.4 Protection: Major Pests, Diseases, and Management Strategies

Buckwheat (*Fagopyrum esculentum*) is a resilient, fast-growing pseudocereal with natural resistance to many pests and diseases (Majaw et al. 2024; Babu et al. 2018). It has long been used as a cover crop due to its rapid growth and strong weed suppression, both during its lifecycle and after termination (Falquet et al. 2015). However, most research on its biocontrol and weed suppression benefits has focused on vegetable cropping systems (Candelaria-Morales et al. 2022). However, when exposed to unfavorable environmental conditions such as high humidity, excessive rainfall, poor soil drainage, or temperature fluctuations, it can become vulnerable to various fungal, bacterial, and viral pathogens. It is affected by viral, bacterial, and fungal diseases at different stages of plant development, depending on pathogen type. These pathogens can significantly affect plant health, leading to reduced growth, seed production, and overall yield (Demchenko and Boyko 2016). Buckwheat pests and diseases primarily affect the leaves (Liu et al. 2022).

Buckwheat (*Fagopyrum esculentum*, *F. tartaricum*, and *F. dibotrys*) faces significant yield and quality losses due to fungal, bacterial, and viral diseases. Fungal pathogens such as *Peronospora fagopyri* (downy mildew), *Erysiphe polygoni* (powdery mildew), *Botrytis cinerea*, *Puccinia fagopyri*, *Rhizoctonia solani*, *Pythium* spp., *Fusarium* spp., and *Sclerotinia sclerotiorum* are major constraints, causing foliar damage, root rot, and seed discoloration (Subedi et al. 2020; Kongari et al. 2024). These infections are exacerbated by environmental factors, with diseases like downy mildew thriving in humid conditions and powdery mildew being prevalent in subtropical climates (Subedi et al. 2020).

In addition to fungal threats, bacterial pathogens such as *Pseudomonas syringae* and *Erwinia amylovora* cause leaf spots and blights, further reducing crop productivity. Viral infections, particularly *Fagopyrum mosaic virus* (FaMV), stunt plant growth and limit yield potential (Kongari et al. 2024). Mycotoxins from *Fusarium* spp. pose additional concerns for food safety, given buckwheat's rising demand in health-conscious markets (Kawanishi et al. 2022).

Managing these biotic stresses requires an integrated approach, including genetic resistance, cultural practices, and fungicidal or bactericidal treatments. Ongoing research on resistance genes in Tartary buckwheat (Schuster 2023) and chemical control measures for fungal and bacterial diseases (Subedi et al. 2022) highlights the need for sustainable disease management strategies to ensure stable buckwheat production. Buckwheat, affected by several pathogens and diseases are being enlisted in Table 4.1.

4.5 Improvement (Genetics and Genomics): Breeding Efforts, Genetic Diversity, and Recent Advancements

Recent advances in genomics have enabled the identification of key genes responsible for its nutritional traits, paving the way for targeted breeding programs. For instance, whole-genome sequencing of *Fagopyrum esculentum* and *Fagopyrum tatarícum* has revealed genetic loci linked to enhanced protein quality, amino acid profiles, and flavonoid biosynthesis (Yasui et al. 2016; Zhang et al. 2021a, b). Marker-assisted selection (MAS) and CRISPR-Cas9 gene editing are now being employed to develop buckwheat varieties with improved nutrient bioavailability and reduced anti-nutritional factors (phytic acid, tannins), making it a more viable functional food (Wijerathna-Yapa and Pathirana 2022). One of the major focuses of buckwheat genomic research is enhancing its flavonoid content, particularly rutin, which has antioxidant, anti-inflammatory, and cardioprotective properties (Kalinová et al. 2018). Additionally, genome-wide association studies (GWAS) have uncovered SNPs associated with elevated micronutrients (zinc, iron) in buckwheat grains, addressing global micronutrient malnutrition (Zhou et al. 2021). These genomic tools accelerate the development of biofortified buckwheat varieties with superior health benefits. Another critical area is improving protein digestibility and amino acid balance in buckwheat. Unlike many cereals, buckwheat contains a nearly complete amino acid profile, including lysine, which is limited in wheat and rice (Zhang et al. 2020). Genomic studies have pinpointed genes regulating storage protein composition (e.g., 13S globulins and 8S albumins), which influence digestibility (Yang et al. 2022). Comparative transcriptomics has identified upregulated genes in high-rutin buckwheat cultivars, such as *FLS* (flavonol synthase) and *F3′H* (flavonoid 3′-hydroxylase), which can be targeted for breeding (Li et al. 2020). Targeted mutation of FtMYB45 using CRISPR/Cas9 successfully raised the flavonoid content of Tartary buckwheat (Wen et al. 2022). This discovery established the groundwork for gene function studies and Tartary buckwheat quality enhancement by proving that the CRISPR/Cas9 system is an effective instrument for accurate genome editing in this crop. Furthermore, CRISPR-based silencing of protease inhibitors (e.g., trypsin inhibitors) has been explored to enhance protein utilization (Zhou et al. 2021). Such advancements ensure that buckwheat can serve as a high-quality plant-based protein source for vegetarian and gluten-free diets. Quantitative trait loci (QTL) mapping has been instrumental in identifying genomic regions associated with key nutritional traits in buckwheat, facilitating marker-assisted breeding for enhanced dietary quality. Several studies have detected QTLs linked to flavonoid biosynthesis, particularly rutin content, in both *Fagopyrum esculentum* (common buckwheat) and *F. tataricum* (Tartary buckwheat). For instance, a major QTL on chromosome 3 of *F. tataricum* was found to regulate rutin accumulation, with candidate genes such as F3H (flavanone 3-hydroxylase) and FLS (flavonol synthase) playing crucial roles (Li et al. 2020). Similarly, QTLs for protein content and amino acid composition have been mapped in common buckwheat, with loci on chromosomes 1 and 5 showing significant associations with lysine and methionine enrichment (Zhang et al. 2021a, b). In addition to flavonoids and proteins, QTLs for mineral nutrients (iron, zinc) and anti-nutritional factors (phytic acid) have been identified, addressing both nutritional enhancement and digestibility. Furthermore, QTLs linked to reduced phytic acid synthesis were detected, which can improve mineral bioavailability in buckwheat-based foods (Yang et al. 2022). Recent advances in genomic technologies have enabled significant progress in understanding the genetic basis of abiotic and biotic stress tolerance in buckwheat (*Fagopyrum* spp.) through GWAS and QTL mapping. GWAS analyses in *F. tataricum* have identified several candidate genes associated with drought and salinity tolerance, including *FtADH1* was found to be a key contributor to drought tolerance in Tartary buckwheat (He et al. 2024a, b), *FtAUR3* can dramatically increase the amounts of components of the reactive oxygen species pathway, such as peroxidase, catalase, and superoxide dismutase, enhancing Tartary buckwheat plant salt tolerance (Lu et al. 2025). Similarly, QTL mapping has revealed genomic regions on chromosomes 2 and 5 linked to aluminum toxicity resistance, a critical trait for acidic soil adaptation (Yang et al. 2022). For biotic stress resistance, GWAS has pinpointed SNPs near *PR* (pathogenesis-related) genes and *RLKs* (receptor-like kinases) that confer resistance to fungal pathogens like *Botrytis cinerea* (Li et al. 2020). The integration of GWAS and QTL data with modern genomic tools such as CRISPR-Cas9 has accelerated the improvement of stress tolerance in buckwheat. For instance, CRISPR-mediated knockout of *ABA8OX* (abscisic acid 8′-hydroxylase) in *F. esculentum* has enhanced drought tolerance by modulating stomatal closure (Zhang et al. 2023a, b). Similarly, transcriptomic studies under cold stress have identified CBF/DREB1 regulatory networks that can be engineered for frost-hardy cultivars (Yasui et al. 2016). For biotic stress, pyramiding multiple QTLs (e.g., *Rfg1* for rust resistance) through genomic selection has shown promise (Kalinová et al. 2018). The integration of QTL mapping with advanced genomic tools, such as CRISPR-Cas9, holds promise for precision breeding, ensuring buckwheat's role as a functional crop for future food security. These strategies, combined with multi-omics data, are paving the way for climate-smart buckwheat varieties with improved yield stability under environmental stresses. These findings provide molecular markers for marker-assisted selection (MAS) to develop stress-resilient buckwheat varieties. These findings enable breeders to develop high-nutrient buckwheat varieties through targeted selection.

4.6 Primary Processing: Postharvest Handling and Processing Techniques

Buckwheat (*Fagopyrum esculentum*) requires careful postharvest handling to ensure high-quality grains and by-products. Although buckwheat consumption has been gradually increasing over the years, limited investment in research and development for postharvest processing and value addition has hindered its large-scale integration into the agricultural sector as a super crop (Chrungoo and Chettry 2021). The primary processing steps after harvesting include drying, cleaning, dehulling, sorting, and storage (Jha et al. 2024; Naik et al. 2022; Radics and Mikóházi 2010). Primary processing refers to the initial handling, treatment, and processing of buckwheat immediately after harvest to maintain quality, reduce spoilage, and prepare the grains for further processing or market distribution. Effective postharvest techniques ensure food safety, enhance shelf life, and reduce losses. Processing buckwheat involves several key steps to ensure the grain is clean, edible, and ready for further use. These steps include drying, cleaning, dehulling, sorting, and storage (Naik et al. 2022).

4.6.1 Drying

Grain typically retains high moisture content at harvest, making drying essential for ensuring proper storage and quality (Radics and Mikóházi 2010). Grain drying is essential for ensuring proper storage and quality but remains an energy-intensive and costly process. Conventional convective dryers, which are widely used for mechanical drying, rely on liquid fuel for air heating and consume substantial power. When processing grain with 17.0–22.0% humidity, these dryers can only reduce moisture by 2.0–3.0% per pass, necessitating multiple drying cycles to achieve the optimal 14.0% moisture level. This prolonged process increases fuel consumption and operational costs, creating economic and environmental challenges in grain processing (Mar'in and Vereshchagin 2016). Moreover, moisture reduction during drying affects the physical properties of buckwheat grains. Quequeto et al. (2018) found that drying increases bulk density, true density, and porosity while causing bulk and true volumetric shrinkage, reducing their values by approximately 14.47% and 14.70%, respectively.

4.6.2 Dehulling

Buckwheat hulls are firm and provide a tight enclosure for the kernels (Zhang et al. 2023a, b; Jha et al. 2024). Before buckwheat can be processed and utilized, its outer hull must be removed. Currently, the primary methods for dehulling hulled grains include mechanical, physical, and chemical techniques. Traditional mechanical dehulling relies on impact or shearing forces to detach the hull from the kernel, but this approach often damages the seeds, leading to a decline in quality (Chu et al. 2022). Physical dehulling involves pretreating soaked grains with techniques such as microwave or hydrothermal processing (Goyal et al. 2010). These methods enhance the ease of hull removal, improving efficiency while preserving kernel integrity. Chemical dehulling, on the other hand, modifies the hull's properties through chemical soaking to facilitate its removal (Dakia et al. 2008). Despite its effectiveness, this method raises concerns regarding food safety and environmental contamination due to the use of chemical agents (Zhang et al. 2023b). Compared to other hulled grains, buckwheat seeds are more susceptible to damage during dehulling, making it difficult to obtain whole groats with high germination rates (Živković et al. 2021). The standard method for producing groats involves thermal dehulling, where seeds are boiled or steamed at high temperatures (130–160 °C) under pressure, then cooled, conditioned, and dried before dehulling. This process causes the endosperm to expand, breaking the husk for easier removal (Janeš et al. 2012; Keriene et al. 2016). However, thermal treatment lowers the seeds' nutritional value and reduces germination potential, making them less suitable for sprout production. Studies have shown that thermal dehulling decreases total phenolic content (TPC), antioxidant activity (AA), and key flavonoids such as rutin and quercetin (Amudha Senthil 2015; Klepacka and Najda 2021). A recent study also found that cold dehulling of buckwheat seeds no negative effects on germination rate and resulted in faster rootlet growth compared to hulled seeds (Živković et al. 2021).

4.6.3 Cleaning and Sorting

Seed cleaning is a crucial step in the storage and processing of grains, ensuring the removal of impurities and improving overall quality (Kolankowska et al. 2021). Over the years, extensive research has been conducted on cleaning and sorting granular materials. Traditional seed cleaning methods rely on differences in aerodynamic and geometric properties to separate impurities from the primary crop. In contrast, modern technologies focus on distinguishing components based on shape variations. Cleaning buckwheat seeds (*Fagopyrum esculentum*) can also be achieved using electrostatic separation techniques. A study by Pozeliene et al. (2010) explored the use of a conveyor-type electro-separator equipped with a flannel belt operating in a high-voltage field for specialized seed cleaning. For instance, buckwheat nutlets can be effectively separated using various advanced techniques, including triangular mesh screens, cylindrical groove separators, grate separators, and indented cylindrical separators (Kolankowska et al. 2021). Additionally, optimizing separation parameters through genetic algorithms has been shown to enhance efficiency and reduce seed loss, outperforming previous suboptimal methods (Konopka et al. 2017). These advancements highlight the continuous improvement in seed cleaning and sorting technologies, leading to more precise and efficient buckwheat grain processing.

4.6.4 Storage

Buckwheat seeds require specific storage conditions to maintain their quality. It is crucial to store them with a moisture content of 14% or less to prevent spoilage and ensure safe storage (Parimala et al. 2013). Among various packing materials, paper bags are recommended as they help retain the physiological and morphological traits of the seeds better than glass, wood, or PVC materials (Tabaković et al. 2019). This is evident from studies where buckwheat seeds stored in paper bags showed higher germination rates and lower dormancy compared to other materials (Tabaković et al. 2019).

The duration of storage also significantly affects the quality of buckwheat seeds. Seeds stored for up to 2 years generally retain good production traits, but beyond 3 years, their quality deteriorates, and they may not be suitable for planting (Jevdjovic and Maletic 2003). Proper storage conditions can extend the shelf life of buckwheat seeds, but it is essential to monitor their quality over time.

4.7 Pre and Post-harvest Technology: Storage, Drying, and Transportation

To preserve quality and reduce losses, buckwheat, a nutrient-rich pseudocereal, needs to be handled carefully after harvest. Postharvest results are greatly impacted by preharvest variables, such as ideal ripeness during harvest. To maintain the quality of seeds during storage the standardization of suitable seed treatments and packaging material is most important because seed treatment is the basic measure to assure adequately healthy crops at emergence and during further growth of plants (Suzuki et al. 2012). Overall, maintaining optimal storage conditions is vital for preserving the viability and quality of buckwheat seeds.

In order to avoid spoiling and microbiological contamination, drying is an essential postharvest procedure. To improve drying efficiency, researchers have explored various drying models. Siqueira et al. (2020) found that the Midilli model effectively represents the drying kinetics of buckwheat grains at 40, 60, and 70 °C, whereas the Approximation of Diffusion model is more suitable for drying at 50 and 80 °C. These findings highlight the potential for optimizing drying parameters to enhance energy efficiency and reduce costs in grain processing. Buckwheat's pest resistance and shelf life are significantly impacted by storage conditions. By lowering oxygen levels, hermetic storage systems such as airtight containers or silo bags effectively prevent the growth of insects and fungi (Atkari et al. 2021). Controlling temperature and humidity is essential; to avoid the formation of mycotoxin, stored buckwheat should be kept below 15 °C and 60% relative humidity (FAO 2021). Although organic markets favor nonchemical techniques such as changed atmospheres, fumigation or natural repellents such as diatomaceous earth may be used in areas with strong pest pressure.

Transportation is a vital aspect of buckwheat postharvest management, as maintaining grain quality during transit from farms to markets or processing units is essential (Chen et al. 2024). In regions such as Mustang, wind-induced shattering during harvest leads to significant grain losses—around 15%—during transport to storage, further exacerbated by historically low yields and inadequate rural infrastructure (Averchev et al. 2021). These logistical challenges not only affect local food security but also hinder broader supply chain efficiency. In Ukraine, overcoming similar barriers is key to reviving buckwheat exports, requiring increased domestic production, improved border logistics, and adaptation to the updated 2023 export regulations, which lifted previous bans but introduced zero quotas for certain goods (Chobitok et al. 2024). Thus, addressing transportation-related losses and infrastructure gaps is crucial to enhancing the overall value chain, improving food security, and expanding global buckwheat trade.

4.8 Utilization Pattern: Traditional and Modern Uses in Food, Medicinal, and Industrial Applications

Traditionally, buckwheat has played a significant role in various cuisines, particularly in Asia and Eastern Europe. Over time, its applications have expanded to modern culinary innovations that cater to health-conscious consumers. Buckwheat seeds are commonly consumed as a cereal after dehusking and are then processed into flour. This flour serves as a key ingredient in various buckwheat-enriched food products, including bread, noodles, pasta, cakes, and biscuits. Additionally, it is used in the production of fermented items such as wine, beer, and vinegar (Suzuki et al. 2020; Jha et al. 2024). In traditional cuisines, buckwheat is commonly used in the form of whole groats, flour, and noodles. In Eastern Europe, it is a staple ingredient in dishes like kasha, a porridge made by roasting and boiling buckwheat groats. Buckwheat noodles are also commonly consumed in northern China, Japan, Korea, and portions of Europe (Wang et al. 2019). Hele, a traditional Chinese noodle, is made entirely from pure common buckwheat flour and exhibits a highly porous structure in its cross-section when observed under a scanning electron microscope. Similarly, soba noodles must contain a minimum of 35% buckwheat to be classified as soba and are appreciated for their high fiber content (Giménez-Bastida et al. 2015; Breslauer et al. 2023). Similarly, in Japan, buckwheat flour is the primary component of soba or "Soba-kiri" noodles (Katsube-Tanaka 2016; Mikami et al. 2018). In France, buckwheat flour is used to make galettes, a type of savory crepe (Skřivan et al. 2023).

With growing interest in gluten-free and plant-based diets, buckwheat has gained popularity in modern food applications (Saha et al. 2024). It is now widely used in gluten-free baking, providing a nutritious alternative to wheat flour in bread, pancakes, and cookies (Breslauer et al. 2023). In recent years, buckwheat has gained popularity in the USA, Canada, and Europe (Giménez-Bastida et al. 2015). It is commonly consumed in various forms, including noodles, pancakes, and muffins, in countries such as China, Ukraine, Japan, Canada, India, and Nepal (Sytar et al. 2016).

4.9 Value Addition and Products: Potential for Creating High-Value Products

Buckwheat, belonging to the genus *Fagopyrum*, is a pseudocereal cultivated for its nutritious grains and bioactive compounds (Nalinkumar and Singh 2020). Buckwheat is a nutrient-rich food packed with essential amino acids, antioxidants, and bioactive compounds such as rutin and quercetin. These contribute to its antioxidant, anti-inflammatory, and potential anticancer properties, making it a valuable addition to a healthy diet (Jose and Selvan 2024). It has been traditionally valued for its high protein content, rich flavonoids, and gluten-free nature (Martínez-Villaluenga et al. 2020). Apart from being a food source, buckwheat is widely used in traditional medicine and has various industrial applications (Zamaratskaia et al. 2024). Over time, its importance has grown due to increasing awareness of its health benefits, sustainability, and functional properties. This study explores the traditional and modern applications of buckwheat in food, medicine, and industry (Table 4.2).

4.10 Marketing: Global Demand, Trade, and Market Trends

Studies have demonstrated the diverse applications of buckwheat by-products across various industries. Shaikhiev et al. (2020) highlighted the effectiveness of buckwheat straw, grain husks, and other waste materials as efficient sorbents for removing harmful substances from aqueous media. Similarly, Jin et al. (2022) emphasized the remarkable potential of buckwheat proteins in food structuring and the development of peptide-based functional foods for disease prevention, reinforcing their value in the food industry (Table 4.1).

Table 4.2 Buckwheat-derived food staffs and bioactive compounds and their beneficial role in health

Food stuffs	Plant parts used/ source	Effects on health
Groats/kasha	Dehulled seeds	Good source of carbohydrates, lowers blood sugar levels, helps in food digestion, and regulate bowel movement
Bread	Seed flour	Intestinal chronic disease, type-2 diabetes and inflammatory bowel diseases (IBDs)
Noodles	Seed flour	Diabetes, obesity, cardiovascular, and chronic disease
Pasta	Seed flour	Celiac disease and type 1 diabetes
Biscuits and snacks	Seed flour	Angiotensin converting enzyme (ACE) inhibitory activity
Honey	Flowers	Antibacterial, antioxidant activities, anti-browning activities, ACE-inhibitory and anti-inflammatory activities, hepatoprotective effects, inhibits DNA damage, effects on antibiotic-resistant pathogens, effects on intestinal microbes and it enhances the growth of probiotics, reducing reactive oxygen species (ROS) activity, wound healing activities, and preventive role in DNA damage
Tarhana	Seeds	Improve nutrient status
Vinegar	Steamed raw seed grain and fermentation	Effects on anti-oxidative activities
Miso and tofu	Seed flour	Improve nutrient status
Wine (soba shochu, tempeh, Chinese rice wine)	Fermentation of malt extracts	Effects on anti-oxidative activities
Tea	Seeds, dried leaves, hulls, and flowers	Effects on anti-oxidative reaction, α-glucosidase inhibitory activities
Bioactive compounds		
Polysaccharides, procyanidin dimer, quercetin, and tannins	Leaf/flower/ seed/rhizome	Antitumor activity
Catechins, coumarins, curcuminoids, mandelic acid, lignans, polysaccharides, phenolic acids, quercetin, rutin, stilbenes, and tannins	Leaf/flower/ seed/rhizome	Anti-oxidant activity
Apigenin, chrysin, hispidulin, hesperidin, isoorientin, isovitexin, luteolin, polysaccharides, quercetin, and rutin	Leaf/flower/ seed/rhizome	Anti-inflammatory activity

(continued)

Table 4.2 (continued)

Food stuffs	Plant parts used/ source	Effects on health
Apigenin, naringenin, polysaccharides, quercetin, quercitrin, rutin, and silymarin	Leaf/flower/ seed/rhizome	Hepatoprotective activity
Chlorogenic acid, epicatechin, hydroxybenzoic acid, luteolin, kaempferol, quercetin, quercitrin, and rutin	Leaf/flower/ seed/rhizome	Antibacterial activity
Quercetin and hesperidin	Leaf/flower/ seed/rhizome	Antifungal activity
Apigenin, catechin, dihydroquercetin, emodin, hesperidin, morin, quercetin, and rutin	Leaf/flower/ seed/rhizome	Antiviral activity
Kaempferol, quercetin, and rutin	Leaf/flower/ seed/rhizome	Anti-ulcer activity
Globulin	Leaf/flower/ seed/rhizome	Anti-fatigue activity
Polysaccharides	Leaf/flower/ seed/rhizome	Hypolipidemic activity
Apigenin, hesperidin, luteolin, polysaccharides, and quercetin	Leaf/flower/ seed/rhizome	Immunoregulatory activity
Galangin, kaempferol, myricetin, N-*trans*-feruloyl tyramine, polysaccharides, and rutin	Leaf/flower/ seed/rhizome	Neuroprotective activity
D-chiro-inositol, isoquercetin, polysaccharides, quercetin, and rutin	Leaf/flower/ seed/rhizome	Antidiabetic activity
Apigenin, cinnamic acid, ferulic acid, gallic acid, isorhamnetin, kaempferol, luteolin, naringenin, quercetin, resveratrol, rutin, and syringic acid	Leaf/flower/ seed/rhizome	Cardioprotective activity
Apigenin, naringenin, nobiletin, phenylpropanoid glycosides, procyanidin dimer, quercetin, and tangeretin	Leaf/flower/ seed/rhizome	Anticancer activity
Quercetin	Leaf/flower/ seed/rhizome	Anti-atherosclerosis activity
Rutin and fagopyritols A1	Leaf/flower/ seed/rhizome	Lipid-lowering, cholesterol-lowering, and blood glucose lowering
Catechins, fisetin, genistein, hydrobenzoic acids, kaempferol, synigric acid, toxifolin, and vanillic acid	Leaf/flower/ seed/rhizome	Antineoplastic activity
Catechins, galangin, kaempferol, and myricetin	Leaf/flower/ seed/rhizome	Antiaging activity

(continued)

Table 4.2 (continued)

Food stuffs	Plant parts used/ source	Effects on health
Kaempferol, myricetin, rutin, and quercetin	Leaf/flower/ seed/rhizome	Antithrombotic activity

Table content have been adopted from Jha et al. (2024), Huda et al. (2021), and Giménez-Bastida et al. (2015)

Expanding on its functional benefits, Zou et al. (2023) explored the bioactive compounds in Tartary buckwheat (TB), highlighting their diverse bioactivities and significant health benefits. The study also addressed safety concerns and industrial applications, paving the way for the commercialization of TB and its derived products. In line with this, Graziano et al. (2022) underscored the great potential of pseudocereals like buckwheat for various food industry applications, further supporting its growing role in functional and commercial food development.

Beyond its nutritional value, buckwheat is widely recognized in the literature as one of the most frequently studied pseudocereals for gluten-free malt and beer production. Over the years, it has demonstrated outstanding results in both productivity and the chemical composition of the final product, making it a valuable ingredient in gluten-free brewing (Brasil et al. 2021; Dabija et al. 2022). Additionally, Gonçalves et al. (2024) highlighted the versatile industrial applications of buckwheat hull and bran, which are widely used in water treatment, biodegradation, fabric production, dye extraction, wine fermentation, and waste management. Buckwheat also plays a crucial role in agro-industry as livestock feed, soil input, and a component in traditional medicinal feeds, while contributing to organic and specialized fertilizers for sustainable agriculture.

These diverse applications highlight buckwheat-growing importance in both industry and agriculture. Its multifunctional properties support food innovation, sustainable farming, and environmental conservation, positioning it as a valuable resource for future developments.

4.10.1 Medicines

Numerous studies have demonstrated that buckwheat provides a wide range of health advantages, including the ability to help fight off a number of chronic illnesses. According to a study by Wieslander (2020), eating buckwheat can help prevent conditions including obesity, cardiovascular disease, high blood pressure, and gallstones. Building on this, research by Huda et al. (2021) and Zhu (2016) demonstrated that buckwheat possesses significant pharmacological properties, including antidiabetic, antitumor, anti-inflammatory, antioxidant, and hepatoprotective effects. Additionally, the flavonoids and phenolic compounds found in buckwheat are widely utilized in treating conditions such as diabetes, hypertension, microbial infections, and cancer (Huda et al. 2021; Jha et al. 2024). Further expanding on these findings (Kreft 2016) highlighted buckwheat's exceptional role in diabetes management, showing that it helps reduce postprandial blood glucose and insulin response. This effect is attributed to bioactive compounds like rutin, which support insulin signaling and metabolic health. Rutin has also been explored for its therapeutic potential in treating Alzheimer's disease, adding to the growing list of buckwheat's medicinal benefits (Kreft 2016).

In traditional medicine, *F. tataricum* and *F. esculentum* were both utilized for a variety of therapeutic purposes. The seeds of *F. esculentum* are said to have the capacity to stimulate the spleen, removing stagnant food and descending qi-flowing, according to Traditional Chinese Medicine (TCM) theory. In Korean folk medicine, it is used as a treatment for fever reduction, detoxification, and anti-inflammation (Kim 2003); in the British Herbal Pharmacopoeia, it is mentioned as an antihemorrhagic and hypotensive medication (Jing et al. 2016). In addition, *F. tataricum* roots were traditionally utilized to regulate qi-flowing for relieving pain, invigorate the spleen and drain dampness, which were commonly used to treat some chronic and incurable diseases, such as rheumatic disorders, cancers, and general debility (Guo et al. 2010; Zheng et al. 2012; Jha et al. 2024). Its roots were therefore called "Qiao ye qi," indicating that it might have similar effects as "San qi" (*Panax notoginseng* roots), which prompted our group to investigate the bioactive metabolites from *F. tataricum* roots and subsequently revealed the abundance of cytotoxic phenylpropanoid glycosides that have potential use in cancer therapy (Jing et al. 2016; Jha et al. 2024). The main flavonol in buckwheat is rutin, which accounts for 90% of its total phenolic content (Sytar et al. 2016). It has proven possible to extract quercetin and rutin from different sections of *F. esculentum*, *F. tataricum*, and *F. cymosum*, which includes seeds, sprouts, leaves, and flowers (Jing et al. 2016). Compared to regular buckwheat, Tartary buckwheat has higher quantities of rutin and quercetin (Zhu 2016; Huda et al. 2021). Extensive research highlights the strong medicinal potential of *Fagopyrum* buckwheats. Among them, *F. dibotrys*, also known as *F. cymosum* (Trev.) Meisn., is a perennial herb with an upright growth habit. Its seeds and leaves, rich in rutin, are not only edible but also contribute to a healthy diet. The leaves can be steamed or boiled and consumed similarly to spinach (Jing et al. 2016). Beyond its nutritional value, *F. dibotrys* holds significant medicinal importance in traditional Chinese medicine (TCM). Its rhizome has long been used for detoxification, improving blood circulation, and promoting wound healing. It has been traditionally employed to treat various ailments, including lung diseases, rheumatism, cancer, dysmenorrhea, inflammation, lower back pain, snakebites, and traumatic injuries, with notable effectiveness in lung cancer therapy (Gao and Meng 1993; Chan 2003). A vital aspect of TCM, golden buckwheat rhizome (FDR) is a particularly valuable kind of *F. dibotrys* with a wide range of medicinal uses. It has been used extensively to treat respiratory conditions, lumbago, cancer, inflammation, dysmenorrhea, and snakebite (Zhang et al. 2022). The "Wei-Mai-Ning" capsule was created as a result of its potent medicinal qualities and is specifically used to cure cancer (Jha et al. 2024). Extracts from golden buckwheat have demonstrated encouraging antiviral qualities in addition to their anticancer potential. Their medical importance has been further expanded by recent investigations that have demonstrated their efficacy against a variety of viral infections, including COVID-19

(Zhang et al. 2021a, b; 2023a, b). Furthermore, a variety of bioactive characteristics are displayed by *Fagopyrum* species, such as *F. cymosum*, *F. tataricum*, and *F. esculentum*. They are useful in both traditional and modern medicine because of their well-known antitumor, antioxidant, anti-inflammatory, hepatoprotective, antihyperglycemic, anti-allergic, antibacterial, and anti-fatigue properties (Jing et al. 2016; Zou et al. 2023).

4.10.2 Conclusion and Future Perspectives

Abundance of nutritional as well as bioactive components of buckwheat leads buckwheat as most valuable crop. Gluten-free nature of buckwheat showed attractive attention to the researchers for its essential roles in human health. In food industries, buckwheat-enriched various bakery, non-bakery food products and fermented food staffs and drinks play an important role in terms of economy. To make different food products from this valuable crop, different food processing techniques are involved. On the other hand, enrichment of buckwheat with different bioactive compounds and other chemical components showed their beneficial effects to get rid of several health diseases. Therefore, to keep in mind the versatile beneficial importance of this crop, future research should be done in such a way that optimum utilization of this crop should be done by providing new by-products with balanced nutrition and therapeutic importance. For this, researchers should select potential genotypes and should perform quality breeding. It is essential to adopt an efficient food processing techniques so that minimal degradation of nutritional and phenolic composition will take place. Till today, least research has been done on toxic effects of this crop, therefore research should be done further to found its all negative effects on human health.

References

Alok A, Sandhya D, Jogam P, Rodrigues V, Bhati KK, Sharma H, Kumar J (2020) The rise of the CRISPR/Cpf1 system for efficient genome editing in plants. *Front Plant Sci* 11:264. https://doi.org/10.3389/fpls.2020.00264

Amudha Senthil SP (2015) Effect of Hydrothermal Treatment on the Nutritional and Functional Properties of Husked and Dehusked Buckwheat. J Food Process Technol 06(07). https://doi.org/10.4172/2157-7110.1000461

Atkari VT, Pawar KR, Patil SC (2021) Hermetic storage technology for food grains: new beginning of organic and safe storage for farmers and consumers: a review. *Bhartiya Krishi Anusandhan Patrika 36*(4):299–303. https://doi.org/10.18805/BKAP278

Averchev, O. V., Kyrylov, Y. E., & Fesenko, G. A. (2021). The current state of buckwheat market in Ukraine. http://hdl.handle.net/123456789/5989

Babu S, Yadav GS, Singh R, Avasthe R, Das A, Mohapatra K, Tahashildar M, Kumar K, Thoithoi Devi M, Rana D, Pande P, Prakash N (2018) Production technology and multifarious uses of buckwheat (Fagopyrum spp.): a review. Indian J Agron 63(4) www.IndianJournals.com

Björkman T (2000) Buckwheat production. Guide to buckwheat production in the Northeast. Available online: http://www.nysaes.cornell.edu/hort/faculty/bjorkman/buck/Buck.html

Boivin N, Fuller DQ, Crowther A (2012) Old world globalization and the Columbian exchange: comparison and contrast. World Archaeol 44:452–469. information: http://www.tandfonline.com/loi/rwar20

Bortesi L, Fischer R (2015) The CRISPR/Cas9 system for plant genome editing and beyond. Biotechnol Adv 33(1):41–52. https://doi.org/10.1016/j.biotechadv.2014.12.006

Bradley D (2011) Proto-Tibeto-Burman grain crops. Rice 4:134–141. https://doi.org/10.1007/s12284-011-9074-y

Brasil VCB, Guimarães BP, Evaristo RBW, Carmo TS, Ghesti GF (2021) Buckwheat (Fagopyrum esculentum Moench) characterization as adjunct in beer brewing. Food Sci Technol 41(suppl 1):265–272. https://doi.org/10.1590/fst.15920

Breslauer R, Nalbandian E, Reinman T, Rezaey M, Ganjyal GM, Murphy KM (2023) Buckwheat production and value-added processing: a review of potential Western Washington cropping and food system applications. Sustainability 15(20):14758. https://doi.org/10.3390/su152014758

Campbell CG, Heller J, Engel JMM (1997) Buckwheat: Fagopyrum esculentum Moench. International Plant Genetic Resources Institute

Candelaria-Morales NP, Grossman J, Fernandez A, Rogers M (2022) Exploring multifunctionality of summer cover crops for organic vegetable farms in the Upper Midwest. Renew Agric Food Syst. https://doi.org/10.1017/S1742170521000545

Chan P-K (2003) Inhibition of tumor growth in vitro by the extract of Fagopyrum cymosum (fago-c). Life Sci 72(16):1851–1858. https://doi.org/10.1016/S0024-3205(03)00013-4

Chen C, Li A (2016) Transcriptome analysis of differentially expressed genes involved in proanthocyanidin accumulation in the rhizomes of Fagopyrum dibotrys and an irradiation-Induced mutant. Front Physiol 7:100. https://doi.org/10.3389/fphys.2016.00100

Chen O, Mah E, Dioum E, Marwaha A, Shanmugam S, Malleshi N, Sudha V, Gayathri R, Unnikrishnan R, Anjana RM, Krishnaswamy K, Mohan V, Chu Y (2021) The role of oat nutrients in the immune system: a narrative review. Nutrients 13(4):1048. https://doi.org/10.3390/nu13041048

Chen YL, Yang KM, Shiao XY, Huang JJ, Ma YA, Chiang PY (2024) Relationship between storage quality and functionality of common buckwheat (*Fagopyrum esculentum* Moench) and Tartary Buckwheat (*Fagopyrum tataricum* Gaertn) at different temperatures. Agric Eng 6(3):3121–3136

Cho KS, Yun BK, Yoon YH, Hong SY, Mekapogu M, Kim KH, Yang TJ (2015) Complete chloroplast genome sequence of Tartary Buckwheat (*Fagopyrum tataricum*) and comparative analysis with common Buckwheat (*F. esculentum*). *PLoS One* 10(5):e0125332. https://doi.org/10.1371/journal.pone.0125332

Chrungoo NK, Chettry U (2021) Buckwheat: a critical approach towards assessment of its potential as a super crop. Indian J Genet Plant Breed (The) 81(01):1–23. https://doi.org/10.31742/IJGPB.81.1.1

Chu Y, Han Y, Jia F, Wang Y, Chen P, Meng X, Zhang J, Li A (2022) Analysis of the impact hulling behaviour of paddy grains. Biosyst Eng 220:243–257. https://doi.org/10.1016/j.biosystemseng.2022.06.005

Clark A (ed.) (2007) Managing Cover Crops Profitably. 3rd ed. Sustainable Agriculture Research and Education (SARE) Handbook Series, bk 9. Sustainable Agriculture Research and Education, College Park, MD

Dabija A, Ciocan ME, Chetrariu A, Codină GG (2022) Buckwheat and Amaranth as raw materials for brewing, a review. Plants 11(6):756. https://doi.org/10.3390/plants11060756

Dakia PA, Blecker C, Robert C, Wathelet B, Paquot M (2008) Composition and physicochemical properties of locust bean gum extracted from whole seeds by acid or water dehulling pre-treatment. Food Hydrocoll 22(5):807–818. https://doi.org/10.1016/j.foodhyd.2007.03.007

Demchenko O, Boyko O (2016) Investigation of the resistance of different varieties of buckwheat to infectious diseases after the pre-sowing treatment of seeds and vegetating plants with biological preparation... February 2017

Econopouly B, Derie ML, Jones SS, du Toit LJ (2017) Downy Mildew of Buckwheat, Fagopyrum esculentum , in the United States caused by Peronospora cf. ducometii. Plant Dis 101(2):391. https://doi.org/10.1094/PDIS-09-16-1362-PDN

Falquet B, GfelleR A, Pourcelot M, Tschuy F, Wirth J (2015) Weed suppression by common Buckwheat: a review. Environ Control Biol 53(1):1–6. https://doi.org/10.2525/ecb.53.1

Fang X, Zhang Y, Zhang Y, Huang K, Yang W, Li X, Zhang Z, Wu K, Xu X, Ruan R, Yuan X, Zhang Z, Yi Z (2019) De novo transcriptome assembly and identification of genes related to seed size in common buckwheat (Fagopyrum esculentum M.). *Breed Sci* 69(3):487–497. https://doi.org/10.1270/jsbbs.18194

FAO (2021) International Network of FOOD Data Systems (INFOODS). http://www.fao.org/infoods/infoods/food-biodiversity/en/. Accessed 15 Feb 2021

Farooq S, Rehman RU, Pirzadah TB, Malik B, Dar FA, Tahir I (2016) Cultivation, agronomic practices, and growth performance of buckwheat. In: Zhou M, Kreft I, Woo SH, Chrungoo N, Wieslander G (eds.) Molecular breeding and nutritional aspects of buckwheat (pp. 299–319). Elsevier. https://doi.org/10.1016/B978-0-12-803692-1.00023-7

Ferreira RJ, Silva EM, Nascimento DD, Santos LB, Trigo VA, Nogueira GP, Vargas PF, Harakava R, Soares PLM (2023) First report of Root-Knot Nematode, Meloidogyne incognita, infecting Buckwheat, Fagopyrum esculentum, in the State of São Paulo, Brazil. *Plant Dis* 107(2):585. https://doi.org/10.1094/PDIS-02-22-0256-PDN

Food and Agriculture Organization of the United Nations (FAO). (2023). Production quantities of buckwheat by country. FAOSTAT Statistical Database. Retrieved June 25, 2025, from https://www.fao.org/faostat/en/#data/QCL

Fukuie Y, Shimoyama H, Morishita T, Tsugama D, Fujino K (2020) A putative AGAMOUS ortholog is a candidate for the gene determining ease of dehulling in Tartary buckwheat (Fagopyrum tataricum). *Planta* 251(4):85. https://doi.org/10.1007/s00425-020-03374-6

Gang Z, Yu T (1998) A primary study of increasing the production rate of buckwheat. In: Campbell C, Przybylski R (eds) Current Advances in Buckwheat Research. Proceedings of the 7th International Symposium on Buckwheat, Winnipeg, Manitoba, Canada, Aug. 12–14, pp. 18–23.

Gao Z, Meng FH (1993) [Effect of Fagopyrum cymosum rootin on clonal formation of four human tumor cells]. Zhongguo Zhong Yao Za Zhi = Zhongguo Zhongyao Zazhi =. China J Chinese Materia Medica 18(8):498–500. 511

Giménez-Bastida J, Piskuła M, Zieliński H (2015) Recent advances in processing and development of Buckwheat Derived Bakery and non-bakery products–a review. Pol J Food Nutr Sci 65(1):9–20. https://doi.org/10.1515/pjfns-2015-0005

Gonçalves MAA, Salvador R, de Francisco AC, Piekarski CM (2024) Value recovery from waste in the processing of buckwheat: opportunities for a circular bioeconomy. Eng Rep 6(7). https://doi.org/10.1002/eng2.12757

Gopi R, Kapoor CHANDAN, Kalita H, Das SK, Avasthe RK (2015) A new report of downy mildew on buckwheat (Fagopyrum esculentum) caused by *Perenospora* sp. in Sikkim. J Mycopathol Res 53(2):95–297

Goyal RK, Vishwakarma RK, Wanjari OD (2010) Effect of moisture content on pitting and milling efficiency of pigeon pea grain. *Food Bioprocess Technol* 3(1):146–149. https://doi.org/10.1007/s11947-009-0255-2

Graziano S, Agrimonti C, Marmiroli N, Gullì M (2022) Utilisation and limitations of pseudocereals (quinoa, amaranth, and buckwheat) in food production: a review. *Trends Food Sci Technol* 125:154–165. https://doi.org/10.1016/j.tifs.2022.04.007

Guo X, Zhu K, Zhang H, Yao H (2010) Anti-tumor activity of a novel protein obtained from Tartary Buckwheat. *Int J Mol Sci* 11(12):5201–5211. https://doi.org/10.3390/ijms11125201

He Y, Zhou M (2022) Buckwheat genome and genomics. In: Chapman MA (ed) Underutilised crop genomes. Compendium of plant genomes. Springer, Cham. https://doi.org/10.1007/978-3-031-00848-1_2

He J, Chen Y, Hao Y, Lai D, Jahan T, Shi Y, Lin H, He Y, Huda MN, Cheng J, Zhang K, Li J, Ruan J, Zhou M (2024a) Combining GWAS and RNA-Seq approaches identifies the FtADH1 gene for drought resistance in Tartary buckwheat. *J Integr Agric 23*(1):123–135

He Y, Zhang K, Shi Y, Lin H, Huang X, Lu X, Wang Z, Li W, Feng X, Shi T, Chen Q, Wang J, Tang Y, Chapman MA, Germ M, Luthar Z, Kreft I, Janovská D, Meglič V, Woo SH et al (2024b) Genomic insight into the origin, domestication, dispersal, diversification and human selection of Tartary buckwheat. *Genome Biol* 25(1):61. https://doi.org/10.1186/s13059-024-03203-z

Hou S, Sun Z, Linghu B et al (2016) Genetic diversity of Buckwheat cultivars (Fagopyrum tartaricum Gaertn.) Assessed with SSR Markers developed from genome survey sequences. *Plant Mol Biol Rep* 34:233–241. https://doi.org/10.1007/s11105-015-0907-5

Hore D, Rathic RS (2002) Collection, cultivation and characterization of buckwheat in north eastern region of India. Fagopyrum 19:11–15

Hsiao H-Y, Ariyawansa HA, Hsu C-C, Wang C-J, Shen Y-M (2022) New records of powdery mildews from Taiwan: Erysiphe ipomoeae comb. nov., E. aff. betae on Buckwheat, and E. neolycopersici comb. nov. on Cardiospermum halicacabum. *Diversity* 14(3):204. https://doi.org/10.3390/d14030204

Huda MN, Lu S, Jahan T, Ding M, Jha R, Zhang K, Zhang W, Georgiev MI, Park SU, Zhou M (2021) Treasure from garden: Bioactive compounds of buckwheat. *Food Chem* 335:127653. https://doi.org/10.1016/j.foodchem.2020.127653

Hughes HD, Hensen ER (1934) Crop production principles and practices. New York

Hunt HV, Shang X, Jones MK (2018) Buckwheat: a crop from outside the major Chinese domestication centres? A review of the archaeobotanical, palynological and genetic evidence. *Veg Hist Archaeobot* 27(3):493–506. https://doi.org/10.1007/s00334-017-0649-4

Janeš D, Prosen H, Kreft S (2012) Identification and quantification of aroma compounds of Tartary Buckwheat (Fagopyrum tataricum Gaertn.) and some of its milling fractions. J Food Sci 77(7). https://doi.org/10.1111/j.1750-3841.2012.02778.x

Jevdjovic R, Maletic R (2003) Effect of buckwheat seed storage duration on its quality. *J Agric Sci* 48(2):135–141. https://doi.org/10.2298/JAS0302135J

Jha R, Yadav HK, Raiya R, Singh RK, Jha UC, Sathee L, Singh P, Thudi M, Singh A, Chaturvedi SK, Tripathi S (2022) Integrated breeding approaches to enhance the nutritional quality of food legumes. *Front Plant Sci 13*:984700. https://doi.org/10.3389/fpls.2022.984700

Jha R, Zhang K, He Y, Mendler-Drienyovszki N, Magyar-Tábori K, Quinet M, Germ M, Kreft I, Meglič V, Ikeda K, Chapman MA, Janovská D, Podolska G, Woo S-H, Bruno S, Georgiev MI, Chrungoo N, Betekhtin A, Zhou M (2024) Global nutritional challenges and opportunities: Buckwheat, a potential bridge between nutrient deficiency and food security. Trends Food Sci Technol 145:104365. https://doi.org/10.1016/j.tifs.2024.104365

Jin J, Ohanenye IC, Udenigwe CC (2022) Buckwheat proteins: functionality, safety, bioactivity, and prospects as alternative plant-based proteins in the food industry. *Crit Rev Food Sci Nutr* 62(7):1752–1764. https://doi.org/10.1080/10408398.2020.1847027

Jing R, Li H-Q, Hu C-L, Jiang Y-P, Qin L-P, Zheng C-J (2016) Phytochemical and pharmacological profiles of three Fagopyrum Buckwheats. Int J Mol Sci 17(4):589. https://doi.org/10.3390/ijms17040589

Jose N, Selvan SS (2024) Buckwheat: biology, functional potential and sustainable utilization, pp 233–258. https://doi.org/10.1007/978-3-031-64237-1_13

Joshi DC, Zhang K, Wang C, Chandora R, Khurshid M, Li J, He M, Georgiev MI, Zhou M (2020) Strategic enhancement of genetic gain for nutraceutical development in buckwheat: a genomics-driven perspective. *Biotechnol Adv* 39:107479. https://doi.org/10.1016/j.biotechadv.2019.107479

Kalinová JP, Vrchotová N, Tříska J (2018) Contribution to the study of rutin stability in the achenes of Tartary buckwheat (Fagopyrum tataricum). *Food Chem 258*:314–320. https://doi.org/10.1016/j.foodchem.2018.03.090

Katsube-Tanaka T (2016) Buckwheat production, consumption, and genetic resources in Japan. In: Molecular breeding and nutritional aspects of Buckwheat. Elsevier, pp 61–80. https://doi.org/10.1016/B978-0-12-803692-1.00005-5

Kaur N, Alok A, Shivani, Kaur N, Pandey P, Awasthi P, Tiwari S (2018) CRISPR/Cas9-mediated efficient editing in phytoene desaturase (PDS) demonstrates precise manipulation in banana cv. Rasthali genome. Funct Integr Genom 18(1):89–99. https://doi.org/10.1007/s10142-017-0577-5

Kawanishi M, Suenaga K, Hirano T (2022) Single-laboratory validation of an analytical method for detarmination of Fusarium toxins in Buckwheat and Job's Tears. Food Hyg Safe Sci (Shokuhin Eiseigaku Zasshi) 63(3):117–121. https://doi.org/10.3358/shokueishi.63.117

Keriene I, Mankeviciene A, Bliznikas S, Cesnuleviciene R, Janaviciene S, Jablonskyte-Rasce D, Maiksteniene S (2016) The effect of buckwheat groats processing on the content of mycotoxins and phenolic compounds. CyTA – J Food 14(4):565–571. https://doi.org/10.1080/19476337.2016.1176959

Ke H, Wang X, Zhou Z, Ai W, Wu Z, Zhang Y (2021) Effect of weimaining on apoptosis and Caspase-3 expression in a breast cancer mouse model. J Ethnopharmacol 264:113363. https://doi.org/10.1016/j.jep.2020.113363

Kim C (2003) Anti-allergic action of buckwheat (Fagopyrum esculentum Moench) grain extract. Int Immunopharmacol 3(1):129–136. https://doi.org/10.1016/S1567-5769(02)00261-8

Klepacka J, Najda A (2021) Effect of commercial processing on polyphenols and antioxidant activity of buckwheat seeds. *Int J Food Sci Technol* 56(2):661–670. https://doi.org/10.1111/ijfs.14714

Kolankowska E, Choszcz DJ, Markowski P, Reszczyński PS, Lipiński AJ (2021) The process of separating Buckwheat and Wheat Grain in a pneumatic cone separator in the context of sustainable agriculture. Processes 10(1):59. https://doi.org/10.3390/pr10010059

Kongari S, Sahi VP, Rahul P, Sandeep N (2024) Cultivating resilience: disease resistance screening of Buckwheat (Fagopyrum spp.) genotypes in India's semi-tropical belt. J Exp Agric Int 46(1). https://doi.org/10.9734/jeai/2024/v46i12300

Konopka S, Choszcz D, Markowski P (2017) Optimization of the separation parameters and indicators of separation efficiency of Buckwheat seeds. Sustainability 9(11):2134. https://doi.org/10.3390/su9112134

Kreft M (2016) Buckwheat phenolic metabolites in health and disease. Nutr Res Rev 29(1):30–39. https://doi.org/10.1017/S0954422415000190

Kumari A, Chaudhary HK (2020) Nutraceutical crop buckwheat: a concealed wealth in the lap of Himalayas. Crit Rev Biotechnol 40(4):539–554. https://doi.org/10.1080/07388551.2020.1747387

Li Y, Zhang K, Liu Y, Li K, Hu D, Zhang X (2020) Identification of quantitative trait loci for rutin content in Tartary buckwheat (*Fagopyrum tataricum*) by association mapping. *Front Plant Sci 11*:553. https://doi.org/10.3389/fpls.2020.00553

Li S, Shen Q, Wang H, He F, Xiao Z, Peng X, Zhou M, Tang X (2021a) First report of Alternaria alternata causing leaf spot of Tartary Buckwheat in China. *Plant Dis* 105(11):3751. https://doi.org/10.1094/PDIS-01-21-0191-PDN

Li S, Zhang K, Khurshid M, Fan Y, Xu B, Zhou M (2021b) First report of Rhizoctonia solani AG-4 HGI causing stem canker on Fagopyrum tataricum (Tartary Buckwheat) in China. *Plant Dis* 105(2):505–505. https://doi.org/10.1094/PDIS-06-20-1253-PDN

Liu M, Huang L, Ma Z, Sun W, Wu Q, Tang Z, Bu T, Li C, Chen H (2019a) Genome-wide identification, expression analysis and functional study of the GRAS gene family in Tartary buckwheat (Fagopyrum tataricum). *BMC Plant Biol* 19(1):342. https://doi.org/10.1186/s12870-019-1951-3

Liu M, Sun W, Ma Z, Zheng T, Huang L, Wu Q, Zhao G, Tang Z, Bu T, Li C, Chen H (2019b) Genome-wide investigation of the AP2/ERF gene family in Tartary buckwheat (*Fagopyrum tataricum*). *BMC Plant Biol* 19(1):84. https://doi.org/10.1186/s12870-019-1681-6

Liu X, Zhou S, Chen S, Yi Z, Pan H, Yao R (2022) Buckwheat disease recognition based on convolution neural network. *Appl Sci* 12(9):4795. https://doi.org/10.3390/app12094795

Logacheva MD, Schelkunov MI, Fesenko AN, Kasianov AS, Penin AA (2020) Mitochondrial genome of Fagopyrum esculentum and the genetic diversity of extranuclear genomes in Buckwheat. *Plants (Basel, Switz)* 9(5):618. https://doi.org/10.3390/plants9050618

Lu X, Zuo Q, Huda MN, Shi Y, Li G, Wang X, Xiao Y, Khurshid M, Jahan T, Dhami N, Gauchan DP, Ali MA, Cheng J, Meng Y, Ruan J, Zhou M (2025) Genome-wide association analysis locates FtAUR3 in Tartary buckwheat that contributes to enhance plant salt resistance. J Integr Agric *24*(4). https://doi.org/10.1016/j.jia.2025.04.008

Majaw J, Bharath J, Prabhasankar P (2024) Effect of pseudocereals on gluten-free pasta: rheological, microstructural, and cooking quality. In: Development of Gluten-Free Pasta. Elsevier, pp 135–156. https://doi.org/10.1016/B978-0-443-13238-4.00007-1

Manandhar HK, Timila RD, Sharma S, Joshi S, Manandhar S, Gurung SB, ... Stahapit BR (2016) A field guide for identification and scoring methods of diseases in the mountain crops of Nepal. NARC, DoA, LI-BIRD and Bioversity International, Nepal

Mar'in VA, Vereshchagin AL (2016) Physical principles of processing off-grade buckwheat. Foods Raw Mater 4(1):51–60. https://doi.org/10.21179/2308-4057-2016-1-51-60

Martínez-Villaluenga C, Peñas E, Hernández-Ledesma B (2020) Pseudocereal grains: nutritional value, health benefits and current applications for the development of gluten-free foods. *Food Chem Toxicol* 137:111178. https://doi.org/10.1016/j.fct.2020.111178

Matsui K, Mizuno N, Ueno M, Takeshima R, Yasui Y (2020) Development of co-dominant markers linked to a hemizygous region that is related to the self-compatibility locus (S) in buckwheat (Fagopyrum esculentum). Breed Sci 70(1):112–117. https://doi.org/10.1270/jsbbs.19129

Maurya KA, Singh AU, John V, Murmu R (2021) Occurrence of Alternaria alternata causing leaf spot in Buckwheat (Fagopyrum esculentum) in Prayagraj area of Uttar Pradesh, India Senior Technical Officer, Regional Plant Quarantine Station, Kolkata. Int J Environ Agric Res (IJOEAR) 7(6)

Michiyama H, Hayashi H (1998) Differences of growth and development between summer and autumn type-cultivars in common buckwheat (Fagopyrum esculentum Moench). Jpn J Crop Sci 67:323–330

Mikami T, Motonishi S, Tsutsui S (2018) Production, uses and cultivars of common buckwheat in Japan: an overview. *Acta Agriculturae Slovenica* 111(2). https://doi.org/10.14720/aas.2018.111.2.23

Mondal KK, Rana SS, Sood P (2002) Sclerotinia root rot: a new threat to Buckwheat seedlings in India. Plant Dis 86(12):1404–1404. https://doi.org/10.1094/PDIS.2002.86.12.1404B

Naik M, Modupalli N, Sunil CK, Rawson A, Natarajan V (2022) Major millet processing. In: Handbook of millets – processing, quality, and nutrition status. Springer, Singapore, pp 63–80. https://doi.org/10.1007/978-981-16-7224-8_4

Nalinkumar A, Singh P (2020) An overview of Buckwheat (Fagopyrum spp)-an underutilized crop in India-nutritional value and health benefits. Int J Med Res Health Sci 2020(7):39–44. www.ijmrhs.com

Olson M (2001) Common buckwheat, agri-facts, agriculture, food and rural management. Alberta, Canada. http://www1.agric.gov.ab.ca/$department/deptdocs.nst/all. agdex103/$file.118_20-2.pdf? Open Element.

Parimala K, Subramanian K, Kannan SM, Vijayalakshmi K, Assistance E, Thooyavathy RA, Subbiah VR, Vijayalakshmi EK (2013) Seed storage techniques-a primer centre for Indian knowledge systems. Chennai Revitalising Rainfed Agriculture Network. www.rainfedindia.org

Penin AA, Kasianov AS, Klepikova AV, Kirov IV, Gerasimov ES, Fesenko AN, Logacheva MD (2021) High-resolution transcriptome Atlas and improved genome assembly of common Buckwheat, Fagopyrum esculentum. *Front Plant Sci* 12:612382. https://doi.org/10.3389/fpls.2021.612382

Pozeliene, A., Lynikiene, S., & Sapailaite, I. (2010). Use of electroseparator in improving the quality of ecological buckwheat seeds. https://cabidigitallibrary.org

Quequeto WD, Siqueira VC, Schoeninger V, Martins EAS, Isquierdo EP, da Silva FP (2018) Physical properties of buckwheat (Fagopyrum esculentum Moench) grains during convective drying. Revista Brasileira de Engenharia Agrícola e Ambiental 22(11):793–798. https://doi.org/10.1590/1807-1929/agriambi.v22n11p793-798

Radics, L., & Mikóházi, D. (2010) The European Journal of Plant Science and Biotechnology Principles of Common Buckwheat Production. http://www.globalsciencebooks.info/Online/GSBOnline/images/2010/EJPSB_4(SI1)/EJPSB_4(SI1)57-63o.pdf

Saha D, Pradhan A, Sarkar A, Tripathy PP (2024) Recent trends in pseudocereal-based foods. In: Pseudocereals. CRC Press, pp 264–291. https://doi.org/10.1201/9781003325277-13

Schuster M (2023) Finding the needle in the haystack: novel multi-omics approach readily identifies candidate resistance genes of Tartary buckwheat against Rhizoctonia solani infection. Plant Cell 35(8):2701–2702. https://doi.org/10.1093/plcell/koad133

Shaikhiev IG, Sverguzova SV, Galimova RZ, Grechina AS (2020) Using wastes of buckwheat processing as sorption materials for the removal of pollutants from aqueous media: a review. IOP Conf Ser: Mater Sci Eng 945(1):012044. https://doi.org/10.1088/1757-899X/945/1/012044

Shen Q, Peng X, He F, Li S, Xiao Z, Wang H, Tang X, Zhou M (2021) First report of Nigrospora osmanthi causing leaf spot on Tartary Buckwheat in China. *Plant Dis* 105(4):1227–1227. https://doi.org/10.1094/PDIS-08-20-1773-PDN

Shi T, Li R, Chen Q et al (2017) De novo sequencing of seed transcriptome and development of genic-SSR markers in common buckwheat (Fagopyrum esculentum). *Mol Breed* 37:147. https://doi.org/10.1007/s11032-017-0743-4

Shimatani Z, Kashojiya S, Takayama M, Terada R, Arazoe T, Ishii H, Teramura H, Yamamoto T, Komatsu H, Miura K, Ezura H, Nishida K, Ariizumi T, Kondo A (2017) Targeted base editing in rice and tomato using a CRISPR-Cas9 cytidine deaminase fusion. *Nat Biotechnol* 35(5):441–443. https://doi.org/10.1038/nbt.3833

Sindarovska YR, Guzyk OI, Yuzvenko LV, Demchenko OA, Didenko LF, Grynevych OI, Spivak MY (2014) Ribonuclease activity of buckwheat plant (Fagopyrum esculentum) cultivars with different sensitivities to buckwheat burn virus. Ukr Biochem J 86(3):33–40

Siqueira VC, Leite RA, Mabasso GA, Martins EAS, Quequeto WD, Isquierdo EP (2020) Drying kinetics and effective diffusion of buckwheat grains. Ciência e Agrotecnologia 44. https://doi.org/10.1590/1413-7054202044011320

Skřivan P, Chrpová D, Klitschová B, Švec I, Sluková M (2023) Buckwheat flour (Fagopyrum esculentum Moench)—a contemporary view on the problems of its production for human nutrition. *Foods* 12(16):3055. https://doi.org/10.3390/foods12163055

Subedi S, Dhami N, Gurung S, Neupane S, Thapa S, Oli L (2020) Assessment of disease resistance and high yielding traits of common buckwheat genotypes in subtropical climate of Nepal. *SAARC J Agric* 18(1):143–152. https://doi.org/10.3329/sja.v18i1.48388

Subedi S, Nepali D, Subedi N (2022) Controlling Downy Mildew in Buckwheat using Fungicide. J Plant Prot Soc 7(01):86–93. https://doi.org/10.3126/jpps.v7i01.47292

Sun W, Ma Z, Liu M (2020) Cytochrome P450 family: genome-wide identification provides insights into the rutin synthesis pathway in Tartary buckwheat and the improvement of agricultural product quality. Int J Biol Macromol 164:4032–4045. https://doi.org/10.1016/j.ijbiomac.2020.09.008

Suzuki T, Mukasa Y, Morishita T, Takigawa S, Noda T (2012) Traits of shattering resistant buckwheat 'W/SK86GF'. *Breed Sci* 62(4):360–364. https://doi.org/10.1270/jsbbs.62.360

Suzuki T, Noda T, Morishita T, Ishiguro K, Otsuka S, Brunori A (2020) Present status and future perspectives of breeding for buckwheat quality. *Breed Sci* 70(1):48–66. https://doi.org/10.1270/jsbbs.19018

Sytar O, Brestic M, Zivcak M, Phan Tran L-S (2016) The contribution of Buckwheat genetic resources to health and dietary diversity. *Curr Genomics* 17(3):193–206. https://doi.org/10.2174/1389202917666160202215425

Taboković M, Simić M, Stanisavljević R, Sečanski M, Živanović L, Štrbanović R (2019) Buckwheat seed quality during the five-year storage in various packing materials. *Plant Soil Environ* 65(7):349–354. https://doi.org/10.17221/237/2019-PSE

Tang Y, Ding M, Tang Y et al. (2016) Germplasm resources of buckwheat in China. In Molecular breeding and nutritional aspects of Buckwheat, Zhou M et al., ed. (Academic). pp. 13–20. https://doi.org/10.1016/C2015-0-00352-5

Wang R, Li M, Chen S, Hui Y, Tang A, Wei Y (2019) Effects of flour dynamic viscosity on the quality properties of buckwheat noodles. *Carbohydr Polym* 207:815–823. https://doi.org/10.1016/j.carbpol.2018.09.048

Weisskopf A, Fuller DQ (2014) Buckwheat: origins and development. In: Smith C (ed) Encyclopedia of global archaeology (pp. 1025–1028). NY: Springer-Verlag

Wen W, Li Z, Shao J, Tang Y, Zhao Z, Yang J, Ding M, Zhu X, Zhou M (2021) The distribution and sustainable utilization of Buckwheat resources under climate change in China. *Plants (Basel, Switz)* 10(10):2081. https://doi.org/10.3390/plants10102081

Wen D, Wu L, Wang M, Yang W, Wang X, Ma W, Sun W, Chen S, Xiang L, Shi Y (2022) CRISPR/Cas9-Mediated targeted mutagenesis of FtMYB45 promotes flavonoid biosynthesis in tartary Buckwheat (*Fagopyrum tataricum*). *Front Plant Sci* 13:879390. https://doi.org/10.3389/fpls.2022.879390

Wieslander G (2020) Buckwheat in human health - a medical review / Ajda in zdravje ljudi– pregledni članek. Folia Biologica et Geologica 61(1):55–60. https://doi.org/10.3986/fbg0067

Wijerathna-Yapa A, Pathirana R (2022) Sustainable agro-food systems for addressing climate change and food security. Agriculture 12(10):1554. https://doi.org/10.3390/agriculture12101554

Wu Q, Bai X, Zhao W, Xiang D, Wan Y, Yan J, Zou L, Zhao G (2017) De Novo assembly and analysis of Tartary Buckwheat (Fagopyrum tataricum Garetn.) transcriptome discloses key regulators involved in salt-stress response. *Gene* 8(10):255. https://doi.org/10.3390/genes8100255

Xu D, Hu M, Wang Y, Cui Y (2019) Antioxidant activities of quercetin and its complexes for medicinal application. Molecules 24:1123. https://doi.org/10.3390/molecules24061123

Yang Y, Li Y, Wang X, Chen J, Zhang Y, Wei X (2022) Genetic dissection of phytic acid content in common buckwheat (*Fagopyrum esculentum*) grains using QTL mapping. *Plant Biotechnol J 20*(3):512–525. https://doi.org/10.1111/pbi.13732

Yao YF, Song XY, Xie G, Tang YN, Wortley AH, Qin F, Blackmore S, Li CS, Wang YF (2023) New insights into the origin of buckwheat cultivation in southwestern China from pollen data. The New phytologist 237(6):2467–2477. https://doi.org/10.1111/nph.18659

Yasui Y, Hirakawa H, Ueno M, Matsui K, Katsube-Tanaka T, Yang SJ, Aii J, Sato S, Mori M (2016) Assembly of the draft genome of buckwheat and its applications in identifying agronomically useful genes. DNA Res 23(3):215–224. https://doi.org/10.1093/dnares/dsw012

Zamaratskaia G, Gerhardt K, Knicky M, Wendin K (2024) Buckwheat: an underutilized crop with attractive sensory qualities and health benefits. *Crit Rev Food Sci Nutr* 64(33):12303–12318. https://doi.org/10.1080/10408398.2023.2249112

Zhang L, Li X, Ma B, Gao Q, Du H, Han Y, Li Y, Cao Y, Qi M, Zhu Y, Lu H, Ma M, Liu L, Zhou J, Nan C, Qin Y, Wang J, Cui L, Liu H, Liang C et al (2017) The Tartary Buckwheat genome provides insights into Rutin biosynthesis and abiotic stress tolerance. Mol Plant 10(9):1224–1237. https://doi.org/10.1016/j.molp.2017.08.013

Zhang L, Ma M, Liu L (2020) Identification of genetic locus underlying easy Dehulling in rice-tartary for easy postharvest processing of Tartary Buckwheat. *Gene* 11(4):459. https://doi.org/10.3390/genes11040459

Zhang K, He M, Fan Y, Zhao H, Gao B, Yang K, Li F, Tang Y, Gao Q, Lin T, Quinet M, Janovská D, Meglič V, Kwiatkowski J, Romanova O, Chrungoo N, Suzuki T, Luthar Z, Germ M, Woo SH et al (2021a) Resequencing of global Tartary buckwheat accessions reveals multiple domestication events and key loci associated with agronomic traits. *Genome Biol* 22(1):23. https://doi.org/10.1186/s13059-020-02217-7

Zhang L-L, He Y, Sheng F, Hu Y-F, Song Y, Li W, Chen J, Zhang J, Zou L (2021b) Towards a better understanding of *Fagopyrum dibotrys*: a systematic review. *Chin Med* 16(1):89. https://doi.org/10.1186/s13020-021-00498-z

Zhang C, Jiang Y, Liu C, Shi L, Li J, Zeng Y, Guo L, Wang S (2022) Identification of medicinal compounds of Fagopyri Dibotryis rhizome from different origins and its varieties using UPLC-MS/MS-based metabolomics. Metabolites 12(9):790. https://doi.org/10.3390/metabo12090790

Zhang M, Zhang X, Pei J, Guo B, Zhang G, Li M, Huang L (2023a) Identification of phytochemical compounds of Fagopyrum dibotrys and their targets by metabolomics, network pharmacology and molecular docking studies. *Heliyon* 9(3). https://doi.org/10.1016/j.heliyon.2023.e14029

Zhang Y, Yang T, Yue D, Shao X, Chen Y (2023b) Cold plasma-assisted buckwheat grain dehulling and farinographical properties of dehulled buckwheat flour. *J Cereal Sci* 112:103716. https://doi.org/10.1016/j.jcs.2023.103716

Zheng C, Hu C, Ma X, Peng C, Zhang H, Qin L (2012) Cytotoxic phenylpropanoid glycosides from Fagopyrum tataricum (L.) Gaertn. Food Chem 132(1):433–438. https://doi.org/10.1016/j.foodchem.2011.11.017

Zhou M, Sun Z, Wang C, Zhang X, Tang Y, Zhu X, Shao J, Wu Y (2021) Genome-wide association study reveals the genetic basis of iron and zinc accumulation in Tartary buckwheat (*Fagopyrum tataricum*). *Theor Appl Genet 134*(10):3453–3465. https://doi.org/10.1007/s00122-021-03909-9

Zhu F (2016) Chemical composition and health effects of Tartary buckwheat. *Food Chem* 203:231–245. https://doi.org/10.1016/j.foodchem.2016.02.050

Živković A, Polak T, Cigić B, Požrl T (2021) Germinated Buckwheat: effects of dehulling on phenolics profile and antioxidant activity of Buckwheat seeds. *Foods* 10(4):740. https://doi.org/10.3390/foods10040740

Zou L, Wu D, Ren G, Hu Y, Peng L, Zhao J, Garcia-Perez P, Carpena M, Prieto MA, Cao H, Cheng K-W, Wang M, Simal-Gandara J, John OD, Rengasamy KRR, Zhao G, Xiao J (2023) Bioactive compounds, health benefits, and industrial applications of Tartary buckwheat (*Fagopyrum tataricum*). *Crit Rev Food Sci Nutr* 63(5):657–673. https://doi.org/10.1080/10408398.2021.1952161

Chobitok, V., Pak, A., & Sofronova, M. (2024). Research of consumer properties of buckwheat porridge at the customs examination. https://repo.btu.kharkov.ua//handle/123456789/61551

Water Chestnut (*Trapa natans* L.): The Aquatic Treasure of Traditional Diets

5

Rima Kumari, Pankaj Kumar, and Anil Kumar Singh

Abstract

Water chestnut is an aquatic plant with significant nutritional and cultural value. Its edible fruits and seeds are rich in essential nutrients, making them a valuable component of traditional diets. This abstract highlights the nutritional benefits, processing and marketing, traditional application as well as potential application in water chestnut. The plant's adaptability to aquatic environments and its role in maintaining ecological balance are also highlighted. With its rich nutrient profile and versatility, water chestnut offers opportunities for sustainable food systems and nutritional security. This aquatic treasure can contribute to diverse and resilient food cultures, particularly in regions where it is traditionally consumed.

Keywords

Water chestnut · Nutrition · Traditional diets · Aquatic plant · Sustainable food systems

R. Kumari (✉)
Department of Molecular Biology and Genetic Engineering, Bihar Agricultural University, Sabour, Bihar, India

P. Kumar
Division of Plant Biotechnology, College of Agricultural Biotechnology, Bihar Agricultural University, Sabour, Bihar, India

A. K. Singh
Directorate of Research, Bihar Agricultural University, Sabour, Bihar, India

A. Kumar et al. (eds.), *Fasting Superfoods*,
https://doi.org/10.1007/978-981-95-2762-5_5

5.1 Introduction

Water chestnut (*Trapa natans* L.) is an annual aquatic angiosperm found in tropical and subtropical regions (Singh et al. 2018). It belongs to the family Lythraceae. In India, it is commonly known as water chestnut in English and *singhara* in Hindi. It is treated as a neglected crop because nowhere in India as well as in the whole world the crop is scientifically taken under the economic cultivation. But nowadays, it has been proved as a crop of immense industrial importance. Further, water chestnut has a vast potential pertaining to its nutritional qualities, which is scientifically proven to be highly beneficial for human beings. Moreover, it leaves all their biomasses in the same field where it grows. Henceforth, it contributes a lot of plant essential nutrients to the soil mass. Water chestnut is mainly grown in Bihar, India. It is distributed in temperate and tropical Asia and central and South-East Europe. It is extensively grown in the all districts of Bihar, namely Araria, Arwal, Aurangabad, Banka, Begusarai, Bhojpur, Bhagalpur, Darbhanga, Buxar, Gaya, East Champaran, Gopalganj, Jamui, Khagaria, Kishanganj, Katihar, Lakhisarai, Kaimur, Madhubani, Jehanabad, Madhepura, Munger, Nalanda, Muzaffarpur, Nawada, Purnea, Patna, Saharsa, Samastipur, Rohtas, Sheikhpura, Sheohar, Sitamarhi, Siwan, Saran, Vaishali, West Champaran, and Supaul. On the basis of morphology of fruit of water chestnut, it is of two types: one is thorny and another one is thorn less. In Bihar, both green and red color thorny fruit type water chestnut is found. Geographical area of production of "Bihar Singhara (water chestnut)" takes place in 38 districts of Bihar. The production area lies between the GPS coordinates of 25° 5′ 45.867″ N and 85° 18′ 47.23″ E. Fresh fruit has a greater production value and is mostly consumed in Asian nations (Jana 2025). Water chestnuts are frequently found in the market in whole and sliced form in canned goods, along with flour and dried nuts, for use in a range of recipes at home and at dining establishments. The water chestnut is regarded as a significant food item throughout the Middle East, Africa, Asia Pacific, North America, Latin America, Europe, Australia, and New Zealand. Asia Pacific consumes the most water chestnuts due to their traditional culinary applications; however, North America is consuming more of them as a result of growing acceptance of multi-culinary cuisine. As demand grows worldwide, the water chestnut market is being driven by its low investment cost and multi-culinary use. Additionally, water chestnut production benefits from the favorable climate in Asian nations, especially India. For more profit, farmers should benefit from its cultivation and remote marketing of both fresh and processed goods. This chapter was intended to provide comprehensive information on the nutritious components, applications, economic significance, cultivation, pest–disease management, processing, and marketing of water chestnut fruit.

5.2 Traditional and Modern Uses of Water Chestnut

5.2.1 Medicinal Value

Folklore medicine has acknowledged the medicinal benefits of the entire fruit and herb as a remedy for a number of illnesses. The whole herb has been reported for hepatoprotective activity, antimicrobial activity, antibacterial activity, antitumor activity, antioxidant activity, and free radical scavenging activity. The fruit has been used as intestinal astringent, anti-inflammatory, anti-diarrhea, antileprotic agent, as well as in urinary discharges, fractures, bronchitis, and anemia (Chaudhary et al. 2012). The fruits of *Trapa bispinosa* has been identified as the Ayurvedic drug Shurangataka. It is also said to have cancer-preventing properties. Sicim juice is used in ophthalmic preparations.

5.2.2 Nutrient Value

The quantities of micronutrients in aquatic plants vary significantly depending on the portion of the plant and the chemical properties of the constituents. Mn typically moves very little from the roots to the leaves. The accumulation pattern of nutrients in the tissues of different part of plant varies a lot. The plant has shown higher concentrations of C, P, K, and Zn in their leaves compared to its root and fruit parts. However, N was found to be contained in appreciably higher amounts in the tissues of fruits. While root parts of the plant were noted to be more efficient in uptake of Fe, Mn, and Cu, the uptake trend of nutrients in *T. natans* was more efficient in absorbing the nutrients, particularly N, P, K, Cu, and Zn, which would help in lessening the pollution load of the polluted water bodies. The major nutrient composition of water chestnut is listed in Table 5.1.

5.2.3 Economic Importance

A nut with both nutritional and therapeutic value is the water chestnut fruit. In Indian street markets, it is offered both boiling and fresh. The fishermen, on an

Table 5.1 Nutrient profile of 100 g of water chestnut (Sundar et al. 2024)

Sl no.	Nutrient	Amount per 100 g
1.	Protein	0.86 g
2.	Carbohydrates	21.46 g
3.	Total fat	0.37 g
4.	Energy	95.6 kcal
5.	Calcium	37.15 mg
6.	Phosphorous	62.83 mg
7.	Iron	0.77 mg
8.	Vitamin C	5.26 mg
9.	Total fiber	3.02 g

average, can easily earn an additional income (by selling raw fruit only) of Rs. 268,800 from cultivation in 1 hectare area. The mature pulp of the fruit, known as milky water chestnut, is used to make meals after it has been boiled and dried, while the immature pulp is consumed raw or cooked. Fresh and boiled water chestnuts are used not only as vegetables but also in tea and in preparing curries. The kernels are dried and sold as nuts and are also ground into flour for bread (*chapattis* or *poories*). The dried nuts (kernels) and flour are sold at Rs. 250 and Rs. 300/kg, respectively. The fruits' nutritional value is comparable to that of wheat. In Japan, water chestnuts are also used to make tea, make wine for sale, and make special dishes for festivals. For certain patients, a paste made by diluting water chestnut dough works well.

5.2.4 Uses of Water Chestnut

Water chestnuts are a versatile ingredient that may be utilized in many different cuisines. They can be eaten fresh, cooked, fried, or dried and ground into flour (Chiang and Ciou 2010; Chandana et al. 2013). Further, singhara is highly valued for its possible medicinal benefits. Traditional medicine systems such as Ayurveda and Traditional Chinese Medicine believe it has cooling, diuretic, and anti-inflammatory properties. Singhara is used to treat fever, sore throat, urinary infections, and digestive problems. Singhara extract is sometimes used in cosmetics for its alleged skin-rejuvenating and moisturizing effects. It is used in a variety of skin-care products, including creams, lotions, and serums, to improve skin texture and hydration. In agricultural operations, singhara plants are an important source of food for animals. The leaves and stems are frequently used as fodder, supplying vital nutrients to cattle and other grazing animals. The list of various uses of water chestnut are shown in Fig. 5.1 (Rajput and Singh 2023).

5.3 Cultivation Practices in Water Chestnut

Water chestnut is cultivated either in water bodies/ponds having water depth of 1.20–1.80 m or in 30–45 cm deep shallow agricultural fields.

5.3.1 Ponds System of Cultivation

Traditionally it is grown in naturally formed high organic matter rich ponds/ditches where seed sowing is not required, since leftover seeds of previous crop serve as planting materials of subsequent crops. However, its cultivation in new water bodies (ponds) requires an addition of FYM @ 15 vha followed by two to three wet ploughings of the field after that the fields must be filled with 45 cm depth of water. The bundle of three rootlets is planted at one point and another bundle of three rootlets is planted at a distance of 3 m and row-to-row distance should also be 3 m.

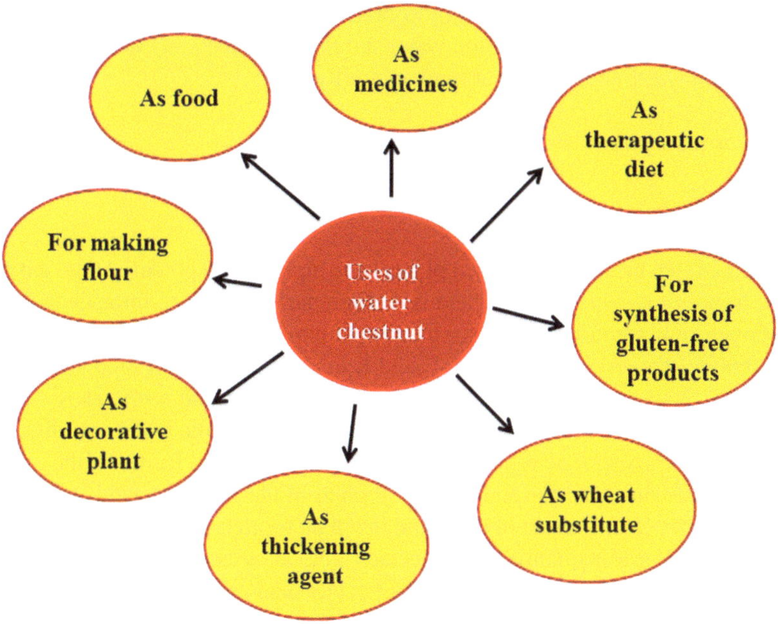

Fig. 5.1 Uses of water chestnut

In ponds, where transplanting of rootlets is not possible, then seeds should be broadcasted @ 69 kg/ha. The transplanting of saplings is done in the month of June–July in agricultural fields. The harvesting of fruit in pond is done with the help of wooden boat. It was recorded that the chemical reaction (pH) of water ranged between 8.2 and 8.6.

5.3.2 Production of Healthy Plant

The following aspects are taken under consideration for producing healthy plant:

(a) Soil and water properties
(b) Improved cultural practices
(c) Cultivation in cropping system mode

(a) *Soil and Water Properties*

The most suitable soil for its cultivation is clayey or silty clay soil type as such soil retains water for longer time. In addition, the soil should be rich in organic matter, iron, manganese, zinc, phosphorus, potassium, and nitrogen. The water chestnut plant is largely very sensitive to nitrogen levels and organic materials. It has also been noticed that the plants growing in nutrient-rich soil are very less affected by any kind of diseases. The irrigation water should not be too salty and acidic; the pH of irrigation water should vary around neutrality.

(b) *Improved Cultural Practices*

Nursery raising

It is well known that nursery-raised plants have higher yield potential than the direct sown crop. For nursery raising, field is well prepared by two to three deep ploughing; however, before ploughing, for the proper nourishment of seedlings, fertilizers @ 120:60:60/ha, respectively, of N, P, and K in combination of 15–20 t/ha compost is applied. Prior to puddling, the soil should also be treated with cakes of neem/mahua/karanj @ 0.8–1.0 t/ha. These cakes help in keeping the soil-borne diseases at way form the rhizospheric region of plants. It would be more fruitful if the mixtures of these cakes (in equal proportion) are applied in the soil. The field is filled with water up to the 45 cm height of bund and the plant rootlets are sown in the month of June or July. For transplanting in 1 hectare area, an area of 182 m^2 is enough for raising the nursery. For transplanting in 1 hectare land area, 3.0 kg seeds are broadcasted in nursery area of 182 m^2. A water level of 45 cm height is maintained throughout the growing period of seedlings, i.e., from June to July. The seedlings are transferred from the nursery plot to the already prepared main field in the first week of August and transplanted at a distance of 3 m × 3 m. The same doses of nutrients are also applied in the main field where transplanting is to be done.

Manures and fertilizers

Being an aquatic plant, its requirement of some essential primary nutrients such as nitrogen (N), phosphorus (P), and potassium (K) is very high. These are the nutrients requirements for the water chestnut crop. The doses of NPK for the crop are @ 120:60:60 kg/ha. One-third of N is applied at the time of puddling and the rest of the fertilizer amount is sprayed on the plant at an interval of 20 days, while the whole amount of P and K is added in the soil during puddling. The organic fertilizer like compost is added @ 15–20 t/ha.

Irrigation

The requirement of irrigation water is very high like any other aquatic plant. As it is transplanted during the monsoon season, i.e., in the month of July, when precipitation starts takes place, in the absence of good precipitation, two to three irrigation is essential for the better development of the crop. The bund of the field is raised up to a height of 2.0 m and irrigation water is field up to a height of 45 cm.

Weed control

The infestation of aquatic weeds is a major problem in filed cultivation. The weed plants that are commonly found in the water chestnut fields are *Marsilea quadrifolia*, *Sagitaria guanensis*, *Ipomoca aquatica* and *Ceratophylum demersum*. The emergence of these weed plants is primarily suppressed by two to three deep puddling with the addition of neem cake @ 0.8–1.0 t/ha, chlorpyriphos weedicide @ 3.0 mL/L solution. Further, the postemergence weedicide such as glyphosate @ 5.0 mL/L is spayed to kill these weeds.

Insect-pest control
The most damaging insect is Singhara beetle (*G. birmanica* Jacoby) (Chrysomelidae: Coleoptera), which generally infests the plant during the grand growth and fruiting stage of the plant. Both adult and grubs scrap the upper surface of green leaves causing severe defoliation. Singhara beetle, *G. birmonica*, has life-cycle of 33–46 days under laboratory conditions. Many synthetic insecticides are being used by farmers to control this pest, which may result in insecticide resistance, a return of the target pest, residues in food, groundwater contamination, negative health effects on humans, and the death of nontarget creatures. The first insecticide resistance in India reported in Singhara beetle against DDT and BHC by Pradhan and his coworkers in 1963. So, the Singhara beetles can be effectively managed by the application of insecticide imidacloprid @ 0.5 mL/L twice at an interval of 15 days or biocontrol agents such as Metarhizium anisopliae IIVR strain (1 × 1010 cfu/g) + Neem oil 1% or *Beauveria bassiana* IIVR strain (1 × 1010 cfu/g) found effective for controlling of Singhara beetle. In temperate climates, aphids are one of the most damaging insect pests to cultivated plants. Aphids and other insects that affect the plant at juvenile growth stage in the nursery block are controlled by application of neem oil @ 3.0 mL/L or imidacloprid @ 0.5 mL/L twice at an interval of 15 days.

(c) *Cultivation in Cropping System Mode*

The disease incidence in water chestnut crop may be decreased up to a certain level through its cultivation in cropping system mode. Water chestnut-based cropping systems such as water chestnut-makhana and water chestnut-berseem. The makhana-water chestnut-berseem has been found most economical and sustainable cropping system in terms of maintaining the soil fertility. Cultivation of water chestnut in cropping system mode with makhana and fish is a good option, especially for resource-poor farmers of north Bihar (Jana et al. 2019).

5.4 Major Pests, Disease, and Their Management Approaches in Water Chestnut

Major pests affecting water chestnut (*Trapa natans* L.) include the water chestnut beetle (*Galerucella birmanica* Jacoby) and aphids. Common diseases include leaf spots, rotting, and yellowing of leaves. Effective management strategies involve a combination of cultural practices, biological control, and chemical control methods to minimize yield losses.

5.4.1 Pests

(a) *Water chestnut beetle (Galerucella birmanica Jacoby)*: This beetle is a significant pest, particularly in wetland ecosystems. Research on pest control in water chestnut identifies that green-colored varieties exhibit some resistance.
(b) *Aphids*: These sap-sucking insects can also cause damage to water chestnut plants.
(c) *Mealybugs*: These are another type of insect pests that can infest water chestnut.

5.4.2 Diseases

(a) *Leaf spot*: Fungal or bacterial infections can cause leaf spots, leading to reduced photosynthetic capacity.
(b) *Rotting*: Waterlogged conditions can promote various rotting diseases affecting roots and other plant parts.
(c) *Yellowing of leaves*: Nutrient deficiencies or disease can cause yellowing of leaves, indicating plant stress.
(d) *Fusarium wilt*: A fungal disease that can cause wilting and death of water chestnut plants.
(e) *Sheath blight*: Another fungal disease that can affect the stems and leaves of water chestnut.
(f) *Blast*: A fungal disease that can cause lesions on leaves and stems.
(g) *Cucumber mosaic virus*: An economically important virus that can affect water chestnut.

5.4.3 Management Approaches

5.4.3.1 Cultural Practices

(a) *Water management*: Ensuring proper water depth and flow can help prevent the buildup of pathogens and reduce pest populations.
(b) *Sanitation*: Removing and destroying infected plant parts can help reduce the spread of diseases.
(c) *Crop rotation*: Rotating water chestnut with other crops can help break disease cycles.
(d) *Resistant varieties*: Using green-colored varieties of water chestnut can provide some resistance to beetle and aphid attacks.

5.4.3.2 Biological Control

(a) *Natural enemies*: Encouraging natural predators and parasites of pests can help control their populations.
(b) *Biopesticides*: Using biological control agents like *Bacillus thuringiensis* can help control certain pests.

5.4.3.3 Chemical Control

(a) *Fungicides*: Approved aquatic-safe fungicides can be used to control fungal diseases.
(b) *Insecticides*: Insecticides may be necessary in severe cases of pest infestations but should be used judiciously to minimize environmental impact.
(c) *Examples of specific chemicals*: Carbaryl (0.05%) and monocrotophos (0.05%) are mentioned for controlling beetles and mealybugs. Bavistin (0.1%) is suggested for fungal diseases.

5.4.3.4 IPM (Integrated Pest Management)

Combining cultural practices, biological control, and chemical control methods provides a more sustainable and effective approach to managing pests and diseases in water chestnut.

Specific examples of IPM:

(a) *For Singhara beetle*: Root dipping of seedlings in chlorpyrifos can help protect the crop during the early stages of growth.
(b) *For fungal diseases*: Spraying with Bavistin (0.1%) at weekly intervals can be effective.
(c) *For leaf-eating beetles and mealybugs*: Spraying with carbaryl (0.05%) or monocrotophos (0.05%) at weekly intervals can be effective.

5.5 Processing and Marketing of Water Chestnut

Water chestnut is a seasonal aquatic crop widely consumed in both raw and processed forms. Its high nutritional value and starch content make it suitable for various food and industrial applications. Efficient processing techniques in water chestnut enhance shelf life, reduce postharvest losses, and open up avenues for commercialization in both domestic and export markets. Integrating mechanization and hygienic practices improves product quality and market potential. Processing and marketing of water chestnut offer immense potential to uplift rural economies, especially in wetland-rich regions (Maiti and Kundu 2015). By promoting value addition, enhancing storage techniques, and connecting producers to wider markets, water chestnut can be transformed from a seasonal aquatic crop into a year-round economic asset. Following are the processing and marketing of water chestnut.

5.5.1 Processing of Water Chestnut

Water chestnut undergoes several postharvest processes (Fig. 5.2) to enhance its shelf life, usability, and market value. The main steps include the following.

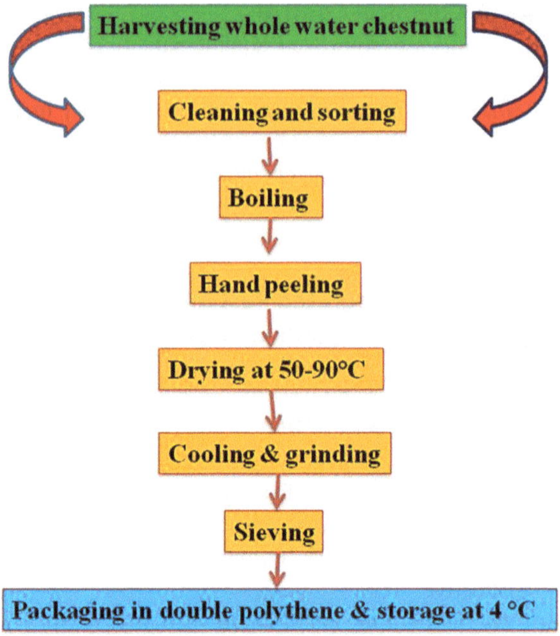

Fig. 5.2 Processing of water chestnut

5.5.1.1 Harvesting

- When the fruits ripen, which is usually in the autumn, water chestnuts are manually collected from ponds or wetlands.
- Boats or hands are used to gather the prickly fruits.

5.5.1.2 Cleaning and Sorting

- To get rid of mud, debris, and aquatic weeds, freshly collected fruits are thoroughly cleaned.
- Fruits that are damaged or immature are separated through sorting.

5.5.1.3 Boiling/Steaming

- The nuts' shells are softened by boiling or steaming them before being consumed fresh or processed further.
- The outer pericarp is also loosened by boiling, which facilitates its removal.

5.5.1.4 Peeling

- The hard outer shell is removed by hand or with the use of mechanical peelers after boiling.
- The white, starchy kernel that results from this is either eaten or processed further.

5.5.1.5 Drying

- To lower moisture content for long-term preservation, peeled kernels can be mechanically dried or sun-dried.
- In India, dried nuts are frequently processed into water chestnut flour, also known as singhara flour, which is extensively utilized during times of fasting.

5.5.1.6 Value-Added Products

- Common processed products include:
  - *Singhara flour* for baking and fasting recipes.
  - *Vacuum-packed kernels* for longer shelf life.
  - *Snacks and sweets* made using boiled or roasted chestnuts.
  - *Pickles and preserves* in certain traditional recipes.

5.5.2 Marketing of Water Chestnut

5.5.2.1 Traditional and Local Markets

- Local markets sell a lot of water chestnuts, particularly during the holidays and fasting seasons.
- Street sellers and small-scale shopkeepers frequently sell fresh, boiled, or roasted nuts.

5.5.2.2 Seasonal Demand

- Farmers' and sellers' profitability is increased by high demand during winter and religious holidays like Navratri.

5.5.2.3 Value Chain and Stakeholders

- Cultivators, regional processors, wholesalers, retailers, and exporters are all part of the value chain.
- Better market connections, storage, and transportation can increase farmers' profits.

5.5.2.4 Export Potential

- Demand for flour and dried water chestnuts is rising in global marketplaces, particularly among gluten-free and health-conscious consumers.
- For export, proper packaging and respect to food safety regulations are essential.

5.5.2.5 Challenges and Opportunities

- *Challenges*: Fresh nuts' perishability, inadequate infrastructure for processing, and a lack of knowledge about value-added applications.
- *Opportunities*: Postharvest handling training, cooperative development, conventional product branding, and e-commerce platform development can all significantly increase market access.

5.6 Conclusion

A fascinating aquatic plant, water chestnuts (*Trapa natans* L.) have long been a wholesome and culturally significant part of traditional meals, especially in Asia. It has many health benefits, such as providing energy, supporting digestion, and possibly reducing inflammation. It is high in carbohydrates, minerals, and antioxidants. In addition to its nutritional benefits, water chestnuts are vital to rural economies because they sustain livelihoods through production, processing, and distribution. In areas with suitable wetland habitats, it is a sustainable food source due to its seasonal growth cycle and resilience to aquatic settings. Water chestnut has enormous potential for agro-biodiversity, cultural heritage preservation, and food security as interest in traditional and functional foods keeps growing. This aquatic gem can be further incorporated into contemporary food systems through strategic research, value addition, and marketing, guaranteeing that its heritage will live on in the diets of future generations.

Acknowledgments The first author wishes to thank the College of Agricultural Biotechnology, Bihar Agricultural University, Sabour, India, for supporting in the preparation of this chapter. This book chapter has been assigned BAU COMMUNICATION NO. 2140/250617.

Conflicts of Interest The authors have no conflicts of interest.

References

Chandana M, Mazumder R, Chakraborty GS (2013) A review on potential of plants under Trapa species. Int J Res Pharm Chem 3(2):502–508

Chaudhary J, Jain A, Bhatiwal S (2012) *Trapa natans* (water chestnut): an overview. Int Res J Pharm 3(6):31–33

Chiang PY, Ciou JY (2010) Effect of pulverization on the antioxidant activity of water caltrop (*Trapa taiwanensis* Nakai) pericarps. LWT Food Sci Technol 43:361–365. https://doi.org/10.1016/j.lwt.2009.08.016

Jana BR (2025) Recent production and marketing scenarios of water chestnut (*Trapa* spp.) in India: a review. Agric Rev 46(2):322–325. https://doi.org/10.18805/ag.R-2555

Jana BR et al (2019) A study on commercial cultivation and storage of water chestnut (*Trapa natans* L.) of wetland ecosystem of North Bihar, India. J Appl Nat Sci 11(2):528–533. https://doi.org/10.31018/jans.v11i2.2105

Maiti AK, Kundu S (2015) Analysis of marketing efficiency of water chestnut (*Trapa natans* L.) in 24-Parganas (North) of West Bengal. J Crop Weed 11(Special Issue):14–18

Rajput JD, Singh SP (2023) Water chestnut (*Trapa natans* L.): functional characteristics, nutritional properties and applications in food industry: a review. J Phytopharmacol 12(2):119–126

Singh IS, Thakur AK, Prakash D (2018) Scientific cultivation of water chestnut crop: a boon for poor fishermen. Indian Farm 68(03):23–27

Sundar S, Dwivedi DH, Verma S, Maurya HK, Kumar B (2024) A review on importance nutritional value and uses of water chestnut (*Trapa natans* var. Bispinosa Roxb.). Int J Plant Soil Sci 36(11):377–385. https://doi.org/10.9734/ijpss/2024/v36i115154

Makhana (*Euryale ferox*): The Super Snack from Water

6

Pankaj Kumar, Rima Kumari, and Anil Kumar Singh

Abstract

Makhana (*Euryale ferox* Salisb.), commonly known as fox nut, has emerged as a nutritionally rich superfood with deep cultural roots in Indian tradition, particularly in the Mithilanchal region of Bihar. As an aquatic crop belonging to the family Nymphaeaceae, makhana thrives in stagnant water bodies and offers a unique combination of agronomic adaptability, socioeconomic relevance, and health benefits. This chapter explores the multifaceted dimensions of makhana—from its botanical characteristics and ecological requirements to traditional cultivation practices, postharvest processing, and evolving market dynamics. The crop's edible seeds are puffed into white kernels that are low in fat, gluten-free, and rich in protein, complex carbohydrates, and essential minerals. Scientific research has substantiated several health-promoting properties of makhana, including its potential for weight management, cardiovascular support, glycemic control, and antioxidant activity. Its culinary versatility further enhances its appeal, with applications ranging from traditional sweets and curries to modern flavored snacks and health products. Despite its growing popularity, makhana production faces critical challenges such as labor-intensive harvesting, lack of mechanization, and postharvest losses. However, concerted efforts in research, government support, and innovations in value addition are paving the way for

P. Kumar (✉)
Department of Molecular Biology and Genetic Engineering, Bihar Agricultural University, Sabour, Bihar, India
e-mail: pankajkumar@bausabour.ac.in

R. Kumari
Division of Plant Biotechnology, College of Agricultural Biotechnology, Bihar Agricultural University, Sabour, Bihar, India

A. K. Singh
Directorate of Research, Bihar Agricultural University, Sabour, Bihar, India

A. Kumar et al. (eds.), *Fasting Superfoods*,
https://doi.org/10.1007/978-981-95-2762-5_6

improved productivity and market access. This chapter underscores its journey from a traditional aquatic plant to a globally acclaimed "Super Snack from Water."

Keywords

Makhana · *Euryale ferox* · Superfood · Aquatic crop · Low glycemic index

6.1 Introduction

In recent years, health-conscious consumers across the world have increasingly turned toward traditional foods for their nutritional value and therapeutic benefits. Among such superfoods, *makhana*, or fox nut, stands out as an emerging health snack rooted in ancient Indian culture. Botanically known as *Euryale ferox* Salisb., makhana is an aquatic crop with unique botanical characteristics and a rich socio-cultural legacy (Ahmed et al. 2015a, b, 2018). Commonly found in stagnant water bodies across parts of India and Southeast Asia, makhana is especially associated with the Mithilanchal region of Bihar, where it has been cultivated and processed for centuries. It is valued for its puffed seeds, which are high in protein, low in fat, gluten-free, and full of antioxidants. It is now being marketed globally as a low-calorie, healthy alternative to fried snacks, capturing attention as a "Super Snack from Water." This chapter delves into the various aspects of *Euryale ferox*—from its botanical background and ecological preferences to its cultivation, processing, nutritional content, health benefits, economic significance, and future prospects.

6.2 Botanical Background

Euryale ferox belongs to the family Nymphaeaceae, which includes aquatic flowering plants such as water lilies. Native to Asia, this plant is known for its floating thorny leaves and attractive purple flowers. The seeds, which develop in spongy fruits submerged in water, are the edible part of the plant and are consumed in various forms across India and China.

- Botanical name: *Euryale ferox* Salisb.
- Kingdom: Plantae
- Division: Penerograms
- Subdivision: Angiosperm
- Class: Dicotyledonous
- Subclass: Polypetalae
- Family: Nymphaceae
- Genus: *Euryale*
- Species: *Ferox*
- Chromosome number: 2n = 58

It is a hardy aquatic plant adapted to survive in muddy, shallow ponds. Its seeds are embedded in mucilage inside large fruits that mature underwater. Once the seeds are harvested and processed, they become the crunchy, white, puffed kernels known as makhana.

6.3 Origin, Distribution, and Historical Significance

Makhana (*Euryale ferox*) is naturally distributed across several Indian states, including Madhya Pradesh, Uttar Pradesh, Assam, Bihar, eastern Odisha, and Jammu & Kashmir (Kumar et al. 2011). However, its commercial cultivation is largely confined to the Mithilanchal region of Bihar—specifically in the districts of Darbhanga, Sitamarhi, Madhubani, Saharsa, Supaul, Araria, Kishanganj, Purnia, and Katihar. Among these, the bulk of processed makhana originates from four key districts: Darbhanga, Madhubani, Purnia, and Katihar. These four districts alone account for approximately 80% of India's total makhana production, covering about 13,000 hectares (Mishra et al. 2003). According to estimates from ICAR-National Research Centre for Makhana, Darbhanga, the total area under makhana cultivation in India spans roughly 15,000 hectares, yielding around 120,000 metric tons (MT) of seeds. After processing, this produces about 40,000 MT of popped makhana. The farm-level value of this production is estimated at 250 crore, while the value at the trading level reaches around 550 crore.

Makhana seeds are typically popped and consumed as roasted snacks or used in the preparation of a variety of sweets and traditional dishes. Valued for their nutritional, medicinal, and ritualistic properties, makhana also supports a thriving cottage industry and provides livelihoods to thousands of fishing families in the region. The seeds are composed of approximately 80% carbohydrates, primarily starch, and 10–12% protein. They are rich in minerals and contain minimal fat of about 0.1%.

6.3.1 Historical Significance

- Makhana has been used in traditional medicine and religious rituals.
- It features prominently in Indian fasting foods due to its "sattvic" (pure) nature.
- It is mentioned in Ayurvedic texts for its rejuvenating and aphrodisiac properties.

6.3.2 Ecological Requirements

Makhana grows in still or slow-moving water bodies and prefers clayey soil with high organic content. The ideal depth for its cultivation ranges between 1.2 and 2.5 m. It thrives in tropical to subtropical climates and is extremely sensitive to changes in water quality and temperature.

6.3.3 Climatic and Soil Requirements

- *Temperature range*: 20–35 °C.
- *Soil pH*: 6.5–7.5.
- *Water depth*: 1.0–2.5 m.
- *Soil type*: Heavy clay with high organic matter.

Makhana is generally grown in natural ponds, wetlands, and low-lying flood-prone fields. It is a rain-fed crop but requires efficient water retention in the field during its entire growth period of about 120–150 days.

6.4 Cultivation, Propagation, and Nursery Management

Makhana is primarily cultivated in stagnant water bodies such as ponds, lakes, and naturally occurring wetlands. There are two main methods of cultivation: the traditional wetland/pond system and the improved field system. In the traditional pond method, makhana is grown in natural or man-made water bodies with depths ranging from 4 to 6 feet. This has been the conventional approach to makhana cultivation for generations. In contrast, the field system is a more recent innovation, wherein makhana is cultivated in shallow agricultural fields with water depths of 1–2 feet. The field system has emerged as a widely adopted and increasingly popular technique among Indian farmers due to its higher yield per hectare and greater time efficiency. Its adoption is steadily rising as it offers better management, ease of harvesting, and suitability for integration into existing cropping systems.

Propagation is usually done using seeds. Mature seeds are collected from the previous crop and stored until the next sowing season. Before sowing, seeds are soaked in water to initiate germination (Fig. 6.1). Seedlings are prepared in a

Fig. 6.1 Mature seed germination

nursery pond or area and transplanted into the main field when they are about 30–40 days old.

6.5 Field Preparation, Transplantation, and Maintenance

Fields are desilted and leveled to ensure uniform water depth. Organic compost or farmyard manure is incorporated to enrich the soil. In natural ponds, aquatic weeds and debris are manually removed. Field should be ploughed 2–3 times and puddle before 3–4 days before transplanting of makhana seedlings. Fertilizer is the key input for realizing the potential yields of makhana. At the time of field preparation, about 20–22 tons of farmyard manure with dose of fertilizers NPK @ 75:45:30 + 40–50 kg lime per ha is recommended, of which entire phosphate and potash are to be given as basal at least 5 days before transplanting. Half dose of nitrogen should be applied after making solution in 4–5 splits and sprayed on leaves after 20–25 days interval. Standing water (minimum 6 inch) should be maintained in main field from transplanting to flower initiation, i.e., up to first fortnight of April. Thereafter, the depth is to be increased to 1 foot and continued to maturity as well as harvesting (buharai) of makhana seed (Gurri), i.e., up to first fortnight of August in field condition. However, in pond system, standing water (minimum 3 feet and maximum 7 feet) should be maintained. This practice helps in good plant establishment and weed control.

Seedlings are manually transplanted at regular intervals (60–70 cm apart). Another approach to planting makhana seeds involves transplanting healthy seedlings during the months of March and April, ensuring a spacing of 1 × 1 m between each row and individual plant. Approximately 2 months later, the plants begin to bloom with purple flowers. About 40–45 days after flowering, the plants produce fruits, which then mature. The makhana plant and its fruit are covered in thorns. When fully mature, the fruits soften, become spongy, and begin to burst. During the initial 2–3 days of this process, the fruits settle at the bottom of the water body. It takes around 4–6 months for the flowers to fully develop and the fruits to mature. In September or October, the makhana seeds are harvested from the bottom of the water bodies. Any remaining seeds that are not collected will automatically germinate in the next growing season. This is a labor-intensive process. During the growing period, the field is maintained at a constant water level. Weed management is crucial, particularly during the early stages after transplanting, as weeds compete with the main crop for nutrients, water, light, and space, thereby affecting the healthy growth of makhana leaves. Depending on the level of weed infestation, a minimum of two to three hand weeding operations are required at intervals of 2–3 weeks, starting around 25–30 days after transplanting. This continues until the water surface is fully covered by the large, spherical leaves of the makhana plant. Once the water surface is completely shaded by these leaves, weed growth is naturally suppressed due to limited light penetration.

Fig. 6.2 Flowering and fruit formation in makhana

6.6 Flowering and Fruit Formation

Under field conditions, makhana plants typically begin bearing flowers 1–2 months after transplanting. However, in the case of direct seeding under the pond system, flowering usually occurs 5–6 months after sowing. Morphological and developmental studies by Mishra et al. (2003) describe the flowers as complete, solitary, and large, with a bright purple color and long pedicel. Flowering and fruiting in makhana generally begin in May and can continue until October–November. Fertilization in makhana flowers occurs at a very early stage of floral development, often when the flower is still hermetically sealed under water. The flowers are bisexual, large, and vividly purple (Fig. 6.2). Makhana is a monotypic genus with limited available genetic variability (CSIR 1952; Sokolov 1952; Mishra et al. 2003; Verma et al. 2010; Das et al. 2006; Kumar et al. 2011). According to Md. Ajaib et al. (2010), the flowers are relatively small about 4 cm across with four erect sepals up to 3 cm long, glabrous inside and prickly outside. The petals are violet inside and green outside, measuring up to 2 cm. The stigma is discoid and depressed, and the ovary is inferior, embedded in a smooth torus. The fruit is conical, measuring up to 11 × 5–10 cm, with a persistent calyx.

6.7 Harvesting

The timing of harvesting is critical for achieving maximum seed yield in makhana cultivation. The optimal harvesting period typically falls in the first fortnight of August. Harvesting and seed processing are still predominantly carried out using traditional methods. Experienced fishermen sweep the entire pond floor to gather sunken seeds into small heaps, which are then collected using a horn-shaped split bamboo implement. Lighter seeds that float on the water surface are retrieved with the help of fine mesh nets. After collection, the seeds undergo a traditional threshing process, wherein they are trampled by foot to remove the outer membranous covering. Under optimal conditions, makhana cultivation can yield approximately 32–35 quintals per hectare of seeds (referred to as guri).

6.7.1 Postharvest Processing

Processing makhana into its edible puffed form involves multiple stages.

6.7.2 Cleaning and Drying

The collected seeds are washed thoroughly to remove the mucilage and are sun-dried for several days. Proper drying is critical for storage and later popping.

6.7.3 Grading and Storage

Seeds are graded based on size and weight. Larger seeds are preferred for puffing. They are stored in cool, dry conditions to prevent mold and insect infestation.

6.7.4 Roasting and Popping

The dried seeds are roasted at high temperatures (around 250–300 °C) in a cast iron pan or rotating drum. After roasting, the seeds are cracked open either manually or using mechanical poppers to extract the puffed kernel.

6.7.5 Final Sorting and Packaging

The popped kernels are sorted by size, color, and quality. They are packed in airtight containers or vacuum-sealed bags to ensure long shelf life.

6.8 Nutritional Composition

Makhana is hailed as a superfood due to its balanced nutritional profile. It is rich in complex carbohydrates and plant-based protein, and low in fat. It is also a good source of minerals such as calcium, magnesium, phosphorus, potassium, and iron. It holds considerable potential for growth in the fields of medicine, nutrition, and economic development. The seeds of this aquatic plant, which have long been used in traditional Chinese medicine and cuisine, are gaining global recognition for their nutritional richness and therapeutic properties. Popped makhana is widely consumed as a healthy dry fruit and is particularly appreciated for its low fat content and high levels of carbohydrates, protein, and essential minerals (Ali et al. 2020).

Table 6.1 Uses of medicinal properties of makhana

Name of benefit	Description
Health benefits	Makhana offers several scientifically validated health benefits
Weight management	Low in fat and high in fiber, makhana promotes satiety and reduces cravings, making it ideal for weight loss diets
Antioxidant properties	The presence of flavonoids like kaempferol gives makhana anti-aging and anti-inflammatory properties
Heart health	It is low in sodium and rich in potassium and magnesium, which help regulate blood pressure and maintain heart function
Diabetic-friendly	Its low glycemic index makes it suitable for people with diabetes. It helps in stabilizing blood sugar levels
Digestive health	High fiber content aids in digestion and prevents constipation
Bone health	Its calcium content helps in maintaining strong bones and teeth

With a protein content of 10–12%, it is considered a nutritious and wholesome food. Studies have shown that makhana surpasses several commonly consumed dry fruits such as almonds, walnuts, cashews, and coconuts in terms of sugar, protein, ascorbic acid, and phenolic compounds (Cormick and Belizán 2019). The kernels are rich in amino acids, containing as many as 16 different types. Both raw and fried forms of makhana are consumed. Makhana seeds are particularly abundant in essential amino acids and have been traditionally used for their therapeutic benefits (Dhua et al. 2001; Ginter et al. 2014; Jana and Idris 2018; Jana et al. 2019; Knawar 2003; Kumar et al. 2018). They are utilized as a stomach tonic and are believed to alleviate joint pain, urinary disorders, and semen loss. Additionally, they are used in treating spleen-related ailments and gonorrhea, and are recommended to help women recover from postnatal fatigue. As a water-based cash crop, makhana is highly nutritious-rich in carbohydrates and proteins, and very low in fat. It has a significant market presence. According to Khadatkar et al. (2020), although only 1.72% of its production is consumed, the crop generates a market surplus of approximately 80%. Nutritionally, makhana is a valuable source of carbohydrates, proteins, and minerals. Unprocessed makhana contains approximately 76.9% carbohydrates, 12.8% moisture, 9.7% protein, 0.9% phosphorus, 0.5% minerals, 0.1% fat, 0.02% calcium, and 0.0014% iron (Leidy 2014; Petruk et al. 2018; Sharma et al. 2008; Sun et al. 2014). In comparison, popped makhana comprises 84.9% carbohydrates, 4% moisture, 9.5% protein, and 0.5% fat. It also offers medicinal benefits and is suggested for use in treating disorders of the respiratory, circulatory, digestive, excretory, and reproductive systems. The energy content is 362 kcal/100 g for raw makhana and 328 kcal/100 g for the popped form (Table 6.1).

6.9 Socioeconomic Importance

Makhana plays a crucial role in rural livelihoods in Bihar, where more than 50,000 families depend on it for income. It involves a labor-intensive value chain including collectors, processors, traders, and vendors.

6.10 Challenges in Makhana Production

Despite its potential, makhana cultivation faces several hurdles:

- *Labor-intensive harvesting*: Manual diving is time-consuming and physically taxing.
- *Lack of mechanization*: Minimal technological intervention in harvesting and popping.
- *Postharvest losses*: Due to poor drying, storage, and lack of grading facilities.
- *Market fluctuation*: Prices are unstable due to demand–supply imbalances.
- *Limited research*: Few improved varieties and agronomic protocols are available.

6.11 Future Prospects

The future of makhana lies in:

- *Scientific cultivation*: Better varieties and agronomic practices.
- *Export growth*: Meeting international food safety standards.
- *Product innovation*: New forms such as drinks, bars, and health supplements.
- *Agri-entrepreneurship*: Involving youth and women in the value chain.
- *Climate resilience*: Developing drought- and flood-tolerant systems.

6.12 Conclusion

Makhana (*Euryale ferox*) represents a compelling example of how a traditional aquatic crop can transform into a modern-day superfood. With its excellent nutritional profile, diverse applications, and high economic value, it has immense potential for sustainable agriculture and food innovation. To harness this potential fully, a multipronged approach involving scientific research, mechanization, policy support, and entrepreneurial ventures is essential. As global demand for functional and health foods rises, makhana is poised to take its rightful place on the international stage as the "Super Snack from Water."

Acknowledgments The first author wishes to thank the ANRF ECRG (File No. 2024/004032) and Bihar Agricultural University, Sabour, India, for their support in the preparation of this chapter.

References

Ahmed D, Kumar V, Verma A, Shukla GS, Sharma M (2015a) Antidiabetic, antioxidant, antihyperlipidemic effect of extract of *Euryale ferox* Salisb. with enhanced histopathology of pancreas, liver and kidney in streptozotocin induced diabetic rats. Springerplus 4:315. https://doi.org/10.1186/s40064-015-1059-7

Ahmed D, Sharma M, Kumar V, Bajaj HK, Verma A (2015b) 2β-hydroxybetulinic acid 3β-caprylate: an active principle from *Euryale Ferox* Salisb. seeds with antidiabetic, antioxidant, pancreas & hepatoprotective potential in streptozotocin induced diabetic rats. J Food Sci Technol 52(9):5427–5441

Ahmed D, Khan MI, Sharma M, Khan MF (2018 Dec) Novel pentacyclic triterpene isolated from seeds of Euryale Ferox Salisb. ameliorates diabetes in streptozotocin induced diabetic rats. Interdiscip Toxicol 11(4):275–288

Ajaib M, Khan Z, Khan N, Wahab M (2010) Euryale ferox Salisb. of The Family Nymphaeaceae: an addition to The Flora of Pakistan. Pak J Bot 42(5):2973–2974

Ali SS, Ahsan H, Zia MK, Siddiqui T, Khan FH (2020) Understanding oxidants and antioxidants: classical team with new players. J Food Biochem 44(3):e13145. https://doi.org/10.1111/jfbc.13145. Epub 2020 Jan 20. PMID: 31960481

Cormick G, Belizán JM (2019) Calcium intake and health. Nutrients 11(7):1606. https://doi.org/10.3390/nu11071606

CSIR (1952) The wealth of India: Raw Materials, CSIR, New Delhi III, p.232

Das S, Der P, Raychaudhary U, Maulike N, Das DK (2006) The effect of Euryale feroxSalisb. (Makhana), an herb of aquatic origin on myocardial ischemic repurfsion injury. Mol Cell Biochem 289(12):55–63

Dhua SR, Sahu RK, Patra BC, Patnaik SSC, Marndi BC, Patnaik A, Dhua U (2001) Annual report 2001–02. CRRI, Cuttack, pp 60–62

Ginter E, Simko V, Panakova V (2014) Antioxidants in health and disease. Bratisl Lek Listy 115(10):603–606. https://doi.org/10.4149/bll_2014_116

Jana BR, Idris M (2018) Anti-aging amino acids in *Euryale ferox* (Salisb.): a review. Adv Plants Agric Res 8(1):39–43. https://doi.org/10.15406/apar.2018.08.00289

Jana BR, Srivastava A, Idris M (2019) New Makhana (*Euryale ferox* Salisb.) processed products for health benefit. J Food Process Technol 10:789. https://doi.org/10.4172/2157-7110.1000789

Khadatkar A, Mehta CR, Gite LP (2020) Makhana (Euryle ferox Salisb.): A high-valued aquatic food crop with emphasis on its agronomic management – A review. Sci Hortic 261:108995. https://doi.org/10.1016/j.scienta.2019.108995

Knawar P (2003) Drudgery perceived by gender in hill farming. Indian J Ext Educ 39(3&4):183–187

Kumar L, Gupta VK, Jha BK, Singh IS, Bhatt BP, Singh AK (2011) Status of makhana (Euryale ferox Salisb.) cultivation in India. Technical Bulletin No. R-32/PAT-21. ICAR Research Complex for Eastern Region, Patna

Kumar N, Rani S, Kuamr G, Kumari S, Singh IS, Gautam S, Choudhary BK (2018) Physiological and biochemical responses of Makhana (*Euryale ferox*) to gamma irradiation. J Biol Phys 45(1):1–12. https://doi.org/10.1007/s10867-018-9511-x

Leidy HJ (2014) Increased dietary protein as a dietary strategy to prevent and/or treat obesity. Mo Med 111(1):54–58. PMID: 24645300; PMCID: PMC6179508

Mishra RK, Jha V, Dharai PV (2003) Makhana (eds.), ICAR, New Delhi, 340p

Petruk G, Del Giudice R, Rigano MM, Monti DM (2018) Antioxidants from plants protect against skin photoaging. Oxidative Med Cell Longev 2018:1454936

Sharma YK, Bangarva GS, Sharma SK (2008) Farming system-based constraints faced by farmers. Indian Res J Ext Educ 8(1):57–59

Sokolov VS (1952) Alkaloid Plants of the USSR. Akademiia Nauk, Moscow

Sun HL, Zhang YQ, Xie XY, Che YY (2014) Studies on chemical constituents from seeds of *Euryale ferox*. Zhong Yao Cai 37(11):2019–2021. Chinese. PMID: 26027125

Verma AK, Banerji BK, Chakarborty D, Dutta SK (2010) Studies on Makhana (Euryale ferox Salisbury). Curr Sci 99:795–800

Barnyard Millet (*Echinochloa* spp.): The Fast-Growing Nutritious Staple

7

Shivam Yadav, Rumana Khan, Mitali Tiwari, Narendra Kumar Singh, Silpa Sahoo, Sushrut Kumar Samanta, and Jitendra Kumar Tiwari

Abstract

Barnyard millet (*Echinochloa* spp.) is an ancient cereal within the Poaceae family, characterized by rapid growth and remarkable adaptability to marginal and drought-prone environments. The genus comprises two cultivated hexaploid species, *Echinochloa frumentacea* and *E. esculenta*, which were domesticated from wild progenitors *E. colona* and *E. crus-galli* in South and East Asia. This underutilized "nutri-cereal" is packed with a remarkable nutritional profile, high in protein, dietary fiber, and important micronutrients such as iron and zinc. It also carries significant cultural value in India as a traditional food and fasting grain. However, genetic improvement has been constrained by the crop's narrow genetic base and polyploid genome, which complicate conventional breeding. Recent advances in germplasm characterization and genomics, including molecular markers and whole-genome sequencing, are facilitating marker-assisted breeding to overcome such limitations. While genomic tools are accelerating cultivar improvement, complementary innovations in post-harvest processing, pest management, and product diversification are equally critical to unlock the full value of this underutilized crop. In light of climate change and increasing global demand for nutrient-dense staple crops, global interest in barnyard millet is increasing, with India at the forefront of its production and with renewed efforts to promote this climate-smart grain for wider food and nutrition security.

Keywords

Sanwa · Tradition food · Minor millets · Climate smart cereal · C4 plant

S. Yadav · M. Tiwari · N. K. Singh
Govind Ballabh Pant University of Agriculture and Technology, Uttarakhand, India

R. Khan (✉) · S. Sahoo · S. K. Samanta · J. K. Tiwari
Rani Lakshmi Bai Central Agricultural University, Jhansi, India

A. Kumar et al. (eds.), *Fasting Superfoods*,
https://doi.org/10.1007/978-981-95-2762-5_7

7.1 Introduction

Barnyard millet (*Echinochloa* spp.), a member of the Poaceae family, exhibits remarkable adaptability to marginal agroecologies, particularly in drought-prone regions with low soil fertility. Originating in Southeast Asia, this short-duration crop has a long history of cultivation and consumption across diverse geographies. In India, locally known as *sanwa*, barnyard millet serves as a staple food in climate-vulnerable regions, providing a reliable harvest under minimal agricultural inputs. India leads global production, cultivating approximately 0.146 million hectares and yielding 0.147 million tons annually, with an average productivity of 1034 kg/ha (IIMR 2018). Barnyard millet, despite being classified as a pseudo-cereal, offers a comprehensive nutritional profile characterized by high protein content, complex carbohydrates, dietary fiber, and essential micronutrients such as iron and zinc, positioning it as a promising alternative to conventional cereals such as rice, wheat, and maize (Singh et al. 2010; Saleh et al. 2013; Chandel et al. 2014). Its agronomic resilience also supports its use as a contingency crop during monsoon failures in rice-based systems (Gupta et al. 2009). Despite its nutritional and agronomic advantages, the crop remains underutilized due to limited awareness and insufficient comprehensive research. Genetic enhancement is constrained by its polyploid genome and restricted genetic diversity among cultivated varieties, which complicate breeding efforts targeting yield and stress resistance. Addressing these constraints will require systematic efforts in germplasm characterization, prebreeding, interspecific hybridization and genomic-assisted breeding approaches to unlocking the full potential of this crop. Given its proven nutritional density, ecological resilience, and low input requirements, barnyard millet is now recognized as a potential "Indian superfood" that aligns with national goals for sustainable food and nutritional security. With growing consumer demand for traditional grains and institutional support through initiatives such as the International Year of Millets (2023), barnyard millet is well-positioned for broader adoption as a health-promoting, climate-smart cereal. Realizing its potential will require an integrated approach spanning research, value chain development, and policy support to mainstream this neglected crop into modern agri-food systems.

7.1.1 Botany

Barnyard millet (*Echinochloa* spp.) is a highly adaptable crop that can thrive at altitudes up to 2000 m above sea level (Gupta et al. 2009). The species exhibits variations in flowering time, inflorescence morphology, spikelet coloration, and overall plant architecture (Obara 1936). It is an annual plant that grows up to 220 cm tall and matures in 45–60 days under ideal circumstances. In northern hill ecosystems, maturation may be prolonged, potentially extending to 3–4 months at higher elevations (Denton 1987; Padulosi et al. 2009; Hulse et al. 1981). The inflorescence is typically an erect terminal panicle, 10–25 cm in length, composed of densely packed racemes, bearing 3–4 mm spikelets (Napper 1965). Spikelets, ranging from green to purple, are arranged on a triquetrous rachis and comprise two florets: a sterile lower floret with a lemma and palea and a bisexual upper floret (Mitich 1990; Gupta et al. 2009). The fertile lemma exhibits a plano-convex morphology, characterized by a smooth surface and an abruptly pointed, with margins inrolled beneath the palea, which demonstrates a comparable surface texture (Napper 1965). Reproductive organs include three stamens and a superior ovary featuring two distinct styles with plumose stigmas (Sundararaj and Thulasidas 1979). Grains, measuring 2–3 mm in length and 1–2 mm in width, are enclosed within a white, hardened lemma and palea. Anthesis initiates at the inflorescence apex and progresses downward over 10–15 days, with most flowers opening between 6.00 a.m. to 7.00 a.m. (Sundararaj and Thulasidas 1979; Jayaraman et al. 1997). Despite being hermaphrodite nature, the species predominantly self-pollinates, with infrequent wind-mediated outcrossing. Late-season florets frequently remain cleistogamous (Seetharam et al. 1990; Maun and Barrett 1986; Potvin 1986, 1991).

Among the 35 *Echinochloa* species, *E. frumentacea* and *E. utilis* (syn. *E. esculenta*) are distinguished by variations in panicle morphology and differences in glume and lower lemma texture (Yabuno 1971, 1987; de Wet et al. 1983). *E. frumentacea* was classified into four races, while *E. utilis* was classified into two races based on inflorescence morphology (Table 7.1). The spikelets of *E. frumentacea* are smaller and awnless with membranous glumes, whereas those of *E. esculenta* are larger and generally awned with chartaceous upper glumes and lower lemmas.

Table 7.1 Taxonomic and agronomic characteristics of cultivated barnyard millet species

Cultivated species	Common name	Subspecies	Race(s)	Key morphological descriptors	Primary geographical distribution
Echinochloa colona (L.) Link	Indian barnyard millet, Sawa millet	*colona*	–	Compact panicle, smaller grains, drought-tolerant, often prostrate growth habit	South Asia (especially India), Tropical Africa
		frumentacea	*stolonifera*, intermedia, robusta, laxa	Larger plant architecture, denser and larger inflorescence, non-shattering	Primarily domesticated and cultivated on the Indian subcontinent
Echinochloa crus-galli (L.) Beauv	Japanese barnyard millet, Hie millet	*crus-galli*	*crus-galli*, macrocarpa	Large, shattering seeds, bristly awns, closely resembles weedy parent form	East Asia, Europe (as a weed and minor forage)
		utilis	*utilis*, intermedia	Non-shattering, plump grains, erect growth habit, awnless.	Japan, Korea, China

7.1.2 Origin

The genus *Echinochloa* comprises approximately 250 annual and perennial species distributed across warmer and temperate regions worldwide (Bajwa et al. 2015), some key *Echinochloa* species and their characteristics are summarized in Table 7.2. Two notable cultivated species are *E. frumentacea* (Indian Barnyard millet) and *E. esculenta* (Japanese Barnyard millet), both belonging to the family Poaceae and subfamily Panicoideae (Clayton and Renvoize 1986; Sood et al. 2015). These species, including their wild ancestors, are hexaploid with a chromosome number of $2n = 6x = 54$, where $x = 9$.

Echinochloa frumentacea, an annual cultivated in India, Central African Republic, Tanzania, and Malawi, emerged through parallel domestication from its wild progenitor *Echinochloa colona*, though the precise domestication timeline remains unestablished (Doggett 1989). In contrast, *Echinochloa esculenta*, cultivated in temperate regions of Japan, Korea, and China, was domesticated approximately 4000 years ago from *Echinochloa crus-galli*, as substantiated by archaeological evidence from Japan's Yayoi and Jomon periods (Watanabe 1970; Nesbitt 2012).

The origins of barnyard millet are complex, involving multiple species and subspecies. Recent molecular studies have enhanced our understanding of genetic diversity within the *Echinochloa* complex. First effort from Hilu (1994) using RAPD markers to assess genomic divergence between cultivated species and their wild relatives showed that *E. frumentacea* has greater genetic variability than *E. utilis*. Later, Nozawa et al. (2004) suggest that *E. esculenta* was likely domesticated from a limited segment of the *E. crus-galli* population, resulting in distinct clusters for *E. esculenta* and *E. crus-galli* using SSR markers on 170 accessions. Further insights were provided by whole-genome genotyping-by-sequencing on a core collection of 89 accessions identified several thousand segregating single-nucleotide polymorphisms (SNPs) and delineated four population clusters within *E. colona* as well as three within *E. crus-galli* (Wallace et al. 2015). Moharil et al. (2016) confirmed extensive molecular diversity and grouped 21 accessions into two morphological races, *E. frumentacea* and *E. colona*, enhancing previous findings.

7.1.3 Phylogeny

The genus *Echinochloa* includes both minor cereal crops and common field weeds; however, the relationships between its species are still not fully understood due to their significant morphological variability. Previous studies by Hilu (1994), Michael (1994, 2001), and Yabuno (1962, 1966, 1984, 2001) have attempted to clarify the phylogenetic relationships within the genus. Controlled crosses have shown that hybrids from certain combinations exhibit normal meiotic behavior, while intergroup crosses often result in abnormalities. It is suggested that the wild hexaploid species, i.e., *E. colona* and *E. crus-galli*, may have been the progenitors of *E. frumentacea* and *E. esculenta*, respectively (Yabuno 1966). Further hybridization

Table 7.2 Comparative morphological, geographic, and domestication-related traits of four key *Echinochloa* species

Trait category	Trait	*Echinochloa colona* (jungle rice)	*Echinochloa frumentacea* (Indian barnyard millet)	*Echinochloa crus-galli* (common barnyard grass)	*Echinochloa esculenta* (Japanese millet)
Taxonomy and origin	Common names	Jungle rice, awnless barnyard grass	Indian barnyard millet, Sawa millet, billion dollar grass	Common barnyard grass, cockspur Grass, water grass	Japanese millet, Japanese barnyard Millet, white millet
	Key synonyms	*Panicum colonum*	*E. colona* var. *frumentacea*	*Panicum crus-galli*	*E. utilis*, *E. crus-galli* var. *utilis*
	Origin	Tropical Asia	Domesticated in India/ Africa	Eurasia	Domesticated in East Asia (Japan)
	Domestication status	Wild progenitor	Domesticated from *E. colona*	Wild progenitor	Domesticated from *E. crus-galli*
Distribution and habitat	Global distribution	Widespread tropical/ subtropical weed	Cultivated in Asia, Africa	Cosmopolitan weed	Cultivated in East Asia, USA (forage)
	Habitat	Moist, disturbed areas, upland rice fields	Marginal lands, rainfed or partially waterlogged fields	Moist, disturbed areas, ditches, wetlands, rice fields	Cultivated in paddy or upland fields
Vegetative morphology	General habit	Tufted, erect, or decumbent	Robust, erect annual	Highly variable, tufted, erect, or reclining	Robust, erect annual
	Culm height (cm)	10–100	60–150	Up to 200	60–150
	Leaf blade width (mm)	3–8	5–22	5–30	7–25
Reproductive morphology	Inflorescence type	Panicle of simple, unbranched, distant racemes	Compact panicle of crowded, simple, incurved racemes	Open to dense panicle with rebranched lower racemes	Dense, compact, lanceolate panicle with incurved racemes
	Spikelet arrangement	Regular, 4-rowed	Densely packed, 4–6 rowed	Irregular rows	Irregular, 4–6 rowed

(continued)

Table 7.2 (continued)

Trait category	Trait	*Echinochloa colona* (jungle rice)	*Echinochloa frumentacea* (Indian barnyard millet)	*Echinochloa crus-galli* (common barnyard grass)	*Echinochloa esculenta* (Japanese millet)
	Spikelet length (mm)	2.0–3.2	2.5–3.5	2.5–4.0 (highly variable)	3.0–4.0
	Spikelet width (mm)	1.1–1.8	1.7–2.0	1.1–2.3	1.9–2.5
	Spikelet color	Pale green to purplish	Whitish or light yellow	Green to purplish	Dark brownish to purplish at maturity
	Spikelet Awns	Awnless (muticous)	Awnless	Highly variable: awnless to 50 mm long	Awnless or rarely short-awned
	Fertile lemma tip	Lacks line of minute hairs	Not well documented, likely similar to *E. colona*	Abruptly transitions to withering tip, with a line of minute hairs	Not well documented, likely similar to *E. crus-galli*
Agronomic traits	Caryopsis (grain) size (mm)	1.2–1.6 long	1.7–2.2 long	1.3–2.2 long	1.2–2.3 long
	Caryopsis color	Whitish to yellowish	Yellowish or white	Brownish or reddish-gray	Brown
	Seed shattering	High (disarticulates at maturity)	Low (persistent on plant)	High (disarticulates at maturity)	Low (persistent on plant)
	Seed dormancy	Low to moderate	Low (selected for rapid germination)	High/variable	Low (selected for rapid germination)

Source: Yabuno (1987); Muldoon et al. (1982); de Wet et al. (1983); Doggett (1989); Bandyopadhyay (1999); Yadav and Yadav (2013); Renganathan and Vanniarajan (2015)

experiments, such as those between *E. crus-galli* and *E. oryzoides*, produced 18 bivalents and nine univalents. This outcome implies that two of the three genomes in *E. crus-galli* are homologous to those in *E. oryzoides* and supports the hypothesis that *E. crus-galli* is an allohexaploid formed via natural hybridization between tetraploid *E. oryzoides* and yet unidentified diploid species of *Echinochloa* (Yabuno 1966, 1984). Based on these crossing experiments and the corresponding meiotic behavior, researchers have classified *Echinochloa* species into several cross-compatible groups, highlighting the roles of polyploidization and subsequent amphidiploidization in the genus's diversification (Yabuno 1984, 1966, 2001).

Conventional taxonomy with a reliance on morphology, reproduction, and ecology is deceptive owing to domestication and adaptive characteristics (Yabuno 1966, 1984, 2001; Yamaguchi et al. 1996). Molecular marker analyses have been utilized to clarify genetic relationships among Eurasian annual *Echinochloa* species in order to overcome such uncertainties. The three main cross-compatible groups that have been shown to consistently emerge from the analyses are the *E. oryzicola* group, the *E. crus-galli* complex, and the *frumentacea* group (Yamaguchi et al. 2005; Yasuda et al. 2002; Ruiz-Santaella et al. 2006). Chloroplast DNA analysis of 30 *Echinochloa* accessions revealed five major phylogenetic groups, including a close clustering of *E. crus-galli*, *E. esculenta*, and *E. oryzoides* within a single complex. Indian barnyard millet (*E. frumentacea*) and its wild relative (*E. colona*) formed a distinct group with five unique substitutions. The study suggests domestication and weed mimicry likely evolved after species divergence in Asian *Echinochloa* supporting these taxonomic groupings (Yamaguchi et al. 2005). Moreover, morpho-agronomic analyses have elucidated these complex evolutionary relationships (Aoki and Yamaguchi 2009). High-throughput genomic methods have found thousands of SNPs more strongly correlated with geographic origin than classical morphological classifications, indicating the role in the development of domesticated genetic variation in barnyard millet populations (Yasuda et al. 2002; Ruiz-Santaella et al. 2006).

7.1.4 Geographical Germplasm Distribution

Barnyard millet germplasm pools are valuable for crop improvement and are mostly preserved in Japan and India. Different institutions in Japan and India such as the Vivekananda Parvatiya Krishi Anusandhan Sansthan, the Indian Institute of Millets Research, and the International Crop Research Institute for the Semi-Arid Tropics in India and the National Institute of Agrobiological Sciences in Japan are scientifically screening these germplasms for agronomic traits, biotic and abiotic stress resistance, grain yield, and nutritional content (Gomashe 2017). Global collaboration has been instrumental in broadening the genetic base available to breeders. Between 1976 and 2007, a series of collaborative missions (Indo-Australian, Indo-Japanese, and Indo-Soviet) resulted in the introduction of 601 exotic barnyard millet accessions into India from countries including Japan, the Russian Federation, the United States, and several African nations (Gomashe 2017). This infusion of foreign germplasm has been crucial for introducing novel traits to enhance productivity and stress resilience (Gomashe 2017). In India, barnyard millet cultivation is concentrated in two distinct agroecological zones: the northern Himalayan region (including Uttarakhand, Sikkim, and Arunachal Pradesh), where its short duration and reliability are highly valued, and the southern Deccan plateau, particularly in states such as Tamil Nadu, Madhya Pradesh, and Maharashtra (Anonymous 2014).

Breeding efforts have been limited, with most varieties being selected from local germplasm (Table 7.3) that are susceptible to diseases and have low yields. Introducing exotic barnyard species has shown potential to improve productivity and resistance to pests and diseases (Yadav et al. 2010), highlighting the need for further improvement initiatives to strengthen the crop's resilience to changing climatic conditions.

7.2 Traditional Uses and Cultural Significance

Barnyard millet (*Echinochloa frumentacea* L.), historically known as Shyamaka in the *Vishnu Purana* (450 AD), has been consistently documented in ancient Indian scriptures as both an agronomically important and ritually significant cereal. The *Yajurveda* refers to this grain as "aanava," and its presence alongside foxtail (priyangava) and black finger millet (shy-aamaka) suggests consumption predating the Indian Bronze Age (circa 4500 BC) (Spengler 2019). During the Mughal era, Abul Fazl's Ain-i-Akbari documents barnyard millet cultivation alongside sorghum, pearl millet, kodo millet, and finger millet. As a kharif crop, it was grown across diverse regions of India, including Malwa, Gujarat, Ajmer, Khandesh, Lahore, Agra, Allahabad, Awadh, and Multan, highlighting its agricultural significance and ecological adaptability (Nene 2005).

Barnyard millet is a prominent ingredient in Himalayan cuisine, featuring in dishes such as "Paleu or chencha" and milk-based kheer. It is frequently blended with other cereals such as amaranth, kodo, and proso millet for diverse culinary applications. The grain can be processed through techniques such as popping and puffing for use in confections and flatbreads (Satyarthi et al. 2018). During religious observances like Navratri, barnyard millet serves as a permissible fasting food, alongside other millets such as amaranth and buckwheat. Moreover, small millets are recognized for their high folic acid content and potential therapeutic value in addressing anaemia among women (Rawat et al. 2021).

7.3 Nutritional Significance of Barnyard Millet

Barnyard millet has earned the moniker "nutri-cereal" due to a nutritional profile that is demonstrably superior to that of many major staple grains (Renganathan et al. 2020; Saleh et al. 2013). It is rich in dietary fiber, essential minerals (iron, zinc, calcium, and magnesium), proteins, vitamins, and essential amino acids (Singh et al. 2010; Saleh et al. 2013; Chandel et al. 2014). The grain contains 51.5–65.5 g of carbohydrates per 100 g, predominantly composed of non-starchy

Table 7.3 Characteristics of key Indian barnyard millet varieties

Variety name	Parentage/genetic background	Developing institute(s)	Maturity (days)	Recommended cultivation zone(s)	Potential grain yield (kg/ha)	Key morphological and agronomic traits	Disease and pest resistance	Notable nutritional/ quality traits
K-1	Pure-line selection from Tenkasi local	TNAU	–	Tamil Nadu	1000	Panicles loose and spear-shaped	–	Grains dull white
Kudiravali K2 (CO (Kv) 2)	Pure-line selection from Kovilpatti local type (parentage: EF 79)	TNAU	95	Tamil Nadu	2650 (rainfed)	High tillering, compact pyramidal panicles	–	Good grain quality, brownish-gray grains
MDU 1	Pure-line selection from Arupukkottai local	TNAU	95–100	Tamil Nadu	1500–1700 (rainfed); 2200–2500 (irrigated)	High tillering (4–9), compact pyramidal panicles	–	High Iron (16 mg/100 g), high milling (70%), good taste, yellowish-gray grains
VL Madira-8	Isolated from local collection	VPKAS, Almora	–	Uttar Pradesh	1800	Grains light gray, smooth	–	White endosperm
Anurag	Derived from genetic collection of Gorakhpur	ARS, Gorakhpur	–	Uttar Pradesh	1700–2000	–	Resistant to pink borer and shoot fly	–
Gujarat banti-1	Selection from segregating material of Banti	AICRIP, GAU, Surat	–	Gujarat	2000	–	–	Grain color is yellowish white

(continued)

Table 7.3 (continued)

Variety name	Parentage/genetic background	Developing institute(s)	Maturity (days)	Recommended cultivation zone(s)	Potential grain yield (kg/ha)	Key morphological and agronomic traits	Disease and pest resistance	Notable nutritional/quality traits
VL Madira-21	Local collection from Garhwal region	VPKAS, Almora	–	Central India	2500–2600	–	Very low incidence of smut and leaf spot	–
VL Madira-29	Location collection from Kumaon region	VPKAS, Almora	–	Central India (except Tamil Nadu)	2500	–	Very low incidence of smut disease	–
VL 172	–	VPKAS, Almora	75–80	Uttarakhand, Uttar Pradesh, Karnataka	2200–2300	–	Tolerant to grain smut	High dehulled grain recovery
VL 207	Cross: VL 172 × GECH 506	VPKAS, Almora	80–90	All states except Gujarat and Tamil Nadu	1800–1900	Pyramidical shaped panicles, improved plant type	Tolerant to grain smut	High harvest index, high dehulled grain recovery
PRJ-1	Derived from germplasm accession EC 346261	GBPUA&T, Pantnagar	–	Uttarakhand, Tamil Nadu	2500	Japanese barnyard millet (*E. esculenta*)	Resistant to grain smut disease	High whole grain yield but lower dehulled recovery than VL series
DHBM-23-3	Cross: VL-13 × IEC-566	UAS, Dharwad	85–90	Andhra Pradesh, Karnataka, Madhya Pradesh, Tamil Nadu	2412 (national avg.)	Erect growth, bold oval gray grains	Resistant to grain smut and shoot fly	High protein (11.7%), zinc (45.1 mg/kg), iron (16.2 mg/kg)

Table 7.4 Nutritional profile of barnyard millet/100 g dry weight in comparison with key cereal grains and millets

Nutrient	Barnyard millet	Rice (white)	Wheat (whole)	Maize (yellow)	Oats (rolled)	Quinoa	Finger millet	Foxtail millet
Proximate composition								
Energy (kcal)	307	359	332	365	389	368	363	331
Protein (g)	12.05	7	12.1	9.4	16.9	14.1	9	12.3
Fat (g)	3.6	1	2	4.7	6.9	6.1	3	4.3
Carbohydrate (g)	65.5	80.3	69.4	74.3	66.3	64.2	75	63.2
Dietary fiber (g)	12.6	1.8	12.1	7.3	10.6	7	9	8
Ash (g)	N/A	N/A	1.5	N/A	1.7	2.4	N/A	3.3
Mineral content								
Calcium (mg)	22	10	33	7	54	47	364	31
Iron (mg)	5	2.3	3.7	2.7	4.7	4.6	4.6	2.8
Magnesium (mg)	82	48	117	127	177	197	137	81
Phosphorus (mg)	280	160	323	210	523	457	283	290
Zinc (mg)	3	2.2	3	2.2	4	3.1	2.3	2.4

Source: FAO (1995), Ramashastri, B.V., & Balasubramanian, S.C. (1999) Narasinga Rao, B.S., Deosthale, Y.G., & Pant, K.C. (Eds.), Gopalan, C., Rama Shastri, B.V., & Balasubramanian, S.C. (2009)

polysaccharides and dietary fiber. The grain is particularly rich in dietary fiber, with values reported as high as 12.6 g/100 g, far exceeding that of polished rice and maize (Ugare et al. 2014). The composition enables gradual sugar release, potentially offering a lower glycemic index for individuals managing type 2 diabetes (Saleh et al. 2013; Ugare et al. 2014; Veena Bharati et al. 2005). Resistant starch in grain mitigates blood glucose, cholesterol, and triglyceride levels (Krishna Kumari and Thayumanavan 1998). Barnyard millet provides metabolic stability through slow-digesting carbohydrates and high fiber content, potentially alleviating gastrointestinal disorders such as constipation, bloating, and abdominal discomfort (Rao et al. 2017). A comparative study of various millets confirmed this, finding barnyard millet's average protein content to be approximately 12.05 g/100 g (Kumari et al. 2024) and notably high iron levels of 4.5–5.6 mg per 100 g (refer Table 7.4) (Saleh et al. 2013; Renganathan et al. 2017; Vanniarajan et al. 2018).

The high fiber content in barnyard millet may be useful for glycemic control and gastrointestinal health, but it can also interfere with the digestibility of other nutrients. Antinutrients such as phytates, polyphenols, tannins, and oxalates bind essential minerals such as calcium, zinc, and iron, reducing their bioavailability (Arora and Srivastava 2002). Despite this limitation, the phytate content (3.30–3.70 mg/100 g) remains within acceptable limits (Gupta et al. 2013) and natural dehulling process enhance mineral bioavailability, rendering it a nutrient-dense food particularly beneficial for individuals at risk of nutritional deficiencies (Panwar et al. 2016). Processing techniques such as polishing, to enhance the looks of the grain, result in the loss of fiber from 11.2 g to 4.5 g, protein from 10.4 g to 6.8 g, and iron from 7.1 mg to 6.2 mg in unpolished grain to polished grain, respectively, thus compromising nutritional quality with consumer choice (Rajeswari and Priyadharshini 2021).

Barnyard millet contains bioactive molecules, including polyphenols, carotenoids, luteolin, and tricin, which demonstrate anticarcinogenic, anti-inflammatory, antioxidant, and antileukemic properties (Kim et al. 2011; Han et al. 2007; Shahidi and Chandrasekara 2013; Rao et al. 2017). Consequently, this grain has been incorporated into diverse food products, from traditional porridges to contemporary processed foods such as biscuits, cakes, snacks, and nutritional formulations (Arora and Srivastava 2002; Bhat et al. 2018; Poongodi et al. 2003). Moreover, its gluten-free composition renders it particularly suitable for individuals with celiac disease and gluten intolerance (Hoshino et al. 2010).

7.4 Cultivation Practices, Climatic, and Soil Requirements

Barnyard millet (Sawan) is cultivated in Indian states including Madhya Pradesh, Maharashtra, and Tamil Nadu. Typically sown during the *Kharif* season in shallow, low-holding soils, it is often intercropped with rice. The crop demonstrates remarkable adaptability, thriving in both temperate and tropical environments, tolerating temperatures across a wide range and elevations up to 2700 m. Optimal growth requires 200–400 mm of annual precipitation. Field preparation involves plowing and harrowing two to three times to achieve fine tilth, with land configured in ridges and furrows (Prabhakar et al. 2017). In Tamil Nadu, rainfed and irrigated seeding occurs during September–October and February–March, respectively. In Uttaranchal and North Eastern states, sowing is conducted in April–May, with dry seeding common in hilly areas preceding the monsoon. The recommended seed rate is 4 kg per acre (10 kg/ha), with pre-sowing Azospirillum treatment at 60 g/kg. Seeds are planted at a 3–4 cm depth, maintaining 30 cm row spacing and 10 cm between plants. Nutrient management involves applying 5 tons of farmyard manure per hectare 2–3 weeks pre-sowing, supplemented with region-specific NPK fertilizers. Andhra Pradesh and other states utilize a 20:20:0 formulation, while Bihar, Tamil Nadu, and Uttar Pradesh employ a 40:20:0 formulation. Although barnyard millet typically does not require supplemental irrigation, additional watering at 25–30 days and during panicle initiation (45–50 days) is recommended during prolonged dry periods (Prabhakar et al. 2017). Effective weed management is critical for controlling both grassy weeds (*Echinochloa colonum* and *Echinochloa crusgulli*) and broad-leaved species (*Celosia argentea* and *Solanum nigrum*). Strategies include postemergence application of 2,4-D sodium salt (1.0 kg a.i./ha), preemergence isoproturon (1.0 kg a.i./ha), and mechanical interventions such as inter-cultivation and hand weeding. Harvesting involves cutting the crop near ground level with sickles at maturity, followed by field-stacking for approximately 1 week before threshing via traditional bullock trampling. This methodology yields approximately 12–15 quintals per hectare for grain and 20–25 quintals per hectare for straw (Prabhakar et al. 2017).

7.5 Genetic Diversity, and Recent Advancements in Barnyard Millet Breeding

7.5.1 Conventional Breeding Efforts

Barnyard millet exhibits a predominantly autogamous reproductive system with occasional cross-pollination events that facilitate gene exchange among populations (Maun and Barrett 1986). This reproductive mode has led to the development of over 20 improved cultivars in regions such as Uttarakhand and Tamil Nadu (Table 7.3). The crop's small floral size, early flowering, brief pollen viability, and transient floral opening make emasculation and artificial hybridization particularly challenging (Nirmalakumari 2009). These biological constraints hinder the success of controlled crosses, thus limiting recombination breeding efforts and impeding the development of genetically diverse lines (Joshi et al. 2021). The hot water emasculation technique (Sood et al. 2015) has enabled hybridization followed by the pedigree method to develop new varieties. Mehta et al. (2005) introduced an interspecific breeding strategy by crossing early-maturing *E. frumentacea* with high-yielding *E. esulenta* to generate segregants that combine rapid maturity with increased yield. Direct crosses between *E. frumentacea* and *E. esulenta* often yield hybrids with chromosomal pairing disruptions or meiotic incompatibilities, resulting in sterility. Conversely, crosses with wild progenitors are fertile and serve as crucial genetic resources for barnyard millet improvement. Incorporating wild progenitors into breeding programs typically introduces broader genetic diversity, including alleles that facilitate chromosomal pairing and potentially restore fertility (Mandelbaum et al. 1995).

Mutation breeding has played an important role in augmenting genetic diversity in self-pollinating crops, especially when it is challenging to hybridize (Sood et al. 2019). Induced mutagenesis, particularly gamma irradiation, has been found to be efficient in enhancing genetic diversity and agronomic performance in small millets. Gamma irradiation of the low-amylase landrace "Nogehie" facilitated the development of stable, full-waxy mutant lines, enabling the release of the glutinous cultivar "Chojuromochi" in Japan. This advancement addressed consumer preferences in Asian markets for a soft, sticky texture, diverging from the coarse texture typical of most barnyard millet genotypes (Hoshino et al. 2010; Goron and Raizada 2015). Gamma irradiation also has been found to enhance variability in critical morphological characters such as tiller number, plant height, head length, and grain yield, thereby increasing the agricultural value of the crop (Mehra et al. 1985).

More recently, Indian breeding programs have marked a significant milestone by successfully employing recombination breeding. Through persistent efforts and the use of techniques such as hot water emasculation, breeders have overcome hybridization barriers. A landmark achievement is the development and release of the variety DHBM-23-3 (also reported as DHBM-93-3 in some initial documents) by the University of Agricultural Sciences, Dharwad (Govanakoppa et al. 2024), which was developed from a cross between the parental lines VL-13 and IEC-566, DHBM-23-3 represents a major leap forward. It is a high-yielding variety, achieving an average grain yield of 24.12 quintals per hectare (q/ha) in national trials, a 13–15% advantage over existing national checks. Crucially, it combines this high yield with a superior nutritional profile, including high protein (11.7%), zinc (45.1 mg/kg), and iron (16.2 mg/kg), along with strong resistance to major biotic stresses such as grain smut and shoot fly. The success of DHBM-23-3 demonstrates that targeted hybridization is now a viable and powerful strategy for developing multi-trait improved cultivars in barnyard millet.

7.5.2 Next-Generation Barnyard Millet Breeding Efforts

Whole-genome sequencing of weedy *Echinochloa crus-galli* resulted in a nearly complete assembly, covering approximately 90.7% of a 1.27 Gb genome at a sequencing depth of 171× revealed the presence of 108,771 protein-coding genes, 785 miRNAs, 514 Mb of repetitive elements, and various noncoding RNAs (Guo et al. 2017). A comprehensive assessment of genomic resources in public databases revealed notable inconsistencies among *Echinochloa* species. The NCBI repository contains 1246 nucleotide sequences, 822 gene sequences, 2468 protein sequences, 105 SRA records, 74 ESTs, and one GEO dataset. *E. crus-galli* dominates with 652 records, while other species such as *E. oryzicola*, *E. phyllopogon*, and *E. colona* have substantially fewer entries. Notably, only 54% of nucleotide sequences correspond to *E. crus-galli*, with cultivated barnyard millet species such as *E. frumentacea* and *E. esculenta* representing merely 4% of the available genomic information. These results highlight the urgent requirement for further genomic studies to improve our knowledge of cultivated barnyard millet for upcoming crop enhancement projects (Guo et al. 2017).

Recent advances in sequencing technologies have enabled the identification of EST-SSR markers, which have proven instrumental in characterizing repeat motifs and elucidating genomic architectures. Initially, RAPD markers were employed to investigate the phylogenetic relationships and genetic diversity of *Echinochloa* species, successfully differentiating between cultivated and wild progenitors (Hilu 1994; Prabha et al. 2012). Yet the restricted polymorphism and RAPD marker reproducibility have been extensively documented (Muza et al. 1995; Rutledge et al. 2000; Ruiz-Santaella et al. 2006) and encouraged scientists to seek out other approaches such as AFLP. AFLP methods have demonstrated enhanced ability to detect alleles with more polymorphic bands per primer and thus produces higher polymorphism information content studies (Danquah et al. 2002; Tabacchi et al. 2006; Kaya et al. 2014).

Later developments have seen the use of sequence-based markers such as simple sequence repeats (SSRs), EST-SSRs, and single-nucleotide polymorphisms (SNPs). The markers are appreciated for their codominant inheritance behavior, ease of replication, and high level of polymorphism, which makes them highly suitable for genetic diversity research (Lin et al. 2011). For instance, studies using five SSR markers have been able to classify 155 accessions of different *Echinochloa* species into clear-cut clusters, which showed differences in the levels of diversity of species such as *E. esculenta* and *E. crus-galli* (Nozawa et al. 2006). In contrast, while there are evidence of the obvious benefits of such markers, sequence-based information from barnyard millet is in an early stage of development. Babu et al. (2018a, b) recently used rice, finger millet, and maize genomic SSRs for cross-transferability analysis among barnyard millet cultivated and wild species for polymorphic marker identification, syntenic region identification, genetic diversity, and population structure analysis. SSRs of all the crops were highly transferable, polymorphic, and capable of differentiating barnyard millet genotypes distinctly. New strategies such as combining Restriction-site Associated DNA (RAD) sequencing with Illumina platforms have also enhanced the mass and rapid discovery of SSR and SNP markers (Chen et al. 2017). In one of the studies, it resulted in more than 4000 contigs of which virtually 5000 putative SSR markers and almost 49,000 likely SNPs were found (Chen et al. 2017). Validation experiments proved that a portion of these SSR markers showed high polymorphism levels, with 3.1–4.8 alleles per locus on average, and reported a genomic GC content of 48.9%, and it is proposed that they are involved in environmental adaptation (Chen et al. 2017; Osuna et al. 2011). Recent genotyping-by-sequencing (GBS) surveys have detected more than 21,000 SNPs in a set of barnyard millet accessions used as a core collection, allowing discrimination of biotypes of *E. colona* and *E. crus-galli* and enabling well-supported population structure analyses (Wallace et al. 2015). These high-throughput approaches have validated the application of high-throughput sequencing in disentangling intricate genetic relationships as well as guiding marker-assisted selection strategies for crop improvement initiatives.

7.5.3 Gene/QTL Mapping

Employment of molecular markers, such as SSRs and SNPs, to accurately define quantitative trait loci (QTLs) controlling key micronutrient profiles and agronomic performance, offers genetic basis of intricate traits, but establishment of an exhaustive genetic linkage map remains unaccomplished relative to other millets such as foxtail and finger millet (Wallace et al. 2015; Chen et al. 2017; Manimekalai et al. 2018; Murukarthick et al. 2019). To date, two mapping studies only have been published on barnyard millet. Ishikawa et al. (2013) identified functional SNP markers for waxy traits and these waxy traits are governed by three loci (EeWx1, EeWx2, and EeWx3), where the presence of functional alleles across all loci results in normal amylose levels, while mutations in one or more loci lead to a reduction in amylose content. Another study used bulked segregant analysis to link an SSR marker (BMESSR 39) to anthocyanin pigmentation (Renganathan et al. 2019). More recent studies have pointed to the identification of putative QTLs or genes for key traits, such as *EcGlb1* for protein content (Kumar et al. 2018), *qDTY3.1* for drought tolerance (Gupta et al. 2020), *qFe7.2* for iron content (Rana et al. 2019), *qEF2.1* for early flowering (Joshi et al. 2021).

7.6 Diseases and Management Strategies in Barnyard Millet

Barnyard millets are prone to various biotic stresses, with fungal pathogens and viruses being particularly significant. These pathogens can substantially diminish the crop's genetic yield potential and reduce overall productivity, especially under changing climatic conditions. While numerous pathogens are present, only a few cause severe diseases, with their impact primarily determined by host susceptibility and environmental factors.

(i) Blast (*Pyricularia grisea*)

Blast disease, caused primarily by *Pyricularia grisea*, poses a significant threat to barnyard millet as a secondary host. Although *P. grisea* is known for its severe impact on finger, proso, and little millets (Viji et al. 2000), its presence on barnyard millet shows its ability to adapt to a wider range of hosts. The pathogen creates water-soaked, elliptical to diamond-shaped lesions that can merge to form large necrotic areas. The life cycle is complex, involving seed-borne inoculum in the pericarp and endosperm (Viswanath and Seetharam 1989), as well as airborne dispersal. Studies have shown yield reductions in related hosts ranging from moderate to severe levels, suggesting that barnyard millet could also suffer significant losses under favorable infection conditions (Nagaraja et al. 2007a, b).

(ii) Leaf Spot/Leaf Blight (*Exserohilum crus-galli*)

Exserohilum crus-galli is the primary causal agent of leaf blight in barnyard millet, particularly in warm, humid cultivated regions. The disease manifests as oval lesions on the leaf lamina, progressively darkening from brown to deep brown. Under severe infection, the plant's neck may fracture, causing the affected tissues to hang. The pathogen additionally targets ear heads, fingers, and grains, resulting in underdeveloped, shriveled kernels and substantial crop yield reduction (Kumar and Singh 2010; Nagaraja et al. 2007a, b). The pathogen can persist in soil for up to 18 months and remains viable on infected debris and seeds, complicating disease management (Vidhyasekaran 1971).

(iii) Smuts (*Ustilago panici-frumentacei*)

Grain smut in barnyard millet is due to the fungal pathogen *Ustilago panici-frumentacei*, which infects floral tissues during grain development. Under conditions of low temperature and high relative humidity, the pathogen forms characteristic velvety green sori. Sori rupture, releasing dark, powdery spores that enable rapid pathogen dissemination (Das 2017; Mundkur and Thirumalachar 1952). Studies by Jain and Tripathi (2007) demonstrated that grain smut significantly reduces little millet yield, with grain production decreasing by 9.80–53.55% and panicle length diminishing by 6.40–38.90%. Phenotypic screening revealed disease resistance, particularly in *E. esculenta* accessions PRB 9402 and PRB 9602, as documented by the All India Coordinated Small Millets Improvement Project (2001).

(iv) Kernel Smut (*Ustilago paradoxa*)

Ustilago paradoxa has been sporadically documented in unprotected cultivation across multiple regions of India, including Bihar, Maharashtra, and Tamil Nadu (Viswanath and Seetharam 1989). During the flowering stage, the fungus induces transformation of floral parts into swollen, greenish structures. These morphological changes result in smut sori formation, typically measuring 1.5–4 mm × 1–2 mm, potentially affecting up to 25 grains per ear (Nagaraja et al. 2016). Chlamydospores, smooth, olive-brown, spherical structures measuring 7–11 μm in diameter, serve as critical morphological indicators for initial disease diagnosis.

(v) *Viral Diseases*

Barnyard millet is susceptible to viral pathogens, including Wheat streak virus, Sugarcane mosaic virus, and Eleusine virus 2. Infected plants typically display chlorotic stripes, streaks, and mosaic patterns on leaves, with transmission occurring through insect and noninsect vectors under favorable agroecological conditions (Nagaraja et al. 2016; Das 2017).

7.6.1 Control Strategies for Major Diseases Affecting Barnyard Millet

Cultural control strategies mitigate pest and disease risks by systematically modifying agricultural environments. Key interventions include crop rotation, optimal plant spacing, precise sowing timing, balanced population density, targeted fertilization, deep summer tillage, crop residue management, field boundary maintenance, infected plant removal, and regulated irrigation. These integrated practices effectively reduce pathogen transmission and enhance crop resilience (refer Table 7.6).

Host Plant Resistance

The most effective strategy for combating this disease, particularly among marginal and resource-constrained farmers, is the development of resistant cultivars. Disease-resistant varieties identified and available for millet-growing regions in India is given in Table 7.5.

7.6.2 Major Pests and Management Strategies in Barnyard Millet

Worldwide, at least 150 insect species are recorded as feeding on millets (Nwanze and Harris 1992). Out of these, 116 species have been recorded from India (Kishore 1996). Most of the pests are common to all species of millet (Gahukar 1989).

Table 7.5 Disease-resistant varieties identified and available for millet-growing regions in India

Cultivar name	Developing institution	Year of release	Pedigree	Maturity (fays)	Grain yield (q/ha)	Key agronomic features	Primary disease resistance Profile	Additional resistances/ tolerances
OLM 203 (Tarini)	OUAT, Odisha	2001	Selection from local cultivar (KL 2)	105–110	10–11	Widely adapted, used as a national check	Blast, grain smut	Drought tolerant genotype
GNV-3 (Gujarat Navsari Vari-3)	NAU, Waghai, Gujarat	2016	Pure line selection from accession WV-125	112–115	9–28*	High yielding, good tillering, erect habit	Blast (moderate), grain smut (moderate)	Moderately resistant to sheath blight
OLM 217	OUAT, Odisha	2009	Selection from Lajigada local	105–110	15–16	Higher yield than OLM 203	Rust, grain smut	–

Table 7.6 Chemical control for different diseases that affect barnyard millet

Disease	Recommended chemical treatment
Downy mildew	Treat seeds before sowing with Ridomyl-MZ at 6 g/kg. Depending on disease severity, apply one or two foliar sprays at 3 g/L
Blast	For seed protection, apply carbendazim at 1 g/kg. Foliar sprays can include Carbendazim (0.2%), Iprobenphos (0.1%), or combinations such as Mancozeb (0.1%) + Carbendazim. Other effective options include Ediphenphos (0.1%), Propiconazole (0.1%), and Tricyclazole (0.1%)
Smut	Before sowing, mix seed with sulfur powder at a rate of 4 g/kg
Rust	Control involves foliar application of Mancozeb (0.2%). Other fungicides such as Hexaconazole (0.1%), Difenoconazole (0.1%), or Propiconazole (0.1%) may also be used
Leaf spot/ leaf blight	Treat seeds using either Carbendazim (2 g/kg) or Mancozeb (0.2%). Additional foliar sprays can be applied based on disease spread

(i) Rhizophagous White Grubs (*Holotrichia* sp., *Anomala dimidiata, Apogonia* sp.)

White grubs lay their eggs in the soil, and the larvae feed on roots for months before pupating and becoming adults. White grubs not only target barnyard millet but also attack sorghum and pearl millet, causing patchy wilting and stunted growth due to root damage, reduces the plant's ability to uptake water and nutrients. Infestation can lead to yield losses of over 20% in affected areas (Sharma and Davies 1988).

Management

Recommended strategies include cultural practices such as crop rotation and deep ploughing to expose larvae, biological control with entomopathogenic nematodes and fungi, and targeted chemical treatments when population thresholds are reached.

(ii) Shoot Flies (*Atherigona falcata, A. pulla, A. simplex, A. soccata, A. oryzae, Atherigona nudiseta*)

Atherigona species represent a group of dipteran pests that inflict damage on various millet crops across Africa and Asia, with each species named for its preferred host. The barnyard millet shoot fly (*Atherigona falcate* Thompson) is one such species, distinguished by its host specificity. In these insects, the complete developmental cycle is approximately 3 weeks, and the pupal stage endures for about 6–11 days, occurring either on the host plant or within the soil matrix. Adult flies exhibit a characteristic morphology, notably three pairs of black spots distributed along the dorsal abdomen (Kalaisekar and Padmaja 2017).

Management

Management of shoot fly infestations incorporates both chemical and cultural practices. Pre-sowing soil treatments with granular formulations of phorate at a dosage of 3 kg active ingredient per hectare have been recommended to reduce pest populations. In addition, agronomic practices such as augmenting the seeding rate followed by the removal of compromised seedlings have been employed to mitigate crop losses. The survival and persistence of these pests are further enhanced by the availability of alternate hosts, such as *Cynodon dactylon*, which support the pest population during off-season periods (Singh and Jotwani 1973; Talati and Upadhyay 1978).

(iii) Armyworm (*Mythimna separata*)

The armyworm, *Mythimna separata*, primarily consumes foliage, a feeding behavior documented by Balasubramanian et al. (1975). These generalist feeders are known to infest multiple cereal crops, such as sorghum, maize, and pearl millet and barnyard millet been observed, with less pubescent varieties exhibiting reduced attractiveness for oviposition by *Spodoptera frugiperda* Smith, suggesting that the absence of leaf pubescence confers a degree of resistance (Leuck et al. 1968, 1977). These insects inflicted extensive defoliation on grasses, while the larvae of *M. albistigma* and *M. frugalis* demonstrated the ability to migrate during daylight hours. Pupal densities averaged 26.75, 18.5, and 15.0 per square meter in fields of finger millet, maize, and grasses, respectively (Rajagopal and Musthak Ali 1992).

7.7 Postharvest Handling of Barnyard Millet

7.7.1 Drying and Storage Technique

Barnyard millet is generally harvested at approximately 24–26% dry basis (db) moisture content. Immediately after harvesting, the grains are dried to a targeted moisture level of about 12% (db) to prevent spoilage and to prepare the grain for further processing (Singh et al. 2009). Most millet varieties have excellent storage properties and can be kept for up to 4–5 years in granaries constructed from reeds and mud walls (as seen with finger millet in Uganda), sealed storage drums, mud straw bins, and earthenware pots or jars, even underground storage has been reported in some regions, which helps protect the grains from insects and environmental moisture (McFarlane 1969).

7.7.2 Primary Processing Technology of Barnyard Millet

Processing millets is crucial due to challenges such as a tough seed coat, unique flavor, and limited availability of processed products. These factors make it harder for millets to be as widely accepted as rice or wheat, which are typically processed to improve edibility. Efficient processing can significantly increase the market value of millets, as unprocessed millets sell for around Rs. 15–20/kg compared to Rs. 80–100/kg for processed products. This presents economic benefits and opportunities for enhancing nutrition and adding value to future food products. Effective dehusking is necessary for processing barnyard millet into a grain suitable for human consumption. Traditional manual methods, such as pounding with a pestle and mortar, can separate the endosperm from the husk and bran, but they are labor-intensive and time-consuming. These approaches tend to produce uneven removal of bran, affecting the uniformity and quality of the resultant grain (Lohani et al. 2012). Mechanized threshers have been designed to break through the limitations of threshing and dehusking barnyard millet. The Vivek Millet Thresher 1 threshes 40–60 kg of millet in an hour and dehusks 5–6 kg of millet per hour with improvised sieves, although it will need several passes (Singh et al. 2003). The dehusking mill developed based on collaboration between Indian and Canadian scientists is 98% efficient with less than 2% grain breakage and produces about 2 kg of clean seed per hour. Specialized dehusking machines, such as the one developed by the Central Institute of Agricultural Engineering in Bhopal, India, can dehusk up to 100 kg/h at 10–12% moisture content and can also be configured for small batches to make the most of efficiency.

7.8 Potential for Creating High-Value Products from Barnyard Millet

Barnyard millet (*Echinochloa frumentacea*) is a nutritionally rich, drought-tolerant cereal with immense scope for creating high-value products. With increasing demand for gluten-free, healthy foods, barnyard millet finds special mention because of its high protein, fiber, and micronutrient composition. The fact that it can grow in adverse conditions makes it an eco-friendly raw material for various uses.

7.8.1 Value-Added Products from Barnyard Millet

Current consumer behavior suggests greater demand for nutritionally strong and functional foods (Kaur and Sharma 2020). Barnyard millet stands out due to its beneficial nutritional profile of highly digestible proteins, high dietary fiber content, and low carbohydrate content, which cumulatively serve to deliver a tempered glycemic effect and augmented metabolic effects (Rao et al. 2017). The inherent characteristics of barnyard millet have spurred its use in value-added food items. Scientists have created composite flours by mixing barnyard millet with conventional cereal flours to improve the nutritional content of foods without compromising flavor (Jaybhaye and Ma 2014; Veena Bharati et al. 2005). Barnyard millet flour is sensitive to moisture and may be utilized in a variety of dietetic products ranging from infant food to snack foods and other nutrition-based foods (Vijayakumar and Mohankumar 2009; Anju and Sarita 2010). A good example is the creation of a ready-to-eat food, where a composite product with 60% barnyard millet, 37% potato mash, and 3% tapioca powder was designed to minimize textural degradation and maximized sensory acceptance with improved nutritional improvement (Jaybhaye and Srivastav 2015).

Barnyard millet is a highly versatile and nutritious food ingredient that has extensive applications in foods from fermented and unfermented breads to snacks, porridges, ready meals, baby food, millet wine, and nutrition powders among other staple foods (Chandrasekara and Shahidi 2010; Nisha and Saravanakumar 2019). It is usually added to recipes in the range of 30–50%, and it is perfect to formulate bakery products such as bread, cookies, cakes, and Khari. Millet food products provide viable alternatives to traditional rice and wheat recipes, introducing new twists to classic dishes such as idli, dosa, idiyappam, rotti, pittu, upma, vermicelli, adai, porridge, khakra, and chapati (Salunke et al. 2019). Surekha et al. (2013) developed cookies by mixing barnyard millet with sago, pulse, and vegetable flours. In the same way, Nazni and Karuna (2016) created bran muffins and rusks by incorporating wheat flour with barnyard millet bran. The said products were developed through sophisticated processing methods such as steam cooking, cold extrusion, and high-temperature short-time (HTST) puffing that provide added texture and sensory attributes to the final ready-to-eat snacks.

7.8.2 Industrial Applications

Amylose-rich barnyard millet starch is gaining up popularity in biodegradable packaging as an antioxidant packaging material (Renganathan et al. 2020). Aqueous extracts of the aerial parts of the *E. colona* plant have been used to synthesize silver nanoparticles (AgNPs), providing a green solution for nanoparticle biosynthesis. Borage seed oil incorporation into barnyard millet starch enhances the flexibility of the starch at the cost of tensile strength, water permeability, and moisture content, making it an ideal choice to produce biofilms (Renganathan et al. 2020).

7.9 Global Demand and Market Trends of Barnyard Millet

Projected population in 2050 is estimated to be 9 billion, requiring an impressive amount of food increase to achieve the goal established by the World Summit on Food Security (Beddington 2010; Tester and Langridge 2010). Nonetheless, the problem is further compounded by climate change, which continues to exert overwhelming pressure on agricultural yields globally. Given the limited availability of arable land, increasing food production will require either a shift in crop consumption patterns or the expansion of cultivation into natural landscapes. Nonetheless, ensuring food security is not solely about boosting production; it is equally crucial to guarantee access to nutritious and healthy food in adequate quantities. This is especially crucial in the developing nations such as India, where poverty and hunger are still prevalent issues. To counter these, it requires an integrated strategy that balances enhanced farming yields with enhanced food distribution and nutritional value. Unlike rice, maize, and wheat, which thrive on fertile lands, barnyard millet is typically cultivated in poor, marginal areas with minimal agricultural attention having immense potential as a staple crop grown in countries such as India, China,

Japan, and Korea. Globally, India is the biggest producer of barnyard millet, both in terms of area (0.146 m/ha) and production (0.147 mt) with average productivity of 1034 kg/ha during the last 3 years (IIMR 2018). In India, barnyard millet is cultivated in states such as Madhya Pradesh, Uttarakhand, Tamil Nadu, Andhra Pradesh, Karnataka, and Uttar Pradesh, covering approximately 66.09 thousand hectares and yielding 85.29 thousand tons annually (Channappagoudar et al. 2010; Yadav et al. 2010). It is the second most important minor *kharif* crop in the hilly regions of the Central Himalayas, where it is a dietary staple and traditionally grown between major crop seasons as a backup against crop failures (Kumar et al. 2007; Trivedi et al. 2017; Manimekalai et al. 2018). Despite its valuable agronomic and nutritional traits, barnyard millet remains one of the most under-researched cereals, with limited genetic resources available for its domesticated species (Trivedi et al. 2017). Research on barnyard millet has largely focused on seed bank diversity and germplasm collection, while genetic studies essential for developing improved varieties remain limited (Sood et al. 2020).

References

Anju T, Sarita S (2010) Suitability of Foxtail Millet (*Setaria italica*) and Barnyard Millet (*Echinochloa frumentacea*) for Development of Low Glycemic Index Biscuits. Malays J Nutr 16:361–368

Anonymous (2014) Status Paper on Coarse Cereals, Directorate of Millets Development, Department of Agriculture and Cooperation, Ministry of Agriculture, Government of India

Aoki D, Yamaguchi H (2009) Oryza sh4 gene homologue represents homoeologous genomic copies in polyploid *Echinochloa*. Weed Biol Manag 9(3):225–233. https://doi.org/10.1111/j.1445-6664.2009.00343.x

Arora S, Srivastava S (2002) Suitability of millet-based food products for diabetics. J Food Sci Technol Mysore 39(4):423–426

Babu B, Rashmi C, Sood S (2018a) Cross transferability of finger millet and maize genomic SSR markers for genetic diversity and population structure analysis of barnyard millet. Indian J Genet Plant Breed 78:364–372. https://doi.org/10.31742/IJGPB.78.3.5

Babu B, Sood S, Kumar D, Joshi A, Pattanayak A, Kant L et al (2018b) Cross-genera transferability of rice and finger millet genomic SSRs to barnyard millet (Echinochloa spp.). 3 Biotech 8:95. https://doi.org/10.1007/s13205-018-1118-1

Bajwa AA, Jabran K, Shahid M, Ali HH, Chauhan BS, Ehsanullah. (2015) Eco-biology and management of *Echinochloa crus-galli*. Crop Prot 75:151–162. https://doi.org/10.1016/j.cropro.2015.06.001

Bandyopadhyay BB (1999) Genotypic differences in relation to climatic adaptation of two cultivated barnyard millet species at Garhwal Hills. Indian J Genet 59:105–108

Balasubramanian R, Seshu Reddy KV, Govinda R, Deviah MA (1975) New record armyworm, Pseudaletia separate Walker (Lepidoptera: Noctuidae) as a pest of ragi in India. J Bomb Nat Hist Soc 72:588–589

Beddington J (2010) Food security: contributions from science to a new and greener revolution. Philos Trans R Soc B Biol Sci 365:61–71. https://doi.org/10.1098/rstb.2009.0201

Bhat N, Wani I, Hamdani A, Gani A (2018) Effect of extrusion on the physicochemical and antioxidant properties of value added snacks from whole wheat (Triticum aestivum L.). Food Chem 276. https://doi.org/10.1016/j.foodchem.2018.09.170

Chandel G, Meena RK, Dubey M, Kumar M (2014) Nutritional properties of minor millets: neglected cereals with potentials to combat malnutrition. Curr Sci 107(7):00113891

Chandrasekara A, Shahidi F (2010) Content of insoluble bound phenolics in millets and their contribution to antioxidant capacity. J Agric Food Chem 58(11):6706–6714. https://doi.org/10.1021/jf100868b
Channappagoudar BB, Hiremath SM, Biradar NR, Koti RV, Bharamagoudar TD (2010) Influence of morpho-physiological and biochemical traits on the productivity of barnyard millet. J Farm Sci 20(3):Article 3. http://14.139.155.167/test5/index.php/kjas/article/view/888
Chen G, Zhang W, Fang J, Dong L (2017) Identification of massive molecular markers in *Echinochloa phyllopogon* using a restriction-site associated DNA approach. Plant Divers 39(5):287–293. https://doi.org/10.1016/j.pld.2017.08.004
Clayton WD, Renvoize SA (1986) Genera graminum. Grasses of the world
Danquah EY, Hanley SJ, Brookes RC, Aldam C, Karp A (2002) Isolation and characterization of microsatellites in *Echinochloa* (L.) Beauv. Spp Mol Ecol Notes 2(1):54–56. https://doi.org/10.1046/j.1471-8286.2002.00144.x
Das IK (2017) Millet diseases: current status and their management. Millets and sorghum: biology and genetic improvement, pp. 291–322
de Wet JMJ, Prasada Rao KE, Mengesha MH, Brink DE (1983) Domestication of mawa millet (*Echinochloa colona*). Econ Bot 37(3):283–291. https://doi.org/10.1007/BF02858883
Denton DC (1987) Food crops for waterfowl. In: Wesley DE, Leitch WG (eds) Fireside Waterfowler: Fundamentals of Duck and Goose Ecology, 352. Stackpole Books, Harrisburg, PA
Doggett H (1989) Small millets—a selective overview. Small Millets Global Agric 22(3):59–70
FAO (1995) Sorghum and Millets in Human Nutrition. FAO Food Nutr Series No. 27. Rome, Italy
Gahukar RT (1989) Insect pests of millets and their management: a review. Int J Pest Manag 35:382–391
Gomashe S (2017) Barnyard millet: present status and future thrust areas: biology and genetic improvement. In: Millets and sorghum: biology and genetic improvement, pp 184–198. https://doi.org/10.1002/9781119130765
Goron TL, Raizada MN (2015) Genetic diversity and genomic resources available for the small millet crops to accelerate a New Green Revolution. Front Plant Sci 6. https://doi.org/10.3389/fpls.2015.00157
Govanakoppa N, Anitha S, Jagadeesh BR, Jakir Hussain KN (2024) High yielding variety bred through recombinant breeding technology in barnyard millet (*Echinochloa Frumentacea*. L). J Adv Biol Biotechnol 27(11):1195–1203. https://doi.org/10.9734/jabb/2024/v27i111705
Guo L, Qiu J, Ye C, Jin G, Mao L, Zhang H, Yang X, Peng Q, Wang Y, Jia L, Lin Z, Li G, Fu F, Liu C, Chen L, Shen E, Wang W, Chu Q, Wu D et al (2017) *Echinochloa crus-galli* genome analysis provides insight into its adaptation and invasiveness as a weed. Nat Commun 8(1):1031. https://doi.org/10.1038/s41467-017-01067-5
Gupta A, Mahajan V, Kumar M, Gupta H (2009) Biodiversity in the barnyard millet (*Echinochloa frumentacea* Link, Poaceae) germplasm in India. Genet Resour Crop Evol 56:883–889. https://doi.org/10.1007/s10722-009-9462-y
Gupta S, Shrivastava SK, Shrivastava M (2013) Study of biological active factors in some new varieties of minor millet seeds. Int J Innov Eng Technol 3:115–117
Gupta SK, Singh R, Varshney RK (2020) Genomic resources and breeding for climate resilience in small millets. Front Genet 11:562
Han X, Shen T, Lou H (2007) Dietary polyphenols and their biological significance. Int J Mol Sci 8(9):950–988
Hilu KW (1994) Evidence from RAPD markers in the evolution of *Echinochloa* millets (Poaceae). Plant Syst Evol 189(3):247–257. https://doi.org/10.1007/BF00939730
Hoshino T, Nakamura T, Seimiya Y, Kamada T, Ishikawa G, Ogasawara A, Sagawa S, Saito M, Shimizu H, Nishi M, Watanabe M, Takeda J, Takahata Y (2010) Production of a fully waxy line and analysis of waxy genes in the allohexaploid crop, Japanese Barnyard Millet. Plant Breed 129(4):349–355. https://doi.org/10.1111/j.1439-0523.2009.01668.x
Hulse JH, Laing EM, Pearson OE (1981) Sorghum and millets: Their composition and nutritive value
IIMR (2018) Annual report 2017–18. Indian Institute of Millets Research, Hyderabad

Ishikawa G, Seimiya Y, Saito M, Nakamura T, Hoshino T (2013) Molecular characterization of spontaneous and induced mutations in the three homoeologous waxy genes of Japanese barnyard millet [*Echinochloa esculenta* (A. Braun) H. Scholz]. Mol Breed 31(1):69–78. https://doi.org/10.1007/s11032-012-9769-9

Jain AK, Tripathi SK (2007) Management of grain smut (Macalpinomyces sharmae) in little millet. Indian Phytopathol 60(4):467–471

Jayaraman N, Suresh S, Nirmala Ganeshan NM (1997) Genetic enhancement and breeding strategies in small millets. In: National seminar on small millets: current research trends and future priorities as food feed and in processing for value addition, 19–21. 23–24 April, 1997, Bangalore, India, (Extended summaries)

Jaybhaye RV, Srivastav PP (2015) Development of barnyard millet ready-to-eat snack food: Part II. Food Sci Res J 6:285–291. https://doi.org/10.15740/HAS/FSRJ/6.2/285-291

Jaybhaye RV, Ma P (2014) Processing and technology for millet based food products: a review. https://www.semanticscholar.org/paper/Processing-and-Technology-for-Millet-Based-Food-A-Jaybhaye-Ma/e50e966715cfe6c910864a3aa775948db43581fd

Joshi D, Meena R, Chandora R (2021) Genetic resources: collection, characterization, conservation, and documentation, pp 19–31. https://doi.org/10.1016/B978-0-12-820089-6.00003-3

Kalaisekar A, Padmaja PG (2017) Insect Pests of Millets and Their Host Plant Relations. In: Patil JV (ed.) Millets and Sorghum (1st ed., pp. 267–290). Wiley. https://doi.org/10.1002/9781119130765.ch10

Kaur H, Sharma S (2020) An overview of Barnyard millet (*Echinochloa frumentacea*). J Pharmacog Phytochem 9(4):819–822

Kaya HB, Demirci M, Tanyolac B (2014) Genetic structure and diversity analysis revealed by AFLP on different *Echinochloa* spp. from northwest Turkey. Plant Syst Evol 300(6):1337–1347. https://doi.org/10.1007/s00606-013-0965-9

Kim JY, Jang K-C, Park B-R, Han S-I, Choi K-J, Kim S-Y, Oh S-H, Ra J-E, Ha TJ, Kang H-W, Seo W-D, Lee JH, Hwang J (2011) Physicochemical and antioxidative properties of selected barnyard millet (*Echinochloa utilis*) species in Korea. Food Sci Biotechnol 20(2):461–469

Kishore P (1996) Evolving management strategies for pests of millets in India. J Entomol Res 27:357–360

Krishna Kumari S, Thayumanavan B (1998) Characterization of starches of proso, foxtail, barnyard, kodo, and little millets. Plant Foods Human Nutrit (Dordrecht, Netherlands) 53(1):47–56. https://doi.org/10.1023/a:1008083020810

Kumar B, Singh KP (2010) Important small millets diseases in India and their management. Plant Pathology Section, College of Forestry and Hill Agriculture, Ranichauri, Tehri Garhwal, Uttarakhand

Kumar J, Kumar B, Yadav VK (2007) Small millets research at G.B. Pant University. GB Pant University of Agriculture and Technology

Kumar S, Stecher G, Li M, Knyaz C, Tamura K (2018) MEGA X: molecular evolutionary genetics analysis across computing platforms. Mol Biol Evol 35(6):1547–1549

Kumari PL et al (2024) A systematic review on millets: a potential nutritional and therapeutic source for the future. Front Sustain Food Syst 8:1324046. https://doi.org/10.3389/fsufs.2024.1324046

Leuck DB, Taliafero CM, Burton GW, Bowman MC (1968) Fall armyworm resistance in pearl millet. J Econ Entomol 61:693–695

Leuck DB, Burton GW, Windstrom NW (1977) Insect oviposition and foliage feeding resistance in pearl millet. J Georgia Entomol 12:138–140

Lin H-S, Chiang C-Y, Chang S-B, Kuoh C-S (2011) Development of Simple Sequence Repeats (SSR) Markers in Setaria italica (Poaceae) and Cross-Amplification in Related Species. Int J Mol Sci 12(11):Article 11. https://doi.org/10.3390/ijms12117835

Lohani UC, Pandey JP, Shahi NC (2012) Effect of degree of polishing on milling characteristics and proximate compositions of barnyard millet (*Echinochloa frumentacea*). Food Bioprocess Technol 5(3):1113–1119. https://doi.org/10.1007/s11947-011-0518-6

Mandelbaum CI, Barbeau WE, Hilu KW (1995) Protein, calcium, and iron content of wild and cultivated species of *Echinochloa*. Plant Foods Hum Nutr 47(2):101–108. https://doi.org/10.1007/BF01089258

Manimekalai M, Dhasarathan M, Karthikeyan A, Murukarthick J, Renganathan VG, Thangaraj K, Vellaikumar S, Vanniarajan C, Senthil N (2018) Genetic diversity in the barnyard millet (*Echinochola frumentacea*) germplasms revealed by morphological traits and simple sequence repeat markers. Curr Plant Biol 14:71–78. https://doi.org/10.1016/j.cpb.2018.09.006

Maun MA, Barrett SCH (1986) The biology of canadian weeds.: 77. *Echinochloa crus-galli* (L.) Beauv. Can J Plant Sci 66(3):739–759. https://doi.org/10.4141/cjps86-093

Mehra HS, Joshi HC, Chikara J (1985) Induced mutations in Japanese millet. Indian J Agric Sci 55:294–295

Mehta H, Tyagi PC, Mohapatra KP (2005) Genetic diversity in Barnyard millet (Echinochloa frumentacea Roxb.) Indian. Indian J Genet 65:293–295

McFarlane JA (1969) Special study of stored products problems in Ethiopia. Preliminary report and recommendations. IAR. Addis Ababa, Ethiopia. 30 pp + xxiv

Michael P (1994) Distribution and taxonomy of *Echinochloa* – a world view with a key to the species occurring in China. Proceedings of the 5th Weed Science Conference of China (Kunming, China, 25–30 November 1994). Weed Science Society of China, Kunming, China, pp 161–166

Michael P (2001) The taxonomy and distribution of *Echinochloa* species (barnyard grasses) in the Asian-Pacific region, with a review of pertinent biological studies. In: Proceedings of the 18th APWSS Conference (Beijing, China, 28 May–2 June 2001). Standard Press of China, Beijing, pp 57–66

Mitich LW (1990) Barnyard grass. Weed Technol 4(4):918–920

Moharil MP, Gawai D, Dikshit N, Dudhare MS, Jadhav PV (2016) Assessment of genetic diversity in Indian Barnyard millet (*Echinochloa* spp. Complex) using morphological and molecular markers. J Appl Nat Sci 8(3):Article 3. https://doi.org/10.31018/jans.v8i3.1016

Muldoon DK, Pearson CJ, Wheeler JL (1982) The effect of temperature on growth and development of *Echinochloa* millets. Ann Bot 50(5):665–672. https://doi.org/10.1093/oxfordjournals.aob.a086408

Mundkur BB, Thirumalachar MJ (1952) Ustilaginales of India. Commonwealth Mycological Institute

Murukarthick J, Manimekalai M, Karthikeyan A, Perumal S, Dhasarathan M, Kandasamy T et al (2019) Transcriptomes of Indian barnyard millet and barnyardgrass reveal putative genes involved in drought adaptation and micronutrient accumulation. Acta Physiol Plant 41:66. https://doi.org/10.1007/s11738-019-2855-4

Muza FR, Lee DJ, Andrews DJ, Gupta SC (1995) Mitochondrial DNA variation in finger millet (*Eleusine coracana* L. Gaertn). Euphytica 81(2):199–205. https://doi.org/10.1007/BF00025434

Nagaraja A, Jagadish P, Ashok EG, Gowda KK (2007a) Avoidance of finger millet blast by ideal sowing time and assessment of varietal performance under rain fed production situations in Karnataka. Journal of Mycopathological Research 45(2):237–240

Nagaraja A, Kumar J, Jain AK, Narasimhudu Y, Raghuchander T, Kumar B, Hanumanthe Gowda B (2007b) Compendium of small millets diseases. Project coordination cell, All India Coordinated Small Millets Improvement Project, UAS, Bangalore, p 80

Nagaraja A, Kumar B, Jain AK, Patro TS, Nageswar Rao TG (2016) Diseases of small millets. In: Diseases of field crops and their management. Indian Phytopathological Society, New Delhi, pp. 295–371

Napper DM (1965) Grasses of Tanganyika with keys for identification

Nazni P, Karuna TD (2016) Development and Quality Evaluation of Barnyard Millet Bran Incorporated Rusk and Muffin. J Food Ind Microbiol 02. https://doi.org/10.4172/2572-4134.1000116

Nene Y (2005) Historical perspectives of grain legumes in food security and sustainable agriculture with focus on South Asia

Nesbitt M (2012) The cultural history of plants. Routledge

Nirmalakumari A (2009) Phenotypic analysis of anther and pollen in diversified genotype of barnyard millet (*Echinochloa frumentaceae*) floral characters. The IUP Journal of Genetics & Evolution 2(3):12–16

Nisha N, Saravanakumar M (2019) Development of reciprocating cutter bar test rig for measurement of cutting force of finger millets. Int J Curr Microbiol App Sci 8:2441–2447. https://doi.org/10.20546/ijcmas.2019.804.284

Nozawa S, Nakai H, Sato Y (2004) Characterization of microsatellite and ISSR polymorphisms among *Echinochloa* (L.) Beauv. Spp. Breed Res 6(4):187–193. In Japan

Nozawa S, Takahashi M, Nakai H, Sato Y-I (2006) Difference in SSR variations between Japanese Barnyard Millet (*Echinochloa esculenta*) and its wild relative *E. crus-galli*. Breed Sci 56(4):335–340. https://doi.org/10.1270/jsbbs.56.335

Nwanze KF, Harris KM (1992) Insect pests of pearl millet in West Africa. Rev Agric Entomol 80:1132–1185

Obara T (1936) On the nutritive value of Japanese barnyard millet. J Agr Chem Soc Japan 12:1049–1058

Osuna MD, Okada M, Ahmad R, Fischer AJ, Jasieniuk M (2011) Genetic diversity and spread of Thiobencarb resistant early watergrass (*Echinochloa oryzoides*) in California. Weed Sci 59(2):195–201. https://doi.org/10.1614/WS-D-10-00124.1

Padulosi S, Mal B, Ravi SB, Gowda J, Gowda K, Shanthakumar G, Yenagi N, Dutta M (2009) Food security and climate change: role of plant genetic resources of minor millets. Indian J Plant Genet Resour 22(01):1–16

Panwar P, Dubey A, Verma AK (2016) Evaluation of nutraceutical and antinutritional properties in barnyard and finger millet varieties grown in Himalayan region. J Food Sci Technol 53(6):2779–2787. https://doi.org/10.1007/s13197-016-2250-8

Poongodi T, Vijayakumar JBM, Nazni P, Rajeshwari M (2003) Value addition for minor millets and its glycemic load among normal and type 2 diabetic subjects. In: Paper presented at the 39th national Nutritional Conference of NSI, Hyderabad

Potvin C (1991) Temperature-induced variation in reproductive success: Field and control experiments with the C4 grass *Echinochloa crus-galli*. Can J Bot 69(7):1577–1582. https://doi.org/10.1139/b91-201

Potvin C (1986) Biomass allocation and phenological differences among southern and northern populations of the C4 grass Echinochloa crus-galli. J Ecol 74:915–923

Prabha D, Negi Y, Khanna V (2012) Assessment of genetic diversity of barnyard millet accessions using molecular markers. Indian J Plant Genet Resourc 25(02):174–179

Prabhakar PC, Ganiger BB, Bhat S, Nandini C, Tippeswamy VK, Manjunath HA (2017). Improved production technology for barnyard millet. Project Coordinating Unit ICAR-AICRP on Small Millets GKVK, Bengaluru-560065, 16

Rajagopal D, Musthak Ali TM (1992) Outbreak of cutworms under dryland conditions. Mysore J Agric Sci 26(1):46–50

Rajeswari N, Priyadharshini VP (2021) Evaluation of nutritional and nutraceutical content of polished and unpolished barnyard millet-an analytical study. Curr Res Nutrit Food Sci 9(3):1067

Rana S, Lemoine E, Granger JP, Karumanchi SA (2019) Preeclampsia: pathophysiology, challenges, and perspectives. Circ Res 124(7):1094–1112

Rao BD, Bhaskarachary K, Christina GDA, Devi GS, Tonapi VA (2017) Nutritional and health benefits of millets. ICAR- Indian Institute of Millets Research (IIMR), Hyderabad, India

Rawat JM, Pandey S, Debbarma P, Rawat B (2021) Preparation of alcoholic beverages by tribal communities in the Indian Himalayan region: a review on traditional and ethnic consideration. Front Sustain Food Syst 5. https://doi.org/10.3389/fsufs.2021.672411

Renganathan V, Vanniarajan SIC (2015) Combining ability for quantitative traits in Barnyard millet (*Echinochloa frumentaceae* (Roxb.) Link). Electronic Journal of Plant Breed 6(2):389–394

Renganathan VG, Vanniarajan C, Senthil N et al (2017) Genetic variability, correlation and path analysis for yield and nutritional traits in barnyard millet (*Echinochloa frumentacea* (Roxb.) Link). Int J Curr Microbiol App Sci 6:2211–2220

Renganathan VG, Vanniarajan C, Ramalingam J (2019) Genetic analysis and identification of molecular markers linked to the anthocyanin pigmentation in barnyard millet [*Echinochloa frumentacea* Roxb (Link)]. In Proceedings of the Neglected and Underutilized crop species for Food, Nutrition, Energy and Environment. NIPGR, New Delhi, p 43

Renganathan VG, Vanniarajan C, Karthikeyan A, Ramalingam J (2020) Barnyard millet for food and nutritional security: current status and future research direction. Front Genet 11. https://doi.org/10.3389/fgene.2020.00500

Ruiz-Santaella JP, Bastida F, Franco AR, De Prado R (2006) Morphological and molecular characterization of different *Echinochloa* spp. and *Oryza sativa* populations. J Agric Food Chem 54(4):1166–1172. https://doi.org/10.1021/jf0520746

Rutledge J, Talbert RE, Sneller CH (2000) RAPD analysis of genetic variation among propanil-resistant and -susceptible *Echinochloa crus-galli* populations in Arkansas. Weed Sci 48(6):669–674. https://doi.org/10.1614/0043-1745(2000)048

Saleh ASM, Zhang Q, Chen J, Shen Q (2013) Millet grains: nutritional quality, processing, and potential health benefits. Compr Rev Food Sci Food Saf 12(3):281–295. https://doi.org/10.1111/1541-4337.12012

Salunke P, Chavan U, Kotecha P (2019) Studies on nutritional quality of barnyard millet cookies. Int J Chem Sci 7:651–657

Satyarthi, K., H. P. State Centre on Climate Change, A. N., Project (India), U.-G.-M. A., & Himachal Pradesh State Biodiversity Board, C. H. (2018). Traditional foodgrain crops of Himachal Pradesh. https://southasiacommons.net/artifacts/5036621/traditional-foodgrain-crops-of-himachal-pradesh/5802253/

Seetharam A, Riley KW, Harinarayana G (1990) Small millets in global agriculture

Shahidi F, Chandrasekara A (2013) Millet grain phenolics and their role in disease risk reduction and health promotion: a review. J Funct Foods 5:570–581. https://doi.org/10.1016/j.jff.2013.02.004

Sharma HC, Davies JC (1988) Insects and other animal pests of millets. ICRISAT, Ptencheru, Andhra Pradesh, India. p. 33 + 28

Singh VS, Jotwani MG (1973) Efficacy of some systematic insecticides against the bajra shoot fly, Atherigona approximate Malloch. Indian J Entomol 35:130–133

Singh KP, Kundu S, Gupta HS (2003) Development of higher capacity thresher for ragi/kodo. In: Recent trend in millet processing and utilization. CCSHAU, Hissar, India, pp 109–116

Singh KP, Mishra HN, Saha S (2009) Sorption isotherms of barnyard millet grain and kernel. Food Bioprocess Technol. https://doi.org/10.1007/s11947-009-0195-x

Singh KP, Mishra HN, Saha S (2010) Moisture-dependent properties of barnyard millet grain and kernel. J Food Eng 96(4):598–606. https://doi.org/10.1016/j.jfoodeng.2009.09.007

Sood S, Khulbe RK, Kumar A, Agrawal PK, Upadhyaya HD (2015) Barnyard millet global core collection evaluation in the submontane Himalayan region of India using multivariate analysis. Crop J 3:517–525. https://doi.org/10.1016/j.cj.2015.07.005

Sood S, Joshi DC, Chandra AK, Kumar A (2019) Phenomics and genomics of finger millet: current status and future prospects. Planta 250(3):731–751. https://doi.org/10.1007/s00425-019-03159-6

Sood S, Joshi D, Pattanayak A (2020) Breeding Advancements in Barnyard Millet (pp. 391–409). https://doi.org/10.1007/978-3-030-41866-3_15

Spengler RN (2019) Fruit from the sands: the silk road origins of the foods we eat, 1st edn. University of California Press. https://doi.org/10.2307/j.ctvh1dx4s

Sundararaj DD, Thulasidas G (1979) Botany of field crops

Surekha N, Naik RS, Mythri S, Devi R (2013) Barnyard millet (*Echinochloa frumentacea* Link) cookies: development, value addition, consumer acceptability, nutritional and shelf life evaluation. IOSR J Environ Sci Toxicol Food Technol 7:1–10

Tabacchi M, Mantegazza R, Spada A, Ferrero A (2006) Morphological traits and molecular markers for classification of *Echinochloa* species from Italian rice fields. Weed Sci 54(6):1086–1093. https://doi.org/10.1614/WS-06-018R1.1

Talati GM, Upadhyay VR (1978) Status of shoot fly, Atherigona approximate (Malloch) as a pest of bajra (*Pennisetum typhoides*) crops Gujarat state. Guj Agric Univ Res J 4:30–35

Tester M, Langridge P (2010) Breeding technologies to increase crop production in a changing world. Science 327:818–822. https://doi.org/10.1126/science.1183700

Trivedi AK, Arya L, Verma SK, Tyagi RK, Hemantaranjan A (2017) Evaluation of barnyard millet diversity in central Himalayan region for environmental stress tolerance. J Agric Sci 155:1497–1507. https://doi.org/10.1017/S0021859617000545

Ugare R, Chimmad B, Naik R, Bharati P, Itagi S (2014) Glycemic index and significance of barnyard millet (*Echinochloa frumentacae*) in type II diabetics. J Food Sci Technol 51(2):392–395. https://doi.org/10.1007/s13197-011-0516-8

Vanniarajan C, Anand G, Kanchana S, Giridhari VVA, Renganathan VG (2018) A short duration high yielding culture—Barnyard millet ACM 10145. Agric Sci Dig. https://arccjournals.com/journal/agricultural-science-digest/D-4574 8

Veena Bharati B, Chimmad V, Naik RK et al (2005) Physico-chemical and nutritional studies in barnyard millet. Karnataka J Agric Sci 18(1):101–105

Vidhyasekaran P (1971) Saprophytic survival of *Helminthosporium nodolosum* and *H. tetramera* in soil. Indian Phytopathol 24:347–353

Vijayakumar PT, Mohankumar JB (2009) Formulation and characterization of millet flour blend incorporated composite flour. Int J Agric Sci 1(2):46

Vijayakumar TP, Mohankumar JB, Jaganmohan R (2009) Quality evaluation of chapati from millet flour blend incorporated composite flour. Indian J Nutr Diet 46:144–155

Viji G, Gnanamanickam SS, Levy M (2000) DNA polymorphisms of isolates of *Magnaporthe grisea* from India that are pathogenic to finger millet and rice. Mycol Res 104:161–167

Viswanath S, Seetharam A (1989) Diseases of small millets and their management in India. In: Seetharam A, Riley KW, Harinarayana G (eds) Small millets in global agriculture. Oxford & IBH Publishing Co. Pvt. Ltd., New Delhi, pp 237–253

Wallace JG, Upadhyaya HD, Vetriventhan M, Buckler ES, Tom Hash C, Ramu P (2015) The genetic makeup of a global barnyard millet germplasm collection. Plant Genome 8(1):plantgenome2014.10.0067. https://doi.org/10.3835/plantgenome2014.10.0067

Watanabe N (1970) A spodographic analysis of millet from prehistoric Japan. J Fac Sci Univ Tokyo, Sect 3:357–379.

Yabuno T (1962) Cytotaxonomic studies on the two cultivated species and the wild relatives in the genus *Echinochloa*. Cytologia 27(3):296–305

Yabuno T (1966) Biosystematic study of the genus *Echinochloa*

Yabuno T (1971) A note on barnyard millet

Yabuno T (1984) A Biosystematic study on *Echinochloa oryzoides* (Ard.) Fritsch. Cytologia 49(3):673–678. https://doi.org/10.1508/cytologia.49.673

Yabuno T (1987) Japanese barnyard millet (*Echinochloa utilis*, poaceae) in Japan. Econ Bot 41(4):484–493. https://doi.org/10.1007/BF02908141

Yabuno T (2001) Taxonomy and phylogeny of the genus *Echinochloa*. In: Yabuno T, Yamaguchi H (eds) Natural history of genus *Echinochloa*, 15—30. Zennokyo Shuppan, Tokyo. (in Japanese)

Yadav R, Yadav V (2013) Comparative performance of Indian and Japanese barnyard millet cultivars under varied fertility conditions for dual use in Indian Central Himalaya. Range Manag Agrofor 34(2):175–178

Yadav VK, Kumar B, Yadav R, Malik N, Kumar J (2010) All India coordinated small millet improvement project

Yamaguchi H, Umemoto S, Masanaga Y (1996) Studies on barnyard grasses, especially [the] non-shattering form of *Echinochloa oryzicola* Vasing., in Yunkui Plateau, China

Yamaguchi H, Utano A, Yasuda K, Yano A, Soejima A (2005) A molecular phylogeny of wild and cultivated *Echinochloa* in East Asia inferred from non-coding region sequences of trnT-L-F. Weed Biol Manag 5(4):210–218. https://doi.org/10.1111/j.1445-6664.2005.00185.x

Yasuda K, Yano A, Nakayama Y, Yamaguchi H (2002) Molecular identification of *Echinochloa oryzicola* Vasing. And E. crus-galli (L.) Beauv. Using a polymerase chain reaction–restriction fragment length polymorphism technique. Weed Biol Manag 2(1):11–17. https://doi.org/10.1046/j.1445-6664.2002.00041.x

Proso Millet (*Panicum miliaceum*): The Prodigious Millet for Diverse Uses

8

C. Nandini, Dinesh Chandra Joshi, Vetriventhan Mani, B. Nandini, B. Niranjan Kumar, and J. Meenakshi

Abstract

The ancient grain proso millet (*Panicum miliaceum* L.) is acknowledged for its tolerance to climatic fluctuations, presenting considerable potential to address food and nutritional shortages. Despite its limited use, it is highly regarded for its capacity to flourish in marginal soils, short growth duration of 65–75 days, minimal fertilizer and water needs, and intrinsic tolerance to biotic and abiotic stressors. Proso millet, traditionally farmed in Asia and Eastern Europe and now grown globally, is genetically enriched in carbohydrates, proteins, important minerals, such as iron and magnesium, and dietary fiber. Its gluten-free composition and low glycemic index render it especially advantageous for health-conscious diets

C. Nandini (✉)
Zonal Agricultural and Horticultural Research Station, Babbur Farm, Hiriyur, KSNUAHS, Shivamogga, Karnataka, India
e-mail: nandinic@uahs.edu.in

D. C. Joshi
Crop Improvement Division, Vivekananda Parvatiya Krishi Anusandhan Sansthan, Almora, Uttarakhand, India

V. Mani
International Crops Research Institute for Semi-Arid Tropics (ICRISAT), Hyderabad, India

B. Nandini
College of Horticulture, Kolar University of Horticultural Sciences, Bagalkot, Karnataka, India

B. Niranjan Kumar
Agricultural and Horticultural Research Station, Kattalagere, KSNUAHS, Shivamogga, Karnataka, India

J. Meenakshi
AICRP on Small Millets, University of Agricultural Sciences, GKVK, Bengaluru, Karnataka, India

A. Kumar et al. (eds.), *Fasting Superfoods*,
https://doi.org/10.1007/978-981-95-2762-5_8

and management of metabolic diseases, including diabetes. This chapter offers a comprehensive and interdisciplinary analysis of proso millet, including its botanical characteristics, historical origins, domestication, traditional uses, and cultural importance in different regions. It details cultivation methods, soil and climatic requirements, issues in pest and disease control, and the use of genomics, molecular breeding technologies, and value-added products. This chapter emphasizes the need to enhance research and breeding initiatives to increase productivity and incorporate it into mainstream agriculture. Through targeted initiatives, proso millet can address global challenges such as food insecurity, hunger, and climate change, positioning it as a sustainable crop for the future.

Keywords

Proso millet · Glycemic index · Climate change · Nutritional quality · Genomics

8.1 Introduction: Origin, Geographical Distribution, and Significance

A deficient supply of micronutrients results in hidden hunger for almost two billion individuals (Chaudhary et al. 2025). In comparison to millets, staple grains, including maize, rice, and wheat, possess lower mineral concentrations (Bekkering and Tian 2019). To attain sustainable agriculture, existing agricultural systems must transition toward crop diversity, prioritizing climate-resilient, underutilized, and nutrient-dense crops. Small millets, such as foxtail, barnyard, brown top, proso, guinea, kodo, and tiny millet, are diminutive grains recognized for their climatic resilience, ecological flexibility, minimal water needs, and insect resistance, rendering them essential for semiarid regions. Despite cultivation on a limited scale, they continuously provide high yields in suboptimal soils.

Proso millet (*Panicum miliaceum* L.) is a tetraploid minor millet with a basic chromosome number of $x = 9$ and a chromosomal configuration of $2n = 4x = 36$. Humans have cultivated these ancient cereals for several millennia. Several appellations have illustrated its extensive use. For instance, in China, it is referred to as broomcorn millet; however, in the United States, it is designated as hog millet. In Germany, it is known as "hersey," whereas in the Asia-Pacific region, including Korea and Japan, it is referred to as common millet. Rajput et al. (2014) designated it as French White in France. The Indian subcontinent is referred to by several names in different languages: Panivaragu in Tamil, Cheena in Bengali, Baragu in Kannada, Variga in Telugu, Vari in Marathi, Bachari-bagmu in Odia, and Cheno in Gujarati (Rajasekaran and Francis 2021). This crop substantially enhances dietary diversity, varies greatly, and has remarkable nutritional benefits (Bhat et al. 2019). Primarily used for human consumption in Asia, it is also used as an animal feed and birdseed in the United States. As a minor millet, it is cultivated in marginal soils with little input. This shallow-rooted plant flourishes in warm, arid climates because of its short growth cycle (70–90 days) and limited water and input needs. It produces seeds with an annual precipitation of 330–350 mm and has remarkable drought resilience, flourishing in semiarid and desert areas (Baltensperger 2002; Lyon et al. 2008).

Proso millet, a crucial dietary crop in ancient civilizations, was first grown 10,000 years ago. Harlan (1975) proposed that domestication occurred in both China and Europe, but Vavilov (1926) identified China as the center of proso millet diversity. Lu et al. (2009) reported that the oldest evidence of proso millet was discovered in the Yellow River valley in Cishan, China, with a chronology ranging from 10,300 to 8700 BP. Additionally, burnt grains and ceramic imprints have been discovered at other Eastern European sites, dating to 7000 cal BP, providing evidence of proso millet (Hunt et al. 2008; Zohary and Hopf 2000). These findings suggest that proso millet was either domesticated separately in Eastern Europe and Central Asia or first domesticated in China before spreading westward over the Eurasian steppe (Hunt et al. 2011; Jones 2004). Currently, it is widely cultivated in South Asia (India, Africa, Russia, the Middle East, Romania, and Turkey), Eastern Europe (Ukraine and Belarus), and East Asia (China and Nepal) (Das et al. 2019). Proso millet was brought to North America by German–Russian immigrants who settled along the eastern Atlantic coast in 1875 (Habiyaremye et al. 2017). The principal regions for proso millet cultivation in North America are the Central Great Plains, namely Colorado, Nebraska, and South Dakota (Lyon et al. 2008).

This grain is gluten-free and rich in carbohydrates and fatty acids. It contains a substantial concentration of key nutrients, including protein (12.5 g), carbohydrates (70.4 g), fat (3.1 g), dietary fiber (14.2 g), minerals (1.9 g), calcium (14 g), phosphorus (206 mg), and iron (10 g), all of which are critical constituents of a standard human diet. In India, proso millet is a crucial superfood for fasting, appreciated for its nutritional advantages, especially its high fiber content, which aids digestion and helps control blood sugar levels, making it ideal for those who fast. It also plays a role in protein biosynthesis and is abundant in essential minerals and vitamins, which boost its nutritional value. During religious events and festivals such as Navratri, meals prepared from amaranth, buckwheat, proso, foxtail, and barnyard millet are allowed on fasting days (Ankita and Seth 2025). This crop is used in several applications such as the preparation of proso grains may be prepared by cooking, steaming, or boiling. They may be savored with milk and honey at breakfast, sautéed with veggies and tofu, or integrated into salads. Fermented drinks made from proso millet in Africa and Asia may be used as bases for distilled spirits and beer (Habiyaremye et al. 2017). Whole grains of proso millet are rich in Mg and are associated with a reduced incidence of type 2 diabetes. Magnesium has advantages for individuals with atherosclerosis and cardiovascular disease, as it may reduce the risk of myocardial infarction and migraine headaches (Shobana and Malleshi 2007; Bhat et al. 2018). This chapter examines the agronomic techniques, nutritional relevance, germplasm collection, and possible applications of proso millet, highlighting how it may improve food security, address nutritional problems, and advance sustainable farming methods.

8.2 History: Traditional Uses and Cultural Significance

Proso millet has a cultivation history spanning over 10,000 years, which significantly influences the cultural, ceremonial, and dietary customs of numerous traditional communities. Its capacity to flourish in dry climates and its short growing period contribute to its early domestication in arid regions, with its origins traced back to Central Asia and Eastern Europe (Dwivedi et al. 2012). Ancient writings, such as the Huangdi Neijing, acknowledge proso millet as one of the "Five Sacred Grains," integral to the agricultural practices of early Chinese society (Lu et al. 2009).

In India, proso millet has long been cherished not only as a staple food but also as an emblem of wealth and fertility. It continues to play a role in various harvest celebrations such as Pongal in Tamil Nadu and Sankranti in Karnataka, where dishes made from millet are presented to deities as a gesture of gratitude for a bountiful harvest (Bhat et al. 2019). Among tribal groups, especially in northeastern and central India, proso millet is incorporated into life-cycle ceremonies such as births and weddings, where its consumption and symbolic significance convey themes of continuity and communal unity.

Proso millet plays a vital role in traditional culinary customs. In the Himalayan areas of Sikkim and Darjeeling, they are used to ferment beverages such as *Tongba* and *Rakshi*, which are saved during religious and social gatherings (Sharma et al. 2019a, b). In countries such as Russia, Germany, and parts of Central Asia, proso millet porridge is a customary breakfast dish. It is often accompanied by milk, fruit, and meat, demonstrating the versatility of grain across different cultural cuisines (FAO 2018). In addition to its culinary use, proso millet is valued for its health benefits. It is abundant in iron, magnesium, phosphorus, vital amino acids, and complex B vitamins. Historically, its high niacin content has helped prevent pellagra among low-income groups reliant on cereal grains (Bhat et al. 2019). Additionally, it was part of a local healing diet for individuals with celiac-like symptoms or gastrointestinal issues long before these conditions were officially recognized by the medical field (Bangar et al. 2021). Thus, proso millet is not only a resilient and nutritious crop but also a preserver of intangible cultural heritage. It uses span rituals, nutrition, sustainable farming, and traditional medicine, underscoring its integrated role in the sociocultural and agricultural systems of various communities.

8.3 Production: Cultivation Practices, Climate and Soil Requirements, and Global Production Statistics

Proso millet exhibits a short cultivation cycle, maturing within 6–12 weeks, and requires minimal water usage (Spengler et al. 2014). It requires annual rainfall of less than 600 mm (Hoffmann-Bahnsen 2004) and a daytime temperature of more than 17 °C during its growth phase (Hoffmann-Bahnsen 2004). This crop requires limited nitrogen fertilizer and demonstrates robust disease resistance (Humphrys 2005). Proso millet can be cultivated at altitudes of up to 3500 m, although it is best suited for plateau regions in the northern areas, akin to other millet varieties

(Baltensperger 1996; Matz 1986). Proso millet, classified as a C4 plant, demonstrates effective carbon fixation, particularly under severe drought conditions characterized by elevated temperatures and restricted availability of carbon dioxide and nitrogen (Habiyaremye et al. 2017).

8.3.1 Cultivation

Proso millet requires a finely tilled, clean seedbed for optimal establishment. However, it does not benefit from deep plowing because of its characteristic shallow root system. Sowing can be performed either by broadcasting or line sowing using seed drills, placing the seeds in furrows at a depth of 3–4 cm. For proso millet cultivation, a spacing of 22.5 × 10 cm is recommended. Line sowing is preferred over broadcasting because it enhances germination, reduces seed rate requirements, and facilitates intercultural operations. For broadcasting, the suggested seed rate was 15 kg/ha, whereas for line sowing it was 10 kg/ha. Since proso millet is a crop with a short growing season, it requires comparatively less fertilizer than other cereals, approximately 40–60 kg N, 30 kg P_2O_5, and 20 kg K_2O/ha. Crop grown in the rainy season do not require irrigation, but during dry spells, one irrigation during tillering is essential. Two to four irrigations are required for summer crops, based on the soil and climate.

A wide variety of weed species are frequently observed in proso millet fields, competing with crops for light, moisture, and nutrients, particularly in the early stages of growth. Among the grassy weed flora commonly encountered *Echinochloa crus-galli* (commonly referred to as sawan), are *Echinochloa colona*, *Dactyloctenium aegyptium* (makra), *Cynodon dactylon* (doob), *Eleusine indica*, *Phragmites karka* (narkul), *Sorghum halepense* (banchari), *Setaria glauca* (banra), and *Cyperus rotundus* (motha). In addition, several broad-leaved weeds, such as *Commelina benghalensis* (kankoua), *Phyllanthus niruri* (hulhul), *Celosia argentea* (chilimil), *Amaranthus viridis* (chaulai), and *Solanum nigrum* (makoi), are frequently observed (Chapke et al. 2020). Effective weed management during the initial growth phase is critical for minimizing crop-weed competition and ensuring optimal plant establishment. Manual weeding during the early crop growth period plays an important role in maintaining crop vigor and enhancing the yield potential in millet cultivation systems.

It typically reaches physiological maturity within 65–75 days after sowing, depending on the cultivar and environmental conditions. Harvesting should be performed carefully to minimize seed loss due to shattering, which commonly occurs in the upper portion of the panicles, where seeds mature earlier than those in the lower parts. The optimal harvest time is when approximately two-thirds of the grains across the panicles reach maturity. Delaying harvest can cause substantial yield losses because of the natural seed-shattering habit.

Harvested crops are commonly threshed using traditional methods, either manually or by employing draught animals, depending on local practices and available resources. Currently, combined harvesters are used for harvesting and trashing purposes. Under optimal agronomic management and improved cultivation practices, proso millet can achieve a grain yield of approximately 20–23 quintals per hectare (q/ha) and straw yield of 50–60 q/ha under irrigated conditions. In contrast, under rainfed conditions, yields are comparatively lower, with grain production ranging between 10 and 15 q/ha and fresh straw yield between 30 and 40 q/ha. These figures underscore the importance of adopting improved packages of practices and providing adequate water to enhance productivity (Chapke et al. 2020).

8.3.2 Climate

It is cultivated during the *kharif* and summer seasons in various agroclimatic regions of India. This species is highly adapted to drought-prone environments and demonstrates significant resilience under conditions of limited and erratic precipitation. Furthermore, it exhibited a moderate tolerance to transient water stagnation. Owing to its harsh nature and rapid phenological development, proso millet completes its life cycle within a short duration, making it an ideal crop for cultivation in marginal, resource-constrained ecosystems (Madhavan et al. 2024).

8.3.3 Soil Requirements

Proso millet demonstrates adaptability to a broad spectrum of soil types, ranging from fertile to relatively low-fertility soils with textures extending from sandy loam to clay-rich black cotton soils. However, its cultivation is not recommended for coarse sandy soils because of poor water retention and nutrient availability. Ideal growth conditions are found in sandy loam soils or well-drained soils that are free of stones and gravel, exhibit a good water-holding capacity, and possess high levels of organic matter. These soil characteristics support robust plant development, improved yield potential, and enhanced resilience under various environmental conditions (Madhavan et al. 2024).

8.3.4 Global Production Scenario

Globally, the area and production of millets are categorized under the single-crop label "millet." India is the leading millet producer of millet worldwide, accounting for 80% of Asia's millet production and 20% of the world's millet production. In 2021, India was responsible for 41% of the global millet output, whereas Niger and China contributed approximately 12% and 8%, respectively. India produced approximately 17.5 million tons of millets in the 2023–24 period, and millets are cultivated over an area of approximately 13.8 million hectares (Anon 2023). In India, Tamil Nadu boasts of the highest millet yield of 769 kg/ha, with Karnataka at 664 kg/ha. In contrast, other states that cultivate millet report yields below 450 kg/ha. Gujarat and Chhattisgarh recorded the lowest productivity, between 165 and 186 kg/ha, respectively (Baruah 2024). Proso millet is grown in approximately 5–ten million acres worldwide. The major producers include India, China, Russia, Nepal, the United States, Romania, Belarus, Turkey, and Ukraine. Comprehensive crop-specific national statistics for proso millet are limited.

8.4 Protection: Major Pests, Diseases, and Management Strategies

Proso millet is naturally resilient to numerous pests and diseases owing to its hardy nature and adaptability. However, recent shifts in climatic patterns, such as rising temperature, irregular rainfall, and increased humidity, have altered pest dynamics and pathogen virulence. This crop is susceptible to several diseases, including sheath rot/banded blight, head smut, grain smut, rust, and leaf spot. *Rhizoctonia solani* Kuhn, the pathogen responsible for sheath rot, was first identified in 1991 at the College of Agriculture in Trivandrum (Rajagopalan et al. 1992). The initial symptoms during the tillering stage include grayish-green lesions on the leaf sheath, which later become grayish-white with brown margins and progress upward, leading to leaf blight. Head smut caused by *Sporisorium destriata* is prevalent in Europe and Asia. The entire inflorescence is affected, turning it into a sorus covered by a grayish-white membrane that breaks open to release dark-brown spores. Carboxin and Benomy are fungicidal seed treatments that can effectively control this disease (Sharma and Sugha 1991). Grain smut, also known as covered or kernel smut, is caused by *Sphacelotheca sorghi*. This disease affects grains and transforms them into white-grayish sacs, called smut sori. These sori, which are slightly oval in shape, contain black powder composed of chlamydospores that persist through the contaminated seeds. Crop rotation for 2–3 years, early collection and burning of diseased ears, and seed treatment are examples of management techniques. Another prevalent illness that has been documented in India, the Philippines, and Ceylon is rust, caused by *Uromyces linearis* Berk. On the upper surface of the leaves, it appears as a large number of tiny, narrow, brown pustules arranged in straight rows. The disease has physiological specialization, meaning that strains from different host plants cannot infect one another, causing little harm to crops. *Bipolaris panici-miliacei* causes leaf spot, a seed-transmitted disease that can result in seed rotting, coleoptile spots, and seedling blight (Lee-DuHyung 1997). The pathogen produces conidia and dark olivaceous brown conidiophores. Growing-resistant varieties such as RAUM-7, is recommended for management. Udbatta disease, caused by *Ephelis oryzae* Syd., transforms the affected panicles into a compact agarbatti-like shape. Similar to other diseases, such as bacterial stripes, first observed in 1971, are characterized by narrow, brown, water-soaked streaks on leaves, sheaths, and culms. The causal bacterium, *Pseudomonas avenae*, is seed-transmitted. Nematode infection in proso millet seeds is the least affected and has been identified as *Aphelenchoides besseyi*, which can be eradicated by presoaking in hydrogen peroxide followed by hot water treatment or exposure to high temperatures (Gokte et al. 1922).

The proso millet shoot fly *Atherigona pulla* poses a major threat to proso millet crops, potentially leading to severe yield reductions. This pest can cause losses of up to 80%, or even complete crop failure (Jagadish et al. 1995). The larvae of the fly consume the central growing shoots, resulting in "dead heart" symptoms where the central leaf withers. Current management strategies during the rainy season include early planting, increased seed rates, removal and destruction of seedlings with dead hearts, and application of carbofuran (furadan) 3G granules at 20 kg/ha in furrows before sowing. Although using granular insecticides offers effective control, it is not cost-effective for most small-scale farmers and its success depends on adequate soil moisture. Traditional methods for controlling shoot flies are neither practical nor affordable for small and marginal farmers. Because the pest is confined to the seedling stage, many farmers may find it challenging to adhere to timely spray schedules, primarily because of the low economic returns from these crops (Ravulapenta Sathish et al. 2017).

8.5 Improvement (Genetics and Genomics): Breeding Efforts, Genetic Diversity, and Recent Advancements

8.5.1 Breeding

This crop, characterized by a chromosomal count of $2n = 36$, is primarily self-pollinated, although natural cross-pollination can exceed 10% (Popov 1946). In affluent nations, it is fed to birds, used as animal feed, and is considered as food in a few regions of Asia (Rajput et al. 2014). Genetic improvement programs predominantly focus on trait improvement, such as high-yielding genotypes, resistance to lodging, resistance to shattering, earliness, and waxiness. This crop is autogamous, rendering breeding methods, including mass selection, pureline selection, pedigree breeding, and mutation breeding appropriate for its enhancement.

The inflorescence of proso millet exhibits a drooping type characterized by multiple open or compact spikelets (Hulse et al. 1980; Nandini et al. 2025) (Figs. 8.1 and 8.2). A pair of lemmas and glumes are present in a single spikelet, with the inner glumes matching the spikelet length and the outer glumes being shorter. The upper

Fig. 8.1 Drooping panicles of proso millet

Fig. 8.2 Proso millet crop showing two types of panicles: (**a**) open type and (**b**) compact type

Fig. 8.3 Blooming in proso millet

lemma was fertile, whereas the lower lemma was sterile. The palea of fertile florets are prominent, whereas those of sterile florets are reduced (Seetharam et al. 2003). Three stamens were present in a single floret, exhibiting a color range from dark brown to tan. The ovary exhibits plumose stigmas and a bifid morphology (Nanda and Agrawal 2008). The panicle exhibited a downward blooming pattern (Sundararaj and Thulasidas 1976) (Fig. 8.3). The anthesis period occurs from 10:00 AM to noon and lasts for 12–15 days (Jayaraman et al. 1997). The stigma receptivity was aligned with the timing of pollen shedding. Anthers retain stickiness until they release minute pollen following the opening of the florets (Nelson 1984). Spiklets should be opened for 10–15 min (Zhuravel 1956). Flowering is favored by bright sunlight, low humidity, and elevated temperature. The seeds exhibited a color variation that included dark gray, black, and golden yellow (Fig. 8.4).

Proso millet breeding employs various crossing techniques, including hot water treatment, contact method, pollination after manual emasculation, USSR method, and SMUASB method, which positions male and female plants in proximity. The manual emasculation technique requires the removal of anthers from female flowers prior to pollination. The hot water treatment method involves immersing the

Fig. 8.4 Variation in seed color of proso millet: (**a**) cream colon and (**b**) black color

panicles in warm water to emasculate the flowers, whereas the Union of Soviet Socialist Republics (USSR) method employs mechanical techniques to open the florets and extract anthers. The Small Millets University of Agricultural Sciences, Bengaluru (SMUASB) method employs ice-cold water to open florets naturally, followed by emasculation and pollination, resulting in a more efficient crossing technique with minimal flower damage (Nandini et al. 2019; Nagaraja et al. 2023; Nandini et al. 2025). Despite these methods, most released varieties have been developed using pure lines or mass-selection techniques (Nandini et al. 2021; Nagaraja et al. 2024) (Fig. 8.5). In India, 63% of the released proso millet varieties are selected from local cultivars and landraces (Vetriventhan et al. 2020). The improved cultivars released for proso millet cultivation are listed in Table 8.1.

8.5.2 Genetic Diversity

The success of breeding initiatives depends on the diversity of the germplasm. Establishing comprehensive germplasm collections for cultivated crops is crucial, especially for under-researched crops, such as proso millet (Upadhyaya et al. 2014) (Table 8.2). A total of 29,308 proso millet accessions have been conserved globally. The largest collection, comprising over 8778 accessions, is held by the N.I. Vavilov All-Russian Scientific Research Institute of Plant Industry (VIR) in Russia, followed by the Institute of Crop Science, Chinese Academy of Agricultural Sciences (ICS-CAAS), which is the second-largest collection of 8451 accessions. The United States, India, and Ukraine have substantial additional repositories. The National Bureau of Plant Genetic Resources (NBPGR) and the All-India Coordinated Research Project on Small Millets (AICRP-Small Millets) oversee the maintenance of 2767 accessions in India. Currently, the International Crop Research Institute for Semi-Arid Tropics (ICRISAT) maintains 849 accessions. NBPGR manages 998 proso millet accessions, whereas AICRP manages 920 accessions.

Fig. 8.5 Field view of proso millet varieties: (**a**) GPUP 25 and (**b**) GPUP 28

The core and minicore collections represent small and diverse subsets derived from extensive germplasm collections. They are essential for crop improvement and offer a wide range of genetic resources that underpin breeding programs. These collections contribute to the preservation of agricultural biodiversity by identifying and protecting unique germplasms. Researchers can enhance the efficiency of breeding programs and streamline genetic diversity studies by selecting core and minicore

Table 8.1 Improved little millet varieties released for diverse agroecological zones of the country (India)

Sl. no.	Name of variety	Pedigree	Maturity (days)	Avg. yield (q/ha)	Area of adaptation
1	TNAU 145	PV 1454 × TNAU 96	70–72	18–20	Tamil Nadu
2	CO(PV) 5 (TNAU 143)	PV 1403 × GPUP 21	70–75	23–25	National
3	TNAU 151	TNAU 96 × PV 1673	72–75	18–20	National
4	TNAU 164	TNAU 137 × CO 4	70–75	18–20	National
5	PratapCheena-1 (PR-18)	Pure line selection	65–70	15–17	National
6	PRC 1	Selection from GPMS 519	70–75	10–12	Uttarakhand hills
7	TNAU 202	PV 1453 × GPUP 16	70–75	18–20	National
8	TNPm-230	TNAU-164 × IPM-19	70–75	21–23	National
9	DHPM-2769	Selection from IPM-2769	70–72	Grain yield: 24.6 q/ha Fodder yield: 4.16 t/ha	Recommended for cultivation in agro-climatic Zone-3 and −8 of Karnataka state
10	PMV-442	GPMS 109 × GPMS 908	70–75	14–16	Andra Pradesh, Karnataka, Madhya Pradesh, Tamil Nadu, Bihar, Telangana, Puducherry

subsets from larger collections (Backiyalakshmi et al. 2021). The representative sets could be evaluated across various locations over multiple years, facilitating the identification of promising germplasms with specific traits for crop improvement. The ICRISAT has 849 proso millet accessions, which are categorized into five distinct groups based on race. Classification was achieved via clustering of 20 morpho-agronomic traits using Ward's method. Approximately 10% of each of the 101 clusters or at least one accession were randomly selected to establish a core collection of 106 germplasms (Upadhyaya et al. 2011) (Table 8.3).

Studies on germplasm characterization in proso millet have yielded significant insights (Table 8.4). Graybosch and Baltensperger (2009) identified that the characteristics of waxy endosperm starch are regulated by duplicate recessive alleles located at two separate loci in 650 accessions. Dikshit and Sivaraj (2013) identified accessions that exhibited elevated protein levels and a range of morpho-agronomic traits. Elangovan et al. (2023) assessed 634 germplasm accessions at the Indian

Table 8.2 The current status of pros millet germplasm resources preserved in various ex situ gene banks around the globe

		Germplasm accessions		
Country	Institute	Cultivated	Wild	Total
Australia	Australian Tropical Crops & Forages Genetic Resources Centre (ATCFC)	228		228
Bangladesh	Plant Genetic Resources Centre, BARI (PGRC, BARI)	198		198
Bulgaria	Institute for Plant Genetic Resources "K. Malkov" (IPGR)	489		489
China	Institute of Crop Science, Chinese Academy of Agricultural Sciences (ICS-CAAS)	8451		8451
Czech Republic	Genebank Department, Division of Genetics and Plant Breeding, Research Institute of Crop Production (RICP)	171		171
Germany	Genebank, Leibniz Institute of Plant Genetics and Crop Plant Research (IPK)	165	1	166
Hungary	Institute for Agrobotany (RCA)	243	1	244
	Station of Ujmajor, Research Institute for Vegetable Crops (UJMAJOR)	50		50
India	AICRP on Small Millets (AICRP-Small Millets)	920		920
	International Crop Research Institute for Semi-Arid Tropics (ICRISAT)	849		849
	National Bureau of Plant Genetic Resources (NBPGR)	994	4	998
Japan	Department of Genetic Resources I, National Institute of Agrobiological Sciences (NIAS)	516		516
Mexico	Estación de Iguala, Instituto Nacional de Investigaciones Agrícolas (INIA-Iguala)	400		400
Poland	Botanical Garden of Plant Breeding and Acclimatization Institute (BYDG)	354		354
	Plant Breeding and Acclimatization Institute (IHAR)	359		359
Russian Federation	N.I. Vavilov All-Russian Scientific Research Institute of Plant Industry (VIR)	8778		8778
Ukraine	Institute of Plant Production n.a. V.Y. Yurjev of UAAS (IR)	1046		1046
	Ustymivka Experimental Station of Plant Production (UDS)	3975	1	3976
The United States	North Central Regional Plant Introduction Station, USDA-ARS, NCRPIS (NC7)	717	4	721

Institute of Millet Research, and identified donors for essential agronomic traits. Vetriventhan and Upadhyaya (2018) conducted an analysis of 200 accessions from 30 countries and identified favorable traits, including high grain yield, large seed size, and improved nutrient levels. Wang et al. (2007) evaluated accessions at the National Centre for Crop Germplasm Conservation in Beijing for morpho-agronomic traits, nutritional content, and disease resistance, identifying potential

Table 8.3 Representative subsets (core collections) of proso millet for the diverse application

Crop	Germplasm subsets	No. of traits used	No. accessions in base collection	No. of accessions in subset	Reference
Proso millet	Core collection	20	833	106	Upadhyaya et al. (2011)

trait donors. Numerous studies have demonstrated considerable variations in morphological and yield-related traits (Natarajan et al. 1978; Manoharan and Sivasubramanian 1982; Hawlader 1991; Prasad et al. 1995; Panwar and Kapila 1992; Reddy et al. 2007; Salini et al. 2010; Sasamala et al. 2011). These studies collectively demonstrate the genetic diversity within the proso millet germplasm and identify potential sources for crop improvement programs.

8.5.3 Recent Advancements

Recent progress in molecular markers for proso millet has greatly enhanced our understanding of the diversity and potential for improvement of this essential crop. Because of limited resources, SSR markers in this crop were examined utilizing related crop species genic information. Hu et al. (2009) conducted research on the genetic diversity of Chinese proso millet using SSR markers of oats, barley, rice, and wheat. Cho et al. (2010) identified 25 polymorphic simple sequence markers from the SSR-enriched library. Hunt et al. (2011) analyzed the distribution of proso millet from the site of domestication to different regions of the world using 16 SSR crop-specific markers. SSR markers of switchgrass were used to analyze the genetic diversity of proso millet by Rajput et al. (2014). The genetic diversity of 88 proso millet accessions was studied by Liu et al. (2016), using 500 SSR primers. The Yanshu-5 cultivar was utilized to develop 8139 EST-SSR-based markers through transcript sequencing by Jiang et al. (2018). Although SNP analysis relies on molecular studies, research on these markers is scarce. The first genetic map was constructed by involving 93 RILs, and eight QTLs were identified for economically important traits. At the waxy locus, six SNPs were identified from 132 accessions by Wang et al. (2018).

In an extensive genetic study of kodo millet, little millet, and proso millet, Johnson et al. (2019) found 1882 SNPs in proso millet, which accurately reflected its population structure. Research on chromosomal assembly and sequencing by Shi et al. (2019) and Zou et al. (2019) demonstrated the potential for gene mining and GWAS applications. Association analysis using 143 AFLP markers identified two markers each for seed yield and fodder yield, with stable and significant associations observed across all environments under saline conditions (Yazdizadeh et al. 2020). Boukail et al. (2021) carried out research on a worldwide collection of 88 proso millet varieties and landraces to examine the seed traits of proso millet and to identify links between genetic markers and phenotypes. They sequenced DNA fragments associated with restriction sites, generating 494 million reads and 2412

Table 8.4 Trait specific accessions of proso millet

Proso millet		
Trait	Germplasm sources	References
Grain yield and large seed size	IPm 9, IPm 390, IPm 361, IPm 366, IPm 673, IPm 2620, IPm 2621, IPm 2783, IPm 2784, IPm 2802, IPm 2824, GR 658, TNPm 471, TNPm 534	Elangovan et al. (2023); Vetriventhan and Upadhyaya (2018); Calamai et al. (2020)
Fe, Zn, Ca, and protein	IPm 2005, IPm 2037, IPm 2053, IPm 2069, IPm 2076, IPm 2077, IPm 2278, IPm 2537	Vetriventhan and Upadhyaya (2018)
Early flowering	TNPM 471, TNPM 615, TNPM 638, TNPM 643, TNPM 669, 336, 347, 348, IC0483576, IC0483602, IC0483867	Elangovan et al. (2023)
Dwarf	TNPm 347, TNPm 348, TNPm 643, TNPm 669, IC0483913	Elangovan et al. (2023)
Basal tillers	IC048165, ICo483718, ICo483859, IC0484177	Elangovan et al. (2023)
1000-grain weight	TNPm 336, TNPm 348, TNPm 471, IC0483602, IC 0484177	Elangovan et al. (2023); Sharma et al. (1993); Zolotukhin et al. (1998); Konstantinov et al. (1986); Ilyin et al. (1993); Konstantinov et al. (1991); Krasavin and Usmanova (1988)
Smut	Kh86, Gorlinka, Barnaulskoye 80, K8763, Saratovskoye 2, Orenburgskoe 9, Saratovskoye 3, Saratovskoye 6, Veselepodolyankoye 632, MS1316, Khar'kovskoe 86, "II" Inovskoe	Elangovan et al. (2023); Sharma et al. (1993); Zolotukhin et al. (1998); Konstantinov et al. (1986); Ilyin et al. (1993); Konstantinov et al. (1991); Krasavin and Usmanova (1988)
Banded leaf and sheath blight	RAUM 8TNAU 137 and GPUP 22	Jain and Tiwari (2013)
Melanosis	K7606, K8740, K8773, K8789, K8790, UNIIZ670, Solnechnoe, Orenburgskoe 9, Krasnoe Toidenskoe 215 (KT215), Khar'kovskoe 57	Konstantinov et al. (1989); Konstantinov et al. (1991); Krasavin and Usmanova (1988)
Shoot fly	RAUm # 1, 2, 3, GPMS # 101, 102, 105, 108, 112, 114, 115, 117, 122, 123, 124, 125, 126, 135, 136, 138, 148, 152, 153, 155, 156, 157, 159, 164, MS 1307, MS1316, MS1437, MS1595, MS4872, PM 29–1, Co1, BR 6	Murthi and Harinarayana (1986)
Rice moth	TNAU 151	Shailaja et al. (2009)
Fall army warm	PI 176653	Wilson and Courteau (1984)

high-quality SNPs. These SNPs were employed in a GWAS to evaluate diversity within the collection, leading to the identification of 13 marker-trait associations for important agronomic and seed traits. This suggests the presence of alleles that could be advantageous for proso millet breeding programs. $F_{2:3}$ mapping populations were employed to detect marker-trait associations for plant height using bulk segregant analysis (Liu et al. 2022c). Jiang et al. (2024) reported SSR marker-trait associations for plant height in proso millet. The proso millet diversity panel at the ICRISAT GenBank was genotyped using GBS-based SNPs to evaluate its genetic diversity and population structure by examining 5621 single nucleotide polymorphisms (SNPs) across 160 distinct accessions. The mean genetic distance among these accessions was determined to be 0.268, with the miliaceum race exhibiting the highest diversity and the ovatum race being the lowest. A total of 40 marker-trait associations (MTAs) were identified, comprising 34 MTAs related to agronomic traits and 6 associated with grain nutrients, with 20 MTAs located within genes (Vetriventhan et al. 2024). Recently, the genome sequence of proso millet has been made available (Zou et al. 2019). Researchers have successfully created and documented a high-quality genome assembly at the chromosome level, which revealed 339 microRNA genes and 55,930 protein-coding genes (Zou et al. 2019). This detailed genome sequence is anticipated to be a crucial resource for proso millet molecular breeding and genomic selection (Santra et al. 2019).

8.6 Transcriptomics, Proteomics, and Metabolomics

Few studies have been conducted on the transcriptomics of proso millet. Hou et al. (2017) were the pioneers in reporting the proso millet transcriptome for the development of SSR markers. Transcription factors (TFs), including those involved in signal transduction and stress responses, are crucial for controlling vital plant growth processes (Riaño-Pachón et al. 2007; Zhang et al. 2011). The NAC family is among the largest groups of plant-specific TFs (Yao et al. 2012), which originate from NAM, ATAF1/2, and CUC2 proteins. NAC genes are essential for drought tolerance. In proso millet, Shan et al. (2020) identified 180 NAC genes in proso millet and studied their evolution, structure, and response to drought stress. Liu et al. (2022a, b, c) conducted transcriptome analysis of two proso millet cultivars that exhibited different levels of nitrogen use efficiency. The study revealed that T184 had a smaller number of differentially expressed genes than S111. The upregulated DEGs in T184 were associated with transporter activity, whereas the downregulated DEGs in S111 were associated with photosynthesis.

Proteomes generally correspond to transcriptomes, and their levels and diversity are not always directly proportional because of posttranslational modifications (Khound and Santra 2020). Mass fingerprinting and 2-D seed protein analyses were performed in four Korean proso millet varieties. Roy et al. (2017) conducted a seed protein analysis on four Korean proso cultivars using 2-DE and mass fingerprinting. From the 1152 differentially expressed proteins, 26 were identified using MALDI-TOF/MS. In all cultivars, two proteins showed increased expression, 11 were downregulated, and 13 proteins were upregulated in two specific cultivars. Proteins with increased expression were associated with transcription, starch metabolism, and disease development, whereas those with decreased expression were associated with glycolysis, stress response, and signal transduction. Yuan et al. (2022) examined proso millet genotypes that were either sensitive or tolerant to salt, examining their responses to salt stress and subsequent rewatering. Under these conditions, salt-sensitive cultivars expressed 38.28% and 44.68% proteins, respectively, whereas salt-tolerant cultivars expressed 43.56% under stress and 39.31% during rewatering, indicating posttranscriptional regulation.

While "metabolomics" is the thorough study of all the metabolites inside an organism at a certain stage, "metabolome" refers to all the metabolites in a biological sample (Diola et al. 2014). Phenolic chemicals found in proso millet grains serve as reducing agents and free-radical scavengers and they have strong antioxidant properties (Shahidi and Chandrasekara 2013). In the early stages of metabolomics research on proso millet, gas chromatography-time-of-flight mass spectrometry was used to assess the grain quality (Kim et al. 2013). This investigation concentrated on three Korean varieties and discovered 48 metabolites, comprising 43 primary metabolites and five phenolic compounds. Ultra-high-performance liquid chromatography coupled with triple quadrupole mass spectrometry (UHPLC-QqQ-MS) is widely employed in metabolomic studies, incorporating both nontargeted and targeted approaches. This technique provides high sensitivity and broad coverage for the analysis of plant metabolites (Cho et al. 2016; Zhu et al. 2018). Li et al. (2021a, 2021b) applied UHPLC-QqQ-MS/MS to investigate proso millet varieties with varying bran colors, identified 672 metabolites, and compared colored grains with white grains. This research uncovered functional components linked to antioxidant activities, enhancing the understanding of metabolite composition in proso millet grains and supporting future breeding efforts.

8.7 Primary Processing: Post-harvest Handling and Processing Techniques

Proso millet possesses excellent grain quality, making it highly suitable for processing. The primary processing steps typically include dehusking, destoning, dehulling, cleaning, pulverizing, and grading. These grains can be used for the preparation of both traditional foods and novel value-added products. Proso millet grains can be prepared in various ways, such as cooking them whole, removing their outer layer, or grinding them into flour, depending on the requirement, using either traditional or modern techniques. Nevertheless, there remains a significant need to explore alternative applications and innovative uses of millets to enhance their value and utilization. When the proso millet was de-hulled, the phytin phosphorus was reduced by 12%. Secondary processing converts initially processed materials into consumable food products, such as ready-to-eat (RTE) and ready-to-cook (RTC) items, which improves convenience and decreases the time required for cooking. Traditional methods, such as popping and flaking, along with modern techniques, such as roller drying and extrusion cooking, can be effectively applied to millets to develop RTE products and promote their wider use. Proso millet is currently utilized in the production of extruded flakes; however, further research is required to develop a wider range of bakery and snack products from proso millet, as has been achieved with other millets (Chapke et al. 2020).

8.8 Pre- and Postharvest Technology: Storage, Drying, and Transportation

Generally, the crop is ready to harvest between 65 and 75 days from the date of sowing. Maturity is not uniform throughout the panicle (Matz 1986) because of its basipetal flowering nature. When more than two-thirds of the seeds in the ear mature, it indicates that the time for harvest arrives. Delaying harvest until full maturity at the field level may result in yield loss due to shattering. Threshing of ear heads is conducted either manually or through trampling by bullocks, followed by winnowing to further cleaning of the harvested seed. Threshing may be postponed until the seed moisture content reaches 13% or lower (Baltensperger et al. 1995). The outer layer of the proso millet, known as the husk, must be removed for consumption. The removal of husks, referred to as dehulling, was conducted using machines. The resulting product is known as hulled proso millet, which is prepared for use. Following dehulling, grading may be implemented to enhance the visual appearance. The grading of produce is conducted according to its size and quality to maintain the desired standards. Before packaging the produce with appropriate materials, it is essential to ensure that the seed is adequately dried, maintaining a moisture content between 11% and 13% for long-term storage, thereby preserving the storage quality of the produce. Sun-drying is the predominant method employed for drying; however, if climatic conditions are unfavorable, oven-drying can be utilized by processing in small batches. Parboiling and subsequent shade-drying improved both the milling and cooking qualities of the proso millet. The drying duration, temperature, seed moisture content, and airflow are critical factors that influence the drying process. Following adequate drying and packaging, the seed is prepared for transport to an appropriate storage location characterized by good ventilation and absence of storage pests and rodents.

8.9 Utilization Pattern: Traditional and Modern Uses in Food, Medicinal, and Industrial Applications

Proso millet has been used in Chinese, German, and Russian cuisine for thousands of years, from 8000 to 10,000 BC. It is used to make both savory and sweet meals in Russia. In savory meals, it is mixed with meat and vegetables, but in sweet preparations, it is mixed with milk and sugar. Proso millet, which is frequently combined with different vegetables such as legumes, potatoes, and squash, was eaten in China without the use of milk or sugar. In Europe, it is added to sweet meals along with honey and apples. Proso millet is ideal for bread manufacturing because of its mild taste, light hue, and gluten-free qualities, particularly for those with celiac disease. Proso millet has the potential to produce bioethanol. According to a 2012 study from the University of Nebraska-Lincoln, the fermentation efficiency of proso millet ranged from 84% to 91%, which is similar to 97% of the total efficacy for highly fermentable corn hybrids (Rose and Santra., 2013).

Proso millet is a grain that can be processed in a number of ways, including roasting and fermentation. Phytic acid levels are decreased by milling, soaking, heating, fermentation, and bread baking. The following are the different ways of utilization:

1. *Fermentation*: Fermentation is a popular technique for increasing the nutritional content and digestibility of proso millet. Microbial activity, either natural or controlled, is used to break down complicated substances and improve flavors.
2. *Roasting*: Proso millet is frequently processed via roasting, which can be performed in a variety of ways, such as dry roasting. The nutty flavor of millet is enhanced by roasting, and it can be used in dishes or as snacks. Proso millet might be ground into flour, followed by roasting to create roasted flour. There are several culinary uses for roasted millet flour, including smoothing soups and stews and creating flat breads.
3. *Cooking*: One of the most popular ways to prepare proso millet ready for eating is to cook it. A variety of recipes can be made using millet grains by boiling, steaming, or cooking them in water. This can serve as a cereal or a side dish.
4. *Baking*: Bread, cakes, and other baked items can be made from proso millet flour. Several dishes, including rice or wheat flour, can be used with other flours.
5. *Germination*: Proso millet grains must be soaked and allowed to sprout in order to germinate. Considered to be more nutrient-dense, germinated millet is frequently added to salads, used as a garnish or as a component in a variety of cuisines (Morah and Etukudo 2017).
6. *Milling*: A popular processing technique involves grinding proso millet grains into flour. According to Bora and Das (2018), millet flour can be used to produce a wide range of foods including bread, porridge, and baked items.
7. *Popping*: The grain of proso millet can be popped like popcorn. Popped millet is a tasty and healthy snack that may also be used in dishes, such as salad cereals or desserts.
8. *Extrusion*: Extrusion is a contemporary processing method used to create a variety of morning dishes and snacks made from millet. The millet mixture was then heated and compressed to produce various textures and shapes. The intended purpose of proso millet, regional preferences, and culinary traditions influence the selection of the processing method. The taste and caloric profile of the finished product can be significantly influenced by the processing method employed, and each approach delivers distinct flavors and textures.
9. *Livestock and poultry feed*: Cows, grazing and slaughtering cattle, dairy cows, lambs, pigs, and poultry can all be fed proso millet, which has nutritional value comparable to other grains used for animal feed. Therefore, whole millet seeds can be sufficiently cracked in a hammer mill equipped with a quarter-inch screen.

8.10 Value Addition and Products: Potential for Creating High-Value Products

Proso millet flour may be used as an alternative to rice flour in the manufacture of diverse snack products. Due to the absence of gluten, it should not be used as an exclusive ingredient in the formulation of baked products. An alternative is to stream it into *idly* as a batter composed of millet mixture. Other products present in the market include puffed and popped millets, extruded items, and flaked breakfast cereals. They mainly serve as supplementary and weaned foods. Proso millet is used in conjunction with other grains to attain a particular texture in gluten-free beer manufactured in the United States.

Composite flour: Proso millet cannot not be used in bread or any bakery item preparation owing to its gluten-free nature. However, it is highly suitable for mixing with any other grain flour and can be used in bakery products, snacks, and healthy drinks. Due to its high nutritional value and high dietary content, many people prefer it to be combined with a variety of flours for the preparation of *rotis*, *chapati*, *idly*, *dosa*, and so on.

Breakfast and snacks: Proso millet can be consumed at every meal, breakfast, lunch, dinner, and snacks. Since it is gluten-free, people intolerant to wheat products prefer to consume on a daily basis. Breakfasts such as *chila*, *idly*, *dosa*, *vegetable khichdi*, *mixed poori*, *chapati*, dessert such as *halwa*, *payassam*, and snacks such as *gatte*, *biscuits*, *namkins* can be prepared.

Ready to cook products: Nowadays, ready-to-cook premixes are gaining popularity due to modern lifestyles. Millet offers nutritional benefits and is helpful in combating lifestyle-related diseases. Premixes such as *upma*, *khichdi*, *dosa*, and *idly batter* are popular.

Alcoholic beverages: Some tribes, including the Tao people of Taiwan's Orchid Island, have long used millets as a vital grains to make millet beer. Using long straws, people share a fermented millet drink that is made in a large pot with hot water. Additionally, millet serves as the foundation for the traditional alcoholic beverage known as *tongba* in both Nepal and India as well as the distilled whiskey known as *rakshi* in Nepal. The fermented beverage *boza* is made from millet in Bulgaria and Romania.

One of the beverage production industries in Europe that is expanding rapidly is gluten-free products. Owing to its distinct flavor and taste, malted proso millet is the most popular gluten-free beer in Germany. A few Great Plains brewing companies in the United States have recently begun using proso millet as a grain source for beer. To produce gluten-free beer for Colorado commercial breweries, Colorado Malting Co. (Alamosa, Colo.) prepared malting millets from proso millet. Proso millet has been used to make beer in Nebraska at the Modern Monks Brewery near Lincoln (Das et al. 2019).

8.11 Marketing: Global Demand, Trade, and Market Trends in Proso Millet

Proso millet has gained global popularity because of its exceptional nutritional value and versatility in various culinary applications. This gluten-free grain, which is rich in proteins and essential amino acids, has become particularly popular in health-conscious markets across North America, Europe, and parts of Asia. In response to this growing demand, a wide range of proso millet-based products, such as flour, snacks, and beverages, have been developed to cater to diverse consumer preferences. Furthermore, grain sustainability, highlighted by its drought resistance and low water requirements, has increased its appeal among environmentally conscious consumers, thereby expanding its market share. The proso millet market worldwide is witnessing remarkable growth, propelled by rising consumer awareness of its health benefits and its gluten-free nature. Global exports have reached unprecedented levels, with the Middle East and Africa becoming prominent importers. The market is anticipated to grow significantly by 2033, driven by health-conscious individuals, sustainable eating trends, and concerns regarding food security. India is a major player in the export market, as the world's largest producer of millet. Because of its versatility, proso millet can be used for purposes other than human consumption, such as animal feed, bird seeds, and possible industrial applications. The Netherlands, the United Kingdom, Canada, and Japan are important export markets and are staple foods in African nations. Marketing campaigns highlight its nutritional benefits, appeal to health-conscious consumers, showcase its adaptability, and support sustainable practices. Trends such as the growing demand for gluten-free products, emphasis on sustainable diets, and consumers' growing nutritional awareness all influence the market. Among the millet species produced worldwide, proso millet is the most important species traded in the world market, with the United States being among the top producers (Castro 2013). The United States generally exports 15–20% of its annual millet production to over 70 countries, primarily as feed. The largest export markets for US proso millet include the Netherlands, the United Kingdom, Canada, and Japan (Powell et al., undated2). Argentina is a major competitor of the export market.

India is the world's leading producer of millet, closely following Nigeria. In 2021–22, India achieved a 5% increase in millet production compared with the previous year, demonstrating its commitment to boosting millet farming. The government aims to establish India as a global center for "Shree Anna," focusing on improving productivity and supplying high-quality, affordable raw materials for value-added millet products. India's millet exports have experienced significant growth, reaching US$ 16.902 million in 2022–23, up from US$ 12.90 million in 2019–20. Major global millet exporters include the United States, Russia, Ukraine, China, the Netherlands, France, Poland, Argentina, and India, with total exports valued at approximately US$ 221.68 million in 2023. In 2020–21, India exported millet worth US$ 27.43 million, with Nepal, UAE, and Saudi Arabia as the primary importers. Other significant markets included Libya, Tunisia, Morocco, the United Kingdom, Yemen, Oman, and Algeria. The United Nations' designation of 2023 as the International Year of Millets, led by India, has heightened the global interest in millets (Banik et al. 2024). This acknowledgment highlights the ancient nutritional benefits of millets and offers India a substantial opportunity to capitalize on its production capabilities and expand its international market presence. To maintain and enhance its competitive advantage, India needs to invest in the genetic improvement, processing, and strengthening of the millet value chain, particularly for proso millet.

8.12 Conclusion and Future Prospects

Proso millet is often underestimated despite its extraordinary capacity to thrive in a variety of climates owing to its low water requirements. This hardy cereal is particularly resilient in harsh environments such as arid regions and nutrient-poor soils. Compared to other major cereals, proso millet offer numerous advantages, although it has not been given much prominence. Nowadays, there is a growing interest in traditional crops, such as millet, as they thrive with limited resources and are crucial for future nutritional food security. Therefore, it is essential to introduce crops, such as proso millet. Recently, there has been a noticeable shift toward millets. Proso millet is an under-researched crop that requires extensive research on genetic resource utilization, the creation of variations through crossing, and the use of genomic tools to identify candidate genes that control important traits. With the advent of next-generation sequencing technologies, several advancements, such as SNP discovery for important traits and identification of SSR and SNP through transcriptomic analysis, have paved the way for proso millet genomic research. Further research should focus on the adaptability of proso millet to diverse environments, development of non-lodging and non-shattering varieties, cultivars suitable for mechanical harvesting, and high-yielding types with quality and nutritional traits in the context of climate change. Although a large number of germplasms are preserved worldwide, these accessions are rarely utilized to exploit the variability in economic traits. Therefore, it is essential to evaluate the agronomic and nutritional traits of these germplasm lines. The crop remains underutilized owing to lack of awareness, insufficient marketing infrastructure, processing challenges, and breeding difficulties. The small size of its flowers complicates crossbreeding, leading to earlier varieties being developed mainly through simple selection methods. Many new techniques for artificial hybridization have been developed. Breeders can use these methods to create variability and incorporate specific traits into existing cultivars. However, its production and yield are lower than those of the other cereal crops. This is largely because many farmers continue to cultivate local varieties despite the availability of numerous high-yielding cultivars. It is crucial to educate farmers on this issue, and the government should take the initiative to provide high-yielding varieties at subsidized rates. Integrating this millet into the public distribution system would make it more accessible to middle-income and low-income farmers. Collaboration among researchers, policymakers, and industry stakeholders is vital for accelerating ,genetic improvements in small millet. Developing nutritionally enhanced and climate-resilient varieties would enhance their role as sustainable food sources. Furthermore, the lack of processing and grading facilities limits its marketability and value addition, whereas a shortage of scientific literature restricts its inclusion in government programs and nutritional policies. To promote this nutrient-dense crop, targeted strategies, such as biofortification, value addition, and awareness campaigns are essential. In conclusion, the proso millet is not just an ancient grain, but it is a viable solution to modern agricultural, nutritional, and environmental challenges. Through focused research, innovative breeding strategies, and supportive policies, proso millet could play a crucial role in ensuring food and nutritional security, promoting sustainable agriculture, and addressing global health concerns.

Author Contributions Statement Nandini C. (N.C.) conceptualized the topic and drafted the initial version of the chapter by incorporating inputs from coauthors D.C.J., V.M., B.N., N.K.B., and M.J. N.C. and D.C.J. were responsible for editing and refining the chapter. B.N. prepared the sections on value-added products and climatology. N.K.B. and M.J. contributed by drafting the introduction and climatology. V.M. helped to enhance the chapter content.

References

Ankita, Seth U (2025) Millets in India: exploring historical significance, cultural heritage and ethnic foods. J Ethnic Food 12(2):1–16. https://doi.org/10.1186/s42779-024-00262-2

Anonymous (2023) Just Agriculture, Just Agriculture 3(5):January 2023

Backiyalakshmi C, Vetriventhan M, Deshpande S, Babu C, Allan V, Naresh D, Gupta R, Azevedo VCR (2021) Genome-wide assessment of population structure and genetic diversity of the global finger millet germplasm panel conserved at the ICRISAT genebank. Front Plant Sci 12:692463

Baltensperger D (1996) Foxtail and Prosomillet. In: Janick J (ed) Progress in new crops. ASHS Press, Alexandria, pp 182–190

Baltensperger DD (2002) Progress with proso, pearl and other millets. Trends in New Crops and New Uses:100–103

Baltensperger D, Lyon D, Anderson R, Holman T, Stymiest C, Shanahan J, Nelson L, Deboer K, Hein G, Krall J (1995) Producing and marketing Prosomillet in the High Plains. EC95-137, Univ Nebraska Coop Ext Serv, Lincoln

Bangar PS, Ashogbon AO, Dhull SB, Thirumdas R, Kumar M, Hasan M, Chaudhary V, Pathem S (2021) Proso-millet starch: properties, functionality, and applications. Int J Biol Macromol 190:960–968. https://doi.org/10.1016/j.ijbiomac.2021.09.064

Banik R, Mansingh B, Bag S, Sunartiya A, Das N, Sow S, Ranjan S, Aziz T (2024) Millets in global food industries: trends, obstacles and possibilities. In: Sunita V, Rekha S, Swarnlata S, Madhulika S (eds) Millets: sustainable grain for a healthy future. ABS Books

Baruah A (2024) Current status of millets in India—trends, prospects and challenges

Bekkering CS, Tian L (2019) Thinking outside of the cereal box: breeding underutilized (pseudo) cereals for improved human nutrition. Front Genet 10:1289

Bhat S, Nandini C, Tippeswamy V (2018) Significance of small millets in nutrition and health – a review. J Dairy Foods Home Sci 37(1):35–40. https://doi.org/10.18805/ajdfr.DR-1306

Bhat S, Nandini C, Srinathareddy S, Jayarame G, Prabhakar (2019) Prosomillet (Panicum miliaceum L.) – a climate resilient crop for food and nutritional security: a review. Environ Conserv J 20(3):113–124. https://doi.org/10.36953/ECJ.2019.20315

Bora P, Das P (2018) Some physical and functional properties of Prosomillet (Panicum miliaceum L.) grown in Assam. Int J Pure App Biosci 6(2):1188–1194

Boukail S, Dell'Acqua M, Ganugi P, Palchetti E, Benedettelli S, Masoni A, Calamai A (2021) Genome wide association study of agronomic and seed traits in a world collection of Prosomillet (Panicum miliaceum L.). BMC Plant Biol 21(1):1–12

Calamai A, Masoni A, Marini L, Dell'acqua M, Ganugi P, Boukail S, Benedettelli S, Palchetti E (2020) Evaluation of the agronomic traits of 80 accessions of Prosomillet (Panicum miliaceum L.) under Mediterranean pedoclimatic conditions. Agriculture (Switzerland) 10(12):1–15

Castro K (2013) Evaluation of mille t(various species) Grain from US as a pathway for regulated weeds. Canadian Food Inspection Agency. Plant Biotechnology Risk Assessment Unit, Ottawa, pp 1–33

Chapke RR, Shyam Prasad G, Das IK, Hariprasanna K, Singode A, Kanthi SBS, Tonapi VA (2020) Latest millet production and processing technologies. Booklet, ICAR–Indian Institute of Millets Research, Hyderabad, p 030

Chaudhary R, Sharma C, Kumar V, Rajput V, Naik B, Prasad R, Sharma S, Kumar V (2025) Millet biofortification for enhanced iron content: roadmap for combating hidden hunger. J Agric Food Res 19:101654. https://doi.org/10.1016/j.jafr.2025.101654

Cho YI, Chung JW, Lee GA, Kim TS, Ma KH, Dixit A, Park YJ (2010) Development and characterization of twenty-five new polymorphic microsatellite markers in Prosomillet (Panicum miliaceum L.). Genes Genom 32:267–273

Cho K, Cho KS, Sohn HB, Ha IJ, Hong SY, Lee H, Kim YM, Nam MH (2016) Network analysis of the metabolome and transcriptome reveals novel regulation of potato pigmentation. J Exp Bot 67(5):1519–1533

Das S, Khound R, Santra M, Santra DK (2019) Beyond bird feed: Prosomillet for human health and environment. Agriculture 9(3):64

Dikshit N, Sivaraj N (2013) Diversity for protein and morpho-agronomical characteristics in Prosomillet germplasm collections of Ratnagiri District, Maharashtra. India Vegetos 26(2):164–170

Diola V, MenezesDaloso D, Antunes WC (2014) Metabolomics. In: Omics plant breeding, pp 81–101. https://doi.org/10.1002/9781118820971.ch5

Dwivedi SL, Upadhyaya HD, Senthilvel S, Hash CT, Fukunaga K, Diao X, Santra D, Cole G, Canci PC (2012) Millets: genetic and genomic resources. Plant Genet Resour 10(3):283–292

Elangovan M, Singode A, Venkatesh K, Amasiddha B, Karthik AS, Pandey S, Pandey CD (2023) Agro-morphological characterization of Prosomillet germplasm for utilization. Int J Plant Soil Sci 35(16):381–392

FAO (2018) Millets and pseudo cereals: traditional foods with modern applications. In: Future smart food – rediscovering hidden treasures. FAO, Rome. http://www.fao.org/3/i8499en/I8499EN.pdf

Gokte N, More SS, Patel JD (1922) Seed-borne nematodes in Prosomillet and their management. Indian J Agric Sci 22(2):140–143

Graybosch RA, Baltensperger DD (2009) Evaluation of the waxy endosperm trait in Prosomillet (Panicum miliaceum). Plant Breed 128(1):70–73

Habiyaremye C, Matanguihan JB, D'Alpoim Guedes J, Ganjyal GM, Whiteman MR, Kidwell KK, Murphy KM (2017) Prosomillet (Panicum miliaceum L.) and its potential for cultivation in the Pacific Northwest, U.S.: a review. Front Plant Sci 7:1961. https://doi.org/10.3389/fpls.2016.01961

Harlan JR (1975) Crops and man. American Society of Agronomy, Madison, Wisconsin, p 295

Hawlader MSH (1991) Genetic variability and correlation study in Prosomillet (Panicum miliaceum L.). Ann Bangladesh Agric 1(2):61–64

Hoffmann-Bahnsen R (2004) Alte Kulturpflanze neu entdeckt—Perspektiven und Möglichkeiten der Wiederinkulturnahme von Rispenhirse (Panicum miliaceum) im ökologischen Landbau. In: On-farm-Erhaltung genetischer Ressourcen von Getreide und Ölpflanzen, pp 91–100

Hou S, Sun Z, Li Y, Wang Y, Ling H, Xing G, Han Y, Li H (2017) Transcriptomic analysis, genic SSR development, and genetic diversity of Prosomillet (Panicum miliaceum; Poaceae). Appl Plant Sci 5(7):1600137

Hu X, Wang J, Lu P, Zhang H (2009) Assessment of genetic diversity in broomcorn millet (Panicum miliaceum L.) using SSR markers. J Genet Genomics 36(8):491–500. https://doi.org/10.1016/S1673-8527(08)60139-3

Hulse JH, Laing EM, Pearson OE (1980) Sorghum and the millets: their composition and nutritional value. Academic Press, New York, p 99

Humphrys C (2005) Raps und Hirse im biologischen Anbau (rape and millet in organic farming). In: Biolandbau (Organic Farming) Congress

Hunt HV, Vander Linden M, Liu X, Motuzaite Matuzeviciute G, Colledge S, Jones MK (2008) Millets across Eurasia: chronology and context of early records of the genera Panicum and Setaria from archaeological sites in the Old World. Veg Hist Archaeobotany 17:5–18

Hunt HV, Campana MG, Lawes MC, Park YJ, Bower MA, Howe CJ, Jones MK (2011) Genetic diversity and phylogeography of broomcorn millet (Panicum miliaceum L.) across Eurasia. Mol Ecol 20(22):4756–4771

Ilyin VA, Zolotukhin EN, Ungenfukht IP, Tikhonov NP, Markin BK (1993) Importance, cultivation and breeding of Prosomillet in Povolzye province in Russia. In: Riley KW, Gupta SC, Seetharam A, Mushonga JN (eds) Advances in small millets. Oxford and IBH Publishing, New Delhi, India, pp 109–116

Jagadish PS, Kumari NG, Reddy VR (1995) Assessment of damage by shoot fly, Atherigona pulla. Prosomillet Indian J Entomol 57(2):150–153

Jain AK, Tiwari A (2013) Evaluation of promising genotypes of Prosomillet against banded leaf and sheath blight disease caused by Rhizoctonia solani Kuhn. Mysore J Agric Sci 47(3):648–650

Jayaraman N, Suresh S, Nirmala A, Ganeshan NM (1997) Genetic enhancement and breeding strategies in small millets. In: National Seminar on small millets, 23–24 April, Coimbatore, India, pp 19–21

Jiang Y, Li H, Zhang J, Xiang J, Cheng R, Liu G (2018) Whole genomic EST-SSR development based on high-throughput transcript sequencing in Prosomillet (Panicum miliaceum). Int J Agric Biol 20:617–620

Jiang Y, Dong L, Li H, Liu Y, Wang X, Liu G (2024) Genetic linkage map construction and QTL analysis for plant height in Prosomillet (Panicum miliaceum L.). Theor Appl Genet 137(4):78. https://doi.org/10.1007/s00122-024-04576-2

Johnson M, Sood S, Gowda CLL, Upadhyaya HD, Sharma S, Bhattacharjee R, Singh S, Senthilvel S, Varshney RK (2019) Genome-wide population structure analyses of three minor millets: kodo millet, little millet, and Prosomillet. Plant Genome 12(3):190021. https://doi.org/10.3835/plantgenome2019.03.0021

Jones M (2004) Between fertile crescents: minor grain crops and agricultural origins. In: Between fertile crescents: minor grain crops and agricultural origins. p 127

Khound R, Santra DK (2020) Omics for Prosomillet genetic improvement. Nucleus 63(3):241–247

Kim JK, Park SY, Yeo Y, Cho HS, Kim YB, Bae H, Park CH, Lee JH, Park SU (2013) Metabolic profiling of millet (Panicum miliaceum) using gas chromatography-time-of-flight mass spectrometry (GC-TOFMS) for quality assessment. Plant Omics 6(1):73–78

Konstantinov SI, Linnik VM, Ya SL (1986) Head smut resistant proso variety Khar'kovskoe 86. Selektsiya i Semenovodstvo (USSR) 5:35–36

Konstantinov SI, Linnik VM, Ya SL (1989) Use of smut-resistant induced mutants in breeding Prosomillet. Selektsiya i Semenovodstvo (Kiev) 66:25–28

Konstantinov SI, Linnik VM, Ya SL, Grigorashchenko LV (1991) Breeding Prosomillet for resistance to diseases. In: Litun PP (ed) Urozhai i adaptivnyi potentsial ekologicheskoi sistemy polya. Ukrainian Academy of Sciences, Kiev, Ukraine, pp 112–117

Krasavin VD, Usmanova RI (1988) Prosomillet variety Orenburgskoe 9 resistant to head smut. Selektsiya i Semenovodstvo (Moscow) 1:54–55

Lee-DuHyung (1997) Morphological characters and seed transmission of Bipolaris panici-miliacei causing leaf spot of common millet. Korean J Pl Pathol 13(1):18–21

Li J, Wang Y, Wang L, Zhu J, Deng J, Tang R, Chen G (2021a) Integration of transcriptomic and proteomic analyses for finger millet [Eleusine coracana (L.) Gaertn.]in response to drought stress. PLoS One 16(2):e0247181

Li W, Wen L, Chen Z, Zhang Z, Pang X, Deng Z, Liu T, Guo Y (2021b) Study on metabolic variation in whole grains of four Prosomillet varieties reveals metabolites important for antioxidant properties and quality traits. Food Chem 357:129791. https://doi.org/10.1016/j.foodchem.2021.129791

Liu M, Xu Y, He J, Zhang S, Wang Y, Lu P (2016) Genetic diversity and population structure of broomcorn millet (Panicum miliaceum L.) cultivars and landraces in China based on microsatellite markers. Int J Mol Sci 17(3):370

Liu C, Yuan Y, Liu J, Wang H, Ma Q, Zhou Y, Liu C, Gong X, Feng B (2022a) Comparative transcriptome and physiological analysis unravel Prosomillet (Panicum miliaceum L.) source leaf adaptation to nitrogen deficiency with high nitrogen use efficiency. Environ Exp Bot 199:104891

Liu J, Yuan Y, Ma Q, Gao Y, Yang Q, Gao X, Feng B (2022b) Cleaner production of Prosomillet (Panicum miliaceum L.) in salt-stressed environment using re-watering: from leaf structural alleviations to multi-omics responses. J Clean Prod 334:130205. https://doi.org/10.1016/j.jclepro.2021.130205

Liu T, He J, Dong K et al (2022c) Genome-wide identification of quantitative trait loci for morpho-agronomic and yield-related traits in foxtail millet (Setaria italica) across multi-environments. Mol Gen Genomics 297:873–888

Lu H, Zhang J, Liu K, Wu N, Li Y, Zhou K, Ye M, Zhang T, Zhang H, Yang X, Shen L, Xu D (2009) Earliest domestication of common millet (Panicum miliaceum) in East Asia extended to 10,000 years ago. Plant Mol Biol Report 27(2):242–251

Lyon DJ, Burgener PA, DeBoer KL (2008) EC08-137 producing and marketing Prosomillet in the Great Plains. University of Nebraska-Lincoln, Lincoln

Madhavan G, Logeshwaran J, Abarajitha P, Saranya S (2024) Organic cultivation practices for Prosomillet. WORLD 944(1449.5):1535

Manoharan V, Sivasubramanian V (1982) Variability studies in Prosomillet (Panicum miliaceum L.). Madras Agric J 69

Matz S (1986) Millet, wild rice, adlay, and rice grass. In: Cereal science. Avi Press, Westport, pp 225–229

Morah FNI, Etukudo UP (2017) Effect of sprouting on nutritional value of Panicum miliaceum (proso millet). Edorium J Nutr Diet 4:1–4

Murthi T, Harinarayana G (1986) Insect pests of small millets and their management in India. In: Seetharam A, Riley KW, Harinarayana G (eds) Small millets in global agriculture. Oxford and IBH Publishing, New Delhi, pp 255–270

Nagaraja TE, Nandini C, Bhat S, Parveen SG (2023) Artificial hybridization techniques in small millets—a review. Front Plant Sci 14:1112117

Nagaraja TE, Gazala Parveen S, Meenakshi J, Murtujasab S, Chetana, Manjunatha M, Bhat S, Nandini C (2024) Panorama of small millets breeding: a review. Plant Breed 143(6):810–827

Nanda JS, Agrawal PK (2008) Botany of field crops, vol 1. Kalyani Publishers, India, p 381

Nandini C, Sujata B, Srinatha R, Jayaramegowda, Prabhakar (2019) Modified crossing (SMUASB) method for artificial hybridization in Prosomillet (Panicum miliaceum L.) and little millet (Panicum sumatrense). Electron J Plant Breed 10:1161–1170

Nandini C, Bhat S, Prabhakar J, Krishnappa M, Reddy SS, Ganiger PC, Vinay N (2021) A new high yielding Prosomillet variety (Panicum miliaceum L.): PMV 442. Electron. J Plant Breed 12(1):9–16

Nandini C, Joshi DC, Nandini B, Nagaraja TE, Bhat S, Gazala Parveen S, Ekbote SD, Dushyanthakumar BM (2025) Advances in hybridization techniques in small millets. In:

Joshi DC, Singh AK, Sood S, Prasad M (eds) Minor millets. Springer, Singapore. https://doi.org/10.1007/978-981-96-4265-6_5

Natarajan US, Raveendran TS, Appadurai R (1978) A path coefficient analysis of yield and yield components in Prosomillet (Panicum miliaceum L.). Madras Agric J 65(7):430–434

Nelson LA (1984) Technique for crossing Prosomillet. Crop Sci 21:205–206

Panwar KS, Kapila RK (1992) Variation and character association in Prosomillet. Crop Improv (India) 19:130–133

Popov GI (1946) The importance of diversity in millet. Agrobiologiya 2:28–43

Prasad GS, Nagaraja TE, Seetharam A, Gowda BTS (1995) Genetic variability and character association studies in Prosomillet. Crop Improv 22(2):225–227

Rajagopalan K, Sreekumar J, Sivaraman K (1992) Occurrence of sheath rot of Prosomillet caused by Rhizoctonia solani in Kerala. College of Agriculture, Trivandrum

Rajasekaran R, Francis N (2021) Genetic and genomic resources for improving Prosomillet (Panicum miliaceum L.): a potential crop for food and nutritional security. Nucleus 64(1):21–32

Rajput SG, Plyler-Harveson T, Santra DK (2014) Development and characterization of SSR markers in Prosomillet based on switchgrass genomics. Am J Plant Sci 5(1):69–75

Ravulapenta Sathish M, Manjunatha K, Rajashekarappa K (2017) Incidence of shoot fly, Atherigona pulla (Wiedemann) on Prosomillet at different dates of sowing. J Entomol Zool Stud 5(5):2000–2004

Reddy VG, Upadhyaya HD, Gowda CLL (2007) Morphological characterization of world's Prosomillet germplasm collection. J SAT Agric Res 3:4

Riaño-Pachón DM, Ruzicic S, Dreyer I, Mueller-Roeber B (2007) PlnTFDB: an integrative plant transcription factor database. BMC Bioinform 8(1):1–10

Roy SK, Kwon SJ, Yu JH, Sarker K, Cho SW, Moon YJ, Jung TW, Park CH, Woo SH (2017) Comparison of protein profiles of Prosomillet (Panicum miliaceum) seeds of various Korean cultivars. Korean J Crop Sci 62(1):40–50

Salini K, Nirmalakumari A, Muthiah AR, Senthil N (2010) Evaluation of Prosomillet (Panicum miliaceum L.) germplasm collections. Electr. J Plant Breed 1(4):489–499

Santra DK, Khound R, Das S (2019) Prosomillet (Panicum miliaceum L.) breeding: progress, challenges and opportunities. In: Advances in plant breeding strategies: cereals. pp 223–257

Sasamala AC, Sahoo LP, Mahapatra KC, Pandey S (2011) Genetic variability and character association in common millet germplasm of Odisha. Indian J Plant Genet Resour 24(1):82–86

Seetharam A, Gowda J, Halaswamy JH (2003) Small millets. In: Chowdhury SK, Lal SK (eds) Nucleus and breeder seed production manual. Indian Agricultural Research Institute, New Delhi, pp 54–67

Shahidi F, Chandrasekara A (2013) Millet grain phenolics and their role in disease risk reduction and health promotion: a review. J Funct Foods 5(2):570–581. https://doi.org/10.1016/j.jff.2013.02.004

Shailaja S, Jagadish PS, Kumar CTA, Nangia N, Gowda J, Nagaraja A (2009) Evaluation of prerelease and released varieties of Prosomillet (Panicum miliaceum L.) to Corcyra cephalonica Stainton (Lepidoptera: Pyralidae) infestation. Environ Ecol 27:445–447

Shan Z, Jiang Y, Li H, Guo J, Dong M, Zhang J, Liu G (2020) Genome-wide analysis of the NAC transcription factor family in broomcorn millet (Panicum miliaceum L.) and expression analysis under drought stress. BMC Genomics 21(1):1–13

Sharma YR, Sugha SK (1991) Chemical control of head smut of Prosomillet. Plant Dis Res 6(1):45–47

Sharma PN, Sugha SK, Panwar KS, Sagwal JC (1993) Performance of Prosomillet (Panicum miliaceum) germplasm in relation to smut (Sphacelotheca destruens). Indian J Agric Sci 63:675

Sharma N, Sharma A, Bhardwaj R, Chauhan R, Thakur N, Singh BP (2019a) Minor millets: a potential food and feed resource for nutritional and livelihood security. In: FAO report—future smart food: rediscovering hidden treasures of neglected and underutilized species for zero hunger in Asia. FAO, Rome, pp 42–47. http://www.fao.org/3/ca7151en/CA7151EN.pdf

Sharma N, Rai RD, Rai SK, Singh SB (2019b) Himalayan food systems: relevance, sustainability and policy issues. Food and Agriculture Organization (FAO)

Shi J, Ma X, Zhang J, Zhou Y, Liu M, Huang L, Sun S, Zhang X, Gao X, Zhan W, Li P (2019) Chromosome conformation capture resolved near complete genome assembly of broomcorn millet. Nat Commun 10(1):1–9

Shobana S, Malleshi NG (2007) Preparation and functional properties of decorticated finger millet (Eleusine coracana). J Food Eng 79:529–538

Spengler R, Frachetti M, Doumani P, Tarasov PE, Hunt HV, Jones MK, Boivin N (2014) Early agriculture and crop transmission among bronze age mobile pastoralists of Central Eurasia. Proc R Soc B 281:20133382

Sundararaj DP, Thulasidas G (1976) Botany of field crops. Macmillan Publishers, India, p 509

Upadhyaya HD, Sharma S, Gowda CLL, Reddy VG, Singh S (2011) Developing Prosomillet (Panicum miliaceum L.) core collection using geographic and morpho-agronomic data. Crop Pasture Sci 62(5):383–389

Upadhyaya HD, Dwivedi SL, Singh SK, Singh S, Vetriventhan M, Sharma S (2014) Forming core collections in barnyard, kodo, and little millets using morphoagronomic descriptors. Crop Sci 54(6):2673–2682

Vavilov NI (1926) Studies on the origin of cultivated plants. Institute of Botany of Applied and Amelioration of Plants, Leningrad

Vetriventhan M, Upadhyaya HD (2018) Diversity and trait-specific sources for productivity and nutritional traits in the global Prosomillet (Panicum miliaceum L.) germplasm collection. Crop J 6(5):451–463

Vetriventhan M, Azevedo VCR, Upadhyaya HD, Nirmalakumari A, Kanepotaka J, Anitha S, Ceasar SA, Muthamilarasan M, Bhat BV, Hariprasanna K, Bellundagi A, Cheruku D, Backiyalakshmi C, Santra D, Vanniarajan C, Tonapi VA (2020) Genetic and genomic resources, and breeding for accelerating improvement of small millets: current status and future interventions. Nucleus 63:217–239

Vetriventhan M, Upadhyaya HD, Deshpande S, Johnson MS, Wallace JG, Victor A, Naresh D, Rayaprolu L, Singh K, Mayes S (2024) Genome-wide assessment of population structure and association mapping for agronomic and grain nutritional traits in Prosomillet (Panicum miliaceum L.). Sci Rep 14(1):21920. https://doi.org/10.1038/s41598-024-72319-w

Wang L, Ang XY, Wen QF, Cao LP (2007) Identification of protein and fat content for Chinese Prosomillet germplasm. J Plant Genet Resour 8:165–169

Wang R, Wang H, Liu X, Ji X, Chen L, Lu P, Liu M, Teng B, Qiao Z (2018) Waxy allelic diversity in common millet (Panicum miliaceum L.) in China. Crop J 6(4):377–385. https://doi.org/10.1016/j.cj.2018.02.004

Wilson RL, Courteau JB (1984) Search of plant introduction Prosomillet for fall armyworm resistance. J Econ Entomol 77:171–173

Yao D, Wei Q, Xu W, Syrenne RD, Yuan JS, Su Z (2012) Comparative genomic analysis of NAC transcriptional factors to dissect the regulatory mechanisms for cell wall biosynthesis. BMC Bioinform 13(15):1–12

Yazdizadeh M, Mohammadi R, Abdul Kader M, Ghaffari V et al (2020) Association analysis between agronomic traits and AFLP markers in a wide germplasm of Prosomillet (Panicum miliaceum L.) under normal and salinity stress conditions. BMC Plant Biol 20(1):1–18

Yuan Y, Wu C, Liu L, Ma Q, Yang Q, Feng B (2022) Unravelling the distinctive growth mechanism of Prosomillet (Panicum miliaceum L.) under salt stress: from root-to-leaf adaptations to molecular response. GCB Bioenergy 14:192–214. https://doi.org/10.1111/gcbb.12910

Zhang H, Jin J, Tang L, Zhao Y, Gu X, Gao G, Luo J (2011) PlantTFDB 2.0: update and improvement of the comprehensive plant transcription factor database. Nucleic Acids Res 39(Suppl_1):D1114–D1117

Zhu G, Wang S, Huang Z, Zhang S, Liao Q, Zhang C, Lin T, Qin M, Peng M, Yang C, Cao X (2018) Rewiring of the fruit metabolome in tomato breeding. Cell 172(1–2):249–261

Zhuravel BN (1956) On the biology of millet blooming under conditions of Belorussian SSR. Sb. Nauch. Tr. Akad Nauk B SSR Inst Sotsialistselkh 4:249–264

Zohary D, Hopf M (2000) Domestication of plants in the Old World: the origin and spread of cultivated plants in West Asia, Europe and the Nile Valley, 3rd edn. Oxford University Press

Zolotukhin EN, Tikhonov NP, Lizneva LN (1998) New cultivar of Panicum miliaceum, 'Il'Inovskoe. Int Sorghum Millet Newsl 39:133–134

Zou C, Li L, Miki D, Li D, Tang Q, Xiao L, Rajput S, Deng P, Peng L, Jia W, Huang R (2019) The genome of broomcorn millet. Nat Commun 10(1):1–11

Part III

Minor Vegetable, Tuber and Root Crops

Kalingda (Citrullus colocynthis): The Desert Melon with Medicinal Power

9

Devendra Upadhyay, Ashish Kumar Tiwari, Sonal Tiwari, Anjali Sharma, and Jitendra Kumar Tiwari

Abstract

Kalingda (*Citrullus colocynthis*), locally known as *Mateera* or *Kalingda* in the Indian arid zones, is a drought-hardy, underutilized cucurbit with immense nutritional, medicinal, and ecological value. Traditionally cultivated in harsh desert climates with minimal inputs, this ancient crop offers a rich source of edible seeds known for their high oil and protein content. The fruit rind also serves as nutritious livestock fodder, aiding in sustainable resource use. As modern diets narrow and climate-resilient food sources gain global attention, *Kalingda* emerges as a vital fasting superfood—nutrient-dense, culturally significant, and ideal for low-calorie, high-impact dietary regimens. Its bioactive compounds, adaptogenic potential, and heritage use in dryland mixed farming systems align it with the emerging paradigm of food security, resilience, and health during fasting periods. This chapter explores the botany, nutritional profile, traditional uses, and modern research advances on *Kalingda*, making a compelling case for its inclusion in the global repertoire of sustainable superfoods.

D. Upadhyay (✉)
Dau Kalyan Singh College of Agriculture and Research Station, Indira Gandhi Krishi Vishwavidyalaya, Bhatapara, Chhattisgarh, India
e-mail: devendra.upadhyay@igkv.ac.in

A. K. Tiwari
Sant Vinova Bhave College of Agriculture and Research Station, IGKV, Marra, Chhattisgarh, India

S. Tiwari
Department of Genetics and Plant breeding, Mahatma Gandhi Udyaniki Evam Vanki Vishwavidyalaya, Sankra-Patan, Chhattisgarh, India

A. Sharma · J. K. Tiwari
Rani Lakshami Bai Central Agricultural University, Jhansi, Uttar Pradesh, India

A. Kumar et al. (eds.), *Fasting Superfoods*,
https://doi.org/10.1007/978-981-95-2762-5_9

Keywords

Kalingda · *Citullus colocynthis* · Desert superfood · Drought hardy crop · Fasting nutrition

9.1 Introduction

With only about 30 plant species contributing 90% of the global food supply, modern agriculture has witnessed a dangerous narrowing of the food basket. This reduced biodiversity not only jeopardizes nutritional security but also compromises our resilience to climate change, pandemics, and ecological shocks. In this context, the resurgence of interest in underutilized crops—often dubbed "future foods"—offers a pathway to rediscover the culinary, cultural, and medicinal treasures embedded in traditional agricultural systems.

Kalingda (*Citrullus colocynthis*), a member of the Cucurbitaceae family, exemplifies such potential. Grown in the arid belts of northwestern India, particularly Rajasthan and Gujarat, this hardy creeper thrives under extreme conditions—shifting sand dunes, searing temperatures, minimal rainfall, and erratic monsoons. Yet, it yields nutrient-rich seeds that are valued both as snacks and functional food ingredients. Unlike commercial watermelon cultivars bred for sweet, fleshy fruit, seed-purpose varieties of *Kalingda* are cultivated for their oil-rich, protein-dense kernels, locally known as "*magaj*." These are used in traditional sweets, savory snacks, and even in *Ayurvedic* and *Unani* medicine systems for their cooling and anti-inflammatory effects.

During fasting periods, especially in Indian and Middle Eastern cultures, nutrient-dense yet light foods are preferred for sustaining energy and promoting detoxification. *Kalingda* seeds, roasted or hulled, are a staple during such rituals, offering concentrated nourishment in a compact form. Beyond fasting, its adaptability to low-input, rainfed conditions and multipurpose value—as oilseed, snack food, livestock fodder, and medicinal ingredient—makes it a cornerstone in the dialogue on sustainable food systems.

In this chapter, we present *Kalingda* as a fasting superfood—anchored in traditional knowledge, validated by modern research, and positioned as a climate-smart crop with immense potential for health-conscious, eco-aware populations. Through its journey from desert margins to the global wellness table, *Kalingda* exemplifies the transformative power of indigenous crops in reshaping the future of nutrition and agriculture.

9.1.1 About *Kalingada (Citrullus lanatus)*

The family Cucurbitaceae consists of about 90 genera and 750 species, out of which 15 species from 8 genera are most commonly cultivated in India. The four species under the genus *Citrullus* are *Citrullus vulgaris* Schrad. (now renamed *C. lanatus* (Thumb.) Matsum. & Nakai), which includes cultivated watermelon, *C. colocynthis* L., *C. ecirrhosus* (Cogn.), and *C. naudinianus* (Sond.) Hook. After seed extraction, green as well as dry rind portion of the fruits locally known as "*khuparia*" can be very well utilized as animal feed. It may substantially replace the costly concentrate and thereby reduce the feeding cost.

Kalingada/Mateera (*Citrullus lanatus* Mansf. Thumb.) is an important extremely drought hardy cucurbitaceous creeper. It is an important crop and as multipurpose values as snack food, oil seed and culinary, fodder, and so on. The arid regions of Rajasthan, Gujarat, Punjab, and Haryana account for 90% of the total hot arid region with an area of 31.7 million hectares of which northwestern Rajasthan (61%) and north Gujarat (19.6%) constitute more than 80% of Indian arid zone. The hostile region is characterized by huge shifting and rolling sand dunes, marred with high velocity wind, high diurnal variation in temperature, intense solar radiation, scarce rainfall, and high rate of evapotranspiration. The average annual rainfall in western Rajasthan is less than 200 mm (Jaisalmer), 250 mm (Bikaner and Barmer), and 350 mm (Jodhpur). The productivity of major rainfed *kharif* crops such as pearl millet (1000 kg/ha), cluster bean (500 kg/ha), moth bean (300 kg/ha), and sesame (350 kg/ha) remain very low due to uncertain monsoon rains and erratic distribution during crop growth period (July–September). Therefore, to minimize the risk of crop failure and uncertain production under rainfed conditions, mixed/intercropping has been in practice in the region since time immortal. Traditionally, cucurbits are preferred choice as a mixed crop with cluster bean/pearl millet in arid regions. Among the cucurbits, watermelon, popularly known as "*Mateera*" or "*Kalinagada*," is an ideal choice as a companion crop due to its adaptive mechanism in mixed cropping in arid parts of Rajasthan.

Watermelon (*Citrullus lanatus* (Thunb.) Matsum. & Nakai) is well-known for its fleshy, juicy sweet fruits cultivated worldwide as cash crop. On the other hand, throughout sub-Saharan Africa and arid regions of India, it is being grown in mixed cropping under resource constraint situations of rainfed agriculture to minimize the risk of crop failure under aberrant weather situations (Fig. 9.1) for its highly priced seeds (Mahla and Choudhary 2013). After harvest of the main crop, farmer family collects the leftover fruits of *Mateera* and extracts the seeds from fruits (Figs. 9.2 and 9.3). Seed price varies from Rs. 5000 to 8000 per quintal (sometimes ≥ Rs. 10,000) in local market, which is remunerative among rainfed kharif crops, thus supporting the livelihood in a big way. The seeds of watermelon used in different preparations are in great demand in local, national, and international markets. The roasted seeds are taken as a common snack, while after removal of hull it is mainly used in sweets and pharmaceuticals. Presently, the kernels (Fig. 9.4) locally called *magaj* are used in restaurants/hotels as source of flavor and thickeners in vegetable preparation (Mahla et al. 2014). Traditionally, green as well as dry rind portion of the fruits locally called as *khuparia* is used as livestock feed. The dried rinds of fruits are stored and supposed to have good keeping quality. On dry weight basis, *khuparia* contains 12–15% crude protein, 20–25% ether extract, and 6–10% ash (Mahla et al. 2014).

Besides, being highly adapted to arid conditions and livelihood supportive role in Indian *Thar Desert*, limited research attention was paid for its seed yield improvement. Keeping in view the importance of watermelon seed in livelihood and oil content in seeds, the concept of seed-purpose watermelon is highlighted in Indian *Thar Desert* (Mahla and Choudhary 2013).

Fig. 9.1 Mixed cropping of seed-purpose watermelon with cluster bean in arid region

Fig. 9.2 Farmer family extracting the seed from watermelon fruit

Fig. 9.3 Dried rind of fruits (left) and storage (right) for its utilization as animal feed

Fig. 9.4 A view of kernel (left) and hull (right) after separation from whole seed.

9.1.2 Evaluation of Seed-Purpose Watermelon Germplasm

Edible seed watermelon has been grown in China, and through systematic breeding efforts, varieties with yield potential of over 2.0 tons per hectare and crude fat content up to 48% have been developed (Kechi 1990). Keeping in view the importance and potential of watermelon as seed-purpose crop rather than fruit or vegetable crop, the breeding work has been initiated at the Central Arid Zone Research Institute, Regional Research Station, Jaisalmer, in 2009. A large number of germplasm accessions including exotic and local landraces collected from various sources were evaluated under rainfed conditions of Jaisalmer (Fig. 9.5). Among the germplasm accessions, a wide range of variability was recorded for quantitative traits (Table 9.1). Further, diversity for fruit size, shape, and stripe pattern, flesh color, texture, and seed size, shape, and color was quite evident (Fig. 9.6). After four selection cycles, the identified promising lines were evaluated in station trial for seed yield potential under rainfed conditions. Genotype CAZJK-13-1 and CAZJK-13-2 (Fig. 9.7) were found promising and gave 783 and 753 kg seed yield

Fig. 9.5 Evaluation of watermelon germplasm under rainfed conditions

Table 9.1 Variability for quantitative traits in seed-purpose watermelon germplasm

Sl no.	Character (s)	Mean ± SEm	Range	Coefficient of variation (CV %)
1	Fruits per plant (no.)	7.6 ± 0.38	2.02–27.25	53.59
2	Fruit yield per plant (kg)	13.2 ± 0.61	1.08–34.50	49.34
3	Fruit weight (g)	1493.8 ± 79.8	464.0–5635.0	57.10
4	Rind weight (g)	986.5 ± 54.8	295.0–3152.0	59.27
5	Fruit diameter (cm)	43.7 ± 0.61	28.80–65.80	14.84
6	Seeds per fruit (no.)	499.8 ± 19.70	156.4–1072.0	42.08
7	Seed yield per fruit (g)	40.6 ± 1.05	18.53–66.49	27.53
8	Seed yield per plant (g)	273.6 ± 13.85	25.0–870.0	54.03
9	1000 – seed weight (g)	93.8 ± 3.60	45.81–191.88	40.93
10	Oil content (%)	28.9 ± 0.38	14.60–44.20	14.02

per hectare with 35.6% and 39.3% oil content, respectively. Consequently, multilocation evaluation of promising genotypes over year resulted in identification of two varieties, namely SKNK-1102 (2016) and CAZJK-13-2 (2018) for cultivation in states of Rajasthan and Gujarat.

Fig. 9.6 Variability in seed size, shape, and color of seed-purpose watermelon germplasm

Fig. 9.7 Improved variety of seed-purpose watermelon CAZJK-13-2

9.1.3 Area of Cultivation and Climate Change

Kalingada is an important crop that is cultivated in dry, hot, and drought-prone rainfed area as sole or intercrop with main crop. This crop is very hardy, require less care as compared to watermelon. Genetic improvement of *Kalingada* is studied at different locations of India under All India Coordinated Research Network on Potential Crops.

Kalingada/Mateera (*Citrullus lanatus* Mansf. Thumb.) is an important extremely drought hardy cucurbitaceous creeper. It is an important crop and as multipurpose values as snack food, oil seed and culinary, fodder, and so on.

There is scope of high potentiality for exploitation of *Kalingada* as nontraditional edible oil producing crops, particularly drought-prone areas of north and north-west Gujarat and worth-west Rajasthan. Further, it would be a better crop of choice as a sole crop in changing climate scenario in arid zone, besides its horizontal expansion in mixed cropping.

9.1.4 Resource Management

The crop can be grown during summer as well as in *kharif* season. It is particularly grown in hot and dry situation on any ordinary soil. The crop thrives best on deep, rich, well-drained sandy loam soil. It is extremely drought hardy cucurbitaceous creeper. It is cultivated in Gujarat (Banaskantha, Patan, Mehsana, and Kachchh districts) and Rajasthan (Jodhpur, Jaisalmer, Bikaner, and Sikar districts). In Gujarat, mainly this crop is grown as sole crop in area of the Vav and Suigam talukas Banaskantha district, while in rest of the district area, it is grown as mixed crop or intercrop with main *kharif* crops (pearl millet, sorghum, Castor, Cluster bean, green gram, etc.).

The main reason for growing *Kalingada* as mixed crop is its fast-growing habit and drought hardy nature. Further, its extensive root system makes it most suitable to tolerate harsh climatic conditions. Second, just after sowing of the crops, prolonged dry spell and high wind speed hinders the germination and proper plant stand of the main crop. The crop is extensively cultivated in *kharif* and *rabi* seasons as an intercrop with pearl millet and sorghum in western regions of India.

9.1.4.1 Mixed Cropping of Sorghum with *Kalingada*

When *Kalingada* was grown as an intercrop with sorghum, the productivity of the main crop, i.e., sorghum, has increased by virtue of soil moisture conservation by *Kalingada* crop, as it covers the surface of the soil. From this experiment, recommendation was made to farmers to sow the *Kalingada* in every sixth line of sorghum as a mixed crop.

9.1.4.2 Inter/Mixed Cropping System in Pearl Millet with *Kalingada*

The farmers of North Gujarat Agro-climatic Zone (AES-1) growing pearl millet crop on light textured soil under rainfed conditions are advised to take *Kalingada* as a mixed crop along with pearl millet in every third row to get more PMEY (3526 kg/ha) as well as higher benefit–cost ratio (2.28).

9.1.5 Maturity and Storage

Seeds continue to mature as the fruit ripens and the rind lightens in color. Seeds will be easier to extract from the fruit if the fruit is held in storage (in the shade or in a seed processing room) for a few days after removing them from the vine. If seeds are left too long in the fruit, they will germinate in situ.

9.1.6 Scope for Area Expansion and Improving Productivity

It is likely that area of potential crops would rise significantly with conscious efforts. However, *Kalingada* have large potential to be cultivated as sole crops and as intercrop in several major crops. For instance, *Kalingada* can be grown as an intercrop with major crop such as bajra, sorghum, castor. Since castor is a widely spaced crop and a major crop of this region, efforts should be made to promote *Kalingada* as component crop with castor. The productivity of underutilized crops can be improved by adopting appropriate measures, which include:

1. Cultivating *Kalingada* on reasonably fertile soil and irrigation at critical period of the crop growth can enhance its productivity.
2. In terms of varieties of *Kalingada*, GK-1 was released for general cultivation and is available to the farmers, but more need to be developed.
3. Creation of awareness among the farmers can rectify the use of imbalance fertilizer nutrients in potential crops. Production technology on fertilizer management and other agronomic aspects has been made available in *Kalingada.*
4. Delayed planting and optimum plant populations are two major constrains. The research finding has indicated that timely sowing with optimum plant population alone increases yield levels by at least 30–50%.

9.1.7 Future Thrust Areas

It has been very well recognized that seed-purpose watermelon *Kalingada* is an integral component of rainfed mixed cropping that supports the livelihood of inhabitant of *Thar Desert* in a big way. However, little attention has been paid for its commercial cultivation due to various obvious reasons. Therefore, the targeted breeding program and in-depth diagnostics of quality parameters of seed will help to identify factors that would boost its production. In the changing climate scenario, concerted scientific efforts on following aspects are very much required to support the arid zone economy (Mahla et al. 2014):

- The exotic germplasm collections as well as indigenous collections of *Kalingada* should be broadened for future needs.
- Development of high seed yield varieties and availability of seeds to farmers.
- Genotypes that can withstand water stress conditions under changing climate should be developed.
- Biochemical analysis for detailed proximate composition of watermelon seeds for multifarious uses, namely food, pharmaceuticals, and industrial applications. Value addition to *Kalingada*-derived products for its wider acceptability.
- Standardization of package of practices for higher seed production in different cropping systems and *nitch* areas. *Kalingada* is an integral component of rainfed mixed cropping that supports the livelihood of desert dwellers in a big way.
- Designing and fabrication of seed extractor for easy and quick seed extraction.
- Evaluation of dry rind as supplement livestock feed to costly concentrate.
- To get remunerative prices and usage of this crop, value addition efforts and to have linkages with entrepreneurs required.

The cultivation of improved genotypes of seed-purpose watermelon in arid region not only increases the income of the farmers but also diversifies the cropping system, which leads to sustainable rainfed production. Further, higher seed production and assured supply of raw material to processing units will provide congenial environment for industries and opportunities for employment generation in the region.

9.2 History: Traditional Uses and Cultural Significance

The historical roots of *Kalingda* (*Citrullus colocynthis*) trace back centuries, embedded in the cultural and agricultural tapestry of arid regions across Africa, the Middle East, and the Indian subcontinent. In India, particularly in Rajasthan and Gujarat, *Kalingda* has been cultivated by desert-dwelling communities since ancient times. Despite its austere growing environment, the crop has been cherished for its multipurpose utility and enduring resilience.

Traditionally known as *Mateera* or *Kalingda* in local dialects, it has been an integral part of indigenous cropping systems practiced by smallholder farmers in desert zones. These farmers often incorporated *Kalingda* as a companion crop in mixed farming systems, especially alongside pearl millet (*bajra*) and cluster bean (*guar*), to reduce the risk of total crop failure under erratic monsoon conditions. This approach, passed down through generations, not only ensured food security but also offered economic returns through the sale of high-value seeds, which were often traded in local village markets.

The seeds of *Kalingda*, locally called *magaj*, have held significant cultural and dietary importance. During religious fasts, festivals, and seasonal detox rituals observed in various Indian traditions, these seeds are consumed as energizing snacks or used in preparing sweets such as *magaj barfi* and *halwa*. Their high nutrient density, pleasant taste, and lightness make them particularly suitable for fasting diets, where minimal yet nourishing intake is desired. The seeds are also believed to have cooled properties, aligning with the *Ayurvedic* concept of balancing bodily heat during summer fasts. In rural households, the dried rind of the fruit, known as *khuparia*, is stored and used as livestock fodder, especially during lean months when green fodder is scarce. The tradition of collecting and sun-drying the rinds for long-term storage is an age-old practice that reflects the community's resourcefulness in ensuring sustainability. In fact, women and children often take the lead in harvesting, seed extraction, and drying activities, making *Kalingda* a crop that also contributes to gender-inclusive rural livelihoods. Beyond its domestic utility, *Kalingda* has also found mention in folk medicine. The seeds and rind have been used in traditional systems for their purported anti-inflammatory, antidiabetic, and digestive properties. While the pulp of *C. colocynthis* is not typically consumed due to its bitterness and toxicity, skilled use of the seeds and other parts in ethnomedicine continues in many rural settings.

As a symbol of survival in some of the harshest climates, *Kalingda* represents the ingenuity of desert agriculture. Its long-standing cultural role, especially during fasting seasons, reinforces its value not merely as a food crop but as a spiritual and ecological ally. The preservation of such traditional knowledge and its integration with modern food systems could be a key strategy for sustainable and culturally rich nutrition.

9.3 Production: Cultivation Practices, Climatic and Soil Requirements, and Global Production Statistics

Kalingda (*Citrullus colocynthis*), known for its rugged resilience and low input demand, thrives under extreme agroclimatic conditions. Its production is largely confined to hot arid regions, where traditional and marginal farming systems dominate. While not widely cultivated on a commercial scale globally, its niche cultivation in India—particularly in Rajasthan, Gujarat, and parts of Punjab and Haryana—serves as a model for climate-smart, resource-efficient agriculture.

9.3.1 Climatic and Soil Requirements

Kalingda is exceptionally drought-tolerant and well-adapted to the harsh conditions of the Indian *Thar Desert*. It grows optimally in regions with low rainfall (150–350 mm annually), high solar radiation, large diurnal temperature variations, and soils ranging from sandy loam to coarse desert sand, often low in fertility. It is sown with the onset of monsoon (July) and matures by late September to early October. The crop's deep-rooted nature allows it to extract moisture from deeper soil layers, making it ideal for rainfed cultivation in wastelands and degraded soils. *Kalingda* prefers slightly alkaline, well-drained sandy soils. Waterlogging is detrimental to its growth, while the crop tolerates salinity to a limited extent, making it suitable for cultivation in saline-prone marginal lands.

9.3.2 Cultivation Practices

Sowing is generally done by broadcasting or dibbling seeds in small basins or pits. A spacing of 1.5–2 m is maintained to allow for vine spreading. As a traditional rainfed crop, *Kalingda* requires minimal external inputs. Irrigation is seldom practiced, unless extreme drought conditions prevail. Fertilizer application is limited to FYM or compost, as the crop is adapted to nutrient-deficient soils. Weeding and thinning are done manually, typically once during early vine growth.

Kalingda is often grown as a mixed or intercrop, most commonly with pearl millet or cluster bean, to optimize land use and reduce the risk of total crop failure. This system benefits from the natural ground cover provided by *Kalingda* vines, which conserve soil moisture and suppress weeds.

9.3.3 Harvesting and Yield

Harvesting is done manually when the fruits attain full maturity. Since the primary economic product is the seed, the fruits are often left in the field for some days post-harvest to allow curing. Seeds are extracted, cleaned, and sun-dried for storage. Under traditional rainfed conditions, seed yields range from 500 to 800 kg/ha, with improved lines such as CAZJK-13-1 and CAZJK-13-2 yielding up to 750–800 kg/ha. Seed oil content ranges between 35% and 40%, making it comparable to several oilseed crops in terms of utility.

9.3.4 Global Production Statistics

While global data on *Citrullus colocynthis* production is limited due to its underutilized and wild nature, its cultivated relative—edible seed watermelon (*Citrullus lanatus*)—has shown growing demand in China, Sudan, Egypt, and parts of West Africa. In India, focused efforts at the Central Arid Zone Research Institute (CAZRI) and its regional stations have helped identify and release seed-purpose varieties for desert farming.

The renewed focus on underutilized and climate-resilient crops under international frameworks such as the FAO's Global Plan of Action on Plant Genetic Resources has opened up possibilities for expanded research and cultivation of *Kalingda*-like species in similar agroecologies worldwide.

9.3.5 Production Constraints

A wide gap of yield levels between farmer's field (achieved yield) and experimental plots (achievable yield) has been observed. Provided right inputs at right time, the yield level of *Kalingada* can be enhanced by manifolds. The major factors that contribute to low productivity on farmers field are mainly attributed to the following:

1. Underutilized crops are specially adapted to particular agroclimatic conditions, hence narrow gene pool (genetic base) is available.
2. These crops are often allotted to poorly maintained soil with low fertility.
3. Nonavailability of high-yielding genotypes (in some crops).
4. Imbalance use of fertilizer nutrients.
5. Delayed planting.
6. Suboptimum plant populations.

However, taking remedial measures such as the use of high yielding genotypes coupled with balance use of fertilizers with proper recommendations practices can enhance the productivity of this crop.

9.4 Protection: Major Pests, Diseases, and Management Strategies

Despite its adaptation to harsh climates and its inherent hardiness, *Kalingda* (*Citrullus colocynthis*) is not immune to biotic stresses. Its cultivation, particularly in resource-constrained arid zones, faces challenges from various pests and diseases (Table 9.2) that can reduce seed yield and affect plant vigor. However, traditional low-input systems and diverse agroecological practices have historically helped in maintaining pest and disease balance.

9.4.1 Integrated Pest and Disease Management (IPDM)

Integrated Pest and Disease Management (IPDM) of *Kalingda* (*Citrullus colocynthis*) involves a combination of cultural, biological, and chemical strategies aimed at minimizing pest damage while preserving the plant's medicinal properties (Table 9.2). *Kalingda* lends itself well to sustainable, ecological approaches to crop protection:

1. *Seed treatment*: With *Trichoderma viride* or *Pseudomonas fluorescens* to enhance seedling vigor and disease resistance.
2. *Intercropping:* With pearl millet or legumes to create a pest-disruptive habitat.
3. *Botanical sprays:* Neem oil, garlic–ginger extracts, and cow urine-based formulations offer eco-friendly alternatives to chemical pesticides.
4. *Soil health improvement*: Regular use of organic matter and compost to encourage beneficial microflora.

Given the increasing interest in climate-resilient, low-input crops, developing resistant varieties and promoting on-farm bio-integrated management practices are crucial for sustainable *Kalingda* cultivation. Modern molecular breeding and plant pathology approaches can also help screen and identify genotypes with durable resistance traits.

Table 9.2 Major insect pests and disease.

Insect	Damaging stage	Symptoms	Management
Pest			
Red Pumpkin Beetle (*Aulacophora foveicollis*)	Both larvae and adults feed on leaves, flowers, and young fruits	Skeletonization of leaves and scarred fruits	(a) Deep summer ploughing to destroy pupae. (b) Use of neem oil or neem seed kernel extract as biopesticide. (c) Manual collection of beetles in small holdings.
Fruit Fly (*Bactrocera cucurbitae*)	Maggots feed inside the fruits, causing rotting	Premature fruit drop and fruit deformation	(a) Field sanitation and timely removal of infested fruits. (b) Use of bait traps (e.g., jaggery + malathion) during fruiting stage. (c) Netting or bagging of fruits to prevent oviposition.
Aphids (*Aphis gossypii*)	All stage	Suck sap from young leaves and transmit mosaic viruses	(a) Spraying with neem-based formulations. (b) Intercropping with repellant plants like marigold or coriander. (c) Encouraging natural predators such as ladybird beetles.
Disease			
Causal organism	*Symptom*	*Impact*	*Management*
Powdery mildew (*Erysiphe cichoracearum*)	White powdery fungal growth on leaves and stems	Reduces photosynthesis, weakens plants, and leads to premature defoliation	(a) Use of sulfur dust or wettable sulfur sprays. (b) Avoid overcrowding and ensure proper aeration.
Downy mildew (*Pseudoperonospora cubensis*)	Yellow spots on upper leaf surface with grayish mold underneath	Leaf drops and stunted growth	(a) Crop rotation with non-cucurbit crops. (b) Spray with metalaxyl-based fungicides at early signs.

(continued)

Table 9.2 (continued)

Insect	Damaging stage	Symptoms	Management
Fusarium wilt (*Fusarium oxysporum*)	Sudden wilting of vines, discoloration of vascular tissue	Wilting	(a) Use of resistant landraces and seed treatments with *Trichoderma* spp. (b) Avoiding water stagnation in fields.
Mosaic virus (cucumber mosaic virus—CMV)	Mottling, leaf distortion, stunted growth	Vector: Aphids	(a) Control of aphid population. (b) Use of virus-free seeds. (c) Rouging of infected plants early in the season.

9.5 Improvement (Genetics and Genomics): Breeding Efforts, Genetic Diversity, and Recent Advancements

Despite its immense potential in arid and semiarid regions, *Kalingda* (*Citrullus colocynthis*) remains underexplored in terms of genetic improvement. Traditionally grown for its hardy nature, seed yield, and multiple ends uses, *Kalingda* has only recently begun to receive attention for systematic breeding and genetic enhancement—particularly in the context of seed-purpose watermelon.

9.5.1 Crop Improvement

9.5.1.1 Germplasm Variability

It requires extensive and detailed assessment of the available germplasm as well as newly evolved promising lines, which could be used in further breeding program or could be directly released as cultivars after proper testing (Figs. 9.8 and 9.9). *Kalingada* is a cross-pollinated crop and offers good potentialities for exploitation of hybrid vigor in improving yield and other qualitative characters.

At present the AICRP on underutilized crops unit maintains 32 accessions collected through exploration. Fifteen morphological descriptors were developed. Using these descriptors the germplasm lines were characterized. The mandate is also to identify and preserve seeds of these precious genetic stocks to NBPGR for long-term preservation (Table 9.3).

9.5.1.2 Genetic Diversity and Germplasm Resources

Kalingda and its close relatives display significant genetic diversity, which is evident in traits such as fruit shape, size, striping patterns, flesh color and texture, seed shape, size, hull color, oil content, and flavor profile. The Central Arid Zone Research Institute (CAZRI), Regional Station Jaisalmer, has played a pioneering role in collecting, conserving, and evaluating germplasm from local landraces and

Fig. 9.8 Variability in fruit shape of *Kalingada*

Fig. 9.9 Variability in seed color and size

Table 9.3 Genetic stocks/germplasm

Sl no.	Entry	IC no. allotted
1	SKNK 7	NGB/RC 645/2001
2	SKNK 4	NGB/RC 646/2001
3	GK-1	NGB/RC 647/2001
4	GK-2	IC 622893

exotic sources. Germplasm evaluation under rainfed conditions revealed wide variability in agronomic and quality traits, providing the foundation for breeding programs.

9.5.1.3 Breeding Objectives

The breeding of *Kalingda* or seed-purpose watermelon has focused primarily on:

- Higher seed yield under rainfed conditions.
- Improved oil content and seed quality.
- Tolerance to biotic and abiotic stresses.
- Suitability for mixed and intercrop systems.
- Uniform maturity and compact growth habit.

9.5.1.4 Selection and Variety Development

Through successive selection cycles and multilocation evaluations, CAZRI identified and released promising varieties:

- CAZJK-13-1: Seed yield 783 kg/ha; oil content 35.6%.
- CAZJK-13-2: Seed yield 753 kg/ha; oil content 39.3%.

SKNK-1102 (2016) and CAZJK-13-2 (2018) were officially released for cultivation in Rajasthan and Gujarat. These varieties have shown superior adaptation to low-input conditions and have supported local livelihoods by offering a dependable source of income.

9.5.1.5 Genomics and Molecular Advances

While molecular studies on *Citrullus colocynthis* are still limited compared to its domesticated cousin (*C. lanatus*), preliminary work has identified important markers and gene sequences related to drought tolerance and root architecture. Seed oil biosynthesis pathways, disease resistance genes (e.g., against CMV and Fusarium), nutritional traits in kernels (magnesium, linoleic acid), advancements in genotyping-by-sequencing (GBS), and transcriptomic analysis in other cucurbits are now being extended to wild and underutilized species like *Kalingda*.

These tools can accelerate:

1. Marker-assisted selection (MAS).
2. Genome-wide association studies (GWAS).
3. Identification of candidate genes for stress resilience and seed traits.

9.5.2 Improved Cultivar

The first-ever variety in *Kalingada*, Gujarat Kalingada-1, having higher seed yield potentiality was registered with NBPGR (Ref. No. NGB/RC 645/2001).

Gujarat Kalingada-1 (GK-1)	
Released year: 2002 Growth habit: surface creeper Length of main vine: 510–525 cm Appearance of first male flower: 44 DAS Appearance of first female flower: 49 DAS Node no. at which first male flower appears: 10 Node no. at which first female flower appears: 21 No of fruits per plant: 8–12 Fruit girth: 41 cm No. of seeds/fruit: 600–650 Avg. seed yield (kg/ha): 1020 Test weight: 6.213 g Seed shape: Oval shape with compressed at one end smooth T.S.S. (%): 3.4; oil (%): 37.1; protein (%): 18.0	
Gujarat Kalingada-2 (GK-2)	
The variety was identified for release by variety identification committee in 27th annual group meet, AICRN on potential crops was held on 2–3/05/2017 at SDAU, Sardarkrushinagar, Gujarat State: Gujarat and Rajasthan Potential seed yield (kg/ha): 750 Crude oil content (g/kg): 390 g/kg Average fruit yield (q/ha): 83. It performed well in different fertility level and spacing Registered in NBPGR: IC 622893	

9.5.3 Constraints in Breeding

The major constraints encountered in genetic improvement of potential crops are outlined as follows:

1. Lack of sufficient variability that is essential for genetic improvement.
2. Utilization of desired germplasm (genetic donors) in breeding programs also lags behind due to inadequate genetic stocks and paucity of information about the existing ones.

3. Considering the large number of crops that are of economic value, number of research organizations and researchers engaged are too few, resulting in dilution effect or research efforts and priorities.
4. Lack of interest by research workers to work with underutilized crops owing to limited utilization.
5. Some of the industrial crops such as Jojoba, Simarouba, and Jatropha have long gestation period also hinders quick genetic improvement.
6. Need of large experimentation fields for evaluation.
7. Difficulties in seed collection and seed distribution.

9.5.4 Future Directions

- Participatory plant breeding (PPB) involving local farmers for trait preference.
- Pre-breeding efforts to introgress useful genes from wild relatives.
- Biotechnological interventions to understand and manipulate oil biosynthesis genes.
- Genomic resource development, including whole-genome sequencing and functional gene annotation.

Given the urgent need to diversify food and nutritional sources under climate change scenarios, targeted improvement of *Kalingda* holds promise not only for food security but also for nutraceutical and industrial applications.

9.6 Primary Processing: Postharvest Handling and Processing Techniques

Postharvest handling and primary processing of *Kalingda* (*Citrullus colocynthis*) play a critical role in maintaining quality, reducing wastage, and enhancing market value especially in arid regions where the crop is primarily grown for its seeds. The fruit, though not consumed fresh like commercial watermelon, is harvested mainly for its oil-rich seeds and livestock-friendly rind, making efficient processing essential for maximizing its value.

9.6.1 Harvesting Practices

9.6.1.1 Maturity Indicators

Kalingda fruits are harvested when the rind color dulls and the tendrils near the fruit dry up.

9.6.1.2 Manual Harvesting

Due to the scattered nature of mixed/intercropped fields, fruits are manually picked and collected in small heaps.

9.6.2 Field Drying

Fruits are sun-dried in open fields for a few days to reduce moisture content, facilitating easier seed extraction.

9.6.3 Seed Extraction

9.6.3.1 Traditional Method

The sun-dried fruits are manually broken open using wooden sticks or tools.

9.6.3.2 Pulp Separation

Seeds are separated from the pulp manually or by trampling and then washed in water to remove any residual pulp and dirt.

9.6.3.3 Drying of Seed

Cleaned seeds are sun-dried on tarpaulin or clean mats to 8–10% moisture content for safe storage. Proper drying enhances seed longevity and reduces fungal infection risks.

9.6.4 Dehulling and Kernel Separation

9.6.4.1 Hulling

In local practices, seeds are manually cracked to remove the tough outer hull. In larger setups, small-scale mechanical dehullers are used.

9.6.4.2 Kernel Yield

Dehulling yields about 60–65% kernels, locally known as *magaj*, which are highly valued for culinary and medicinal use.

9.6.5 Oil Extraction (Optional in Villages)

9.6.5.1 Cold Pressing

Traditional oil expellers or ghani systems are sometimes used in rural areas to extract oil from dried seeds.

9.6.5.2 Modern Pressing

In commercial setups, screw press or hydraulic press machines extract oil more efficiently, followed by filtration.

9.6.6 Rind Utilization

9.6.6.1 Green Rind (Fresh)

The fresh rinds, not consumed by humans, are chopped and fed to cattle as a juicy fodder.

9.6.6.2 Dry Rind (*Khuparia*)

Dried rinds are stored and used as a protein-rich livestock feed, particularly during lean periods. Rind contains ~12–15% crude protein and up to 25% ether extract (on dry weight basis).

9.6.7 Storage Methods

9.6.7.1 Local Storage Methods

Seeds and kernels are stored in earthen pots or jute bags lined with neem leaves or ash to prevent pest infestation. Airtight containers in urban areas or for commercial trade to maintain quality and prevent oil rancidity.

Kalingda's primary processing is inherently low-cost, eco-friendly, and fits well into rural agroecosystems with limited access to modern infrastructure. With proper handling and basic tools, farmers can ensure seed quality, value addition, and extended shelf life.

9.7 Pre- and Postharvest Technology: Storage, Drying, and Transportation

Efficient pre- and postharvest management is vital for reducing losses, preserving quality, and enhancing the commercial viability of *Kalingda* (*Citrullus colocynthis*) seeds and by-products. Since this desert crop is largely cultivated under low-input, rainfed systems, practical and cost-effective technologies are essential for its value chain development—especially in arid and semiarid zones.

9.7.1 Preharvest Considerations

9.7.1.1 Harvest Timing

Timely harvesting is crucial to avoid over-ripening, which can cause seed deterioration or fungal infestation.

9.7.2 Field Preparation for Drying

Before harvest, farmers often prepare a clean, elevated space for sun-drying the fruits and seeds, ensuring protection from livestock and moisture accumulation.

9.7.3 Drying Technology

9.7.3.1 Sun-Drying

It is a widely adopted method due to low cost and effectiveness in arid climates. Fruits are spread on clean surfaces (tarpaulins, mats, concrete floors) and sun-dried for 3–5 days before seed extraction. Post-extraction seeds are further sun-dried to safe moisture levels ($\leq$10%) to maintain viability and prevent fungal attack.

9.7.3.2 Improved Drying Practices

The use of solar dryers or polythene-sheet-covered structures can protect seeds from dust, rain, and birds while ensuring uniform drying. Controlled drying improves seed quality and minimizes aflatoxin contamination.

9.7.4 Postharvest Storage

9.7.4.1 Seed Storage

Properly dried seeds are stored in gunny bags, jute sacks, or metal bins lined with neem leaves, ash, or turmeric powder to deter storage pests. Airtight containers for high-quality seed lots intended for market sale or oil extraction. For long-term storage, maintaining cool, dry, and well-ventilated conditions is key.

9.7.4.2 Kernel Storage

Dehulled kernels (*magaj*) are more prone to rancidity due to their high oil content. Stored in airtight jars, vacuum-sealed pouches, or food-grade plastic containers. The addition of natural antioxidants such as asafetida or neem oil is practiced in some regions.

9.7.4.3 Rind Storage (*Khuparia*)

Dried rinds are bundled and stored under shade or in simple hay-roofed structures. They retain nutritional value for several months and serve as a reliable livestock feed during feed-scarce periods.

9.7.5 Transportation and Packaging

9.7.5.1 Seeds and Kernels

For local markets, it is transported in jute bags or sacks. For distant/national markets, it is packed in moisture-proof polyethylene or laminated pouches to maintain hygiene and shelf life.

9.7.5.2 Bulk Movement

Larger consignments of seeds or kernels may be transported using tractors, pickup vans, or trucks. Proper cushioning and stacking during transport prevent mechanical damage and maintains quality.

9.7.6 Challenges and Opportunities

- Lack of modern drying and dehulling facilities in rural areas leads to variability in product quality.
- Need for low-cost community-level processing units for seed extraction, drying, and oil expelling.
- Opportunities exist for entrepreneurship development and farmer–producer organizations (FPOs) focusing on seed value chain enhancement.

With minimal investment, improved postharvest and storage techniques can significantly enhance the quality and marketability of *Kalingda*-based products, contributing to farmer income and nutritional security in dryland regions.

9.8 Utilization Pattern: Traditional and Modern Uses in Food, Medicinal, and Industrial Applications

Kalingda (*Citrullus colocynthis*), known locally as *Mateera* or *Khuparia*, holds immense traditional and modern utility, especially in arid and resource-limited regions. While the fleshy part of the fruit is bitter and not eaten, the seeds, oil, and rind serve multiple roles in food, health, and livestock nutrition. Its adaptability to extreme climates and multipurpose nature makes it an important fasting superfood, especially during lean seasons in arid regions.

9.8.1 Quality

9.8.1.1 Watermelon as Oilseed Crop

Among nine oilseeds, soybean (39%), groundnut (26%), rapeseed, and mustard (24%) contributes more than 88% of total oilseeds production in the country. However, in terms of vegetable oil production, mustard, soybean, and groundnut contribute 31%, 26%, and 25%, respectively (Table 9.4). Consequently, we are dependent on only three oilseed crops and fulfilling 60% (14 million tons) of our domestic demand of vegetable oil through imports with a total value of Rs. 69,586 crores during 2016–17, which includes largest share of palm oil (63%) followed by soybean (25%) and sunflower (12%) oil (Anonymous 2016). Therefore, it is high time to explore the alternate sources of vegetable oil, which are indigenous in nature and potential for commercial cultivation on large scale.

Seed oil yields of *Citrullus* species were reported as early as in 1940s (Dhingra and Biswas 1945; Sao and Potts 1952; Das et al. 2002) and recently (Jarret and Levy 2012; Mahla et al. 2018). Watermelon seed contains 20–40% oil besides other minerals in appreciable quantity (Ibeanu et al. 2012; Jarret and Levy 2012; Samia et al. 2012). Its seed oil contains 70.0% unsaturated fatty acids, with linoleic acid (18:2) being the dominant fatty acid (Jarret and Levy 2012; Mahla et al. 2018) and is used for oil production at the subsistence level in West Africa and the Middle East (Baboli

Table 9.4 Physicochemical properties and fatty acid composition of watermelon seed oil

Properties	Watermelon seed oil						Sunflower oil[7]	Soybean oil[7]
	1	2	3	4	5	6	[7]As per codex standards	
Refractive index (25 °C)	1.471	1.35	1.466	1.468	–	1.466	1.467–1.469	1.466–1.1467
Saponification value (mg KOH/g oil)	200	220.2	212.6	198.0	205.3	199.8	188–194	189–195
Iodine value (g I/100 g oil)	156	114.9	128.8	107.5	–	97.1	110.143	120.143
Acid value (mg KOH/g oil)	2.4	7.09	5.05	–	6.10	–	4.0	4.0
Peroxide value (mequiv. O_2/kg oil)	3.2	20.0	3.40	1.31	2.80	2.90	10 0.0	10
Oil yield (g/100 g dry seed)	50	57.3	–	–	–	–	40	21
Palmitic (C16:0) acid (%)	11.4	10.6	9.84	15.5	10.6	15.0	3–10	7–14
Stearic (C18:0) acid (%)	7.0	8.3	6.36	12.6	8.3	11.2	1–10	3–5.5
Oleic (C18:1) acid (%)	13.2	13.7	10.8	20.5	13.7	21.2	14–35	18–26
Linoleic (C18:2) acid (%)	68.3	62.1	72.6	50.8	62.1	51.1	55–75	50.57
Total saturated fatty acids (%)	18.4	–	–	–	–	–	4–22.5	10–20.6
Total unsaturated fatty acid (%)	81.5	–	–	–	–	–	69–95	73.5–93.5

[1]Source: Baboli and Kordi (2010), [2]Edidiong and Ubong (2013), [3]Neuza (2015), [4]Azeem (2015), [5]Egbuonu et al. (2015), [6]Raziq et al. (2012), [7]Kirk (1991)

and Kordi 2010; Jarret and Levy 2012). To meet out the challenges of sustainable rainfed production, native crop species with diversified economic products can play a vital role. Seed-purpose watermelon fits well in the traditional system of Indian arid zone and provides substantial livelihood support to desert dwellers. The physiochemical properties and fatty acid profile of watermelon seed oil is comparable with sunflower and soybean and therefore could be used successfully as a source of edible oils for human consumption (Sabahelkhier et al. 2011; Baboli and Kordi

Table 9.5 Quality of *Kalingada* of Gujarat and Rajasthan

Sl no.	Sources of seed	Seed color	Kernel (*magaj*) recovery (%)	Oil content (%)
1.	Banaskantha (Guj) (*Kalingada*)	Black	39–42	20–35
2.	Barmer/Balotral, Jodhpur (Raj) (*Mateera*)	Black/ brown	35–39	17–18
3.	Nokha, Lunkaransar, Nohar Bhadra, Churu (Raj)	White/ brown	35–36	14–15

2010; Navaratne and Subasinghe 2014) and industrial applications (Falade and Obuseh 2014; Essien and Eduok 2013) and can be commercially exploited (Das et al. 2002). This may not only provide livelihoods but also reduce import and promote industries in the arid zone.

Recently, the "*Kalingada beej*" is in great demand in the local market due to its protein-dense and oil-rich kernels, popularly known as "*magaj*." The kernels contain 35–50% crude protein, 28–40% oil and minerals in appreciable quantity (Table 9.5). Further, the oil contains more than 80% unsaturated fatty acids with linoleic acid being the dominant fatty acid. A seed sample collected from *Bhabhar* (Gujarat) has oil content to the extent of 27% in seed. It is well comparable with oil content in rape seed and mustard, which is an irrigated crop.

The proximate composition and other quality parameters of the watermelon seeds have been reported to be comparable with sunflower and groundnut oil and hence seed-purpose watermelon has a great potential not only as oilseed crop but also in livelihood and nutritional security under resource constraint situation of Indian *Thar Desert*. *Kalingada* seeds are used for oil production at the subsistence level in several African countries, and in the Middle East Asia. They can be bruised and rubbed up with water to form an emulsion, which can be used to cure catarrhal infection, disorders of the bowels, urinary passage, and fever. Watermelon seed oil is light, penetrating, and rich in essential fatty acids. Das et al. (2002) reported 28% crude fat and 23% crude protein in *Kalingada* seeds, while in kernel the corresponding values observed were 49% and 40%, respectively. Recently, the "*Kalingada beej*" is in great demand in the local market due to its protein- and oil-rich kernels, popularly known as "*magaj*." The kernels contain 35–50% crude protein and 28–40% oil and minerals in appreciable quantity. Further, evaluated composition of watermelon seed oil for fatty acid profile and reported 81.6% unsaturated fatty acids with linoleic acid (18:2) being the dominant fatty acid (68.3%).

9.8.2 Uses

This seed-purpose watermelon, locally known as *Kalingada* in Gujarat and Rajasthan, is grouped under potential crops with great potential for oil- and protein-rich seed yield in Indian *Thar Desert*. The *Kalingada* has the potential to provide seed yield up to 10 quintals per hectare with more than 30% crude protein and oil.

The immature fruits of *Kalingada* are used as green vegetable. The mature fruits are fed to the livestock. The dehulled kernels are used for garnishing sweets and culinary preparation. The roasted and salted seeds are directly consumed as such or by mixing with other cereal grains. Most importantly, however, seed is used for extraction of vegetable oil as well as good source of protein. It is a source of energy to farmers as refreshing fruit with health benefits. It is a source of livestock feed and livelihood security.

Presently, seeds of *Kalingada* and *Mateera* are being used by "*Magaj*" industries located in Sidhpur and Unjha in Gujarat and in Nokha, Jodhpur, and Bikaner in Rajasthan. The *magaj* recovered from the seed is used in sweets and *mukhwas* (mouth freshener), *thundai*, and to some extent in *Unani* medicine and it is also used in the preparation of gravy for Punjabi sabji/dishes in restaurants. In the years of high production, the seed is used for oil extraction for Vanaspati ghee making by large and other small vanaspati ghee industries.

A wide variability in kernel (*magaj*) recovery and oil content has been observed in the collection from different sources as given below.

9.8.2.1 Traditional Uses

9.8.2.1.1 As Fasting Food

The kernels (*magaj*) are rich in fats and proteins and are widely used during fasting periods. Roasted or raw, they are mixed into traditional sweets such as laddus, halwa, and panjiri during fasting rituals.

9.8.2.1.2 Medicinal Applications

In *Ayurveda* and folk medicine, the dried pulp and seeds are used for purgative and digestive properties, treating constipation, intestinal worms, and inflammation. The seed oil is used externally for joint pain and skin diseases.

9.8.2.1.3 Animal Feed

The rind or *khuparia* is a reliable fodder source for livestock in drought-prone areas. Used in both green and dried forms, it supplements cattle diets, especially in lean months.

9.8.2.2 Culinary Uses

9.8.2.2.1 Kernels (*Magaj*)

Snack food: Roasted and salted as a healthy snack.

Flavor enhancer: Used in cooking gravies and traditional dishes for thickening and flavor.

Bakery and sweets: Incorporated in cakes, energy bars, and sweets for texture and nutrition.

Fasting recipes: Popular in *Navratri* and other fasting cuisines across India.

9.8.2.2.2 Seed Oil

Extracted oil is edible and used for cooking in some local traditions. High in unsaturated fats (~35–48%), making it comparable to other edible oils such as sunflower or safflower oil.

9.8.2.3 Modern Uses

Nutraceuticals and Health Foods Kernels are increasingly marketed as superfoods due to their high protein and essential fatty acids. Its antioxidant properties are suitable for vegan, gluten-free, and low-carb diets.

9.8.2.4 Pharmaceutical and Cosmetic Uses

Seed oil is used in skincare formulations for its moisturizing and anti-inflammatory properties. Extracts are being studied for antidiabetic and anti-inflammatory activities.

9.8.2.5 Industrial Uses

Residual seed cake (after oil extraction) is used in making organic fertilizers or animal feed. It is a potential raw material for biofuels and biodegradable packaging due to oil content.

9.8.3 Livelihood and Food Security

Income generation: High-value seed sales support rural households in arid zones.

Food security: As a dependable dryland crop, *Kalingda* fills dietary gaps in lean seasons.

Climate resilience: Its drought tolerance makes it ideal for cultivation under changing climate scenarios.

In essence, *Kalingda* bridges tradition and modernity—providing nutrition during fasting, income during scarcity, and utility across industries. It stands out as a model crop in the fasting superfood domain, especially for regions where food diversity and resilience are paramount.

9.9 Value Addition and Products: Potential for Creating High-Value Products

Kalingda (*Citrullus colocynthis*), often viewed primarily for its role in subsistence farming and traditional uses, has immense untapped potential for value addition. Through modern processing techniques, the different parts of the plant—especially the seeds, oil, and rind—can be converted into high-value products that cater to both local and international markets. The multipurpose nature of the crop enhances its commercial viability, making it a fasting superfood with diverse applications.

9.9.1 Seed-Based Products

9.9.1.1 Edible Seed Kernels (*Magaj*)

1. *Nutrient bars and snacks*: Roasted, salted, or spiced kernels can be packaged as ready-to-eat snacks. This market is growing, especially among health-conscious consumers looking for high-protein, low-carb snacks.
2. *Energy supplements*: Used in energy bars or smoothies, catering to fitness enthusiasts and people on special diets such as ketogenic, vegan, or gluten-free.
3. *Flour production*: Dehulled and ground kernels can be processed into flour, which can be used in baked goods or as a substitute for wheat flour in gluten-free recipes.

9.9.1.2 Seed Oil

1. *High-quality cooking oil*: With its high unsaturated fat content, watermelon seed oil is comparable to other oils such as sunflower and safflower. It is suitable for cold-pressed oil extraction.
2. *Cosmetic and personal care products*: Extracted oil is used in skincare and haircare products due to its moisturizing and anti-inflammatory properties. Its use in high-end creams, lotions, and soaps is gaining popularity.
3. *Nutraceutical products*: The oil can be marketed as a functional food ingredient, containing essential fatty acids and antioxidants, suitable for dietary supplements.

9.9.2 Rind-Based Products

9.9.2.1 Dried Rind for Livestock Feed

Premium animal feed from dried rind or *khuparia* can be packaged as premium fodder for cattle, poultry, and small ruminants, especially in arid regions where other feed resources are scarce. The high protein content in dried rinds makes it a good supplement.

9.9.2.2 Organic Fertilizer

Rinds, along with seed cakes, can be converted into organic manure, contributing to sustainable agriculture practices in drought-prone regions.

9.9.2.3 Natural Fiber Extraction

Rind fiber for textile industry: The outer layer of the rind has potential for fiber extraction, which could be used in the manufacture of biodegradable textiles or eco-friendly packaging materials.

9.9.2.4 Craft Products

In local markets, the dried rinds can be used in handicrafts, making baskets, mats, or decorative items.

9.9.3 Functional and Medicinal Products

9.9.3.1 Pharmaceutical Formulations

Anti-inflammatory and digestive products with increasing interest in traditional medicine and functional foods, the seeds and rind of *Kalingda* can be processed into herbal supplements for managing digestive health, blood pressure, and cholesterol.

9.9.3.2 Antioxidant Supplements

The antioxidant-rich compounds in the seeds, rind, and fruit pulp can be extracted and marketed as nutraceuticals targeting wellness and antiaging markets.

9.9.3.3 Herbal Tea (*Kalingda* Seed Tea)

The seeds can be roasted and used as a base for herbal tea, marketed as a calming or detoxifying beverage, appealing to the growing demand for natural wellness drinks.

9.9.4 Bio-Based Products

9.9.4.1 Biofuels

With the oil content in *Kalingda* seeds, there is potential to explore their use in biofuel production, especially in rural or off-grid areas, contributing to renewable energy sources and reducing dependency on fossil fuels.

9.9.4.2 Biodegradable Packaging

Extracted fibers from the rind can be used in the production of biodegradable packaging, which is gaining momentum as an eco-friendly alternative to plastic in the global market.

9.9.5 Marketing and Export Potential

9.9.5.1 Export Markets

Due to the increasing demand for natural health products and sustainable raw materials, *Kalingda*-based products, especially seeds and oil, have significant export potential. Countries with growing health-conscious populations, such as the United States and European nations, are key markets.

9.9.5.2 Local and National Demand

The traditional food and medicinal properties of *Kalingda* cater to a niche market in India, especially in arid and semiarid regions. At the same time, modern value-added products can tap into a broader, health-conscious consumer base.

The various uses of *Kalingda* in sectors such as food, health, and industry present numerous opportunities for adding value. By improving processing methods, there is significant potential to create a variety of high-value products that can aid in economic development while also promoting sustainability and enhancing food security in arid areas.

9.10 Marketing: Global Demand, Trade, and Market Trends

Kalingda (*Citrullus colocynthis*) is emerging as a significant fasting superfood, with increasing attention on its health benefits and versatile applications. Despite its traditional and regional significance, there is growing global demand for *Kalingda*-based products in the food, medicinal, and industrial sectors. As markets evolve, the potential for this resilient crop to drive economic growth and sustainable agriculture is becoming more apparent.

9.10.1 Global Demand for *Kalingda*-Based Products

9.10.1.1 Health-Conscious Consumers

1. *Superfood trend*: The rising trend of consuming superfoods and natural products has fueled interest in *Kalingda*, especially its seeds and oil, which are rich in essential fatty acids, antioxidants, and protein.
2. *Nutraceutical market*: The demand for functional foods, dietary supplements, and herbal teas is expanding globally, particularly in North America, Europe, and Asia, where consumers are seeking natural alternatives to traditional pharmaceutical products.

9.10.1.2 Cosmetic and Personal Care Industry

The growing awareness of the anti-inflammatory and moisturizing properties of *Kalingda* oil is contributing to its rise in the cosmetics industry. It is being integrated into skin creams, lotions, and hair care products, positioning it as a premium ingredient in the natural beauty market.

9.10.1.3 Animal Feed and Agricultural Products

As global concerns about sustainable farming and feed security grow, *Kalingda*'s rind and seed cake have garnered interest in the livestock feed and organic fertilizer sectors. The plant's drought tolerance makes it an attractive option for farmers looking for low-input, high-output solutions.

9.10.2 Trade and Export Opportunities

9.10.2.1 Seed and Seed Oil Exports

India, the largest producer of *Kalingda*, has a well-established domestic market for watermelon seeds, but there is vast untapped potential in international markets. China, the Middle East, and Africa are also key players in watermelon seed exports, and *Kalingda* could be promoted as a specialty product in these regions due to its medicinal and nutritional value.

9.10.2.2 Global Packaging and Industrial Use

As the demand for eco-friendly materials increases, *Kalingda*'s potential as a source for biodegradable packaging can offer a unique export opportunity. This aligns with global movements toward sustainable practices in packaging, making *Kalingda* a prime candidate for the green economy.

9.10.2.3 Market Penetration

Through strategic partnerships, branding, and targeted marketing, the export of *Kalingda* products can expand globally. Indian exporters can look to international trade fairs, online platforms, and health-food markets to introduce these products to a broader audience.

9.10.3 Market Trends

9.10.3.1 Health and Wellness Boom

As health consciousness increases globally, consumers are seeking natural and organic alternatives. *Kalingda*'s high protein content, healthy fats, and functional health benefits align with the current shift toward plant-based diets, gluten-free products, and low-carb eating. Plant-based protein is becoming more desirable, and *Kalingda* seeds are a rich source of plant-based protein, making them an attractive addition to this growing market segment.

9.10.3.2 Sustainability and Climate Resilience

Given the growing global concerns about climate change, *Kalingda*'s drought-tolerant nature and ability to thrive in marginal lands position it as an ideal crop for sustainable farming in arid and semiarid regions. The focus on sustainable agriculture and resource-efficient crops further elevates *Kalingda*'s relevance in modern agricultural markets.

9.10.3.3 Rise in Functional Foods

As consumers demand more functional foods, *Kalingda*-based products are set to grow in popularity. The market for functional snacks and dietary supplements is expanding, and *Kalingda* can serve as a core ingredient in these categories due to its nutrient density and medicinal properties.

9.10.4 Challenges and Future Outlook

9.10.4.1 Limited Awareness and Research

Despite the growing interest, awareness about *Kalingda*'s potential is still limited, especially outside of India. Increased research into its genetics, processing, and health benefits will be key to unlocking its full commercial potential. Governments, research institutions, and the private sector need to collaborate to promote *Kalingda* as a high-value crop.

9.10.4.2 Infrastructure for Processing and Marketing

Efficient processing infrastructure for *Kalingda*-based products is still underdeveloped in many regions. Investment in modern processing techniques, such as cold-pressing for oil and advanced packaging for seeds, will boost the quality and shelf-life of these products. Supply chain improvements will be essential for maintaining consistent product quality and meeting international demand.

Kalingda is poised to make a significant impact on the global market as a fasting superfood, thanks to its nutritional benefits, sustainability, and multiuse potential. With proper research, value addition, and marketing strategies, it can become a key player in the global health-food, cosmetic, and agricultural industries (Table 9.6).

Table 9.6 Nutritional, traditional, modern uses, and market potential of *Kalingda* (*Citrullus colocynthis*)

Plant part/ product	Nutritional/ medicinal benefit	Traditional use	Modern/industrial use	Emerging market opportunities
Seeds/ kernels (*magaj*)	High in protein, amino acids, essential fatty acids, vitamins	Roasted as snack, festive and ritual use	Functional food (snack bars, flour), dietary supplements, oil extraction	Health-conscious consumers, nutraceuticals, plant-based protein
Seed oil	Rich in unsaturated fats, antioxidants, anti-inflammatory agents	Medicinal skin use (folk medicine)	Cooking oil, cosmetic ingredient (creams, lotions), pharmaceutical base	Natural cosmetics, personal care, herbal supplement industries
Rind (*khuparia*)	12–15% crude protein, 20–25% ether extract	Fodder for livestock, dried for off-season use	Organic manure, livestock feed, potential source of biodegradable packaging fiber	Sustainable agriculture, eco-friendly packaging materials
Fruit flesh	Hydrating, rich in water and antioxidants	Seasonal food in arid zones	Ingredient in health drinks, pickles, or fermented foods	Functional foods, fasting diets, hydration-based wellness foods
Whole plant (crop)	Grows in degraded soils with minimal inputs	Mixed cropping with pearl millet, cluster bean	Low-input crop for arid farming, cover crop, soil stabilizer	Climate-resilient agriculture, rainfed farming solutions

References

Anonymous (2016) Directorate of Vanaspati, vegetable oil and fats (DVVOF) and Department of Commerce, Kolkata

Azeem MW (2015) Analytical characterization of pure and blended watermelon (*Citrullus lanatus*) oil: impact of blending on oxidative stability. Pak J Anal Environ Chem 16(1):52–58

Baboli ZM, Kordi AAS (2010) Characteristics and composition of watermelon seed oil and solvent extraction parameters effects. J Am Oil Chem Soc 87:667–671

Das M, Das SK, Suthar SH (2002) Composition of seed and characteristics of oil from karingda [*Citrullus lanatus* (Thumb) Mansf]. Int J Food Sci Technol 37:893–896

Dhingra DR, Biswas AK (1945) Component fatty acids of oil of *Citrullus vulgaris* Schrad. (water melon) seed. J Indian Chem Soc 22:119–122

Edidiong AE, Ubong ME (2013) Chemical analysis of *Citrullus lanatus* seed oil obtained from southern Nigeria. Elixir Org Chem 54:12700–12703

Egbuonu ACC, Aguguesi RG, Samuel R, Ojunkwu O, Onyenmeri F, Uzuegbu UE (2015) Some physicochemical properties of the petroleum ether-extracted watermelon (*Citrullus lanatus*) seed oil. Asian J Sci Res 8:519–525

Essien EA, Eduok UM (2013) Chemical analysis of *Citrullus lanatus* seed oil obtained from southern Nigeria. Elixir Org Chem 54:12700–12703

Falade OS, Obuseh V (2014) Evaluation of physicochemical properties and total phenol contents of watermelon (Rothmas and sugar baby) seed oils. IFE J Sci 16:257–264

Ibeanu VN, Onyechi UA, Ugwuanyi GU (2012) Nutrient and dietary fibre profile of dehulled and undehulled seeds of sweet princess watermelon (*Citrullus lanatus*) consumed in Nigeria. Int J Basic Appl Sci 12:249–252

Jarret RL, Levy IJ (2012) Oil and fatty acid contents in seed of *Citrullus lanatus* Schrad. J Agric Food Chem 60:5199–5204

Kirk SR (1991) Pearson's composition and analysis of foods, 9th edn. Longman, London, pp 617–620

Kechi M (1990) Nutrients in seeds of edible watermelon (*Citrullus lanatus* (Thunb.) Matsum. & Nakai). Cucurbit Genet Coop Rep 13:43–44

Mahla HR, Choudhary BR (2013) Genetic diversity in seed purpose watermelon (*Citrullus lanatus*) genotypes under rainfed situations of Thar Desert. Indian J Agric Sci 83:300–303. https://epubs.icar.org.in/index.php/IJAgS/article/view/28345

Mahla HR, Singh JP, Roy MM (2014) Seed purpose watermelon in arid zone. Central Arid Zone Research Institute, Jodhpur

Mahla HR, Rathore SS, Venkatesan K, Sharma R (2018) Analysis of fatty acid methyl esters and oxidative stability of seed purpose watermelon (*Citrullus lanatus*) genotypes for edible oil. J Food Sci Technol 55(4):1552–1561

Navaratne SB, Subasinghe DJS (2014) Determination of fatty acid profile and physicochemical properties of watermelon and soursop seed oils. Eur Int J Appl Sci Technol 1:26–32

Neuza J (2015) Physicochemical characterisation and radical-scavenging activity of Cucurbitaceae seed oils. Nat Prod Res 29(24):2313–2317

Raziq S, Anwar F, Mahmood Z, Shahid SA, Nadeem R (2012) Characterization of seed oils from different varieties of watermelon [*Citrullus lanatus* (Thunb.)] from Pakistan. Grasas Aceites 63:365–372

Sabahelkhier MK, Ishag KEA, Sabir Ali AK (2011) Fatty acid profile, ash composition and oil characteristics of seeds of watermelon grown in Sudan. Br J Sci 1:76–80

Samia E-SF, Salem RH, Abd El-Ghany ME (2012) Chemical and nutritional evaluation of different seed flours as novel sources of protein. World J Dairy Food Sci 7:59–65

Sao TCM, Potts WM (1952) The analysis and characterization of the oil from the seed of *Citrullus vulgaris*. J Am Oil Chem Soc 29:444–445

Sweet Potato (*Ipomoea batatas* L.): The Nutrient-Rich Root of Resilience

10

Devendra Upadhyay, Padmakshi Thakur, Oshin Pali, and Ashish Kumar Tiwari

Abstract

Shakarkandi, or sweet potato (*Ipomoea batatas* L.), is a nutrient-dense root crop that is well-known for its cultural significance, adaptability, and resilience. It is grown all over the world in tropical and subtropical climates and is an essential source of vitamins, minerals, dietary fibre, carbohydrates, and bioactive substances. With a total national productivity of 11.67 MT ha^{-1}, Odisha, Uttar Pradesh, West Bengal, and Madhya Pradesh are the top-producing states in India. The crop is valuable for both human nutrition and animal feed because sweet potato tubers are high in starch (27%), sugars (3–6%), and carotenoids (up to 20 mg/100 g), while the leaves are high in protein (4%) and micronutrients. Its high vitamin A content, low glycaemic index, and antioxidant qualities support the immune system, enhance vision, and a lower chance of developing chronic illnesses. In addition to its role in food security, sweet potatoes have a variety of uses in processed foods, the production of starch and bioethanol, and traditional fasting customs. In Hindu culture, in particular, they are associated with prosperity and good health. As a climate-resilient crop that supports sustainable agriculture and nutritional security objectives, it can flourish in marginal soils in drought-prone and resource-constrained environments. The biological, nutritional, and cultural aspects of sweet potatoes are highlighted in this review, reinforcing their reputation as 'The Nutrient-Rich Root of Resilience'.

D. Upadhyay (✉) · O. Pali
Dau Kalyan Singh College of Agriculture and Research Station, Indira Gandhi Krishi Vishwavidyalaya, Bhatapara, Chhattisgarh, India
e-mail: devendra.upadhyay@igkv.ac.in

P. Thakur
S.G. College of Agriculture and Research Station, IGKV, Jagdalpur, Chhattisgarh, India

A. K. Tiwari
Sant Vinova Bhave College of Agriculture and Research Station, IGKV, Marra, Patan, Chhattisgarh, India

A. Kumar et al. (eds.), *Fasting Superfoods*,
https://doi.org/10.1007/978-981-95-2762-5_10

Keywords

Shakarkandi · Sweet potato · Nutritional security · Biofortification · Functional food · Carotenoids · Low glycaemic index

10.1 Introduction

Sweet potato (*Ipomoea batatas* L.) is a vital starchy food crop that is often overlooked, yet it is recognized as a nutritional powerhouse and symbol of resilience in tropical and subtropical regions (Tedesco et al. 2023). The sweet potato is a versatile, nutrient-rich root crop that has long been cherished across cultures for its delicious taste, high productivity, and exceptional health benefits (Gelaye 2024). Commonly recognized for its natural sweetness and vibrant orange flesh, the sweet potato is increasingly gaining attention worldwide as a '*super food*' due to its dense nutritional profile, functional food properties, and ability to contribute to food and nutritional security (Islam 2024). In the southern United States, it is commonly referred to simply as potato, while what many know as potato is identified as White Potato or Irish Potato (Laurie et al. 2024).

10.1.1 Origin and Botany

The sweet potato does not occur in a wild form (Rodamilans et al. 2021). South America is regarded as the origin of the sweet potato (Bach et al. 2021). In prehistoric eras, it served as a food crop in both tropical America and Polynesia (Cooley 1951; Leon 1976). It spread into Polynesia by Spanish expedition that started from Peru in the sixteenth century (Leon 1976). Sweet potato was cultivated in the coastal area of Peru which was introduced possibly from the north, the coastal area of Ecuador, and Columbia (Martin et al. 1976). The spread of sweet potato into Africa, North America, Europe, India, China, Japan, and the Philippines was the result of Spanish, Portuguese, and British trade, exploration, and colonization (O'Brien 1972; Juras 2021).

Sweet potato (*Ipomoea batatas*) belongs to the family Convolvulaceae. The family Convolvulaceae comprises 45 genera and 1000 species distributed throughout the tropical and warm temperate regions of the world (Jiang et al. 2022b). The genus *Ipomoea* comprises approximately 400 species, primarily found in tropical regions. Among them, *Ipomoea batatas*, widely referred to as the sweet potato, represents the only noteworthy edible root crop within this genus (Jiang et al. 2022a, b). Sweet potato is a hexaploid species with somatic chromosome number of 90. Its origin is likely linked to amphidiploidy, derived from a tetraploid ($2n = 60$) and a diploid ($2n = 30$), which produced a triploid form ($2n = 45$) that later underwent chromosome doubling, resulting in hexaploidy ($2n = 90$) (Yan et al. 2022; Obata et al. 2022). It is believed that sweet potato originated from hexaploid *Ipomoea trifida* (Nishiyama 1961). Purseglove (1965) also considers *Ipomoea trifida* as a progenitor of sweet potato. Nishiyama (1971) confirmed *Ipomoea trifida* as the direct ancestor, while its diploid predecessors are thought to be *Ipomoea leucantha* and *Ipomoea litoralis*. Martin (1972) supported Nishiyama's findings regarding *Ipomoea trifida* and further identified two additional closely related diploid species, *Ipomoea lacunose* and *Ipomoea trichocarpa*. According to Nishiyama et al. (1973), the evolutionary sequence of sweet potato is *Ipomoea leucantha* (BB), *Ipomoea litoralis* (BBBB), *Ipomoea trifida* (BBBBBB), and *Ipomoea batatas* (BBBBBB).

Sweet potato is an herbaceous perennial but cultivated annually (Chauhan et al. 2021). The stem is a branching vine with milky juice. The vines develop roots at the joint during later part of the season. The vines are light green to purple in colour, usually angular in shape, and slender to moderate in thickness. The triangular leaf is borne on a short to long petiole. Leaves are more or less heart-shaped but may be shouldered, toothed, entire, deeply cleft, or variously lobed. Leaves vary in colour usually green to purple.

Tubers of different cultivars vary greatly in form, size, and colour of skin. The shape of tubers is fusiform and globular with smooth or ridged surface. The skin colour is commonly white, yellow-reddish, or purple. The inflorescence of sweet potato is characterized by a cyme that typically features one or a few flowers. The corolla is funnel-shaped, composed of five lobes, and displays a purple hue with a darker throat and lighter margins. There are five unequal stamens that are fixed near the base of the corolla. The pistil consists of a style measuring 2 cm in length, which has a stigma that is divided into two lobes. The seeds of sweet potato are found within a capsule, which can yield up to four seeds, although normally only one or two of these seeds fully mature. The seeds are angular, black in colour, and possess a hard outer coat.

10.1.2 Geographical Distribution

In India, the crop is commonly referred to as '*Sakarkand*'. Its remarkable ability to thrive in a variety of climatic conditions accounts for its extensive distribution, which spans from tropical regions in America, India, and the Pacific Islands to temperate areas in Japan, China, Korea, and North America (Islam 2024). The leading states in terms of area and production are Odisha, Uttar Pradesh, West Bengal, and Madhya Pradesh. Odisha is the largest producer of sweet potato in India (Chauhan et al. 2025) (Table 10.1).

10.1.3 Significance

Sweet potatoes are characterized by a significant starch content, which is notable for its smaller size and resemblance to corn starch granules rather than the larger granules typically found in tubers. The quality of sweet potatoes is largely determined by their sugar content, which ranges between 3% and 6%, varying according to the specific cultivar and environmental conditions. The main type of sugar in sweet

Table 10.1 State-wise area, production and productivity of sweet potato in India (2023–2024)

S.No.	State	Area '000' Ha	Production '000' MT	Productivity MT/Ha
1	Odisha	34.57	335.56	9.71
2	Uttar Pradesh	19.49	269.33	13.82
3	West Bengal	16.33	175.66	10.76
4	Madhya Pradesh	8.36	131.31	15.71
5	Karnataka	7.94	91.24	11.49
6	Jharkhand	3.22	58.88	18.31
7	Chhattisgarh	4.63	53.97	11.65
8	Bihar	2.40	44.77	18.68
9	Jammu & Kashmir	1.37	41.11	30.10
10	Assam	4.43	30.85	6.97
11	Meghalaya	4.97	17.98	3.62
12	Andhra Pradesh	0.50	14.92	30.01
13	Maharashtra	1.47	12.85	8.74
14	Nagaland	1.00	11.72	11.68
15	Tamil Nadu	0.44	7.40	16.87
16	Rajasthan	0.22	2.36	10.76
17	Kerala	0.15	2.16	14.38
18	Telangana	0.13	1.85	14.45
19	Mizoram	0.10	1.00	10.00
20	Andaman and Nico. Islands	0.07	0.71	10.49
21	Goa	0.09	0.28	3.11
22	Manipur	0.01	0.08	16.00
	India	**111.89**	**1306.00**	**11.67**

Source: Ministry of Agriculture & Farmers Welfare, Govt. of India (ON4008)

potatoes is sucrose, accompanied by lesser amounts of reducing sugars (Bhutto et al. 2024; Laurie et al. 2024). Other sugars, including maltose, mannose, galactose, and pentose, are present in very small quantities. Research indicates that the sugar concentration in sweet potato tubers increases during both storage and cooking processes (Islam 2024). Besides sugars, sweet potatoes are also rich in vitamins A, B, and C. Some cultivars can have carotene levels ranging from 5.4 to 20 mg per 100 g, with higher concentrations typically found in the yellow and orange-fleshed varieties, which are significant sources of vitamin A precursors (Bhutto et al. 2024) (Table 10.2).

Sweet potato leaves have a protein content of approximately 3.2% when measured by fresh weight and 12.6% when assessed on a dry weight basis. Additionally, the leaves are abundant in carotene and vitamin C. The green tops provide 4% protein and are notable sources of calcium, phosphorus, and iron (Singh 2023). In terms of fresh tubers, sweet potatoes are primarily composed of carbohydrates, proteins, fats, sugars, various vitamins, and a significant amount of dietary fibre. These nutrients play a crucial role in enhancing the nutritional value of Shakarkandi, establishing it as a valuable food option (Purohit et al. 2024).

Table 10.2 Average chemical composition of tuber and vine

Constituent	Tuber (%)	Vine (%)	Reference
Moisture	70	80.7	Purseglove (1977)
Carbohydrates	27	9.7	
Protein	1.5–2.0	4.2	
Fat	0.2	0.8	
Fibre	1.0	2.4	
Sugar	3–6	–	
Minerals	Mg/100 g		
Calcium	46	360	Maini (1976)
Phosphorus	49	60	
Iron	0.8	10	
Magnesium	24	–	
Potassium	373	–	
Sodium	13	–	
Chlorine	85	–	
Sulphur	26	–	
Vitamins			
Thiamine	0.08	0.07	
Riboflavin	0.04	0.24	
Niacin	0.70	1.70	
Ascorbic Acid	24.0	27	
Carotene	0–18	4	
Energy (K cal)	120	63	

Sweet potato provides various health advantages, including better eye health, boosted immunity, and a lowered likelihood of chronic conditions like diabetes and cardiovascular diseases. Moreover, the crop's adaptability to marginal soils and resilience under drought-prone and resource-limited environments make it an ideal candidate for climate-resilient agriculture and sustainable farming systems, especially in developing countries (Tedesco et al. 2023). Beyond its role in human nutrition, sweet potato also serves as a valuable raw material in industrial food processing, animal feed, and bioethanol production, underscoring its multifaceted importance. With growing interest in functional foods and holistic health, sweet potato is rightly earning its place as a superfood of the future (Gelaye 2024). The main uses of sweet potatoes are for human consumption. It is usually eaten after boiling, baking, and frying and may be candied with syrup or used as a 'puree' (Nyombaire 2023). Tubers are utilized for canning, dehydrating and flour manufacture and are also an important source of starch, glucose, pectin, sugar syrup, and industrial alcohol, and sweet potato is the cheapest source of calories (Histifarina et al. 2023). Considering the crop duration, the approximate food calories production in sweet potato is the highest among the tuber and root crops.

10.2 History: Traditional Uses and Cultural Significance

Indians eat sweet potatoes (*Shakarkandi*) during fasting (*vrat*) primarily due to their high nutritional value, ease of digestion, and sattvic (pure) nature in traditional Hindu dietary practices. The sweet potato is recognized for its value as a natural food source, providing substantial energy, being easily digestible, rich in nutrients, and having a natural sweetness. This aligns with the principles of Sattvic food in Ayurveda. Its hydrating and cooling attributes make it particularly appealing in warmer climates, and it can be used in a wide variety of culinary applications, leading to diverse food preparations. These qualities contribute to its classification as a *Superfood*.

In the rich tapestry of cultural traditions, the sweet potato stands out as a symbol of significance, especially during periods of fasting and joyous occasions like the revered Sakat Chauth. Embracing this fast with genuine devotion is thought to attract the gracious blessings and goodwill of the exalted Lord Ganesha. This day serves as more than merely the passage of time; it represents an opportunity for immense happiness and fulfilment in life. The rituals linked to narrating the fasting story during worship are believed to bring blessings to the child, contributing to a long and successful life, while also fostering a sense of joy and calm within the home. Furthermore, the practice of offering water to *Chandra Dev* (Moon) on this auspicious day is believed to protect the child from illnesses and to attract good fortune. Consequently, the sweet potato, with its nourishing attributes and profound symbolic significance, represents much more than a mere aliment; it embodies a harbinger of well-being and prosperity, intricately woven into the very fabric of life's celebrations and rituals.

Sweet potato is an excellent source of vitamin A, dietary fibre, and several vital trace minerals, including manganese, potassium, and magnesium. Regular consumption of sweet potato can yield numerous health advantages, such as promoting healthy skin, eyes, and bones, enhancing immune function and cognitive health, and stimulating collagen synthesis. Additionally, sweet potatoes are advantageous for individuals with diabetes, as their consumption does not result in an abrupt increase in blood glucose levels, attributed to their low glycaemic index.

10.3 Production

10.3.1 Cultivation Practices

Growth and development of sweet potato (vines and tuber) are dependent on several factors such as environmental conditions, soil quality, and management practices. Optimal temperature, light exposure, and moisture levels play a crucial role in the successful cultivation of sweet potatoes, influencing both vine growth and tuber development. Mellow (1946) observed that the best growth and yield of tubers occurred when the climatic conditions during the growing season included high temperatures both during the day and at night (Dumbuya et al. 2021).

10.3.1.1 Land Preparation

Land preparation for sweet potato differs from cereal crops. The tuberous roots of sweet potato develop better in loose soils. Generally, land preparation depends on the soil structure, fertility, time of planting, and crop sequence followed. For sweet potato planting, the land is loosened cross-wise to a depth of 15–20 cm, depending upon soil texture with the help of cultivator or plough 3–4 weeks before planting.

10.3.1.2 Variety/Cultivar

Most of the existing cultivars are from clonal selection and are self-incompatible. A number of cultivars have been obtained from mutation as chance seedlings or deliberate hybridization. There are large variations in the skin and flower colour, shape, and size of the tubers, the morphological characters, the depth and period of rooting amongst the cultivars. Sweet potato varieties grown in India are mostly popular local cultivar or state-preferred varieties. These are usually grouped either as white or red type. Table 10.3 shows some Indian cultivars/varieties of sweet potato released/recommended for release through AICRP on Tuber Crops.

In Bangladesh, two varieties from exotic germplasms (*Kamalasundari* and *Tripti*) and one from the local germplasm (Daulatpuri) were selected and released for commercial cultivation. In the United States, sweet potatoes are classified as food types and feed types. Usually, the feed type produces higher tuber yield and high starch. Bosewell (1950) divided food types into firm-fleshed and soft-fleshed based on the characteristics of cooked sweet potato. According to Thompson and Kelly (1979), soft-fleshed types have greater commercial importance than firm-fleshed types. The most important soft-fleshed cultivars are Puerto Rico, Unit I, and Cleitt Bunch Puerto Rico.

In Hawaii, there are dozens of good native sweet potato cultivars (Tanaka and Sekioka 1976). The important best-yielding moist cultivar is Cona B, and others are Iliula and Onolena.

There are many Cuban sweet potatoes grown in Florida. The most important cultivar is Verde, which is white-fleshed, sweet, cold tolerant, and has red skin. The important cultivars grown in Japan for industrial use are Koganesengan and Minamiyutaka. Both produce higher tuber yields along with high starch content. Yulmi, a sweet potato cultivar released in S. Korea, selected from the cross Chinmi × M178001-15 (Jeong et al. 1991). Figure 10.1 shows different morphological features of released varieties from India for commercial cultivation.

10.3.2 Soil and Climate

10.3.2.1 Soil

The best soil for sweet potato is sandy loam and clay subsoil (Key 1973; Martin et al. 1976). It can be grown in loamy to clay loam soil also. Laterite soils of good depth and better fertility can also be considered suitable for its cultivation. Heavy clay soil, which becomes hard after drying, checks the development of the tubers, and high sand soil causes the development of long, cylindrical, and thin tubers.

Table 10.3 Some Indian cultivars/varieties[a]

Year of release/ recommendation	Name of the originating centre	Name of variety/cultivar	State/National
1976	TNAU	CO 1	Tamil Nadu
1980	TNAU	CO 2	Tamil Nadu
1980	Dr. RPCAU	Rajendra Sakarkand-5	National
1982	TNAU	CO 3	Tamil Nadu
1983	ICAR-CTCRI	Varsha	Maharashtra
1987	ICAR-CTCRI	Sree Nandhini	Kerala
1987	ICAR-TCRI	Sree Vardhini	Kerala
1987	APHU	Samrat	Andhra Pradesh
1990	Dr. RPCAU	Rajendra Sakarkand-35	National
1990	Dr. RPCAU	Rajendra Sakarkand 43	National
1993	Dr. RPCAU	Rajendra Sakarkand-47	National
1994	APHU	Kiran	Andhra Pradesh, Bihar
1996	ICAR-CTCRI	Sree Bhadra	National
1996	ICAR-CTCRI	Sree Rethna	Kerala
1998	ICAR-CTCRI	Gouri	Odisha
1998	ICAR-CTCRI	Sankar	Odisha
1999	TNAU	COCIP-1	Tamil Nadu
2000	Dr. BSKKV	Konkan Ashwini	Maharashtra
2001	NDUAT	Narendra Shakarkand-9	Uttar Pradesh
2001	Dr. RPCAU	Rajendra Sakarkand-92	Bihar
2004	ICAR-CTCRI RC	Kalinga	National
2005	ICAR-CTCRI RC	Goutam	Odisha
2005	ICAR-CTCRI RC	Kishan	Odisha
2005	ICAR-CTCRI RC	Sourin	Odisha
2006	IGKV	Indira Madhur	Chhattisgarh
2006	IGKV	Indira Naveen	Chhattisgarh
2007	IGKV	Indira Nandini	Chhattisgarh
2008	BCKV	Kamala Sundari	West Bengal
2011	ICAR-CTCRI	CO 3–4	National
2011	IGKV	Indira Priya	National
2011	IGKV	Indira Narangi	National
2011	TNAU	IGSP-14	Tamil Nadu
2011	BCKV	Bidhan Jagannath	West Bengal
2013	TNAU	CO-5	Tamil Nadu
2015	CAU	NFSP-1	Manipur
2017	ICAR-CTCRI RC	Bhu Kanti	Odisha
2017	ICAR-CTCRI RC	Bhu Ja	Odisha
2017	ICAR-CTCRI RC	Bhu Swami	Odisha
2017	ICAR-CTCRI RC	Bhu Sona	Odisha
2017	ICAR-CTCRI RC	Bhu Krishna	Odisha
2019	BCKV	BCSP 10	Bihar

[a]Anonymous (2019)

Fig. 10.1 (**a**) Vines, (**b**) flowers, and (**c**) tubers of different varieties

Fig. 10.1 (continued)

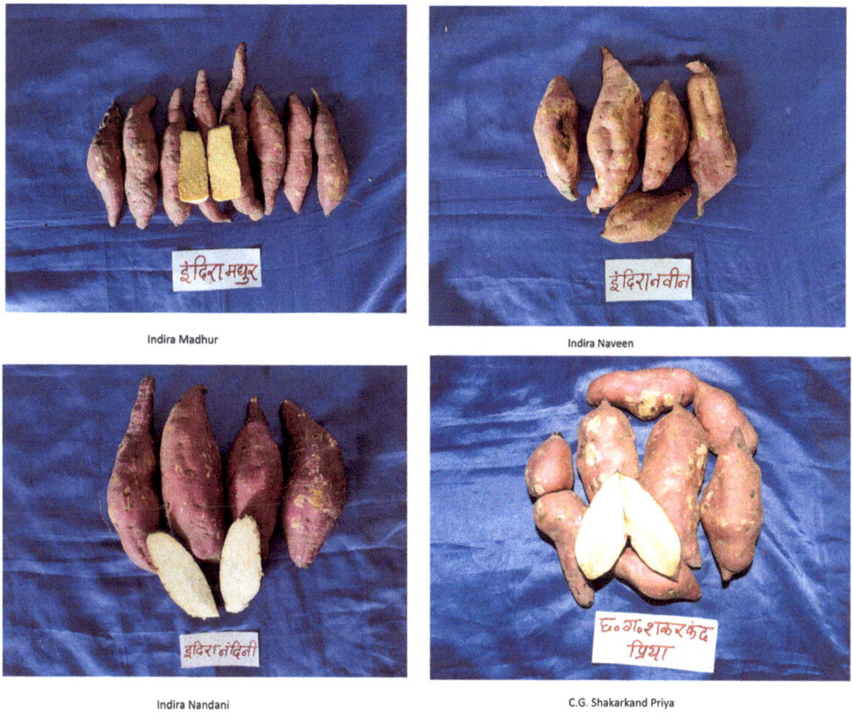

Fig. 10.1 (continued)

Compacted soils often lead to reduced tuber yields due to insufficient aeration (Watanabe 1969). For optimal sweet potato growth, soil pH should ideally be maintained between 5.2 and 6.7. When pH levels fall between 3.5 and 4.5, root development suffers, and high salinity levels can further hinder the growth of both stems and roots (Eguchi et al. 1994). Neutral and high pH may cause pox and scurf disease (Steinbauer and Kushman 1971; Key 1973). Scurf injury was found to be reduced in slightly acid soils (Martin et al. 1976). High fertile soil promotes vigorous vine growth, but it is unfavourable to tuber development. The presence of heavy clay is undesirable and can lead to harvesting difficulties. Additionally, sweet potatoes are relatively sensitive to conditions of saline and alkaline soil.

10.3.2.2 Climate

The cultivation of sweet potato is best suited to regions where a moderately warm climate is present during the 4-month growing period, with temperatures typically ranging from 21.0 to 26.7 °C (Martin et al. 1976). Optimal growth occurs at temperatures above 22.5 °C (Nair et al. 1984). Research by Spence and Humphries (1972) indicated that fibrous roots appeared at 15 °C, while tuber formation increased at temperatures between 25 and 30 °C. Excessive rainfall and extended photoperiods can enhance vine growth but may decrease tuber yield. Longer daylight periods (16 hours) promote vine elongation, whereas shorter periods (9 hours)

10.3.3.2 Vegetative Propagation

Medium-sized healthy tubers free from disease, pests, and bruises are selected for planting in the nursery. Usually, two nurseries, namely, primary and secondary, are needed for producing planting materials. A well-prepared and manured nursery having good drainage helps in producing better planting materials. Planting in nursery is done 3 months ahead of planting in the main field. The selected tubers are planted 5–10 cm deep at a spacing 25 cm in rows, 60 cm apart. In order to promote growth of vine, nursery is maintained properly and the beds irrigated when necessary. The sprouts are cut after about 40–45 days and planted in secondary nursery for further growth. The cuttings 20–30 cm in length are planted in this nursery at a distance of 60 cm between rows and 25 cm in the row. Once the nursery vines attain an adequate length, cuttings are executed and subsequently planted in the field (Choudhury 1976). A mature tuber produces between five to seven sprouts, and 100 kg of tubers may yield approximately 7000 shoots. In regions with cold winters, the same techniques are followed in propagating frames, and 250 kg seed tubers are used for planting in 1 ha (Ustimenko-Bakumovsky 1983). For each hectare of planting, a primary nursery of 100 m^2 and a secondary nursery of 500 m^2 are required (Mandal 1976; Anonymous 1983). Vine cuttings after harvest of tuber or those raised in the nursery are used for planting. The cuttings obtained from apical and middle portion of vis have been found to produce larger numbers of sprouts and higher yield for tubers. Vegetative propagation followed by primary and secondary nursery bed preparation is presented in Fig. 10.2.

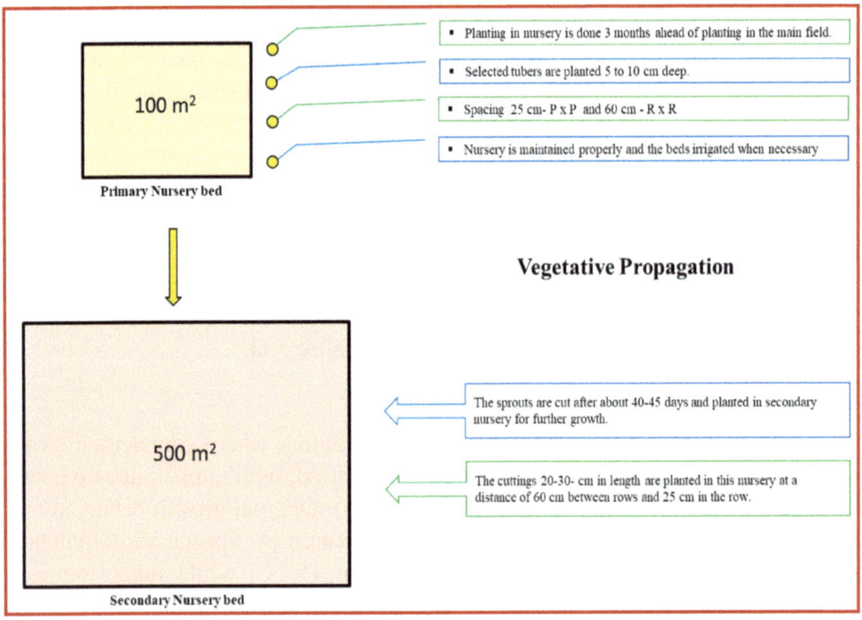

Fig. 10.2 Vegetative propagation followed primary and secondary nursery bed

Mukhopadhyay et al. (1990) reported that top and middle vines were equally efficient to produce more tubers than bottom vines. In general, 20–30 cm long cuttings gave higher yield than shorter cuttings. Eronico et al. (1981) also observed higher leaf area index, marketable tuber, and total yield by planning apical cuttings than middle and basal cuttings.

10.3.3.3 Micropropagation

A comprehensive review by Unnikrishnan and Sheela (2001) on biotechnology of sweet potato provides scientific and applied aspects, including advances in micropropagation. Shoot-tips along with a leaf primordium, as well as meristem, cultured on MS medium with GA_3 and kinetin produced plantlets which were tested as virus-free. GA_3 and kinetin or GA_3 alone were effective for faster development of meristem culture (Unnikrishanan et al. 1991).

10.3.4 Planting Time

The time of planting vine cutting and growing season has a marked effect on the yield of sweet potato. In the tropics and subtropics, sweet potato is traditionally planted at the beginning or end of the rainy season. In Cuba, December-planted crop yields good (Norman et al. 1989) whereas in Thailand, March-planted crop yields more (Sajjapongse and Wu 1989).

In India, planting generally depends on the availability of soil moisture. In North Bihar and parts of Uttar Pradesh, it is planted in two seasons, i.e., spring (February–March) and autumn (September–October). In Tamil Nadu, planting is usually done in September and May, while in Kerala, the vest time for planting is from September to December (Singh and Mandal 1970; Mohankumar and Mandal 1972).

10.3.5 Method of Planting and Spacing

Nair et al. (1984) recommended a planting distance of 60 × 20 cm for optimum yield. At CTCRI, the highest tuber yield was reported by planting at a distance of 45 × 15 cm. Some Ridge and furrow, bed and mound methods are also practiced in different parts of India (Table 10.4).

Table 10.4 Ridge and furrow, bed and mound methods are practiced in different parts of India

Method of planting	State	References
Ridge and furrow	Uttar Pradesh and Bihar	Anonymous (1980)
Ridge and furrow	Tamil Nadu	Anonymous (1983)
Flat bed	North Bihar	Anonymous (1980)
Planting on mound	Keral (CTCRI)	Anonymous (1983)
Vertical planting	U.P. and par of Bihar	–
Horizontal planting	Panjab and Gujarat	–

10.3.6 Manuring and Fertigation

The nutrient requirement of sweet potato is fairly high and depends mainly on the types of soil and its fertility status. Studies on the uptake of NPK and Ca at different stages of growth revealed that the above nutrients are utilized more by the plants at the early stage, up to about 60 days after planting than at a later stage. N and K are more effective than P (Table 10.5). Organic manures in the form of farmyard manure or compost help to improve the soil condition for proper development of tubers.

Nitrogen application in optimum quantity produces a high tuber yield and high starch content in tubers. High rates on N, on the other hand, often cause excessive top growth at the expense of root growth. The application of 60 kg N/ha increased the total and marketable tuber yields of sweet potato (Biswas et al. 1980).

Though the response to phosphorus application is less pronounced in comparison to N and K, about 50 kg P_2O_5/ha or less has been recommended for the cultivation of sweet potato in several States of India.

The role of potassium in the promotion of tuber growth through the accelerated translocation of photosynthates from leaves to tubers has been reported. It is well known that potassium influences the root yield of sweet potato more than other nutrients by increasing the photosynthetic efficiency. The beneficial effect of split application of K was reported by Hossain et al. (1987). The highest growth and yield were recorded at 75 kg K_2O/ha when applied in two equal splits (Mukhopadhyay et al. 1992).

10.3.6.1 Micro Nutrients

Application of $MgSO_4$ @ approx. 20 kg/ha after top dressing, Borax @ approx. 1.5 kg/ha after top dressing, dipping the cuttings in 2–4% zinc sulphate for 15 minutes before planting and foliar spraying of 1.5% zinc sulphate hepta hydrate solution after top dressing and application of Mg, B, S, Zn either alone or in combination may be recommended in locations based on soil requirement to release maximum tuber yield from sweet potato.

Table 10.5 Effect of potassium on tuber yield at different days after planting (DAP)

	Tuber yield (t/ha)		
K_2O kg/ha	90 (DAP)	120 (DAP)	150 (DAP)
Basal application			
0	4.62	9.32	10.40
25	5.98	11.70	12.91
50	6.99	13.39	14.71
75	7.68	14.82	16.54
100	7.82	14.93	15.95
Split application			
25	6.79	13.06	14.45
50	8.13	14.83	16.51
75	8.97	16.40	18.16
100	8.74	16.07	17.79

Mukhopadhyay et al. (1992)

10.3.6.2 Biofertilizers

The effect of biofertilizers like *Azotobacter* (AZT) and *Azospirillum* (AZP) on the nitrogen use efficiency as well as growth and yield of sweet potato was found to be positive. The use of *Azospirillum* @ 2 kg/ha as vine dipping +10 kg/ha as soil application reduced the application of nitrogen up to 2/3rd of the recommended dose and recorded significantly higher yield over control (Anonymous 2001).

10.3.7 Irrigation

Planting may be followed by irrigation or planting in September–October before the cessation of monsoon will ensure establishment. The critical stage of moisture supply is 40 days after planting (DAP) or earthing up (35 DAP) may be followed by an irrigation. Indira and Kabeerathuma (1990) observed that the tuber yield of sweet potato was significantly reduced by water stress at the initial phase (10–30 DAP) but stress during the tuber development phase (30–50 DAP) resulted in slight increase in yield, whereas stress during the tuber maturity phase (75–95 DAP) showed yield reduction in sweet potato.

10.3.8 Interculture; Weed Management

In the early stage of crop growth, i.e., up to 2 months after planting, weeds are a problem for sweet potato. But later, the weeds are suppressed due to the rapid growth of the crop which covers and smother the weeds. Inter-culture operation to suppress weeds at early stage of growth is an essential requirement for obtaining high yield. Earthing up of the soil is helpful to suppress weeds besides improving its physical condition. In a trial at the CTCRI, India, Ravindra and Mohankumar (1987) observed that earthing up at 15 or 30 days after planting resulted in tuber yield of 12.68 and 16.55 t/ha, respectively, compared to 12.84 t/ha under control (no weeding and earthing up). The minimum tillage technique with mulching increased soil temperature and water content and improved growth and also decreased the labour requirement (Yang 1990). Application of pendimethalin @ 1.0 kg/ha (pre-emergence) along with quizalofop-p-ethyl @ 75 g/ha (post-emergence) demonstrates effective control of weeds, especially in areas where grassy weeds are prevalent.

10.3.9 Harvesting

The time of harvesting of sweet potato differs with cultivars, which may extend from 120 to 180 days (Table 10.6). Before harvest, the vines are cut and tubers are dug up by using spade or fork. Sweet potato is usually harvested when the leaves turn yellow and begin to shed. Maturity can be ascertained by cutting tuber and the latex from mature tuber dries up without turning black (Purseglove 1977; Key

Table 10.6 Maturity period of varieties

S.No.	Variety	Maturity period
1	Sree Kanaka	75–85 days
2	Rajendra Sakarkand-92	75–120 days
3	Sree Bhadra, Sree Arun, Sree Varun	90 days
4	COCIP-1, Indira Naveen, Indira Narangi	90–100 days
5	Sree Rethna	90–105 days
6	Sree Nandini, Sree Vardhini	100–105 days
7	Indira Priya, Bhu Krishna	100–110 days
8	Samrat	100–120 days
9	Konkan Ashwini	105 days
10	Kalinga, Goutam, Sourin, CO-3, Bhu Kanti, Bhu Ja, Bhu Swami, Bhu Sona	105–110 days
11	Rajendra Sakarkand-5	105–120 days
12	Rajendra Sakarkand-35	110 days
13	Rajendra Sakarkand-43, Gouri, Kishan, Indira Madhur, Kamla Sundari	110–120 days
14	Kiran	110–130 days
15	Sankar, Indira Nandini, H-41, H-42, Varsha	120 days
16	Rajendra Sakarkand-47, Narendra Sakarkand-9	120–130 days
17	CO-5, CO-1	135 days

1973). Irrigating the field 4–6 days prior to harvest is helpful when soils become compact. While digging out tubers, care should be taken to avoid mechanical injuries which would affect the storage and quality (Kushman 1967).

10.3.10 Yield

The yield fluctuates with cultivars, season of planting, soil types, and fertility. Under good management practice, yield can range from 35 to 40 t/ha in irrigated and 8 to 10 t/ha under rainfed conditions. The average yield in India is 6–7 t/ha (Varma 1984; Ghosh and Nair 1984).

10.3.11 Storage

Highly perishable nature of tubers of sweet potato does not permit it to be stored for a long period in the tropics (Key 1973). Sweet potato tubers are usually stored after curing. Uncured tubers get spoiled quickly at temperatures of 10 °C or below (Lutz and Simons 1972). Sweet potatoes can successfully cure at relative humidity of 85–95% and 30–33 °C for 5–7 days (Banda 2023). Curing healed wounds inflicted during harvest (Walter Jr. and Schadel 1982). After curing, sweet potato may be stored at 10–12 °C with a relative humidity of 85–90%.

10.4 Protection

10.4.1 Diseases

10.4.1.1 Field Disease

10.4.1.1.1 Wilt or Stem Rot

It is an important field disease in India, Japan, and the United States caused by *Fusarium oxysporum* f. *batatas*, and the fungus is carried through infected tubers as well as soil where it can survive for several years. The disease may appear in seed bed where the leaves become yellow along with discolouration of vascular bundles. The infected tubers show shrunken spots.

Control measures:

(i) Use of clean, healthy planting material.
(ii) Crop rotation where sweet potato should be grown once in 2–3 years.
(iii) Growing of resistant cultivars.
(iv) *Fusarium oxysporum* on sweet potato can be controlled by cross-protection with a non-pathogenic strain (Komada 1990).

10.4.1.1.2 Black Rot

The disease is caused by *Certocytis fimbriata* and can occur both in field and in the storage. The visible symptoms are yellow sickly appearance of the foliage and black spot on the underground part of the stem. Control measures are similar to the stem rot.

10.4.1.1.3 *Cercospora* Leaf Spot

It is caused by the fungi *Cercospora batatae* and *Cercospora bataticola* (Anonymous 1983). The disease manifests as dark brown spots on leaves, each featuring a lighter centre. Warm and humid conditions significantly harm the plant's foliage. Three sprayings of 0.25% carbendazim at fortnightly interval are effective to suppress the disease.

10.4.1.1.4 Scurf

It is caused by the fungus *Monilochaetes infuscans* and occurs in the United States, Hawaii, Japan, and Australia. The disease produces black spots of varying sizes and shapes on the tubers. Practising long rotation or applying organic matter at least 1 year before planting is recommended. Good drainage and application of sulphur resulted in reduction of disease (Chupp and Sherf 1960).

10.4.1.2 Virus Disease

About 20 kinds of sweet potato viruses have been reported in the world, but only three of them have been found in China: sweet potato feathery mottle virus (SPFMV), sweet potato latent virus (SPLV), and sweet potato caulimo-like virus (SPCLV). On average, sweet potato yield in China loses over 20% due to sweet potato-virus disease.

10.4.1.3 Storage Disease

10.4.1.3.1 Sof Rot

Soft rot is a very serious disease of sweet potato in storage caused by *Rhizopus nigricans*. The disease starts as a soft watery rot on tuber, and the fleshy tissue becomes soft and stringy, and water exudes from affected skin.

10.4.1.3.2 Ring Rot and Collar Rot

In ring rot and collar rot forms of soft rot, the decay forms rings or collar around tubers. Control measures consist of minimizing injury during harvest and storage. The fungus enters into the tubers through wounds. For healing up the wounds in storage, a temperature of 26.7–29.4 °C and relative humidity of 85–90% should be maintained.

10.4.2 Pests

More than 80 species of insect pests and a few species of mites and nematodes cause damage by feeding on roots, tubers, and foliage (Pillai et al. 1978).

Weevil

The most important and common pest of sweet potato in India is *Cylas formicarius*. It makes hole at the basal portion of the vine and lays eggs. The collar region gets the maximum infestations through which weevil enters into the tubers. Large number of weevils enter into tubers and make them unsuitable for marketing. The infestation begins for 45 days after planting and the intensity is maximum during crop harvest. Maximum tuber damage was observed in February–May when temperature was high and rainfall was low (Rajamma and Premkumar 1984). It has been observed that pests become active with the rise of temperature in both spring and autumn planted crops.

IPM (Integrated Pest Management) Package Against Sweet Potato Weevil

Integrated pest management practices involve:

1. Healthy planting material: Planting should be done with weevil-free planting materials.
2. Clean cultivation: Collect and destroy the weevil-infested tubers, vines, and alternate hosts.
3. Bio pesticides/Microbial agents: Application of entomopathogenic fungi *Beauveria bassiana* (1.5 WP bio powder) @ 6.75 kg/ha may be recommended to reduce sweet potato weevil damage.
4. Barrier crops: Marigold either as a border crop or as inter crop may be planted with sweet potato to reduce the weevil damage in tubers.
5. Intercropping models: It was recommended to grow coriander either in 1:1 or 2:1 ratio or garlic in 1:1 ratio as intercrop with sweet potato to minimise sweet potato weevil damage.

6. Chemical method: Vine dipping with 0.02% chlorpyriphos (20 EC) for 10 minutes, spraying chlorpyriphos 0.02% at 30 and 60 DAP and Nanma at 45 and 75 DAP may be followed for effective management of sweet potato weevil.
7. Sex pheromone traps: Installation of sex pheromone trap @ one per 100 m^2 at 10 DAP and changing the lure at 60 DAP is effective for the control of sweet potato weevil damage.

10.5 Crop Improvements

Sweet potato (*Ipomoea batatas* (L.) Lam) is usually propagated by vegetatively. Only breeding work is propagated by seed. The sexual reproduction of sweet potato is restricted for two reasons: (i) its tendency not to bloom, and (ii) the low fertility of its flowers. Sweet potato is characterized by its high level of heterozygosity and possesses a chromosome count of $2n(6\times) = 90$, which is classified as a natural hexaploid (Amritha et al. 2024).

The major breeding methods in sweet potato include:

(a) Introduction
(b) Germplasm evaluation and clonal selection.
(c) Polyploid: as interspecific hybridization and selection.
(d) Mutation: X-ray irradiation has been used for creating new variabilities.
(e) Another culture has also been used recently in its improvement programme.

Due to the bisexual nature of the flowers, the process of emasculation becomes necessary. The flowers when mature open before dawn, after 2.00 AM onwards and remain open for a few hours after dawn. Majority of the improved varieties have been developed from the progenies from open-pollinated seeds. The plant has a relatively strong sporophytic self-incompatibility system so that self-pollination usually occurs at low frequency. New genotypes are developed by recombining highly heterozygous hexaploid genotypes. Incompatibility alleles result in specific cross combinations being difficult to achieve, and seeds from controlled sweet potato crossings have especially high value. Controlled crosses were made especially for developing biofortified and stress-tolerant varieties in sweet potato.

The major breeding objectives in sweet potato are:

(a) High yield, resistance to pests and diseases particularly against sweet potato weevil.
(b) Better culinary quality and increase in nutritional value (high carotene and protein contents).
(c) Early maturity (<90 days).
(d) Compact bushy plant type, shade tolerance, and adaptability to intercropping systems,
(e) Improved tuber quality with low sugar and high starch,

(f) High starch content for industry, dual-purpose clones which can be used as leafy vegetable fodder crop with reasonably high leaf yield.
(g) Good fried chip quality, high keeping quality, wider adaptability, and photo insensitivity.

An introduction and collection of both native and foreign genetic materials have been conducted. In India, the Central Tuber Crops Research Institute (CTCRI) and the All India Coordinated Research Improvement Project for Tuber Crops primarily oversee the extensive collection of sweet potato germplasm. CTCRI is currently preserving over 1827 genetic stocks of sweet potato sourced from various regions in India and other nations. The initial phase of the improvement program involved the development of Pusa Safed and Pusa Red cultivars from selected varieties.

10.6 Primary Processing

Sweet potatoes (*Ipomoea batatas*) are a highly adaptable root vegetable that is rich in nutrients and widely cultivated in tropical and subtropical regions. The main processing methods applied to these harvested tubers focus on enhancing their shelf life, flavour, and compatibility for various value-added products. This processing transforms the sweet potatoes into forms that are more suitable for consumption, storage, or further industrial applications.

10.6.1 Harvesting

Harvesting, which is the initial stage of sweet potato processing, has an enormous effect on its quality. Depending on the variety and region, sweet potatoes are normally harvested 90–120 days after planting, or when the tubers achieve their ideal size and maturity (Woolfe 1992).

Timing	Harvesting occurs when vines begin to yellow, indicating tuber maturity. Delaying harvest could boost production, but there's a chance of pest damage or over-maturation, which would alter flavour and texture	Scott and Maldonado (1999)
Method	Manual harvesting with hoes or forks is common in small-scale farming to avoid tuber damage. Mechanized harvesting, used in large-scale operations, requires careful calibration to minimize bruising	FAO (2013)
Considerations	Harvesting during dry weather decreases soil adhesion and microbiological contamination, and handling tubers gently can minimise cuts or bruising that could cause spoiling during storage	Kays (1985a)

10.6.2 Cleaning

In order to improve appearance and lower the microbial load, post-harvest washing eliminates soil, debris, and surface pollutants (FAO 2013). Loose soil is removed by brushing or shaking the tubers.

Wet Cleaning Tubers are washed with clean water, using manual scrubbing or mechanized systems with high-pressure sprays or rotating brushes in commercial settings.

Sanitization A mild disinfectant, such as a chlorine solution (50–100 ppm), may be applied to reduce surface pathogens, followed by rinsing with potable water (Scott and Maldonado 1999).

Drying Tubers are air-dried or dried with forced air to prevent moisture-related spoilage.

10.6.3 Curing

The curing process aids in the repair of wounds, prolongs the shelf life of products, and enhances sweetness through the conversion of starches into sugars. This procedure is essential for achieving optimal storage conditions and preserving high quality (Kays 1985a, b).

Process Tubers are maintained in a regulated environment with temperatures between 29 and 32 °C and relative humidity levels of 85–90% for a duration of 4–7 days. Conventional curing methods consist of arranging tubers in shaded areas that have good ventilation, whereas industrial practices utilize specialized curing chambers (Woolfe 1992).

Benefits Heals cuts and bruises by forming a protective suberized layer. Reduces water loss during storage. Enhances flavour through enzymatic starch breakdown.

Over-curing can lead to sprouting or weight loss, while under-curing may result in poor wound healing (FAO 2013).

10.6.4 Sorting and Grading

Sorting and grading ensure uniformity and quality for market or further processing (USDA 2005).

Sorting Damaged, diseased, or undersized tubers are removed, either manually or with automated systems using conveyors and optical sorters.

Grading Tubers are categorized based on size, shape, and quality (e.g., market-grade, processing-grade, or reject). For example, large, uniform tubers are preferred for fresh market sales. Smaller or irregular tubers are directed to processing, such as for flour or puree (Scott and Maldonado 1999).

Standards Grading follows regional or international standards, such as those set by the USDA or local agricultural authorities (USDA 2005).

10.6.5 Initial Preparation

Initial preparation makes sweet potatoes ready for consumption or secondary processing, depending on the intended use (FAO 2013).

Peeling Manual peeling with knives or mechanized peelers (abrasive or steam peelers) removes the skin. Steam peeling is efficient for large-scale operations but requires energy-intensive equipment (Woolfe 1992).

Cutting/Slicing Tubers are cut into chunks, slices, or cubes for cooking, drying, or freezing. Uniform cuts ensure consistent processing outcomes.

Blanching For processed products like frozen or canned sweet potatoes, tubers are blanched in hot water (85–95 °C for 2–5 minutes) or steamed to inactivate enzymes, preserve colour, and improve texture (Kays 1985a, b).

Packaging Processed or fresh tubers are packed in breathable bags, crates, or vacuum-sealed containers to maintain quality during transport and storage.

10.6.6 Challenges in Primary Processing

Mechanical Damage Improper handling during harvesting or cleaning can cause bruising, reducing market value (Scott and Maldonado 1999).

Resource Constraints Small-scale farmers may lack access to curing facilities or mechanized equipment, limiting shelf life (FAO 2013).

Varietal Differences Varieties differ in skin thickness, sugar content, and storage potential, requiring tailored processing methods (Woolfe 1992).

Post-Harvest Losses Inadequate curing or storage can result in significant losses, especially in developing regions (FAO 2013).

10.6.7 Applications of Primary Processed Sweet Potatoes

- *Fresh Market*: Cleaned, cured, and graded tubers are sold directly to consumers (USDA 2005).
- *Food Industry*: Sliced or blanched sweet potatoes are used for fries, chips, or canned products (Scott and Maldonado 1999).
- *Industrial Use*: Processed sweet potatoes serve as raw material for flour, starch, or bioethanol production (Woolfe 1992).

10.7 Pre and Postharvest Technology of Sweet Potato

Sweet potato is a vital tuber crop globally, valued for its nutritional content (rich in carbohydrates, beta-carotene, vitamins, and minerals) and versatility in food, feed, and industrial applications. Post-harvest technology aims to minimize losses (20–40% globally due to improper handling), extend shelf life, and maintain quality. The following sections outline key practices.

Pre-Harvest Practices

- Foliar application of maleic hydrazide (1000 ppm) 14 days before harvest reduces sprouting (Woolfe 1992).
- Pre-harvest curing (removing foliage 10–14 days prior) toughens skin, reducing post-harvest losses by 30–40% (Banda 2023).

10.7.1 Curing

Curing of harvested sweet potato tubers is done at 29–32 °C temperature and 85–90% relative humidity for 4–7 days, depending on variety and damage extent. Adequate airflow must be maintained to prevent CO_2 buildup and rotting of tubers. Curing is a process that promotes the healing of wounds caused by harvesting, minimizes water loss, and improves flavour through the conversion of starches into sugars. Additionally, it fortifies the periderm, which helps to lower the risk of microbial invasion. Curing reduces weight loss by 10–15% and decay by 20–30% during storage (Ray and Ravi 2005).

10.7.1.1 Types of Curing

10.7.1.1.1 Field Curing

In tropical areas, tubers are typically placed in hips under shaded conditions for a duration of 3–5 days, often being covered with leaves or tarps.

10.7.1.1.2 Controlled Curing

In commercial environments, curing rooms equipped with heating devices and humidifiers are utilized. In the United States, curing chambers are kept at a temperature of 30 °C with a relative humidity level of 90% (Rahman et al. 2023).

10.7.2 Cleaning

Tubers are washed with clean water to remove soil and debris. Chlorinated water (100–150 ppm) reduces microbial load (Picha 1986). Gentle brushing avoids skin damage.

10.7.3 Sorting and Grading

- Sorted by size, shape, and quality.
- Damaged, diseased, or undersized tubers are separated for immediate use or processing.
- Grading enhances market value; uniform tubers fetch 10–20% higher prices (FAO 2008).

10.7.4 Storage

After harvesting sweet potato tubers are stored at 13–15 °C temperature to prevent chilling injury (<10 °C) or sprouting (>16 °C) and 85–90% RH to minimize shrinkage. At this condition, storage can be done for 6–12 months for cured tubers and uncured tubers last 1–2 months.

10.7.4.1 Storage Methods

10.7.4.1.1 Traditional Storage

In tropical regions, tubers are stored in pits, heaps, or baskets lined with straw or leaves. Sand or ash layering reduces moisture loss (Yego 2022).

10.7.4.1.2 Modern Storage

Refrigerated warehouses with controlled temperature and humidity are used in developed countries. Modified atmosphere storage (low O_2, high CO_2) extends shelf life by 2–3 months (Kays 1985a, b).

10.7.5 Packing and Transportation

Ventilated crates, mesh bags, or cardboard boxes are used as packaging material for sweet potatoes to ensure airflow. Plastic bags are avoided due to moisture buildup. Loose packing prevents compression damage (20–25 kg per crate is standard storage quantity). For transportation, use refrigerated trucks (13–15 °C) for long distances to maintain quality. Avoid rough handling as mechanical damage increases rot by 15–25% (Bao et al. 2021). In developing countries, tubers are transported in open trucks, requiring shade and cushioning (e.g., straw) to reduce bruising.

10.8 Utilization Pattern

10.8.1 Sweet Potato: As Super Food

Fasting is a practice widely embraced across various cultures and religions in India and is recognized for its substantial benefits on physical, mental, and spiritual health. It facilitates the process of detoxification, enhances digestion, and promotes metabolic health by providing the body with time to rest and rejuvenate, a concept supported by research on intermittent fasting. From a mental perspective, fasting encourages discipline, mindfulness, and emotional strength, aiding individuals in developing self-control and mental clarity. Spiritually, it plays a crucial role in traditions such as *Navratri* and *Ekadashi*, symbolizing concepts like purification, devotion, and gratitude. In a social context, fasting fosters community connections through shared practices and charitable activities. By combining health advantages with cultural and spiritual relevance, fasting continues to be an essential practice in India, contributing to a holistic approach to well-being.

Fasting includes several dietary approaches such as Calorie Restriction (CR), Intermittent Fasting (IF), Periodic Fasting (PF), Long-Term Fasting (LF), Fasting-Mimicking Diet (FMD), and Very Low-Calorie Diet (VLCD). These methods initiate a metabolic transition from using glucose to utilizing ketones, referred to as the G-to-K switch, which is a key feature of the physiological impacts of fasting. Following a period of fasting, refeeding leads to a reversal of this process known as the K-to-G switch (Anton et al. 2018). Sweet potato holds a unique importance during festivals in India. Women observing fasts often present it as food and *prasad* while also consuming it during their fasting practices. Its significance is heightened by the variety of dishes prepared using sweet potato, which are integral to Indian cuisine. As a result, it features prominently in nearly every festival, earning it a reputation as a superfood. Furthermore, sweet potato is recognized for its numerous medicinal properties, making it a vital component of traditional dietary practices.

In India, sweet potatoes (*Shakarkandi*) are consumed during fasting (*vrat*) because they provide significant nutritional benefits, are easy to digest, and align with the *sattvic* (pure) principles of traditional Hindu dietary customs. Here's why sweet potatoes are a preferred choice during fasting is presented in Table 10.7.

Table 10.7 Sweet potatoes: preferred choice during fasting

Particular	Content	References
Rich Source of Energy	During fasting, people avoid grains and certain staple foods. Sweet potatoes are rich in complex carbohydrates, which provide sustained energy and prevent fatigue.	Bartels (2021) and Rose et al. (2024)
Easily Digestible	Unlike heavy grains, sweet potatoes are light on the stomach, making them easier to digest while still being filling.	Abdullahi (2023)
Nutrient Dense	They are packed with essential vitamins and minerals like Vitamin A (beta-carotene), Vitamin C, potassium, and fibre, which help maintain good health during fasting.	Abewoy et al. (2024) and Islam (2024)
Natural Sweetness	Since refined sugar is often avoided during fasting, sweet potatoes offer a naturally sweet taste, satisfying cravings without the need for added sugar.	Avena (2023)
Sattvic Food in Ayurveda	In Ayurveda, fasting serves as a method for detoxification of the body. Sattvic foods, known for being pure, light, and nourishing, are recommended. Sweet potatoes are categorized as sattvic and assist in balancing the body's energies.	Bandekar and Namdeo (2024), Jaitwar et al. (2024), and Shirhatti et al. (2023)
Hydrating and Cooling	Sweet potatoes have high water content, keeping the body hydrated and preventing dehydration, which can occur during fasting.	Rashid et al. (2022) and Elgabry et al. (2025)
Versatility in Cooking	Sweet potatoes can be roasted, boiled, or made into desserts like *shakarkandi halwa* or *chaat*, making them a delicious and flexible fasting option.	–

In India, sweet potatoes hold a special place during fasting periods. Their nutritional benefits are well-recognized, and they are an integral part of various spiritual ceremonies that mark these sacred times.

10.8.2 Why Sweet Potato Is Super Food?

Sweet potatoes are loaded with essential nutrients, making them an excellent food source for all age groups considered as super food specially for fasting people. Key nutrients include:

- Carbohydrates and Fibre: Sweet potatoes are rich in complex carbohydrates, providing sustained energy. Additionally, their high fibre content aids digestion and improves gut health (Bovell-Benjamin 2007; Smith 2023; Rose et al. 2024).
- Vitamins: Sweet potatoes are an excellent source of vitamin A (as beta-carotene), vitamin C, and several B vitamins, all of which play crucial roles in immunity, vision, and metabolism (Teow et al. 2007; Mourtala et al. 2023; Abewoy et al. 2024; Noreen et al. 2024).

- Minerals: It contains essential minerals like potassium, iron, magnesium, and calcium, which contribute to cardiovascular health and bone strength (Laurie et al. 2012; Ranteallo et al. 2023; Vicente et al. 2022).
- Antioxidants: The presence of anthocyanins, flavonoids, and carotenoids in different coloured varieties provides antioxidant benefits, protecting against chronic diseases (Nguyen et al. 2021; Ooi et al. 2021; Chintha et al. 2023).

10.8.3 Comparative Nutritional Assessment

Sweet potatoes offer greater nutritional advantages when compared to other staple crops like white potatoes, rice, and wheat. In contrast to white potatoes, which possess a high glycaemic index, sweet potatoes gradually release sugar into the bloodstream, making them a great choice for those with diabetes (Noreen et al. 2024; Elgabry et al. 2023). Additionally, sweet potatoes are much richer in vitamin A than rice and wheat, both of which frequently lack this essential nutrient (Gelaye 2024).

Furthermore, biofortified varieties of sweet potatoes have been developed to combat malnutrition in vulnerable populations. These include the orange-fleshed sweet potato (OFSP), which has been widely promoted for its high beta-carotene content (Mwanga et al. 2021; Mir et al. 2024).

10.8.4 Health Benefits of Sweet Potato

Vitamin A deficiency is a major public health issue in India, particularly affecting children and pregnant women (Kundu et al. 2021). The beta-carotene in orange-fleshed sweet potato (OFSP) is highly bioavailable and can significantly improve vitamin A levels (Low et al. 2007; Girard et al. 2021; Abewoy et al. 2024). Several community-based interventions in African and Asian countries have demonstrated the effectiveness of promoting OFSP to address vitamin A deficiency (Girard et al. 2021).

1. Glycaemic Control and Diabetes Management

Despite being a carbohydrate-rich crop, sweet potatoes have a low glycaemic index (GI), meaning they release glucose into the bloodstream gradually (Bartels 2021). Studies have shown that sweet potato consumption helps regulate blood sugar levels and enhances insulin sensitivity (Ludvik et al. 2004; Kinoshita et al. 2023). This is particularly beneficial in the Indian context, where the prevalence of type 2 diabetes is on the rise.

2. Gut Health and Digestive Benefits

Sweet potatoes contain high levels of dietary fibre and resistant starch, which act as prebiotics, fostering a healthy gut microbiome and preventing constipation (Zhang et al. 2013; Nolte Fong 2021; Smith 2023). Resistant starch has been found to improve digestion and support the growth of beneficial gut bacteria, reducing the risk of gastrointestinal disorders (Chen et al. 2024).

3. Heart Health and Blood Pressure Regulation
The potassium content in sweet potatoes helps maintain blood pressure levels and supports cardiovascular health by reducing the risk of hypertension (Bovell-Benjamin 2007; Stone et al. 2021; Chintha et al. 2023). Additionally, the fibre content in sweet potatoes contributes to cholesterol reduction, further lowering the risk of heart disease (Chang et al. 2021; Qiao et al. 2021).

4. Role in Cancer Prevention
Emerging research suggests that the antioxidants present in sweet potatoes, particularly anthocyanins in purple varieties, have potential anti-cancer properties (Jiang et al. 2022a, b; Tang et al. 2023). These bioactive compounds help neutralize free radicals and reduce inflammation, thereby lowering the risk of certain types of cancer, including colorectal and breast cancer (Panchal et al. 2022).

10.9 Value Addition and Products

Due to its high level of carbohydrates, dietary fibre, vitamins (particularly vitamin A precursors), and minerals, sweet potatoes (*Ipomoea batatas*) is a nutrient-dense, adaptable root crop with substantial potential for value addition. In order to increase economic returns, promote food security, and satisfy a range of consumer demands, value addition entails turning sweet potatoes into high-value items.

Importance of Value Addition
Value addition in sweet potato processing increases shelf life, enhances marketability, and creates diverse products for local and global markets. Key benefits include:

- Economic Gains: High-value goods fetch higher prices, which helps processors and farmers (Woolfe 1992).
- Reduced Post-Harvest Losses: Processing reduces spoiling, particularly in areas with inadequate infrastructure for storage (FAO 2013).
- Nutritional Enhancement: Fortified or functional products address dietary deficiencies, such as vitamin A deficiency (Scott and Maldonado 1999).
- Market Diversification: Value-added products cater to urban consumers, health-conscious markets, and industrial applications (Ray and Sivakumar 2009).

10.9.1 Value-Added Products

Sweet potato's versatility allows for a wide range of products, from traditional to innovative, meeting consumer preferences for convenience, taste, and nutrition (Table 10.8).

Table 10.8 Value-added products prepared from the sweet potato

S.No.	Value-added products	Product description	Process	Market potential	References
1.	*Snacks and convenience foods*				
	(a) Sweet Potato Chips and Fries	Crispy chips or fries are made by frying, baking, or air-drying thinly sliced sweet potatoes. Adding natural flavours or spices to food improves its appeal. Compared to potato-based snacks, these items are well-liked for their inherent sweetness and health advantages.	Wash, peel, slice, blanch (to retain colour), and fry or bake. Vacuum frying reduces oil content for healthier options.	Wash, peel, slice, blanch (to retain colour), and fry or bake. Vacuum frying reduces oil content for healthier options.	Ray and Sivakumar (2009)
	(b) Extruded Snacks	Sweet potato flour is mixed with cereals (e.g., corn or rice) to create puffed snacks or breakfast cereals through extrusion cooking.	Grind tubers into flour, blend with additives, extrude, and dry or fry.	Appeals to children and urban consumers seeking fortified, ready-to-eat snacks.	Woolfe (1992)
2	*Bakery and confectionery products*				
	(a) Sweet Potato Flour-Based Products	Sweet potato flour is used as a partial substitute for wheat flour in bread, cakes, cookies, and pastries, improving nutritional value and reducing gluten content.	Peel, slice, dry (solar or mechanical dryers), and mill tubers into flour. Blend with wheat flour (10–30% substitution) for baking.	Gluten-free and nutrient-enriched products target health-conscious and celiac consumers.	Scott and Maldonado (1999)
	(b) Jams and Purees	Sweet potato puree is cooked with sugar, pectin, and preservatives to produce jams, spreads, or pie fillings.	Steam or boil tubers, mash into puree, cook with additives, and package in sterilized containers.	High demand in confectionery and dessert industries.	FAO (2013)

(continued)

Table 10.8 (continued)

S.No.	Value-added products	Product description	Process	Market potential	References
3.	(a) Fermented Products	Sweet potato can be fermented to produce alcoholic beverages like wine or beer, or non-alcoholic drinks like probiotic beverages.	Mash tubers, add yeast or starter cultures, ferment under controlled conditions, and filter or pasteurize.	Growing interest in functional beverages with health benefits.	Ray and Sivakumar (2009)
	(b) Lactic Acid Fermented Foods	Similar to sauerkraut, sweet potatoes can be fermented with lactic acid bacteria to produce tangy, shelf-stable products.	–	Appeals to niche markets seeking probiotic-rich foods.	Woolfe (1992)
4.	*Industrial products*				
	(a) Sweet Potato Starch	Sweet potato's high starch content and bioactive compounds make it a valuable raw material for industrial applications.	Wash, grate, slurry, filter, and dry to obtain starch. Wet milling ensures high purity.	Competitive with corn and potato starch due to cost-effectiveness and functionality.	–
	(b) Modified Starch	Chemically or physically modified sweet potato starch is used in processed foods, pharmaceuticals, and biodegradable plastics.	–	Growing demand for eco-friendly materials.	Ray and Sivakumar (2009)
	(c) Bioethanol	Sweet potato's high carbohydrate content makes it a viable feedstock for bioethanol production, especially in regions with surplus production.	Hydrolyze starch into sugars, ferment with yeast, and distil to produce ethanol.	Renewable energy markets and government incentives for biofuels drive demand.	Woolfe (1992)
5.	*Nutraceutical and functional products*				
	(a) Nutraceutical	Sweet potatoes' bioactive compounds, such as beta-carotene, anthocyanins (in purple-fleshed varieties), and phenolic compounds, offer opportunities for high-value nutraceutical products.			

(continued)

Table 10.8 (continued)

S.No.	Value-added products	Product description	Process	Market potential	References
	(b) Beta-Carotene Supplements	Orange-fleshed sweet potatoes are processed into powders or extracts for vitamin A supplements to combat deficiency.	Dry tubers, grind into powder, or extract beta-carotene using solvents or supercritical CO_2.	High demand in developing countries and health supplement markets.	Scott and Maldonado (1999)
	(c) Anthocyanin Extracts	Purple-fleshed sweet potatoes are processed to extract anthocyanins for use as natural food colourants or antioxidants.	Extract with ethanol or water, purify, and concentrate.	Growing preference for natural over synthetic colourants.	Ray and Sivakumar (2009)
	(d) Functional Flours	Flour from coloured sweet potatoes is marketed as a functional ingredient for its antioxidant and anti-inflammatory properties	–	Appeals to health-conscious consumers and food manufacturers	Zhang et al. (2018)

10.9.2 Technologies for Value Addition

Advanced technologies enhance efficiency and quality in sweet potato processing:

- *High-Pressure Processing (HPP)*: Preserves nutritional quality and extends shelf life of purees and juices (FAO 2013).
- *Vacuum Frying*: Produces low-fat snacks with better colour and texture (Ray and Sivakumar 2009).
- *Spray Drying*: Converts purees into powders for easy storage and transport (Woolfe 1992).
- *Extrusion Technology*: Enables production of fortified snacks and cereals with consistent quality (Scott and Maldonado 1999).

10.9.3 Potential for High-Value Products

The global market for sweet potato-based products is expanding due to:

- Health Trends: Demand for natural, nutrient-rich, and gluten-free products drives innovation in snacks, flours, and supplements (Ray and Sivakumar 2009).

- Sustainability: Sweet potatoes' low input requirements and industrial applications (e.g., biodegradable plastics, biofuels) align with sustainability goals (Zhang et al. 2018).
- Export Opportunities: Processed products like starch, flour, and snacks have strong export potential to developed markets (FAO 2013).
- Local Empowerment: Small-scale processing units can create jobs and improve livelihoods in rural areas (Scott and Maldonado 1999).

10.10 Marketing: Global Demand, Trade, and Market Trends

The sweet potato (*Ipomoea batatas* L.) is an adaptable root vegetable acknowledged for its significant nutritional benefits, its capacity to thrive in various agro-climatic environments, and its numerous applications in the food, feed, and industrial sectors. On a global scale, sweet potato is ranked among the five leading food crops, subsequent to rice, wheat, maize, and cassava (Scott 2021). There is a consistent increase in worldwide demand for sweet potato, attributed to its designation as a 'Superfood' that contains significant levels of beta-carotene, dietary fibre, and antioxidants, while also having a low glycaemic index. This trend corresponds with a heightened consumer focus on health and wellness. Additionally, orange-fleshed sweet potato varieties are gaining popularity for their pro-vitamin A content, especially in efforts to combat malnutrition in developing nations.

Sweet Potato Production, 2003 to 2023 (source: FAO (2025) via Our World in Data) presents a 20-year global and regional trend in sweet potato production, measured in million tonnes. The data from Table 10.9 and Fig. 10.3 revealed that worldwide production has decreased from around 130.97 million tonnes in 2003 to approximately 93.52 million tonnes in 2023. The steepest decline occurred between 2003 and 2006, after which global production stabilized with minor fluctuations. China, a historically dominant producer, shows a consistent and sharp decline, from about 107.56 million tonnes in 2003 to less than 51.40 million tonnes in 2023. This trend largely contributes to the overall decline in Asia's and global production. Asia mirrors the downward trend of China due to China's major contribution. After 2015, Asia's production levels off at around 59.58–58.19 million tonnes. In contrast to Asia, Africa shows a gradual but consistent increase in sweet potato production. It rises from around 12.09 million tonnes in 2003 to over 31.16 million tonnes in 2023, making Africa a growing player in sweet potato agriculture. India's plays marginal role, production remains very low and almost flat, indicating minimal contribution to global sweet potato output.

The global decline in sweet potato production is largely driven by reduced cultivation in China, possibly due to changing food preferences, urbanization, or a shift to higher-value crops. Africa's increasing production highlights the crop's growing importance for food security, nutrition, and climate-resilient farming systems on the continent. India has not yet capitalized on the nutritional and economic potential of sweet potato, suggesting room for policy and research interventions.

Table 10.9 Global production scenario of sweet potato from last two decades

Year/country	World	Asia	China	Africa	India
2003	130.97	115.26	107.56	12.09	1.13
2004	130.77	113.34	105.66	13.75	1.18
2005	127.80	110.16	102.54	14.00	1.18
2006	106.73	88.51	81.04	14.58	1.07
2007	101.29	83.12	75.60	14.47	1.07
2008	103.30	85.71	78.23	13.93	1.09
2009	96.29	77.49	69.81	14.90	1.12
2010	94.03	73.03	65.60	16.78	1.09
2011	94.63	72.21	64.64	17.97	1.05
2012	90.88	67.70	59.79	18.95	1.07
2013	90.37	64.83	56.89	21.21	1.13
2014	93.80	63.62	55.78	25.55	1.09
2015	91.50	62.04	54.20	24.88	1.23
2016	90.55	59.58	51.39	26.34	1.45
2017	92.93	59.14	51.45	28.97	1.46
2018	91.81	60.81	53.01	26.62	1.50
2019	89.59	57.28	50.16	27.68	1.16
2020	91.57	58.65	51.54	28.57	1.14
2021	92.34	58.47	51.57	29.44	1.12
2022	92.21	58.33	51.09	29.62	1.18
2023	93.52	58.19	51.40	31.16	1.29

Source: Food and Agriculture Organization of the United Nations (2025)

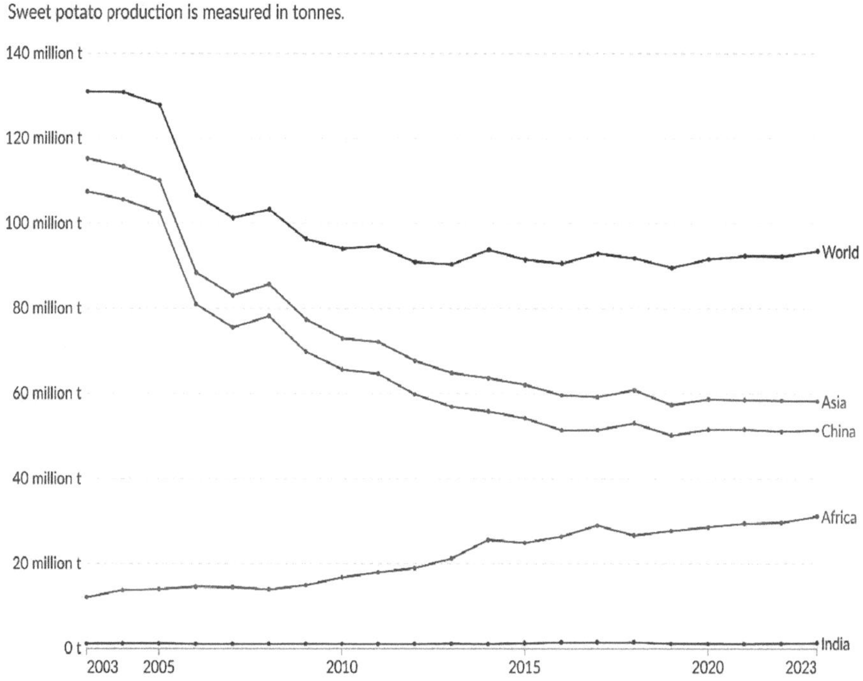

Fig. 10.3 Sweet potato production 2003–2023

This shift in production patterns indicates a rebalancing of global sweet potato agriculture, with Africa emerging as a key region in future trade and food security initiatives.

References

Abdullahi NG (2023) Health risk assessment of heavy metals in sweet potatoes harvested across the six geopolitical zones of Nigeria. Kwara State University

Abewoy D, Megersa HG, Banjaw DT, Lemma DT (2024) Major nutritional content of orange fleshed sweet potato (OFSP) and its importance. Glob Acad J Agric Bio Sci 6:1–7

Amritha MS, Dhanya OG, Nair AG, Vidya P, Muthusamy SK, Mohan C (2024) Genetic diversity studies between Ipomoea wild species and cultivated sweet potato (*Ipomoea batatas* (L.) Lam.) using SSR markers. S Afr J Bot 169:452–463

Anonymous (1980) Achievement at a glance 1969–1980, All India Coordinated Research Project on tuber crops (other than potato). ICAR, Tirhut College of Agriculture, Dholi

Anonymous (2001) Sixth group meeting, AICRP on tuber crops

Anonymous (2019) Standard operating procedures for AICRP on tuber crops. CTCRI, Trivandrum

Anonymus (1983) Two decades of research. CTCRI, Trivandrum

Anton SD et al (2018) Flipping the metabolic switch: Understanding and applying the health benefits of fasting. Obesity 26(2):254–268

Avena NM (2023) Sugarless: a 7-step plan to uncover hidden sugars, curb your cravings, and conquer your addiction. Union Square & Co

Bach D, Bedin AC, Lacerda LG, Nogueira A, Demiate IM (2021) Sweet potato (*Ipomoea batatas* L.): a versatile raw material for the food industry. Braz Arch Biol Technol 64:e21200568

Banda K (2023) Postharvest physiology of sweet potato (*Ipomoea batatas* L.) storage roots: the effects of curing, and ethylene and 1-methylcyclopropene on quality attributes. Cornell University ProQuest dissertations and thesis, p 30813531

Bandekar SS, Namdeo AG (2024) Eating and healing: traditional food as medicine. In: Role of herbal medicines: management of lifestyle diseases. Singapore, Springer Nature Singapore, pp 479–4950

Bao G, Wang G, Wang B, Hu L, Xu X, Shen H, Ji L (2021) Study on the drop impact characteristics and impact damage mechanism of sweet potato tubers during harvest. PLoS One. plos.org

Bartels A (2021) The glycaemic index and load of three ghanaian staple foods: sweet potato, taro and ripe plantain. University of Cape Coast

Bhutto RA, Khanal S, Wang M, Iqbal S, Fan Y, Yi J (2024) Potato protein as an emerging high-quality: Source, extraction, purification, properties (functional, nutritional, physicochemical, and processing), applications, and challenges using potato protein. Food Hydrocoll 157:110415

Biswas D, Sen H, Som MG, Jana PK (1980) Effect of irrigation and nitrogen levels on the yield, consumptive use and consumptive use efficiency of sweet potato. Veg Sci 7:1–5

Biswas J, Sen H, Mukhopadhyay SK (1989) Effect of light hours on growth and tuber yield in sweet potato. J Root Crops 15:123–125

Bosewell VR (1950) USDA Farmers Bull., 2020

Bovell-Benjamin AC (2007) Sweet potato: a review of its past, present, and future role in human nutrition. Adv Food Nutr Res 52:1–59

Chang HH, Lan YC, Chung SD, Chien CT (2021) Sweet potato leaf feeding decreases cholesterol, oxidative stress and thrombosis formation in syrian hamsters with a high-cholesterol diet. Life 11(8):802

Chauhan VBS, Behera S, Pati K, Bansode VV, Nedunchezhiyan M (2021) Breeding for drought tolerance in sweet potato (*Ipomoea batatas* L.). In: Recent advances in root and tuber crops. Brillion Publishing, pp 65–87

Chauhan DS, Vashisht P, Bebartta RP, Thakur D, Chaudhary V (2025) Jerusalem artichoke: a comprehensive review of nutritional composition, health benefits and emerging trends in food applications. Compr Rev Food Sci Food Saf 24(1):e70114

Chen Z, Liang N, Zhang H, Li H, Guo J, Zhang Y, Chen Y, Wang Y, Shi N (2024) Resistant starch and the gut microbiome: Exploring beneficial interactions and dietary impacts. Food Chem: X 21:101118

Chintha P, Sarkar D, Pecota K, Dogramaci M, Hatterman-Valenti H, Shetty K (2023) Phenolic bioactive-linked antioxidant, anti-hyperglycemic, and antihypertensive properties of sweet potato cultivars with different flesh color. Hortic Environ Biotechnol 64(5):877–893

Choudhury B (1976) Vegetable. National Book Trust, India

Chupp C, Sherf AF (1960) Vegetable diseases and their control. The Ronald Press Company, New York

Cooley JS (1951) The sweet potato—Its origin and primitive storage practices. Econ Bot 5:378–386. https://doi.org/10.1007/BF02984804

Dumbuya G, Alemayehu HA, Hasan MM, Matsunami M, Shimono H (2021) Effect of soil temperature on growth and yield of sweet potato (*Ipomoea batatas* L.) under cool climate. J Agric Meteorol 77(2):118–127

Eguchi T, Kitano M, Eguchi H (1994) Effect of root temperature on sink strength of tuberous root in sweet potato plants (Ipomoea batatas Lam.). Biotronics 23:75–80

Elgabry RM, Sedeek MS, Meselhy KM, Fawzy GA (2023) A review on the potential health benefits of sweet potato: insights into its preclinical and clinical studies. Int J Food Sci Technol 58(6):2866–2872

Elgabry RM, Fawzy GA, Meselhy KM, Ramadan NM, Younes NF, Al-Taweel AM, Praveen S, Sedeek MS (2025) Phytochemical analysis and anti-aging efficacy of two Egyptian sweet potato cultivars in a d-galactose-induced skin aging model. Future J Pharm Sci 11(1):8

Eronico CA, Escalada RG, Trenucla RM (1981) Effects of different portions and length of storage of cuttings on the growth and yield of sweet potato. Ann Trop Res 3:144–149

Food and Agriculture Organization (FAO) (2008) Post-harvest management of sweet potato. Food and Agriculture Organization

Food and Agriculture Organization (FAO) (2013) Post-harvest handling and storage of sweet potatoes. FAO Agricultural Services Bulletin

Food and Agriculture Organization of the United Nations (2025) With major processing by our world in data. "Sweet potato production – FAO"

Gelaye Y (2024) Sweet potato: a versatile solution for nutritional challenges in Ethiopia. Technol Agron 4:e014. https://doi.org/10.48130/tia-0024-0011

Ghosh SP, Nair RG (1984) Tuber crops research and development. Indian Fmg 53:5–8

Girard AW, Brouwer A, Faerber E, Grant FK, Low JW (2021) Orange-fleshed sweet potato: Strategies and lessons learned for achieving food security and health at scale in Sub-Saharan Africa. Open Agric 6(1):511–536

Histifarina D, Purnamasari NR, Rahmat R (2023) Potential development and utilization of sweet potato flour as a raw material for the food industry. IOP Conf Ser: Earth Environ Sci 1230(1):012006

Hossain MM, Siddique MA, Choudhury B (1987) Yield and chemical composition of sweet potato as influnenced by timing of N-K fertilizer application under different levels of irrigation. Bangladesh J Agriculture 12:181–188

Indira P, Kabeerathuma S (1990) Physiometabolic Changes in Sweet potato grown under different level of soil moisture. J Root Crops 16:28–32

Islam S (2024) Sweetpotatoes [*Ipomoea batatas* (L.) lam]: the super food of the next century? An intensive review on their potential as a sustainable and versatile food source for future generations. CyTA-J Food 22(1):2397553. https://doi.org/10.1080/19476337.2024.2397553

Jaitwar AP, Jatav R, Verma R (2024) Holistic healing: the principles and practices of ayurvedic medicine. Guru Publishing House

Jeong BC, Oh SK, Park KY, Rho SP, Chung KB, Chung DH, Chin MS, Hong EH, Rark RK, Jung JW, Cha YH, Lee HB (1991) A new good eating quality sweet potato variety "Yulmi". Research reports of the rural development administration, upland and industrial crops. 33:22–28

Jiang T, Ye S, Liao W, Wu M, He J, Mateus N, Oliveira H (2022a) The botanical profile, phytochemistry, biological activities and protected-delivery systems for purple sweet potato (*Ipomoea batatas* (L.) Lam.): an up-to-date review. Food Res Int 161:111811

Jiang F, Wang S, Wang H, Wang A, Xu D, Liu H, Yang B, Yuan L, Lei L, Chen R, Li W, Fan W (2022b) A chromosome-level reference genome of a Convolvulaceae species *Ipomoea cairica*. G3 12(9):jkac187

Juras S (2021) Understanding the prehistoric presence of the sweet potato in polynesia: an investigation on seed floatability and viability. The Ohio State University

Kays SJ (1985a) Postharvest physiology of sweet potato storage roots. Hort Sci 20(6):1005–1009

Kays SJ (1985b) Postharvest physiology of perishable plant products. In: Postharvest technology of horticultural crops. University of California, pp 123–148

Key DE (1973) Root crops. Tropical Products Institute, London, pp 144–159

Kinoshita A, Nagata T, Furuya F, Nishizawa M, Mukai E (2023) White-skinned sweet potato (*Ipomoea batatas* L.) acutely suppresses postprandial blood glucose elevation by improving insulin sensitivity in normal rats. Heliyon 9:e14719

Komada H (1990) Biological control of soil borne plant pathogens. CAB International, Wallingford, pp 65–75

Kundu S, Rai B, Shukla A (2021) Prevalence and determinants of Vitamin A deficiency among children in India: findings from a national cross-sectional survey. Clin Epidemiol Glob Health 11:100768

Kushman LJ (1967) USDA Marketing Bull., 38

Laurie SM, Faber M, van Jaarsveld PJ, Laurie RN, Duchateau L, Modisane PC (2012) Beta-carotene content of selected sweet potato (*Ipomoea batatas*) varieties from South Africa. J Food Compos Anal 27(2):92–100

Laurie SM, Mulabisana J, Sutherland R, Sivakumar D, Pofu K, Mphela WM, Truter M, du Plooy I, Araya N, Araya H, Nyathi M, Kistnasamy A, Cloete M, Nkosi B, Shimelis H, Laing MD, Malebane ME, Bairu MW (2024) Seventy years of sweet potato [*Ipomoea batatas* L.(LAM)] *Research in South Africa*. Crop Sci 64(3):1112–1128

Leon J (1976) Proceedings of the fourth symposium of the International Society for Tropical Root Crops, Columbia, pp 20–36

Low JW, Arimond M, Osman N, Cunguara B, Zano F, Tschirley D (2007) A food-based approach introducing orange-fleshed sweet potatoes increased vitamin A intake and serum retinol concentrations in young children in rural Mozambique. J Nutr 137(5):1320–1327

Ludvik B, Neuffer B, Pacini G (2004) Efficacy of *Ipomoea batatas* (caiapo) on diabetes control in type 2 diabetic subjects treated with diet. Diabetes Care 27(2):436–440

Lutz JM, Simons JW (1972) USDA Prod Rsh Rept, 147-1-17

Maini SB (1976) Quality aspect of sweet potato, I. Short specialise training course on sweet potato production tech. CTCRI, Trivandrum

Mandal RC (1976) Short specialized training course on sweet potato production tech. CTCRI, Trivandrum

Martin FW (1972) USDA Prod Rsh Rept, 147-17

Martin JH, Leonard WJ, Stamp DL (1976) Principles of field crop production, 3rd edn. Macmillan Publishing Co., Inc, New York

Mellow DA (1946) Cultivation of the sweet potato. Ceres 7:54–60

Mir MA, Gul N, Ashraf Bhat M, Ganai SA, Malik FA, Rather AR, Shah A (2024) Biofortification of sweet potato: enhancing nutritional quality for improved human health. In: Harnessing crop biofortification for sustainable agriculture. Singapore, Springer Nature Singapore, pp 193–202

Mohankumar CR, Mandal RC (1972) Two decades of research. CTCRI, Trivandrum

Mourtala IZM, Innocent NM, Habibou M, Oselebe H (2023) Recent progress in breeding for beta-carotene, dry matter content and sugar in sweet potato [Ipomoea Batatas (L.) Lam]-A review. Eur J Agric Food Sci 5(1):6–13

Mukhopadhyay SK, Sen H, Jana PK (1990) Effect of planting materials on growth and yield of sweet potato. J Root Crops 16:119–122

Mukhopadhyay SK, Sen H, Jana PK (1992) Effect of K on growth and yield of sweet potato. J Root Crops 18:10–14

Mwanga ROM, Swanckaert J, da Silva PG, Andrade MI, Makunde G, Grüneberg WJ, Kreuze J, David M, De Boeck B, Carey E, Ssali RT, Utoblo O, Gemenet D, Anyanga MO, Yada B, Chelangat DM, Oloka B, Mtunda K, Chiona M, Koussao S, Laurie S, Campos H, Yencho GC, Low JW (2021) Breeding progress for vitamin A, iron and zinc biofortification, drought tolerance, and sweetpotato virus disease resistance in sweet potato. Front Sustain Food Syst 5:616674. https://doi.org/10.3389/fsufs.2021.616674

Nair GM, Ravindran CS, Prasad M (1984) Cultural practices for tuber crops. Indian Fmg 33:29–33

Nguyen HC, Chen CC, Lin KH, Chao PY, Lin HH, Huang MY (2021) Bioactive compounds, antioxidants, and health benefits of sweet potato leaves. Molecules 26(7):1820

Nishiyama I (1961) Proceedings of the 10th pacific science congress, Hawaii, pp 119–228

Nishiyama I (1971) Evolution and domestication of the sweet potato. Bot Mag 84:377–387

Nishiyama U, Miyazuki T, Sakamito S (1973) Evolutionary sequence of the sweet potato, Ipomoea batatas (L.) Lam. Genetics 74:197

Nolte Fong J (2021) Influence of resistant starch from potatoes on glycemic response in healthy, overweight females

Noreen S, Niazi MK, Shehzadi S, Ikram A, Arshad MT, Gnedeka KT (2024) Extend the emphasis to encompass sweet potatoes health advantages, industrial applications, and nutritional qualities. CyTA-J Food 22(1):2390994

Norman MC, Rodrjgues N, Cuadra A (1989) Performance of six early sweet potato cultivars on alluvial soil. Cienc Tec Agric 12:49–57

Nyombaire G (2023) Conversion of sweet potato starch to syrup by acid hydrolysis. Tuskegee University

O'Brien PJ (1972) The Sweet Potato: Its Origin and Dispersal. Am Anthropol 74:342–365

Obata N, Tabuchi H, Kurihara M, Yamamoto E, Shirasawa K, Monden Y (2022) Mapping of nematode resistance in hexaploid sweetpotato using a next-generation sequencing-based association study. Front Plant Sci 13:858747

Ooi SF, Sukri SAM, Zakaria NNA, Harith ZT (2021) Carotenoids, phenolics and antioxidant properties of different sweet potatoes (*Ipomoea batatas*) varieties. IOP Conf Ser: Earth Environ Sci 756(1):012077

Panchal SK, John OD, Mathai ML, Brown L (2022) Anthocyanins in chronic diseases: the power of purple. Nutrients. mdpi.com

Picha DH (1986) Postharvest handling and storage of sweet potatoes. Hort Sci 21(3):267–270

Pillai KS, Lal SS, Palaniswami MS (1978) Some new records of insects and mites as pests on sweet potato. J Root Crops 4:45–47

Purohit SR, Sharma V, Kumari M, Muthukumarappan K, Kane-Potaka J (eds) (2024) Future crops and processing technologies for sustainability and nutritional security. CRC Press, pp 203–218

Purseglove JW (1965) In: Baleer HG, Stebbins GL (eds) The genetics of colonizing species. Academic Press

Purseglove JW (1977) Tropical crops-dicotyledons. Longaman Group Ltd, London, pp 78–88

Qiao H, Shao H, Zheng X, Liu J, Liu J, Huang J, Zhang C, Liu Z, Wang J, Guan W (2021) Modification of sweet potato (*Ipomoea batatas* Lam.) residues soluble dietary fiber following twin-screw extrusion. Food Chem 335:127522

Rahman SA, Khan SA, Iqbal S, Rehman MM, Kim WY (2023) Eco-friendly, high-performance humidity sensor using purple sweet-potato peel for multipurpose applications. Chemosensors. mdpi.com

Rajamma P, Premkumar (1984) Sweet potato weevil. Indian Fmg 33:62–63

Ranteallo Y, Ahmad M, Syam A, Nilawati A (2023) Identification and quantification of minerals and vitamins of purple sweet potato (*Ipomoea batatas*) leave. IOP Conf Ser Earth Environ Sci 1230(1):012134

Rashid MT, Liu K, Jatoi MA, Safdar B, Lv D, Li Q (2022) Energy efficient drying technologies for sweet potatoes: Operating and drying mechanism, quality-related attributes. Front Nutr 9:1040314

Ravindra CS, Mohankumar CR (1987) Cultural practices of sweet potato. CTCRI, Technical Bulletin, Trivandrum

Ray RC, Ravi V (2005) Post-harvest spoilage of sweet potato in tropics and control measures. Crit Rev Food Sci Nutr 45(7–8):623–644

Ray RC, Sivakumar PS (2009) Traditional and novel fermented foods and beverages from tropical root and tuber crops: review. Int J Food Sci Technol 44(6):1073–1087

Rodamilans B, Casillas A, García JA (2021) P1 of sweet potato feathery mottle virus shows strong adaptation capacity, replacing P1-HCPro in a chimeric plum pox virus. J Virol 95:10–128

Rose BA, Edwin MR, Merrin RS, Bharathi VY (2024) Sweet potatoes: a nutrient-rich powerhouse for enhancing athletes' performance with optimized carbohydrate consumption—a review. In: Revitalizing health through humanities. Taylor & Francis, pp 94–100

Sajjapongse A, Wu MH (1989) Effect of some climatic factors on sweet potato yield. J Nat Sci 23:86–92

Scott GJ (2021) A review of root, tuber and banana crops in developing countries: past, present and future. Int J Food Sci Technol 2021(56):1093–1114

Scott GJ, Maldonado L (1999) Sweet potato post-harvest handling and processing. International Potato Center (CIP)

Shirhatti VR, Marathe SJ, Shah NN, Singhal RS (2023) Traditional food systems: Going backwards to move forward towards finding solutions to nutritional problems. In: Nutrition science, marketing nutrition, health claims, and public policy. Academic Press, pp 95–117

Singh AK (2023) Horticultural practices and post-harvest technology. Academic Guru Publishing House

Singh KD, Mandal RC (1970) Two decades of research. CTCRI, Trivandrum, 1983

Smith JEW (2023) Impact of daily sweet potato ingestion on alterations to the gut microbiome, cholesterol, and exercise performance. Mississippi State University

Spence JA, Humphries EC (1972) Effect of moisture supply, root temperature, and growth regulators on photosynthesis of isolated rooted leaves of sweet potato (*Ipomoea batatas*). Ann Bot 36:115–121

Steinbauer CE, Kushman LJ (1971) Sweet potato culture and diseases. USDA Hand Book 388:1–74

Stone MS, Martin BR, Weaver CM (2021) Short-term RCT of increased dietary potassium from potato or potassium gluconate: effect on blood pressure, microcirculation, and potassium and sodium retention in pre-hypertensive-to-hypertensive adults. Nutrients 13(5):1610

Tanaka JS, Sekioka TS (1976) Proceedings of the fourth symposium of the International Society for Tropical Root Crops, Columbia, pp 150–151

Tang C, Han J, Chen D, Zong S, Liu J, Kan J, Qian C, Jin C (2023) Recent advances on the biological activities of purple sweet potato anthocyanins. Food Biosci 53:102670

Tedesco D, de Almeida Moreira BR, Júnior MRB, Maeda M, da Silva RP (2023) Sustainable management of sweet potatoes: A review on practices, strategies, and opportunities in nutrition-sensitive agriculture, energy security, and quality of life. Agric Syst 210:103693

Teow CC, Truong VD, McFeeters RF, Thompson RL, Pecota KV, Yencho GC (2007) Antioxidant activities, phenolic and beta-carotene contents of sweet potato genotypes with varying flesh colours. Food Chem 103(3):829–838

Thompson C, Kelly WC (1979) Vegetable crops. Tata McGraw-Hill Publishing Co. Ltd, New Delhi, pp 405–430

Unnikrishanan M, Mukherjee A, Nair NG (1991) National symposium on genetics & biotechnology for crop improvement. Osmania University Hyderabad, Symp. Abst, p 61

Unnikrishnan M, Sheela MN (2001) In: Parthasarathy VA, Bose TK, Deka PC (eds) Biotechnology of horticultural crops, vol 2. Naya Prokasha, Calcutta

USDA (2005) United States standards for grades of sweet potatoes. United States Department of Agriculture

Ustimenko-Bakumovsky GV (1983) Plant growing in the tropics and subtropics. MIR Publishers, Moscow, pp 166–179

Varma SP (1984) Tuber crops research under the co - ordinated projects. Indian Fmg 34:29

Vicente AR, Manganaris GA, Darre M, Ortiz CM, Sozzi GO, Crisosto CH (2022) Compositional determinants of fruit and vegetable quality and nutritional value. In: Postharvest handling. Academic Press, pp 565–619

Walter WM Jr, Schadel WE (1982) A Rapid Method for Evaluating Curing Progress in Sweet Potatoes. J Am Soc Hor Sci 107:1129–1138

Watanabe K (1969) Effects of soil physical conditions on the growth and yield of crop plants. 9. Mechanism of the excessive vegetative growth of sweet potatoes from the viewpoint of ecological characteristics of the roots. Proc Crop Sci Soc Japan 38:652–656

Woolfe JA (1992) Sweet potato: an untapped food resource. Cambridge University Press

Yan M, Nie H, Wang Y, Wang X, Jarret R, Zhao J, Wang H, Yang J (2022) Exploring and exploiting genetics and genomics for sweet potato improvement: status and perspectives. Plant Commun 3(5):100332, New York

Yang ZG (1990) The effect of minimum tillage on sweet potato production with herbicide application and plastic film mulch. J South China Agri Univ 11:64–71

Yego RC (2022) Production, storage and quality characteristics of sweet potato varieties. M.Sc. thesis the School of Agriculture and Biotechnology, Department of Family and Consumer Sciences, University of Eldoret, Kenya

Zhang L, Li X, Li Z (2013) Gut microbiota regulates resistant starch metabolism and its impact on gut health. Food Res Int 54(1):93–101

Zhang L, Zhao L, Wang Q (2018) Sweet potato starch and its derivatives: characteristics and applications. Starch-Stärke 70(5–6):1700251

Sago (Tapioca) *Manihot esculenta*: The Starchy Sustainer of Energy

11

Sasmita Priyadarsini Dash, Prachi Pattnaik, Nitish Kumar Jena, Padmakshi Thakur, and Jitendra Kumar Tiwari

Abstract

Cassava (*Manihot esculenta Crantza*), a part of Euphorbiaceae family, with chromosome number 2n = 36, is the most crucial crop, being sixth in position in the world. It comes fourth in position of major staple food after rice, followed by wheat and maize in the world. Cassava is basically a perennial woody shrub bearing tuberous roots as the economically important part. The roots are inexpensive source of starch which have enormous demand for industrial uses as well as for animal feed, primarily in Asia. Cassava is a perennial plant, although it is popularly farmed in tropical and subtropical areas as an annual crop consumed for its starchy root tuber, which is delicious as well as a key source of carbohydrates. Most cassava is eaten in cooked form, but large amounts are also used to extract starch that may be utilized in food, feed of animals, and industry. Processed products made up of tapioca are also popular among consumers like chips, sago, and vermicelli. Tapioca pearls or sabudana are small, translucent globules; and have become an important part during fasting seasons in India. They have immense importance particularly during religious festivals like Navratri, Ekadashi, Shivaratri, and Saawan season. Various staple popular dishes are prepared using tapioca such as khichdi, thalipeeth, upma, kheer or payasam, and vada, aside from crispy snacks of papad and javvarisi vadaam. There's a potential gap between demand and supply in the industrial sector, which might require expanding cassava farming to new regions. Promoting value-added pro-

S. P. Dash (✉) · P. Pattnaik · N. K. Jena
IAS, Siksha 'O' Anusandhan deemed to be University, Bhubaneswar, Odisha, India

P. Thakur
S.G. College of Agriculture and Research Station, IGKVV, Jagdalpur, Chhattisgarh, India

J. K. Tiwari
Rani Lakshmi Bai Central Agricultural University, Jhansi, Uttar Pradesh, India

A. Kumar et al. (eds.), *Fasting Superfoods*,
https://doi.org/10.1007/978-981-95-2762-5_11

cessing of cassava can enhance profitability and create new market opportunities.

Keywords
Cassava/tapioca · Carbohydrate · Sagu · Value-added product

11.1 Introduction

Cassava (*Manihot esculenta Crantza*), a part of family Euphorbiaceae, with chromosome number 2n = 36, is the sixth most crucial crop in the world (Schaal et al. 2006). More than 500 million people worldwide consume it as a staple food, serving mostly the poor people in developing countries. It is the major source of calories in African countries, where it is grown primarily for its nutritious roots, although it can also serve as a leaf crop. It comes fourth in position of major staple food after rice, followed by wheat and maize in the world (Ceballos et al. 2004). The roots are inexpensive source of starch which have enormous demand for industrial uses as well as for animal feed, primarily in Asia (Schaal et al. 2006). Cassava is basically a perennial woody shrub bearing tuberous roots as the economically important part. A fully grown plant can grow up to height ranging from 1 to 2 m, although it may vary in some cultivars reaching a height up to 4 m (Fig. 11.1). Branching can initiate from as low as 20 cm, while it may be absent in some of the cultivars (Fig. 11.3). The plant is comprised simple leaves arranged spirally around the stem (Fig. 11.2). It is monoecious in nature with staminate flowers developing near the tip and pistillate flowers developing near the base of the inflorescence.

Cassava is a perennial plant, but it is widely adapted as an annual crop in tropical and subtropical areas for its delicious starchy tuber (Figs. 11.4 and 11.5). Most cassava is consumed after boiling, but they are also used largely to extract starch that may be utilized in food, animal feed, and industry. The plant is like a xerophyte,

Fig. 11.1 Fully grown plant of cassava

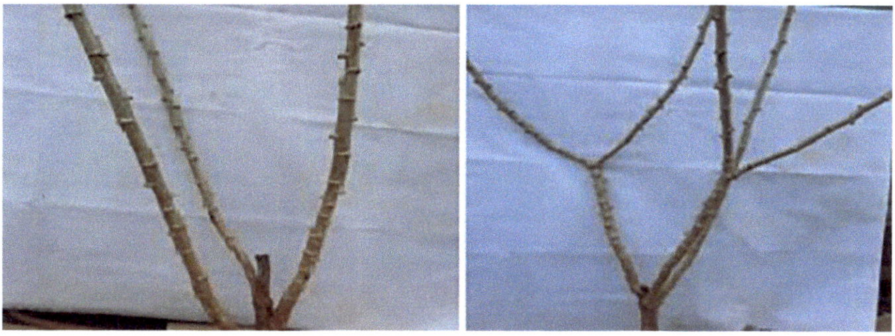

Fig. 11.2 Branching pattern in cassava

Fig. 11.3 Variations in leaves of cassava

having tolerance to drought, which can even sustain on poor soils. The Portuguese brought tapioca, a major starchy root crop produced in the tropics, to India in the seventeenth century. Tapioca is better known as the 'King of Tropical Tuber Crops' (Upadhyay and Pandey 2020) mostly grown in Kerala, Tamil Nadu, and Karnataka in southern peninsular India.

Cassava roots and leaves constitute major parts of the plant, having nutritional benefits. Roots consist of minute amounts of sucrose, glucose, fructose, and

Fig. 11.4 Young stem colour in cassava

Fig. 11.5 Different tuber shapes in cassava

maltose. Essential amino acids such as arginine, glutamic acid, and aspartic acid are found in abundance, whereas methionine, cysteine, and tryptophan have low content in the root (Bayata 2019). Cassava is crucial due to its demonstrated antibacterial, anti-inflammatory, anticancer, antidiabetic, antidiarrheal, and wound-healing properties.

11.2 History

Tapioca is identified by the presence of lactiferous vessels composed of secretory cells. Other crucial crops belonging to the Euphorbiaceae family include castor oil plants (*Ricinus comunis*), rubber trees (*Hevea brasiliensis*), and ornamental plants (*Euphorbia* spp.). It was an introduction by the Portuguese into Africa and Asia in the fifteenth century (Lokko et al. 2007). Cassava's origin was initially pinpointed to Central America—encompassing Colombia, Guatemala, Venezuela, and South Mexico—due to the abundance of wild varieties in that region. Manihot species present in Central America were distantly related to cassava. Thus, over time, Central America was recognized as the 'minor' centre of origin, while Brazil was tagged as the 'major' centre of origin. Cassava is believed to be a result of crossing between two wild Manihot species, followed by clonal propagation and selection of the hybrid. Presently, the plant is widely adapted and cultivated in the Caribbean, Africa, Asia, and Latin America, possessing food security benefits. Africa is the largest cassava-producing country in the world.

11.3 Area and Production

Asia stands second in cassava production among all the continents known for cassava production in the world. Globally, cassava production is highest in Nigeria, followed by DRC, Thailand, Ghana, and Indonesia. In India, the highest production is in Tamil Nadu, followed by Kerala and Andhra Pradesh. In India, the total area and production were recorded at 134 hectares and 4742 MT, respectively (NHB 2021–22). Major exporters of cassava are Thailand and India.

11.4 Botany

Tapioca pertaining to genus Manihot and species esculenta approximately consists of about 98 species in the genus, which range from subshrubs to shrubs and trees. It is diploid with chromosome number 2n = 36, but polyploids are also present with 2n = 54 or 72. Wild tapioca species have slender, fibrous root systems. The progenitors of tapioca are wild species, including *M. flabellifolia* and *M. peruviana*. Cassava is a perennial, thin, with branching shrub that grows 2–7 m tall, producing 5–10 tubers per plant, which are cylindrical in shape (Fig. 11.1). The stem of plant is woody, cylindrical, thick, and palmately lobed, and the segments above are glabrous. Tapioca plant has sympodial branching system that divides further into di/tri/or tetra-chotomously, producing secondary branches (Fig. 11.2). It also has palmately compound leaves, leaving noticeable leaf scars on the stem. Generally, uneven number of lobes may be present, ranging from 3 to 9 (Alves 2001). Leaves present near the inflorescence are smaller in size and lobe number. The petiole's length varies from 5 to 30 cm but may also reach upto 40 cm. Roots of this plant are the main storage organ (Figs. 11.3 and 11.4).

When the plants are sexually propagated through seed, the tap root system develops like dicots. The adventitious roots developing from the taproot later become storage roots. The adventitious roots of the vegetatively propagated plants develop from the cut surface of the stake present at basal end and result in fibrous root system. Anatomically, cassava root is a true root, so it cannot be used for asexual propagation. Fully grown storage root of cassava has three significant tissues, including bark (periderm), peel (cortex), and parenchyma.

Cassava is monoecious plant bearing both male and female flowers on the same plant. Inflorescence rises in axillary racemes near the ends of branches. Male flowers are found at the top of the monoecious inflorescence, whereas the lower part of inflorescence bears female flowers, which are fewer in number than the male flowers. The size of female flowers is greater than that of male flowers. Cassava exhibits protogyny condition in which pistillate flowers open 1–2 weeks in advance to staminate flowers. The flower consists of yellow, reddish, or purple tepals, and it includes an indefinite structure called perianth or perigonium and is devoid of calyx or corolla. Stigma becomes receptive from 6.30 a.m. in the morning and it continues up to 2.30 p.m. Ten stamens,

Fig. 11.6 Variations in rind colour of cassava

alternately long and short, are arranged in two circles, converging and diverging, supporting the anthers. The pollen is orange or yellow in colour, fleshy, and ranging from 122 to 148 μm in size. The pistillate flower consists of ten-lobed basal disk, which is less lobulated than staminate flower. Ovary is superior and tricarpellary, situated on a glandular disc with ten lobes, three loculi, and six ridges. A single ovule is present in each locule, and the styles are surrounded by three-lobed stigmas. The fruit is trilocular capsule has nine longitudinal plicate wings and is ovoid or globular in shape and 1–1.5 cm in diameter (Figs. 11.5 and 11.6). Tapioca is cross-pollinated and highly heterozygous, mostly pollinated by insects. Fruit reaches maturity in about 75–90 days after pollination (Ghosh et al. 1988). The shape of the seed is ovoid or ellipsoidal having dark brown smooth seed coat. All organs of cassava contain cyanogenic principles except the seed. Linamarin is also found in cassava tissues, which is produced in leaf and transferred to storage roots.

11.5 Classification of Cassava

Cassava can be classified based on genotype, duration, and taste (Fig. 11.7). Based on genotype it is divided into three types viz., bunching type, semi bunching type and non-bunching type whereas based on duration and taste it is classified in two categories; short (6–8 months) and long duration (>10 months) and bitter (high cyanide content, requires extensive processing before consumption) and sweet taste (low cyanide content, edible after simple cooking).

11.6 Varieties Cultivated

Varieties vary in colour of rind and flesh, tuber size, stem attributes, leaf and petiole colour, pattern of branching, crop duration, and resistance to pest and disease. High percent of cross-pollination persists, which classifies it as heterozygous in nature. Asexual method of propagation resulted in development of several polyploidy varieties and hybrids. Most of the improvement works on cassava crop are done at the Central Tuber Crops Research Institute (CTCRI), Thiruvananthapuram. Important varieties are given in Table 11.1.

Other popular cultivars: Co2, Co3, CO (TP) 4, CO (Tp) 5, MVD 1, H-165, Sree Vishakam (H.1687), Sree Prabha, Sree Athulya, Sree Apoorva, Sree Pavithra, Sree

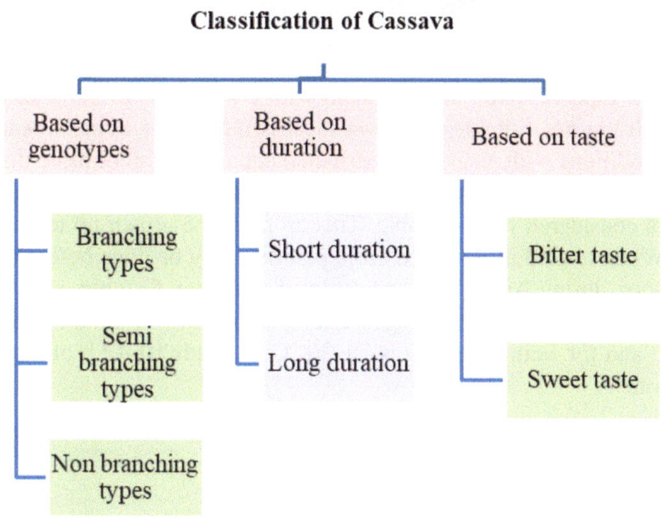

Fig. 11.7 Classification of cassava

Table 11.1 Important varieties of cassava

Varieties	Features
Sree Padmanabha	Resistant to cassava mosaic disease. Peak growing season is from October to November and in kharif June to July. Yield is moderate (38 t/ha). It has upright posture with top branches. It produces cylindrical white flesh tubers having low cyanogenic glucoside. Crop duration—270–300 days
Sree Sahaya	It is multiple hybrid, non-branching with yield capacity 35–40 t/ha. Crop duration—10–11 months
Sree Prakash	Dwarf plant, non-branching and early maturing with yield capacity 35–40 t/ha. Highly tolerant to Cercospora leaf spot. Crop duration—7–8 months
Sree Harsha	Triploid clone, high starch (38–41%)
Sree Vijaya	Clonal selection, crop mature—6–7 months Suitable for low-land cultivation
Sree Jaya	Clonal selection, crop duration—6–7 months Suitable for low-land cultivation
H-226	Semi-branching hybrid with characteristic green coloured leaves; skin is cream with purple patches; tuber rind is light purple; susceptible to cassava mosaic disease (CMD) Crop duration—10 months
Nidhi	Tolerant to mosaic, greyish white stem, white petiole with red shade, skin light pink. Crop duration—5–6 months

Reksha, and Sree Kaveri, H—97. Recent varieties in cassava include drought-tolerant varieties like Sree Kaveri, as well as varieties with enhanced cooking attributes and reduced NPK requirements, such as Sree Annam and Sree Manna, released by the Central Tuber Crops Research Institute (CTCRI), having low cyanogen content.

11.7 Soil and Climate

Cassava can be cultivated in all types of soil. However, tapioca grows well in well-drained soil, especially red lateritic loam with a pH range of 5.5–7.5. Saline and alkaline soil are not suitable. It grows successfully under tropical, warm, humid climatic conditions with an evenly distributed annual rainfall of more than 100 cm. As an irrigated crop, it can be cultivated all-round the year but December–January planting is considered most suitable. This crop can be grown up to an altitude of 1000 m. When treated as a rainfed crop, planting may be done before the onset of SW monsoon during April–May and during September–October, which coincides with NE monsoon. The ideal range of temperature for optimum germination is 25–35 °C, and for stem sprouting is 28–30 °C. It needs 10–12 h of day length for tuberization.

11.8 Field Preparation

To get a fine tilth, land is ploughed 4–5 times. Land is loosened by ploughing properly to a depth of 20–25 cm. At least 30 cm of soil depth should be present, and ridges and furrows should be formed with appropriate spacing. Ridges are made in sloppy land of length varying from 25 to 30 cm for a rainfed crop and in levelled for an irrigated crop. Ridges are established across the slope, while the flat beds are made in level lands having good drainage.

11.9 Propagation

11.9.1 Vegetative Propagation

For planting, setts are prepared from the central portion of the stem. Their length ranges from 25 to 30 cm long having 8–10 nodes. Mechanical damage should be avoided when handling and preparing the setts. Planting of setts is done vertically in beds, mounds, or ridges with the buds pointing upward at a depth of 5 cm. Before planting, setts are soaked in 1 g of carbendazim in 1 l water for 15 min. For planting one hectare of land, 17,000 sets are required. The setts may be treated with a solution of KCL (5 g/l) and ZnSO4 and FeSO4, each at 0.5% for 20 min. The sets are submerged in 30 g/l each of Azospirillum and Phosphobacteria for 20 min. Multiplication ratio of cassava is 1:10.

11.9.2 Mini Sett Technique

It is a technique in which small setts of size 3–5 cm consisting of two nodes and two buds are raised in nursery, 1 month prior to transplanting. Transplanting should be done on onset of monsoon in case of rainfed areas but under irrigated conditions, it can be done at any time of the year. The multiplication ratio under this technique is 1:60. This is developed in ICAR-CTCRI. Seed rate: 50,000 mini sett per ha.

11.9.3 True Cassava Seeds (TCS)

True cassava seeds technique has been developed at CTCRI, Thiruvanathapuram. These seeds are mainly utilized in breeding programme. Seeds remain viable for 8 months, while they remain dormant for 3–4 months. Seed rate: 1.5 kg/ha.

11.10 Planting Time and Spacing

The rainfed crop should be planted in April. Robust, healthy, disease-free plants are selected while taking planting supplies. The setts may be planted either vertically, slanted, or horizontal. Deeper planting can result in swelling of mother stem, which leads to subsequent reduction in size.

- Under rainfed condition: 60 × 60 cm (27,777 setts).
- Under irrigated condition: 75 × 75 cm (17,777 setts) and 90 × 90 cm (12,345 setts).
- Under Kanyakumari circumstances: 90 × 90 cm (12,345 setts).

11.11 Water Management

Sufficient moisture should be available in the initial stage of growth (20 DAP) for proper establishment of cassava stakes. First irrigation is given at the time of planting. This should be followed by a second irrigation on the third day and then continued every 7–10 days up to the third month, and every 20–30 days till eighth month. Moisture stress at any stage may result in increasing concentration of HCN in tuber.

A drip system may be installed for timely and optimum irrigation for cassava. In the lateral system, the drippers could be spaced at 60 cm apart for 4 l/h and 50 cm apart for 3.5 l/h. The laterals should be positioned in the middle of each raised bed, spaced at 120 cm wide and 30 cm apart.

11.12 Nutrient Management

For irrigated crops, 25 t/ha FYM should be incorporated during the ploughing process. Application of 45:90:120 kg of NPK/ha could be done as a basal dose and 45:120 kg of N:K /ha 90 DAP while preparing the soil.

However, in case of rainfed crop, 12.5 t/ha FYM is added to soil along with 50 kg N, 65 kg P, and 125 kg K/ha. After occurrence of rain for 30–60 days after planting, 2 kg of Azotobacter is sprayed into the soil (2.0 kg Azotobacter + 20 kg FYM + 20 kg soil per hectare). Generally, for long-duration varieties, NPK @

100:50:100 kg/ha dose is recommended, and 75:50:75 kg/ha NPK dose is considered optimum for short-duration varieties. Fertigation may also be done at the rate of 90:90:240 kg of NPK per hectare throughout the cropping time, which is applied once every 3 days intervals. In case of micronutrient deficiency, application of $MgSO_4$ @20 kg/ha, $ZnSO_4$ @12.5 kg/ha, and Borax @10 kg/ha may be done.

11.13 Intercultural Operations

Intercultural operations include gap filling, weeding, and thinning. Gap filling may be done within 20 days after planting. Depending on the weed's aggressiveness, first weeding should be done after 20 days of planting, and subsequent weeding may be carried out once in every month for up to 5 months. Thinning should be done by retaining two shoots per plant on the 60th day after planting. Grow aggregatum onions, coriander, short-duration legumes, and vegetables as intercrops may also be done during 60 days of planting.

11.14 Protection

11.14.1 Insects and Pests

List of major insects significantly affecting the yield of casava is presented in Table 11.2.

Integrated Pest Control Practices:

- Eliminate Abutilon indicum, an alternative weed host.
- Place 12 yellow sticky traps per hectare.
- Make smart use of nitrogen.
- Prevent overwatering.
- Spray 3% neem oil, 25 g/l of fish oil rosin soap, or 25 EC/2 ml of methyl demeton. For better contact with the foliage when using neem oil, teepol should be applied

Table 11.2 List of major insect pest and their management

Pest	Characteristic
Mites	Damage is high in summer including blotching and yellow specks on leaves; later curling, drying, and shedding of leaves occur Dicofol 18.5 EC 2.5 ml/l may be sprayed between the third and fifth month to control mites
White fly	Vector of cassava mosaic virus; initially appearance of chlorotic spots on the leaves which latter coalesce forming irregular yellowing of leaf tissue; development of sooty mould;severe infestation results in premature defoliation
Spiralling white fly	Yellowing of leaves, dropping of leaves; appearance of sooty mould in lower leaves are the major damaging symptoms

at 1 ml/l. Apply methyl demeton in the early phases of crop growth and phosphate in the late stages.
- Avoid using artificial pyrethroids.
- Don't let the crop grow over its allotted time.

11.14.2 Diseases of Cassava

List of major diseases significantly affecting the yield of casava is presented in Table 11.3.

11.14.3 Plant Physiological Disorders

Iron deficiency: When a deficiency symptom first appears, ferrous sulphate at the rate of 2.5 g/l to 3–4 times over a weekly period can be applied. Lime should be added in an equal amount to neutralise the solution.

Chlorosis: Foliar spraying of 1% FeSO4 + 0.5% ZnSO4 at 60 and 90 DAP can be given to alleviate micronutrient deficit.

11.15 Harvesting and Yield

Crop harvesting begins from 9 to 11 months after sowing. The time of harvesting depends upon the cultivar. As the tuber matures, leaves turn yellow and half of them dry out and fall off. Staggered harvesting is practiced in cassava. Normally short

Table 11.3 List of major diseases and their management

Disease	Characteristics
Mosaic	White or pale-yellow or pale-green patches appear on infected leaves. Leaves are often twisted, distorted, and stunted. Whitefly (*Bemisia tabaci*) is the carrier of the virus, and transmission of the disease is through their feeding habits. The infected plant is distorted and dwarfed
Cercospora leaf spot	Severe spotting and premature defoliation; presence of circular to angular spots, having greyish colour and velvety appearance Control: Mancozeb can be sprayed twice at intervals of 15 days, at a concentration of 2 g/l
Tuber rot	Caused by the fungus *Phytophthora palmivora* There is no external symptom. Symptoms are visible only after uprooting the tuber. Water-soaked brown lesions in the tuber first starts and finally the tuber will be rotten and exhibit a foul smell Control: Avoid stagnant water Provision of efficient drainage systems Apply *Trichoderma viride* across the soil at a rate of 2.5 kg/ha as a base and at the third and sixth months after planting Spot drench should be done with copper oxychloride 2.5 g/l

duration cultivars mature for harvest in 6–7 months after planting while long-duration cultivars take 10–11 months to mature. A fork or crowbar is used to uproot the tubers. The yield of cassava depends on the soil, climate, variety, and management practices. The average yield of tubers under rainfed conditions is 20–25 t/ha while under irrigated conditions is 40–50 t/ha.

Over the ideal period of 12–15 months after planting, the edible quality of roots—particularly their starch content—improves; beyond which lignification occurs, which primarily causes quality reductions. Cassava usually sheds its leaves during the dry season. Additionally, the quality of the roots changes when it rains, most likely as a result of the starch being remobilized for the production of new leaves. At this stage, boiled cassava loses its mealy texture making the roots unsuitable for this purpose.

Cassava root harvesting is usually done by hand, and it is easier in sandy soil or during the rainy season. When harvesting, it is typically required to dig around the roots to loosen the soil around them, especially in heavier soils or during dry spells. Usually, plants are trimmed down between 30 and 50 cm above the ground so that they may be readily lifted. The roots of the uprooted plant are lifted out of the earth using its projecting stem. Avoiding breaking the roots while lifting is crucial because if they are not removed from the soil, damaged roots might result in losses and pollution that could cause spoiling. Once the land has been cleared, harvesting is the most labour-intensive operation, and research programs are being carried out by agricultural engineers to find ways to mechanize it. There is a need for mechanical harvesting of cassava due to the non-uniform geometry of the roots in the ground. Nevertheless, certain cassava harvesters have been created and are in use at the moment, mostly by large-scale farmers. Mechanical harvesting is quite expensive for farmers with limited resources. In terms of generating economic flow, cassava's young leaves and shoots may be just as important as the roots because they are both edible and eaten as vegetables. On the other hand, root yield may get affected if too many leaves are harvested.

11.16 Post-Harvest Handling in Cassava

Cassava post-harvest handling focuses on preserving the quality and reducing losses of harvested cassava roots, which are highly perishable. This includes storage, processing, packaging, and transportation of both fresh and processed cassava products. The rapid deterioration of cassava roots after harvest makes it crucial to employ effective handling and storage techniques.

11.16.1 Primary Processing of Cassava

The first cultivators of cassava, the Amerindians, developed various processing methods to enhance its taste, prolong its shelf life, and reduce its cyanogenic potential. Today, cassava consumption across different regions of the world involves a wide variety of processing methods which consist of a combination of the following processing steps.

11.16.1.1 Peeling

Firstly, cassava roots are frequently peeled, which significantly reduces the raw material's cyanogenic potential because the peel makes up just 15% of the root's weight and has a cyanogen level that is usually 5–10 times higher than that of the root parenchyma. However, linamarase, an enzyme crucial to cassava detoxification during processing, is also abundant in the peel. The finest results are obtained when peeling by hand with a knife; this method is labour-intensive and sluggish, requiring an average of 25 kg every man-hour.

11.16.1.2 Boiling

Cassava as a vegetable is often consumed after boiling for 15–45 min. Certain types of cassava produce roots that are tender, mealy, and simple to mash after boiling. The roots of these kinds of variations are boiled and then ground into a smooth paste known as fufu in several parts of Africa. The first stage in making farinha (Brazil, Caribbean) and gari (West Africa) is pulping.

11.16.1.3 Pressing

Pressing the grated pulp to lower its moisture content is often the next processing step after cassava grating. The 'tipiti', a long thin basket weave tube created by the Amerindians, has been used for processing cassava for two millennia. The cassava mash would be poured into the tipiti, which would then be stretched from the bottom and hung from a tree branch. This would cause the mash's volume to decrease and the water to be squeezed out. From 60% to 70%, the mash's moisture level drops to roughly 50%. The pressing time can be as short as 15 min with the hydraulic press or as long as 4 days or more when stones are relied upon. The cake obtained after pressing needs to be broken down into granules. This can be done manually or mechanically by passing it again in a grating machine. The powdery granules obtained can then be further processed into the desired products.

11.16.1.4 Drying

The short shelf-life of cassava roots can be greatly extended by reducing their moisture content to the point where all physiological processes and microbiological development are blocked. This stage in cassava is at 14% moisture content, which is equivalent to 0.70 water activity. You can either dry the cassava roots in the sun or in an oven to remove the moisture. Sun-drying is the most popular technique for drying cassava; the moisture content is typically reduced to 8–12%. A drying surface that is exposed to sunlight is utilized for spreading out the cassava chips or granules from a grater. The drying rate will decrease with the number of chips on a drying surface. Compared to thicker chips, thin chips dry more quickly. It should be mentioned that a shorter drying time results in higher-quality chips (e.g., starch content, white colour). However, the longer the drying time, the lower the cassava's cyanogenic potential. Determining the residual cyanogen level of the dried cassava thus depends on drying parameters that influence the drying rate, particularly the loading rate (weight of drying material per unit area of drying surface).

11.16.1.5 Packaging

In any food system, packaging is crucial because it lowers percentage of loss, adds value, prolongs shelf life, preserves a product's quality and wholesomeness, raises market standards, and ensures food safety (Opara and Mditshwa 2013). Poor or insufficient packaging frequently leads to postharvest losses and unstable shelf life. Different factors such as storage conditions, moisture content, and the water activity of the product have a direct impact on the shelf life of packaged cassava products.

Water activity affects the microbial load and promotes physicochemical degradation because a high level of water activity indicates a higher relative humidity and moisture content. The majority of cassava products are hygroscopic, meaning they can absorb gases and moisture from the atmosphere, which promotes the growth of microorganisms.

Several packaging materials are in use for storage of cassava products such as paper, plastic, and sack bags or a combination of one or two of the materials. Paper materials can be coated or laminated with aluminium, and it is superior over other materials because they can be recycled at low cost (Opara and Mditshwa 2013). Plastic films, which include low-density polyethylene (LDPE), high-density polyethylene (HDPE), laminated aluminium foil (LAF), polyethylene (PE), polypropylene (PP), and plastics containers, are widely used for packaging.

11.16.1.6 Storage of Cassava

Due to the perishability nature of cassava roots, it is essential to develop storage schemes that alleviate the problem. Cold storage can significantly slow the decay of cassava. Internal discolouration in cassava may be avoided when they are kept below 4 °C. It was discovered that putting cassava roots in polyethylene bags might keep them fresh for around 2 months.

(a) *Traditional Storage*

In-ground storage is one of the most simplest conventional way of extending the shelf life of root and enhancing food security. Cassava roots have the benefit of prolonged in-ground storage due to their ideal harvesting age and harvest flexibility. However, prolonged storage might cause the storage root to lignify, which degrades some of its distinguishing characteristics. Additionally, rats or some infections (such as *Fusarium solani* and *Phaeolus manihotis*) may infest the root. Coating roots with loamy soil is another age-old technique. However, the root's shelf life was only prolonged by a few days using the conventional techniques. It has been demonstrated that while the root's cyanide, moisture, and starch contents decrease with storage time, its ash, sugar, crude fibre, and acidic content rise with increase in storage days.

(b) *Advanced Technology*

In collaboration with the National Resources Institute (NRI) in Chatham, UK, the Centro International de Agriculture Tropical (CIAT) in Palmira, Colombia, has developed several sophisticated techniques for cassava storage (Fadeyibi 2012). These techniques include paraffin wax waxing, freeze drying, and chilling at lower temperatures between 0 and 5 °C. These methods are used for export purposes. High temperatures and low relative humidity during storage promote the curing process, although careful handling and root selection are necessary before storage.

However, none of these methods can extend the shelf life of root while maintaining the quality attributes.

11.17 Processing of Cassava Root

Cassava root can be detoxified, its shelf life stabilized, and its quality improved by processing it into various food forms. Furthermore, processing can enhance or lessen the quality of modified products. Additionally, processing can also increase or decrease the quality attributes of the processed products. Drying and milling of cassava starch is the major ingredient for most of the value-added product of cassava (Fig. 11.8).

11.18 Tapioca Pearl/Sagu

Tapioca pearls, known by various regional names across India like sabudana in Hindi, javvarisi in Tamil, sabu in Bengali, and chavvari in Malayalam, hold a remarkable place in Indian kitchens and traditions. These translucent, tiny pearls are especially cherished during fasting periods, often making an appearance during festivals like Navratri, Shivaratri, Ekadashi, and the Saawan season. From comforting dishes like sabudana khichdi, thalipeeth, upma, and kheer or payasam, to crunchy delights like papads and javvarisi vadaam, sabudana is a go-to ingredient for many. It's particularly popular during Hindu festivals such as different vrat, Navratri, and Diwali, where light, flavourful dishes made with sabudana are a staple after fasting. Interestingly, sabudana isn't just loved for its taste—Ayurveda also praises it for its natural cooling properties, using it as a remedy to help lower body heat. So, whether it's for tradition, taste, or wellness, sabudana continues to be a beloved part of Indian culinary and cultural life. A sabudana preparation method is presented in Fig. 11.9.

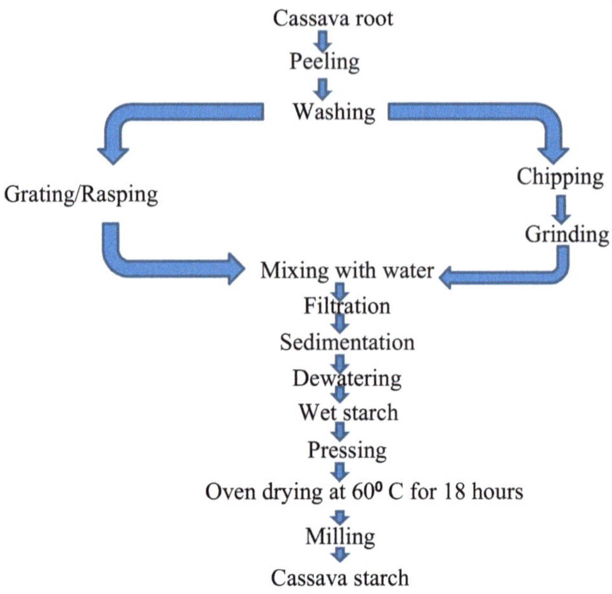

Fig. 11.8 Flow chart for milling cassava starch

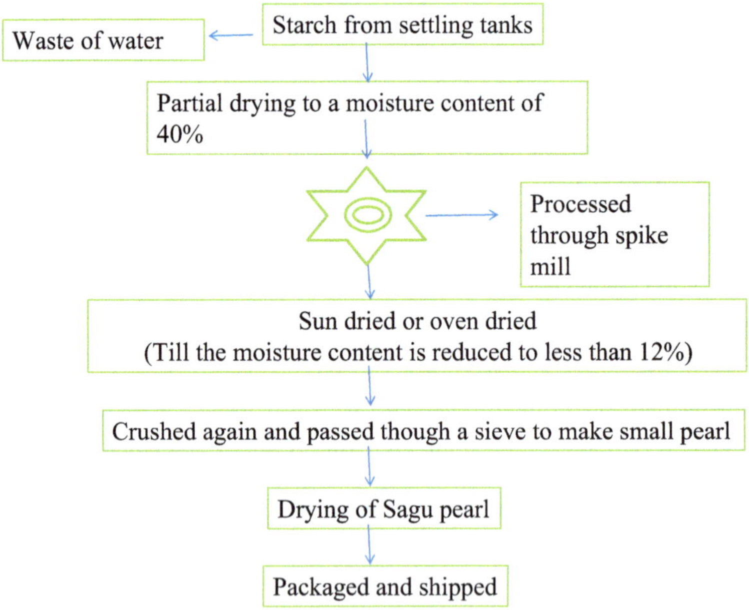

Fig. 11.9 Flowchart for preparation of sagu/tapioca pearl

11.19 Benefits of Sabudana

11.19.1 Source of Ample Energy

Sabudana is rich in starch and sugars, which the body quickly breaks down into glucose, fuelling our cells and keeping vital functions running smoothly. That's why it's a great choice after long hours of fasting or even after a tough workout. It helps recharge your energy levels and can ward off tiredness, light-headedness, and those pesky post-fast headaches.

11.19.2 Supports Gluten-Free Diet

Many young adults and older folks develop a sensitivity or intolerance to gluten—the protein found in grains like wheat, which is a common part of many Indian meals. That's where sabudana comes in. Since it's naturally gluten-free, it makes a great alternative to wheat. You can use it to make chapatis, dosas, or even traditional sweets, making it a safe and tasty option for people with celiac disease or gluten intolerance.

11.19.3 Maintains Bone Density

Sabudana is a great natural source of calcium, making it especially beneficial for strengthening bones in growing children. For older adults, it helps maintain healthy bone density and can ease symptoms of osteoporosis. While younger people can enjoy it regularly, those who are middle-aged or older should stick to moderate portions to support bone health without risking digestive or kidney issues.

11.19.4 Supplies Essential Amino Acids

Sabudana contains some important amino acids, which make it an essential plant-based source of good-quality protein. It provides methionine, which is rich in sulphur and helps nourish your skin and hair. It also has valine and isoleucine, both of which play a key role in healing and repairing muscle tissues. Plus, it includes threonine, which supports healthy teeth and enamel formation.

11.19.5 Checks Blood Sugar Levels

While sabudana is packed with calories and carbs that provide a quick energy boost, it also contains beneficial plant compounds like phytates, tannins, and polyphenols. These slow down digestion, which can help keep blood sugar levels in check—especially useful for people with diabetes. Thanks to its easy-to-digest nature and fibre content, having small portions of sabudana occasionally can also support the management of lifestyle conditions like obesity and diabetes.

11.19.6 Treats Anaemia

Iron deficiency anaemia is a common issue for many men, women, and children in India, often causing constant tiredness and reduced energy levels. Sabudana, being a good source of iron, can be a real help for those with low hemoglobin. Including it in the diet can support healthy blood levels and play a role in managing and improving anaemia.

11.19.7 Boosts Nervous System Function

Including small, controlled portions of sabudana in your daily diet can support better brain and mental health. Thanks to its high levels of the amino acid tryptophan, it helps boost nerve function, sharpen memory, and calm the mind. Tryptophan also helps balance serotonin the feel-good brain chemical making sabudana helpful in managing anxiety, improving mood, and promoting restful sleep.

11.19.8 Augments Heart Health

Sabudana is totally free of cholesterol, making it a heart-friendly ingredient, especially for those dealing with cardiac conditions. It's also rich in dietary fibre and B vitamins, which work together to boost good cholesterol and lessen bad LDL cholesterol. This helps prevent the buildup of plaque and fat in the arteries, supports smoother heart function, and promotes overall cardiovascular health.

11.19.9 Ensures Healthy Pregnancy and Nourishment for Babies

Consuming overnight soaked sabudana can offer great benefits for pregnant and breastfeeding women. Packed with iron and calcium, it's perfect for boosting milk production and helping balance hormones in expectant mothers and new moms. Sabudana's rich nutritional profile makes it an excellent choice for supporting the growing needs of children. Thanks to its starchy content, it helps promote healthy weight gain, supporting their overall development.

11.20 Different Types of Sabudana

Sabudana comes in various forms, each with its own distinct characteristics. The most common types are small, medium, and large pearls. Small pearls are perfect for dishes that cook quickly, while medium and large pearls are better suited for recipes that need a longer cooking time, like desserts or puddings. Besides the size, the pearls can also vary in colour and texture depending on how they're processed.

Fig. 11.10 Traditional and modern sabudana dishes

11.20.1 Traditional and Modern Sabudana Dishes

Traditional dishes like sabudana khichdi, sabudana vada, and sabudana kheer really bring out the unique flavours and textures of sabudana. These dishes are enjoyed across various regions, often served during religious festivals or as breakfast options. With a little creativity in the kitchen, sabudana can be transformed into modern dishes like sabudana biryani, sabudana waffles, or sabudana pancakes (Fig. 11.10).

11.21 Cassava Improvement

The breeding objectives for cassava can be briefly interpreted as (Teeken et al. 2021):

(a) Increase yield of dry matter per unit area.
(b) Increased resistance to major diseases like cassava mosaic disease (CMD), cassava brown streak disease (CBSD), and cassava bacteria blight (CBB).
(c) Resistance against pests such as cassava green mites (CGM) and whiteflies.
(d) Enhanced quality and quantity of starch.
(e) A low production for hydrogen cyanide.
(f) Enhanced nutritional value or biofortification.
(g) Adaptation to varying environmental conditions.
(h) Enhanced plant type and structure for easy mechanization.
(i) Enhanced consumer characteristics, including processing, cooking, and organoleptic properties.

Through conventional breeding, there has been an availability of improved cultivars leading to a remarkable yield increase in most major crops. The traditional method of breeding cassava uses phenotype-based recurrent selection. This method involves generating full-sibling and/or half-sibling progenies and then undergoing several clonal selection stages, such as single-row trials, preliminary advanced trials, and uniform yield trials. Conventional breeding methods have been employed to develop and launch several cassava varieties (Malik et al. 2020).

Due to the heterozygous genetic make-up of cassava, conventional breeding approaches result in considerable challenges. The long breeding cycle of 12 months in tapioca necessitates the need for long duration to develop improved varieties. The efficiency of conventional breeding could be enhanced by new tools and technologies, particularly when multiple traits are being selected simultaneously. To achieve more efficient agriculture—particularly in light of climate change, dwindling resources, land shortages, and rising food demand—it is crucial to modernize breeding programs with innovative tools. Innovative genomic techniques and biotechnology may enhance genetic gain, hasten the creation of superior cultivars, and have an impact on smallholder farmers' lives. Breeding program for enhancement of provitamin A carotenoids in cassava roots is also a high priority. Positive correlation has been found between colour of the cassava storage root parenchyma and total carotenoids present in the tuber.

In recent years, genomic resources for cassava have grown significantly. Using enriched genomic DNA libraries and expressed sequence tags, thousands of simple sequence repeat (SSR) markers have been produced. Numerous sequence polymorphisms have been discovered through re-sequencing of several accessions and the chromosome-scale reference genome of cassava. Methylation polymorphisms, structural variations, insertions-deletions (Indels), and single-nucleotide polymorphisms (SNPs) have all been identified. A haplotype map for cassava has been created, containing 25.9 million SNPs and 19 million indels (Ramu et al. 2017). Researchers can now examine tens of thousands of SNPs at a lower cost, thanks to high-throughput genotyping platforms (e.g., Golden Gate assay). It is now possible to perform rapid and precise high-density fingerprinting with SNP markers, thanks to the advent of inexpensive genotyping methods that utilize multiplex sequencing platforms (like genotyping by sequencing). These resources put forward efficient tools that have assisted genetical research and molecular breeding work in cassava.

11.22 Global Cassava Market Overview

With a compound annual growth rate (CAGR) of 4.2% from 2025 to 2034, the worldwide cassava market is projected to increase from its 2024 valuation of USD 198.90 billion to USD 299.62 billion by 2034. Major factors driving the market for cassava is the demand for its starch in the food and beverage industry, its rising use in the production of sweeteners, and its growing role in snacks and cereals are major drivers fuelling market growth.

In India, the cassava market is divided into segments like food and beverages, animal feed, and others. The food and beverages sector is the largest contributor to revenue, as more consumers are becoming aware of and seeking healthier additives. Animal feed also represents a significant portion of the market, as cassava is widely used in animal feed production. The market for cassava in India is projected to grow at a CAGR of 5% from 2025 to 2030.

11.23 Market Trends and Opportunities

1. Food and beverages: The popularity of cassava in food and beverage products is on the rise, thanks to its healthy reputation and adaptability in different recipes.
2. Industrial sector: The industrial demand for cassava is growing, driven by its diverse uses in products like adhesives, corrugation gums, and paper conversion, as noted by ResearchGate.
3. Animal feed: Cassava is increasingly being recognized as a valuable ingredient in animal feed, and its demand in this area is expected to rise.
4. Exports: India exports cassava to various countries, with key markets including the UK, UAE, and Qatar.
5. Ethanol production: The cassava-to-ethanol market is expanding, especially in countries like Thailand, China, and the Philippines.

11.24 Challenges and Opportunities

11.24.1 Demand-Supply Gap: There's a Potential Gap Between Demand and Supply in the Industrial Sector, Which Might Require Expanding Cassava Farming to New Regions

The growing industrial demand for sago starch has led to a discernible gap between demand and supply, particularly in key sectors like adhesives, textiles, and paper manufacturing. Current production levels, largely concentrated in traditional cassava-growing regions, are increasingly insufficient to meet projected industrial requirements. Studies indicate that bridging this demand-supply gap will necessitate the expansion of cassava cultivation into new, non-traditional areas. For example, meeting projected industrial sector needs may require hundreds of thousands of additional hectares under cassava cultivation, especially in regions outside the established production hubs. This expansion effort would not only help stabilize supply for sago-based industries but also benefit local farmers through crop diversification and increased income opportunities. The strategic introduction of improved cassava varieties and sound agricultural practices in new regions is considered essential for sustaining the industry's growth and securing its vital role in the agro-industrial sector.

11.24.2 Market Access: Ensuring Fair Pricing for Farmers and Improving Their Access to Markets Is Essential

Ensuring fair pricing and improving market access for sago farmers are critical steps towards sustaining the sector's growth and fostering rural livelihoods. Many sago producers, especially smallholders, often contend with limited access to lucrative markets, the dominance of middlemen, and price volatility, reducing their bargaining power and profits. Establishing transparent pricing mechanisms, promoting farmer cooperatives, and leveraging digital platforms for direct sales can help guarantee fair remuneration. Additionally, investments in rural infrastructure—such as better roads, storage facilities, and market information systems—are essential to connect farmers with processors and end-users more efficiently. These measures not only secure better incomes for sago producers but also contribute to developing more resilient and inclusive agri-food value chains.

11.24.3 Value Addition: Promoting the Processing of Cassava into Value-Added Products Could Increase Profitability and Open Up New Opportunities in the Market

Promoting the processing of cassava into value-added sago products holds significant potential for boosting profitability and unlocking new market opportunities. By diversifying beyond raw starch into items such as sago pearls, flour, noodles, and convenient ready-to-eat snacks, farmers and processors can capture higher margins and cater to evolving consumer demands. The development of branded, quality-assured value-added products can also attract niche markets, including health-conscious consumers and export segments. Furthermore, facilitating technology transfer, quality control, and access to packaging solutions will support the growth of sago-based enterprises. This shift towards value addition not only enhances income prospects for cassava growers but also stimulates rural employment, strengthens local economies, and contributes to the long-term sustainability of the sago industry.

References

Alves AAC (2001) Cassava botany and physiology. In: Cassava: biology, production and utilization. Wallingford, Cabi, pp 67–89

Bayata A (2019) Review on nutritional value of cassava for use as a staple food. Sci J Anal Chem 7(4):83–91

Ceballos H, Iglesias CA, Pérez JC, Dixon AG (2004) Cassava breeding: opportunities and challenges. Plant Mol Biol 56:503–516

Fadeyibi A (2012) Storage methods and some uses of cassava in Nigeria. Cont J Agric Sci 5:12–18

Ghosh SP, Ramanujan T, Jos JS, Moorthy SN, Nair RG (1988) Tuber crops. Oxford and IBH Publishing Co., New Delhi, pp 3–146

Lokko Y, Okogbenin E, Mba C, Dixon A, Raji A, Fregene M (2007) Cassava. In: Pulses, sugar and tuber crops. Springer, Berlin, pp 249–269

Malik AI, Kongsil P, Nguyễn VA, Ou W, Srean P, López-Lavalle LAB, Ishitani M (2020) Cassava breeding and agronomy in Asia: 50 years of history and future directions. Breed Sci 70(2):145–166

Opara UL, Mditshwa A (2013) A review on the role of packaging in securing food system: adding value to food products and reducing losses and waste. Afr J Agric 8:2621–2630

Ramu P, Esuma W, Kawuki R, Rabbi IY, Egesi C, Bredeson JV et al (2017) Cassava haplotype map highlights fixation of deleterious mutations during clonal propagation. Nat Genet 49(6):959–963

Schaal BA, Olsen KM, Carvalho LJ (2006) Evolution, domestication, and agrobiodiversity in the tropical crop cassava. In: Darwin's harvest: new approaches to the origins, evolution, and conservation of crops. Columbia University Press, New York, pp 269–284

Teeken B, Agbona A, Bello A, Olaosebikan O, Alamu E, Adesokan M et al (2021) Understanding cassava varietal preferences through pairwise ranking of gari-eba and fufu prepared by local farmer–processors. Int J Food Sci Technol 56(3):1258–1277

Upadhyay A, Pandey P (2020) Cultivation of cassava/tapioca (*Manihot esculenta* Crantz) in India. Just Agric 1(2):161–164

Tikhur (East Indian Arrowroot) (*Curcuma angustifolia* Roxb.): The Starchy Root of Versatile Nutrition

12

Padmakshi Thakur, Sasmita Priyadarsini Dash, Omesh Thakur, Vikas Ramteke, Upendra Naik, Devendra Upadhyay, Rohit Maurya, and Jitendra Kumar Tiwari

Abstract

Curcuma angustifolia is part of the ginger family, *Zingiberaceae*, which also includes familiar spices like ginger and turmeric. It is native of Indian subcontinent. It has around 50 types and 1000 species and is distributed throughout central, southern, and eastern India, but most commonly reported from the Northeast and Western Coastal Plains and hills of India. *Curcuma angustifolia* is a relatively obscure yet significant source of starch. It is also known as white turmeric or narrow-leaved turmeric or East Indian arrow-root. *Curcuma angustifolia* is propagated via rhizomes, which serve as the main reservoir for its nutritional and medicinal properties. In Chhattisgarh, it is locally known as '*tikhur*'. Tribal people in Chhattisgarh, Orissa, and north-eastern India cultivate *Curcuma angustifolia* for starch. The rhizomes and starch of *Curcuma angustifolia* (Tikhur) possess medicinal properties, including acting as an appetizer, alleviating burning sensations and stomach discomfort, assisting in kidney stone removal, and providing

P. Thakur (✉) · V. Ramteke · U. Naik
S.G. College of Agriculture and Research Station, IGKVV, Jagdalpur, Chhattisgarh, India

S. P. Dash
IAS, Siksha 'O' Anusandhan Deemed to be University, Bhubaneswar, Odisha, India

O. Thakur
CoA & Research Station, IGKVV, Gariyaband, Chhattisgarh, India

D. Upadhyay
Dau Kalyan Singh College of Agriculture and Research Station, Indira Gandhi Krishi Vishwavidyalaya, Bhatapara, Chhattisgarh, India

R. Maurya
KVK Turki, DRPCAU, Muzaffarpur, India

J. K. Tiwari
Rani Lakshmi Bai Central Agricultural University, Jhansi, Uttar Pradesh, India

A. Kumar et al. (eds.), *Fasting Superfoods*,
https://doi.org/10.1007/978-981-95-2762-5_12

relief for ulcer patients. It has been reported that Tikhur starch is a rich source of carbohydrates, lipids, proteins, essential nutrients, and minerals. When mixed with water or milk, the starch powder serves as a nourishing, filling meal, especially beneficial for newborns and growing children. Additionally, beverages containing Tikhur starch can act as a nutritious supplement alongside breast milk. It is consumed during fast and festive occasions. Consumption increases during summer season in the form of herbal drinks due to their cooling effect, which prevents body from heat strokes. The starch of *Curcuma angustifolia* is widely used in food production due to its excellent thickening, gelling, and water-retaining characteristics. The rhizome, easily digestible and highly valued in dietary use, has yet to be fully explored or utilized for industrial applications. Despite its potential, the starch of *Curcuma angustifolia* has not received significant attention in the food industry. This chapter deals with the nutritional, medicinal importance, and production potential of this crop.

Keywords

Curcuma · Starch · Starch extraction · Nutrition and yield

12.1 Introduction

Tikhur (*Curcuma angustifolia* Roxb.) is a rhizomatous herbaceous plant within the Zingiberaceae family, characterized by a chromosome number of $2n = 42$. Commonly referred to as white turmeric, narrow-leaved turmeric, or East Indian arrowroot, *Curcuma angustifolia* is one of over 80 species in the genus *Curcuma*, which includes other well-known species such as ginger and turmeric. It is a tropical plant native to India, primarily found in the eastern and southern regions. It is propagated through rhizomes, which serve as a rich source of both starch and fiber (Misra and Dixit 1983) (Fig. 12.1). Ancient Indian communities used it as an important food source due to its high starch content and digestibility. Its easy cultivation in coastal and forested regions made it a staple for indigenous tribes and rural populations. In Chhattisgarh, it is considered as minor tuber crops categorized under forest produce and generally cultivated as an intercrop in orchards.

12.2 Taxonomy

The taxonomy of *Curcuma angustifolia* is classified as follows: Kingdom: Plantae, Class: Liliopsida, Subclass: Commelinidae, Order: Zingiberales, Suborder: Zingiberanae, Family: Zingiberaceae, Genus: *Curcuma* L., and Species: *angustifolia*. In various regions, it is known by several vernacular names, such as Tavakshira and Tavaksheera in Sanskrit, Tekhur and Tikhur in Hindi, East Indian arrowroot in English, and several others across languages like Kannada, Telugu, Tamil, and Malayalam (Bapalal and Nighantu 1985).

Fig. 12.1 Healthy plant (**a**), flower (**b**), rhizome (**c**), and starch (**d**) of *Curcuma angustifolia*

12.3 Area, Production, and Distribution

Curcuma angustifolia is distributed throughout central, southern, and eastern India, though it is most commonly reported from the Northeast and the Western Coastal Plains and hills (Srivastava et al. 2006; Sharma 2012). It thrives as wild growth in the hilly regions of Maharashtra, Madhya Pradesh, Andhra Pradesh, Himachal Pradesh, Odisha, Chhattisgarh, Central Himalayas, Karnataka, Bihar, West Bengal, Tamil Nadu, and Kerala. Furthermore, it is cultivated and widespread in countries such as Sri Lanka, Burma, Laos, Nepal, Pakistan, various regions of Asia and Africa, Cambodia, Northern Australia, and China (Ravindran et al. 2007; Nahak and Sahu 2011; Tiwari and Patel 2013). *C. angustifolia* is frequently found at the edges or in the clearings of forests. The total area of *tikhur* in Chhattisgarh is 93.48 ha with 2047.13 tonnes production and 18.04 t ha^{-1} productivity (Shankar and Singh 2018). In Chhattisgarh, *Curcuma angustifolia* is found in abundance in the hilly tracts and forests of districts such as Bastar, Dantewada, Bijapur, Narayanpur, Kanker, Rajnandgaon, Kawardha, Dhamtari, Bilaspur, Raipur, Korba, Korea, and Surguja. The total annual collection of tikhur rhizomes, classified as a minor forest produce, amounts to 190.00 tonnes. The Bastar and Bilaspur divisions are identified as the key areas with the highest potential for tikhur production in the state (Anonymous 2005).

Two distinct varieties of *Curcuma angustifolia* are found in the Bastar division: one with creamy white flowers and the other with light pink-colored flowers (Singh et al. 1999). Historically, tikhur was widely distributed across the Sal forests of

Chhattisgarh. However, unregulated harvesting and overexploitation have led to a decline in its population, resulting in its occurrence being limited to specific patches. In response, large-scale cultivation of tikhur has now been adopted by farmers in the Bastar division.

12.4 Botany

It is a perennial flowering plant with modest and small spiked inflorescence of three or four pink or white, funnel-shaped flowers within tufts of pink terminal bracts. The flowers of *Curcuma angustifolia* are bisexual, epigenous, and zygomorphic. The calyx is typically tubular, thin, and split on one side, sometimes resembling a spathe, with a 3-toothed or lobed apex. The corolla is tubular at the base and 3-lobed at the tip. The plant features six stamens or staminodes. The ovary is inferior, initially 3-loculed, and becomes 1- or 3-loculed upon maturation. The style is single, very thin, and located in a furrow in the filament between the anther locules (Annonymous 2001). The flowers generally appear at the beginning of the rainy season (July to August) before the leaves have fully developed, and they continue to bloom even after the leaves have fully matured.

The leaves grow to a length of 36–37 cm and a width of 8–10 cm, with a suborbicular or lanceolate to narrowly strap-shaped blade that may be glabrous or hairy. The prominent midvein and numerous pinnate, parallel lateral veins are accompanied by entire leaf margins. These leaves have a scent and taste similar to turmeric. The plant's rhizomes, which can grow up to 1.5 m in length, are the primary source of its nutritional and medicinal benefits, with their aroma attributed to the presence of Curcumene.

12.5 Nutritive Value and Uses

12.5.1 Nutritive Value

Curcuma angustifolia is cultivated by tribal communities in Chhattisgarh, Orissa, and northeastern India primarily for its starch, which serves as both food and medicine. The starch is highly nutritious and can be consumed directly, making it a valuable food supplement and a substitute for arrowroot powder. According to Kashyap and Shahi (2021), *Curcuma angustifolia* is a rich source of carbohydrates, lipids, proteins, nutrients, and minerals. Starch powder is often mixed with water or milk to provide a filling meal, particularly for newborns and growing children. Furthermore, beverages containing *Curcuma angustifolia* starch are considered a healthy alternative to breast milk (Sahoo et al. 2022). Despite its nutritional and medicinal value, the use of *Curcuma angustifolia* remains limited mainly to tribal areas and traditional medicinal preparations. Physicochemical composition of *tikhur* starch (g/100 g): It has been reported that for optimal shelf life, the moisture content of high-quality tikhur starch should be maintained within the range of

10–13.5%. Additionally, the water solubility index and swelling powers of tikhur starch increase as the temperature rises from 50 to 100 °C.

Phytochemical constituents: The rhizomes of *Curcuma angustifolia* have been identified to contain a range of secondary metabolites, including alkaloids, flavonoids, terpenoids, phenols, tannins, saponins, curcumin, steroids, glycosides, and oils (Rajkumari and Sanatombi 2017). Additionally, they are found to contain starch, glucose, sugars, sesquiterpenoids, curcuminoids (curcumin, demethoxycurcumin, and bisdemethoxycurcumin), curcumol, zederone, fyranodiene, pyrocurzerenone, procurcumenol, Curcumanolide A & B, as well as gum and fats (Kulkarni and Ansari 2004; Itokawa et al. 2008) (Table 12.1).

12.5.2 Medicinal Uses

The rhizome of *Curcuma angustifolia* (Tikhur) is mildly bitter, demulcent, nonirritating, nutritive, and fragrant. It is renowned for its medicinal value and is employed in the treatment of various ailments. The fresh rhizome pulp serves as a remedy for fever, joint pain, and leucorrhoea (Hemadri and Rao 1984). It is also used as an appetizer to alleviate burning sensations and stomach pain, aid in kidney stone removal, and assist ulcer patients (Sharma 2003). Additionally, the rhizome pulp is utilized for treating headaches and provides a cooling effect (Nag et al. 2006).

Table 12.1 Physico-chemical composition of *tikhur* starch (g/100 g)

S.No.	Constituents	Deshpande (2008)	Kumari et al. (2017)	Chandel et al. (2018)	Tripathy et al. (2023)	Kaur et al. (2023)
1	Crude gat (%)	0.2	0.62	0.39	0.8	0.54
2	Bulk density (g/cc)	–	0.51	–	0.48	0.84
3	Crude protein (%)	1.6	1.02	0.39	1.8	0.92
4	Crude fiber (%)	3.9	–	–	–	–
5	Ash (%)	0.9	0.3	0.9	1.12	0.52
7	Starch recovery (%)	25–30	–	–	11.65	13.79
8	Total carbohydrate (%)	–	–	85.3	83.45	–
9.	Sugar and dextrins (%)	2.1	–	–	–	–
10.	Amylose (%)	–	32.3	–	–	31.9
11.	Amylopectin (%)	–	67.7	–	–	–
12.	Calorific value (cal/g)	–	3499	–	–	–
13.	Moisture (%)	–	2.72	12.53	13.4	5.97

Externally, powdered rhizome mixed with honey is applied to the mucous membranes in cases of stomatitis and facilitates the healing of stomach ulcers (Elias et al. 2015). Essential oil extracted from the rhizomes is used to combat tapeworms (Banerjee and Nigam 1978) and for antifungal treatments. Furthermore, the starch is employed in the treatment of jaundice (Hemadri and Rao 1984), while the boiled water extract of the rhizome has been reported to aid in diabetes management (Sheikh et al. 2015). The starchy flour is also used as a weaning food known as shotti in North Bengal (Sharma 2012).

Essential Oil Composition of Tikhur Rhizome: Rhizome oil contains *ar*-curcumene, β-Pinene, α-Terpineol, Camphor, Zingiberol, Borneol (Banerjee et al. 1980), curzerenone, furanodienone, isofuranodienone (Nguyen et al. 2001), xanthorrhizol isomer, methyl eugenol, palmitic acid, germacrone, isoborneol, curdione, and 1,8-cineole (Srivastava et al. 2006; Thongkhwan et al. 2017). *Leaf oil contain*: curzerenone (33.2%), 14-hydroxy-δ-cadinene (18.6%), and γ-eudesmol acetate (7.3%) (Jena et al. 2017).

12.5.3 Use of Starch

The starch derived from the rhizomes of *Curcuma angustifolia* is highly nutritious and easily digestible, making it a suitable food source for infants, frail children, and invalids. It is a rich energy source and is commonly consumed during fasting periods. The starch is used in the preparation of various sweet dishes and herbal foods such as halwa, barfi, and jalebi, particularly during fasting (Vrata, Upwas). Farmers also prepare a cooling herbal drink called 'sarbat' using tikhur starch during summer. The use of tikhur starch is prevalent among major tribes in Chhattisgarh, such as the Kamar, Gond, and Halba, though its demand is increasingly spreading across the state.

As a gluten-free alternative to wheat flour, it can be used in place of maize starch as an excipient in medicinal tablets with considerable success, owing to its superior binding and non-crumbling properties (Shahi and Deepak 2018). However, tikhur starch is often adulterated with starches from tapioca, sweet potato, and rice flour (Annonymous 2001).

East Indian Arrowroot is valued for its high starch content, making it useful in various industrial applications beyond its traditional food and medicinal uses. It is utilized across various industries, including food, pharmaceuticals, cosmetics, textiles, and biodegradable products.

In Chhattisgarh, starch of *tikhur* is utilized for the preparation of many dishes. It is consumed during fast and festive occasions. Consumption increases during summer season in the form of herbal drinks due to their cooling effect which prevents body from heat strokes. Various dishes prepared from *tikhur* starch are presented in Fig. 12.2.

Fig. 12.2 Sweet dishes viz., *tikhur barfi* (1), *tikhur halwa* (2), *tikhur jalebi* (3), and drink (*tikhur sharbat (4)*) prepared by *tikhur* starch

12.6 Comparison with Other Arrowroots

East Indian Arrowroot is different from West Indian Arrowroot (*Maranta arundinacea*), which is another plant used for arrowroot starch production. While both produce starch that is easily digestible, *Curcuma angustifolia* is more closely related to turmeric and has additional medicinal properties.

12.7 Varieties Cultivated

At All India Coordinated Research Project, IGKV, Jagdalpur, Chhattisgarh, collected locally available genotypes of *tikhur* from different district of Chhattisgarh. Some high-yielding genotypes like IGSJT-10-1, IGSJT-10-2, IGBT-10-4, and IGBLT-10-1 were collected and identified during 2011–12. During 2018–19, one variety of *tikhur*, *C.G. Tikhur-1* developed from SGCARS, Jagdalpur through clonal selection and was released from SVRC for commercial cultivation in whole Chhattisgarh. Average rhizome yield of this variety is 24.58 t/ha with 14.46% average starch recovery.

12.8 Soil and Climatic Requirement

Curcuma angustifolia (Tikhur) flourishes best in warm, humid tropical and subtropical climates, with an optimal temperature range of 25–35 °C. Sufficient rainfall is crucial for its growth, development, and rhizome yield, with annual rainfall between 95 and 130 cm distributed over a 5–6-month period being ideal. The most critical phase of moisture requirement occurs during the first 4 months after planting, particularly during the growth and rhizome formation stages. The plant thrives in deep, loose, friable soil with high organic matter and adequate nutrient content, which supports healthy rhizome yield. Proper drainage is essential, as stagnant water may negatively affect the rhizome yield. Slightly acidic to neutral soils (pH 5.5–7.0) are preferred for optimal growth. Partial shade is more suitable for growth and yield of *tikhur* rhizomes compared to open field.

Fig. 12.3 *Tikhur* cultivation during at S. G. CoA & RS, IGKV, Jagdalpur

Fig. 12.4 Demonstration of *tikhur* cultivation as inter crop in cashew and mango orchard at Bastar district of Chhattisgarh

12.9 Planting Season

Planting time of rhizomes and growing season have marked effect on the yield of *tikhur* rhizomes and starch recovery. Second fortnight of May to first week of June is the best planting time of *tikhur* in Chhattisgarh. Planting must be completed before monsoon (Figs. 12.3 and 12.4).

12.10 Land Preparation and Planting Technique

Field preparation involves ploughing and turning of soil followed by planting by anyone of the methods, viz., flat bed and ridge-furrow, depending on the soil type. Among these methods, furrow planting is superior to others due to better soil moisture retention. Planting spacing 60 × 20 cm or 60 × 30 cm and planting depth 6–7 cm have been found optimum.

12.11 Seed Rate

Rhizome finger or mother corm of 15–25 g weight is considered best as seed or planting material. For planting in 1 ha, 15–18 quintals *tikhur* rhizomes are required.

12.12 Weeding, Mulching, and Irrigation

Consistent weeding is essential for ensuring a high rhizome yield. Non-chemical weed control practices begin with deep ploughing and implementing diverse crop rotations. Deep cultivation, coupled with soil exposure to sunlight prior to transplanting, aids in the destruction of weed seeds. Maintaining a weed-free field for the first 4–5 weeks after planting is crucial, as this period is vital to prevent weed competition, which can adversely affect yield. Hand weeding is an efficient method for managing weeds between rows of plants. In the main field, a pre-planting application of 1 L of fluchloralin (as basalin 2 L/ha) dissolved in 500 L of water just before transplanting is very effective chemical control method to suppress weeds to emerge. Mulching in a layer of dried leaves or straw is applied to retain soil moisture and prevent weed growth. Plastic mulch can also incorporate in between the row planting. During *kharif* season irrigation is needed in dry spells situation.

12.13 Manures and Fertilizer

According to Sahu et al. (2016), during field preparation, 20 tonnes per hectare of farmyard manure (FYM) or 10–15 tonnes per hectare of vermicompost should be incorporated into the soil. The crop requires 60 kg/ha of nitrogen (N), 40 kg/ha of phosphorus (P_2O_5), and 60 kg/ha of potassium (K_2O) during the crop season. The full dose of FYM is thoroughly mixed into the plots at the time of field preparation. For chemical fertilizers, half of the nitrogen dose and the full doses of phosphorus and potassium are applied as a basal dose before planting. The remaining half of the nitrogen dose is applied 45 days after planting, during the intercultural operation and earthing-up.

Organic production is best practice for cultivation of *tikhur* to maintain and conserve the nutritional and medicinal properties of the crop.

12.14 Duration of Crop, Harvesting

Tikhur rhizomes mature and become ready for harvest approximately 5–6 months after planting. Early harvest results in poor quality of rhizome, which can not be stored for longer period. Harvesting is done manually by *khurpi* or *kudali*. For extraction purposes, it can utilized 5–10 days after harvesting. For seed purposes, rhizome should be placed for curing for about 10–15 days just after harvesting in shade places. Extraction of starch processed by manual or wet grinding machine.

12.15 Yield

The fresh rhizome yield of Tikhur ranges from 180 to 200 quintals per hectare. The crop is considered mature when the leaves begin to yellow and collapse, eventually leading to the drying of the entire plant.

12.16 Post-Harvest Processing of *Tikhur*

Tikhur is primarily cultivated for its starch content. The traditional method of starch extraction involves several labour-intensive steps: cutting and peeling the fresh rhizomes, rubbing them on a rough surface or sieve, soaking in water, decanting, and drying (Patel et al. 2015). From 1 kg of fresh rhizomes, only about 125–140 g of starch is typically extracted, indicating a significant loss of material along with high time and labour costs (Das et al. 2015; Tiwari and Patel 2013). The pure starch obtained is white, odourless, tasteless, and insoluble in cold water and organic solvents (Radley 1953).

12.16.1 Time of *Tikhur* Processing

Best time of *tikhur* processing starts from after harvesting, i.e. from November to March. December to February is ideal for *tikhur* processing due to winter season as low temperature helps in better starch settling.

12.16.2 Methods of *Tikhur* Processing

From ancient time, it is processed by traditional method, which involves rubbing or crushing of rhizome by stone and also grating of rhizome with the help of sieve. In the Bastar district of Chhattisgarh, this procedure is still followed by tribal people (Figs. 12.5 and 12.6).

Procedure of *Tikhur* Processing

Traditional Method	Advance Method
1. Rubbing rhizome on rough surface of stone or crushing by stone. 2. Rubbing or grating of rhizomes on sieves having rough surface.	1. Cutting of rhizome into small pieces and grinding in wet pulveriser machine of 1 HP, 2 HP or 3 HP. It is also known as mechanized starch extraction technique.

Processing Procedure

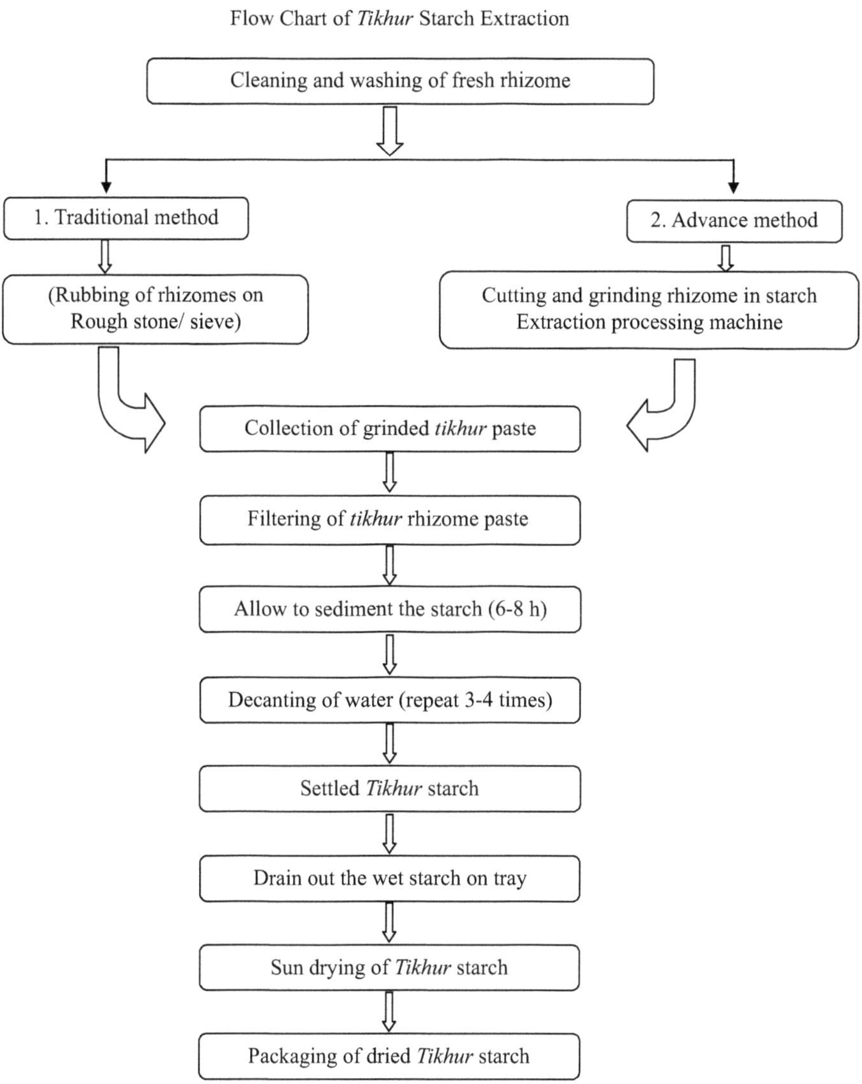

Fig. 12.5 Flow Chart of *tikhur* starch extraction

1. Cleaning and washing of fresh rhizome by removal of root and tuberlets and dipping in water for 1 hour for better removal of dust and stone particles from rhizome.
2. Pasting and grading of rhizome by traditional method and advance method. In case of advance method, cutting of rhizome into small pieces is necessary so that rhizome finger can easily pass into the wet pulverizer machine. Cutting of rhizomes is also necessary when it is crushed or rubbed by rough surface of stone. While in case of grating on sieve, whole rhizome fingers will be used to

Fig. 12.6 Process of *tikhur* starch extraction

ease the process for rubbing into rough surface on sieve. Water is added during the procedure when required. In wet pulverizer machine, paste is collected through 25–30 mm sieve in a pot.

3. The paste was added with water in a *1:2 ratio* to prepare a starch solution. This solution was filtered through *muslin cloth.*
4. The *supernatant part retained on the cloth* was removed separately.
5. The *filtered solution* was collected in an *earthen pot.*
6. The pot was kept undisturbed for *6–8 hours* to allow starch particles to settle.
7. A *starch mass sedimented* at the bottom of the pot.

8. *Decanting of water* was done after 6–8 hours, once the starch had fully settled.
9. The process of decanting was repeated 3–4 times till the bitterness taste was not experienced.
10. On an average 10–15 kg rhizome can process by traditional method in 1 day. While in advance method through wet pulveriser machine 20–25 kg rhizome can be process in a hour, depending capon capacity of machine.

Advantages of mechanized starch extraction technique over traditional method of processing

- It is enhanced or improved starch extraction technique.
- Delivers *20× higher efficiency* than the conventional method. All procedure remains same only grinding of rhizome done by gravy machine.
- It saves labour and time and also improves efficiency of process.
- On an average 3–4 q rhizomes can process by this method in 1 day.
- Initial cost is more but not required any specific setup as it can also perform in a small room with the facilities of electricity and water.

12.17 Economics

- The average input cost for *Tikhur* production is Rs. 120,000–140,000/- ha^{-1} in Chhattisgarh and average rhizome yield is 15–18 t ha^{-1} and average starch recovery obtained from *tikhur* rhizome is 12–16.0%.
- On the basis of average rhizome yield 18 t ha^{-1} produces 1.8 tonne starch. Processing cost of rhizome is Rs. 1000/- per quintal.
- The total production and processing cost is Rs. 2.84 lakhs. The market rate of *tikhur* starch varies between 400/- and 600 Rs per kg.
- The cost-benefit ratio of *tikhur* production and processing for 1 ha is 1:2.35 in intercrop and 1:2.38 in sole crop.

References

Annonymous (2001) The wealth of India, vol 2. National Institute of Science Communication Council of Scientific and Industrial Research, New Delhi, p 261

Anonymous (2005) Chhattisgarh Rajya Laghu Vanopaj, Bajar Sarvekshan Prativedan, CGMFPFED, pp 16–17, 42

Banerjee A, Nigam SS (1978) In vitro antihelmintic activity of the essential oils derived from the various species of genus Curcuma Linn. Sci Cult 44(ii):503–504

Banerjee A, Kual VK, Nigam SS (1980) Riv. Ital. Essenze. Profumi, Piaute Off., Aromat., Syndets Saponi Cosmet. Aerosols 62(2):75–76

Bapalal G, Nighantu A (1985) Chaukhambha Bharati Academy, Varanasi, 1st ed, p 579

Chandel N, Kumar V, Ram DS (2018) Physicochemical and functional properties of *Curcuma angustifolia* (*Tikhur*): an underutilized starch. J Pharmacogn Phytochem 7(6):2438–2440
Das D, Jha S, Kumar KJ (2015) Isolation and release characteristics of starch from the rhizome of Indian palo. Int J Biol Macromol 72:341–346
Deshpande DJ (2008) A handbook of herbal remedies. Agribios Pub, Jodhpur, pp 403–404
Elias L, Pradeep, Harini A, Hegde PL (2015) *Curcuma angustifolia* Roxb (Tavaksheeri): a review. J Pharmacogn Phytother 4(2):241–243
Hemadri K, Rao SS (1984) Jaundice tribal medicine. Anc Sci Life 3:209–212
Itokawa H, Shi Q, Akiyama T, Morris-Natschke SL, Lee KH (2008) Recent advances in the investigation of curcuminoids. Chin Med 11:3
Jena S, Ray A, Banerjee A, Sahoo A, Nasim N, Sahoo S (2017) Chemical composition and antioxidant activity of essential oil from leaves and rhizomes of *Curcuma angustifolia* Roxb. Nat Prod Res 31(18):2188–2191
Kashyap KK, Shahi S (2021) A theoretical approach view on water and their chemistry. EEO 20(6):2479–2486. https://doi.org/10.17051/ilkonline.2021.06.229
Kaur P, Broadway AA, Alam T, David J, Chattree A (2023) Extraction and characterization of *Tikhur* starch (*Curcuma angustifolia*) and its utilisation in the development of edible films. Eco Environ Cons 29(January Suppl. Issue):220–229
Kulkarni PH, Ansari S (2004) The ayurvedic plants. Indian Medical Science Series No 132. Sri Satguru Publications, Delhi, pp 126–261
Kumari R, Shrivastava SL, Mishra HN, Murlidhar M (2017) Pharm Innov J 6(7):114–119
Misra SH, Dixit VK (1983) Pharmaceutical studies on starches of some Zingiberaceous rhizomes. Indian J Pharm Sci 45(5):216–220
Nag JL, Shukla N, Pararey PM, Soni VK, Netam CR, Pandey DK (2006) Effect of extraction methods on production of edible *tikhur* (*Curcuma angustifolia* Roxb.). In: Abstracts book, National Seminar on medicinal, aromatic & spices plants perspective and potential. IGKV, TCB, CARS, Bilaspur, Chhattisgarh, p 185
Nahak G, Sahu RK (2011) Evaluation of antioxidant activity in ethanolic extracts of five Curcuma species. Int Res J Pharm 2(12):243–248
Nguyen, TBT, Trinh DC, Do DR, Nguyen XD (2001) Vietnam Hoa Hoc Va Cong Nghiep Hoa Chat 5:11–14 [CAN 136: 180650]
Patel S, Tiwari S, Pisalkar PS, Mishra NK, Naik RK, Khokhar D (2015) Indigenous processing of Tikhur (*Curcuma angustifolia* Roxb.) for the extraction of starch in Baster, Chhattisgarh. Indian J Nat Prod Resour 6:213–220
Radley JA (1953) Starch and its derivatives, vol 1, 3rd edn. Wiley, New York, pp 369–401
Rajkumari S, Sanatombi K (2017) Nutritional value, phytochemical composition, and biological activities of edible *Curcuma* species: a review. Int J Food Prop 20(S3):S2668–S2687
Ravindran PN, Nahar L, Sarker SD, Skornickova J, Rehse T, Sabu M, Ravindran PN (eds) (2007) Turmeric: the genus Curcuma. CRC Press, Taylor and Francis Group, London, New York, pp 4, 10–11, 43, 72, 458
Sahoo S, Gayakwad T, Shahi S (2022) Medicinal value of edible mushrooms: a review. Int J Health Sci 6(S2):8760–8767
Sahu MK, Shankar D, Kanwar RR, Singh DP (2016) Yield performance of tikhur (*Curcuma angustifolia* Roxb.) genotypes in Narayanpur district of Chhattisgarh. J Plant Dev Sci 8(10):505–508
Shahi S, Deepak D (2018) Separation and structure elucidation of novel decasachharide "oviasose" from Gaddi Sheep's Milk. Eurasian J Anal Chem 13(5)
Shankar D, Singh J (2018) Tuber crops of Chhattisgarh, tropical tuber crops potential and prospects. Westville Publishing House, pp 252–254
Sharma R (2003) Medicinal plants of India – an encyclopedia. Daya Publishing House, Delhi, p 75
Sharma A (2012) Traditional processing of Shotti (*Curcuma angustifolia* Roxb.) – a rhizome based ethnic weaning food. Indian J Tradit Knowl 11(1):154–155
Sheikh Y, Maibam BC, Biswas D, Laisharm S, Deb L, Talukdar NC, Borah JC (2015) Antidiabetic potential of selected ethno-medicinal plants of north east India. J Ethnopharmacol 171:37–41

Singh J, Sharma RB, Singh R (1999) Improved cultural practices for cultivation of medicinal herb – *Tikhur*. In: Health care and development of medicinal plants. Indira Gandhi Agricultural University, pp 319–324

Srivastava S, Chitranshi N, Srivastava S, Dan M, Rawat AKS, Pushpangadan P (2006) Pharmacognostic evaluation of *Curcuma aeruginosa* Roxb. Nat Prod Sci 12:162–165

Thongkhwan P, Chaibunga T, Kwanboonjan H, Theanphong O (2017) Essential oil constituents of the fresh root and rhizome of *Curcuma angustifolia* Roxb. from Thailand. Bull Health Sci Technol 15:52–53

Tiwari S, Patel S (2013) A comparative study of Tikhur traditional and partial mechanical processing and cost economics. Int J Agric Eng 6(1):213–215

Tripathy P, Jena PK, Dandasena TK, Shahi S (2023) Medicinal value about Thikur: review. Eur Chem Bull 12(1):143–150

Part IV

Minor Fruits and Trees

Aonla (Indian Gooseberry) (*Emblica officinalis* Gaertn.): The Vitamin C Powerhouse for Immunity

13

Neetu Saroj, Manoj Kundu, Sukanya Misra, and Ranjit Pal

Abstract

Aonla (*Indian gooseberry*) is a popular fruit crop since time immemorial. As per the ancient Indian mythology, it was the first fruit crop grown on the Earth. It is said to be a cure for deadly diseases due to its high nutritional values. It is also a storehouse of different biologically active substances that have different health benefits. Nearly every part of aonla tree viz. roots, bark, leaves, flowers, fruits, and seeds is used in Ayurvedic and Unani medicine for improving the digestive process, lowering fever and providing relieve against cough and asthma. It is also helpful for the purification of blood and increases the immunity against heart disease, and many more. The abundant polyphenol and ascorbic acid (vitamin C) present in mature aonla fruit contribute significantly to improving the antioxidant activity of the fruit. The anticancer, anti-inflammatory, anti-hyperlipidaemia, antidiabetic, and neurological protective properties are the additional health advantages for daily consumptions of anola fruits. Its distinctive qualities, which are attributed to its bioactive constituents, have expanded the potential applications for improving human nutrition and health on a global scale. Therefore, the purpose of writing this chapter is to present important information by highlighting the nutritional features, health-promoting benefits as well as its other uses.

Keywords

Amla · Antioxidant activity · Ayurveda · Vitamin C

N. Saroj · M. Kundu (✉) · S. Misra · R. Pal
Department of Fruit Science, College of Horticulture & Forestry, Rani Lakshmi Bai Central Agricultural University, Jhansi, Uttar Pradesh, India

A. Kumar et al. (eds.), *Fasting Superfoods*,
https://doi.org/10.1007/978-981-95-2762-5_13

13.1 Introduction

Aonla (*Phyllanthus emblica* or *Emblica officinalis*) is an important indigenous and underutilized fruit crop of Euphorbiaceae family (Bulo et al. 2024). It is also popular known by the name of amla or amalaka (Tomar and Pradhan 2024). It is pondered as a wonder fruit for health. It is believed to be the world's first grown tree (Singh et al. 2019). It grows both under cultivated and in wild forms throughout the tropical as well as subtropical regions of India (Muzaffar et al. 2022). India is the leading aonla-producing country in the world, followed by Brazil, Sri Lanka, and Vietnam. In India, Uttar Pradesh, Madhya Pradesh, and Tamil Nadu account for approximately 75.88% of total aonla production of the country (NHB Database 2021).

In Ayurvedic medicine, aonla is one of the most important components because of its copious amount of nutrients, bioactive compounds and extensive health benefits (Mehere et al. 2024). Due to its peculiar nutraceutical, functional, and medicinal properties, aonla has gained international recognition. It contains around 160 and 20 times more vitamin C than apples and oranges, respectively (Tomar and Pradhan 2024). Out of the natural compounds which play a crucial role in the disease prevention (atherosclerosis, cancer, inflammation, heart disease, hypertension, neurological disorders, as well as osteoporosis) along with few other lifestyle diseases and various epidemic disorder, aonla extract is found to rank close to the top (Muzaffar et al. 2022). In addition, aonla is considered as a therapeutic and functional fruit that encourages proper growth and development of human beings. In Indian traditional medicine, aonla also plays a significant role to prevent burning sensation in eyes and skin and reducing anxiety and anaemic problem. It also helps to increase the functionality of male reproductive system, facilitating digestion, enhancing hepatic function (liver function), and also having a tonic impact on the circulatory system (Gul et al. 2022). In addition, it contains various kind of secondary metabolites such as phenol, flavonoids, tannins, terpenoids, and glycosides). They are also abundant in the extracts obtained from aonla fruit (Singh et al. 2019).

13.2 History

Traditional Uses and Cultural Significance

The aonla fruit is known as the 'Earth Mother' in Hindu mythology because it provides nourishment for humans. To cure various illness, aonla has been used in Ayurveda, Siddha, and Unani systems in India, Tibet, Sri Lanka, and China. It is revered as a rejuvenator (rasayana). Traditionally, it is used to cure jaundice, inflammation, and diarrhoea. An eyewash infusion is made by soaking the dried seeds or fruits in water for whole night (Mehjabin et al. 2024). After fruit is harvested, juice that exudes can also be utilized as an eyewash as well as to treat the inflamed eyes. Oils and charred aonla seeds have been used in herbal compositions against several skin-related issues. It has been reported that asthma, bronchitis, diabetes and fever can also be treated by aonla seeds. It is good for healing wound and can be used to treat snakebites and scorpion bites. In traditional formulations, it is also used as tonic for hair to improve hair growth and pigmentation because it possesses fixed oil which are very useful for hair growth (Saini et al. 2022). Additionally, it is a key element for Triphala and Chavanprash (Avinash et al. 2024). Furthermore, it is also helpful for improving the immunity system of the body and has digestive, carminative, and ophthalmic properties (Mehjabin et al. 2024).

13.3 Utilization Pattern

13.3.1 Nutritional Composition

Aonla is renowned for its nutritional value and has several different chemical components (Table 13.1). The biochemical makeup of aonla fruit is varied due to cultivar difference, management techniques adopted during cultivation, and environmental variabilities (Muzaffar et al. 2022). Aonla juice has 20 times vitamin content C than orange juice and 120 times more vitamin C along with 3 times more protein as compared to apple (Pareek and Kitinoja 2011).

13.3.2 Bioactive Components of Aonla

Vitamins A, C, E, K and minerals such as calcium, potassium, and magnesium along with dietary fibre, and phytochemicals (phenolic acids, tannins, flavonoids, lignans, carotenoids, and stilbenes) are more abundant bioactive substances present in aonla fruits. These bioactive compounds make this crop more popular among human beings (Variya et al. 2016). In traditional medicine, aonla has been used due to its different therapeutic properties, which are directly related to the availability of physiologically active secondary metabolites in the fruit (Saini et al. 2022). The primary bioactive compounds present in aonla fruit are polyphenols (ellagic acid, gallic acid, chebulinic acid, chebulagic acid, quercetin, apeigenin, leutolin, and corilagin) and ascorbic acid. Aonla fruit has also been found to contain sugar-substituted phenolics and other bioactive substances and tannins (emblicanin A, emblicanin B, punigluconin, and phyllaemblicin B) at significantly higher amount (Muzaffar et al. 2022; Khan et al. 2019).

13.3.3 Therapeutic Properties of Aonla

Every part of aonla fruit is beneficial due to its medicinal and pharmaceutical qualities. Different studies indicate the potentiality of aonla fruits in managing several diseases (Muzaffar et al. 2022), which are mainly due to its potential antioxidant, anti-inflammatory, anticancerous, anti-diabetic, antimicrobial and immunomodulator properties (Variya et al. 2016).

Table 13.1 Nutritional constituents of different aonla varieties

Variety	Attributes					
		On dry basis				
	Moisture (%)	Protein (%)	Fat (%)	Ash (%)	Fiber (%)	Carbohydrate (%)
Desi	81.26–82.01	2.12–2.99	0.48–0.50	3.08–3.88	22.35–23.14	73.75–74.60
Banarasi	82.05–83.33	3.17–4.30	0.36–0.45	2.37–2.75	07.18–7.63	87.12–87.79
Hathijhool	86.31–86.98	3.06–3.81	0.30–0.40	2.18–2.90	14.4–15.00	80.06–80.77
Chakaiya	87.5–88.00	4.51–5.11	0.46–0.51	2.88–3.11	14.4–15.22	80.06–80.64
Balwant	86.67–87.11	3.98–4.52	0.35–0.39	2.08–2.65	12.64–13.10	80.95–81.33
Kanchan	83.36–84.02	2.66–3.00	0.42–0.49	2.24–2.83	17.04–17.87	77.62–78.11
NA-7	87.04–87.77	3.80–4.21	0.39–0.47	2.65–3.00	15.98–16.05	77.18–77.31
NA-9	85.12–86.21	3.68–4.07	0.21–0.26	2.57–3.04	15.42–15.74	78.12–78.99
NA-10	84.89–85.51	3.02–3.94	0.40–0.47	2.97–3.22	11.68–11.90	81.93–82.22

Data sources: Muzaffar et al. (2022) and Tewari et al. (2019)

13.3.3.1 Antioxidant Activity

Oxidative reactions within the cell are responsible for the damage of the individual cell. To tackle oxidative damage, nature has given us with complex antioxidant defence systems (vitamin C) and several enzymes (glutathione, catalase, superoxide dismutase, etc.) to battle against oxidative damage (Variya et al. 2016). Aonla is a rich source of phenolic compound and tannins, which mainly functions as antioxidant agent. The antioxidant activities of phenolic components are due to their redox potentiality, which includes the ability to donate hydrogen, scavenge free radicals, quench single oxygen, and form stable radical intermediates. Thus, these substances may aid in preventing oxidation at the physiological and cellular levels (Gul et al. 2022). High antioxidant-rich foods reduce the risk of illness and morbidity as well as mortality rate and also improve the immunity system of the human body.

13.3.3.2 Anticancer Activity

Various nonclinical and clinical studies have revealed that plant-derived polyphenols enhance cancer prevention (Lorenzo et al. 2018). Specifically, polyphenols prevent oxidative stress, produce pro-inflammatory chemicals, prevent DNA from damage, and promote apoptosis by various mechanisms. Notably, the consumption of aonla extracts helps to elevate the activity of caspase-3, 7, and 8 and up-regulate the Fas protein in the HeLa cell line, suggesting the activation of death receptor pathway for apoptosis (Gul et al. 2022). As per the in vitro matrigel invasion study, aonla made MDA-MB-231 cells less invasive, but normal lung fibroblasts (MRC5) did not exhibit any cytotoxicity. Similarly, the polyphenol pyrogallol, which is present in aonla, was reported to prevent human lung cancer (Yang et al. 2009; Gul et al. 2022). Furthermore, studies (Huang and Zhong 2011; Purena et al. 2018) revealed that gallic acid extracted from aonla leaves caused apoptosis in a hepatocellular cancer cell line (BEL-7404).

13.3.3.3 Antimicrobial Property

Infectious diseases are a leading cause for the death as well as morbidity throughout the world particularly, in most of the developing and impoverished nations. Numerous reports indicate the potential antimicrobial properties of aonla, which are important for the production of different antimicrobial drugs at reasonable costs (Kumar et al. 2011; Variya et al. 2016). The methanolic extract of aonla fruit was used for the extraction of alkaloids, which were then subjected to assess for their antibacterial properties. A considerable antimicrobial activity against different gram-positive and gram-negative pathogenic bacteria as well as fungi pathogens such as *Aspergillus niger*, *Candida albicans*, *Sacharomyces cerevisiae*, etc. was observed in this methanolic extract of aonla fruits (Almatroodi et al. 2020).

13.3.3.4 Antidiabetic Activity

Diet plays a significant role in diabetes management and associated metabolic problems (Variya et al. 2016). Aonla possesses a naturally occurring compound, which plays significant role to combat diabetes problem. As per in vitro study, the main phytochemicals (ascorbic acid and ellagic acid) in aonla decreased the activity of important enzymes associated with the digestion of glucose (notably glucosidase and amylase), which ultimately helps to inhibit the breakdown of carbohydrates to glucose (Nampoothiri et al. 2011). A research was conducted in mice to assess the effectiveness of aonla against arsenic-induced hyperglycaemia. Finding indicates that arsenic exposure changes the glucose homeostasis and dramatically reduces hepatic glucose-regulating enzymes, pancreatic inflammation indicators such as IL-1β, TNF-α (Almatroodi et al. 2020). Additionally, exposure to arsenic considerably reduced serum insulin and the level of c-peptide protein in the body. However, regular consumption of aonla (500 mg kg^{-1} body weight) maintains the blood sugar level by activating the hepatic glucose-regulating enzyme activity. Furthermore, it also helps in considerable reduction of glucose-6 phosphatase and phosphoenol pyruvate carboxykinase activity. In addition, it considerably increases the serum insulin concentration and the level of c-peptide protein as compared to arsenic-treated one (Singh et al. 2020).

13.3.3.5 Anti-inflammatory Activity

Aonla fruit is found to be effective against inflammation (Gul et al. 2022). Research on aonla demonstrated that the methanolic extract of aonla leaves had anti-inflammatory properties. As per Li et al. (2020), RAW 264.7 cells, treated with aonla extract (having high concentrations of corilagin, gallic acid, and ellagic acid) had significantly lower levels of inflammatory markers. Furthermore, it also helps to inhibit the production of tumour necrosis factors-like TNF-α, interleukin-1β (IL-1β), and interleukin-6 (IL-6), which ultimately helps to prevent tumour-like growth in the human body. Besides, Granuloma formation is also reduced when hydroalcoholic extract of aonla is used at doses of 300–700 mg kg^{-1} body weight. It also significantly reduces rat paw oedema when administered against all phlogistic agents. At higher dose (700 mg kg^{-1}), aonla extract produced profound anti-inflammatory responses and demonstrated comparable effects to different recommended anti-inflammatory drugs (Golechha et al. 2014).

13.4 Production

13.4.1 Origin and Distribution

Initially, the maharaja of Kashi took the effort to domesticate aonla in Varanasi, Uttar Pradesh, India. In the nearby Vindhyan hills, wild aonla trees are widespread. From this, later on, a superior genotype known as 'Banarasi' was selected. Furthermore, the genotypes were dubbed from Gujarat as 'Francis' and afterwards as 'Hathijhool' due to their drooping branches. Eventually, these varied introductions were crucial for the creation of the high-yielding cultivars from Pratapgarh, Uttar Pradesh, India. Later on, these selections were taken for commercial cultivation throughout the country (Singh et al. 2019).

In India, anola trees are generally found on hill slopes up to 200 m MSL. Furthermore, it also widely spread to tropical, subtropical, and coastal belt of the country. It is also grown in the hilly regions of Kashmir valley and plains (Gantait et al. 2021). However, within the country, aonla is commercially cultivated in Uttar Pradesh, Uttarakhand, Gujarat, Maharashtra, Rajasthan, Tamil Nadu, Andhra Pradesh, Karnataka, Bihar, Haryana, Madhya Pradesh, Rajasthan and West Bengal (Saini et al. 2022). Furthermore, it is a well-suited crop for the wasteland with minimal nutrient and water supply. Hence, the area coverage under aonla has increased dramatically nationwide during the last two decades.

13.4.2 Soil and Climatic Requirement

Aonla can be cultivated both in light and heavy soil due to its hardiness. Furthermore, the typical features of the tree such as deep root system, reduced size of leaves, summer dormancy of fertilized fruitlets (during April–June), followed by maximum growth and development of fruits during monsoon period makes this crop most suited for commercial cultivation under arid as well as semi-arid environments of rainfed ecosystems. Marginal degraded lands and saline/sodic to acidic soils (ESP-35; EC-6–9 ds m^{-1} and pH up to 9.5) are most suitable for aonla cultivation (Singh et al. 2019). The ideal annual rainfall range for optimum growth and development of aonla tree is between 630 and 800 mm (Gantait et al. 2021). The mature aonla tree can tolerate a temperature as high as 48 °C with low temperature up to freezing point. However, under the hot arid climate of western Rajasthan, frost can cause significant harm to the plants during the winter months. A warm temperature appears to be favourable for the flower buds induction, but sufficient atmospheric humidity is necessary for the proper growth and development of dormant fruitlets during the month of July and August. During flowering and fruit set, sudden fluctuations in temperatures have a negative impact on aonla fruiting. Dry spells just after fruit setting may cause heavy fruit drop and also delay in initial growth of the fruits (Singh et al. 2019).

13.4.3 Area and Production

India leads globally in the production and cultivation of aonla. Around 95,000 ha area is currently covered under aonla cultivation with around 1,107,000 metric tons annual production in the country (NHB database 2021). The global market for aonla extract is expected to increase at 4.9% rate per year, reaching a projected value of USD 49.34 billion by the end of 2025 (Muzaffar et al. 2022). Hence, the area under aonla cultivation is also expected to increase in the near future.

13.4.4 Important Cultivars

India is rich in varietal wealth in aonla. From the wide germplasms available in Gujarat, three promising lines have been selected and released as variety under the name of Anand 1, Anand 2, and Anand 3 by Anand Agricultural University (AAU), Anand, Gujarat. Furthermore, the most promising cultivars of aonla for commercial cultivations are NA-25, NA-26, NA-10, NA-9, NA-8, Neelum (NA-7), Amrit (NA-6), Krishna (NA-5), and Kanchan (NA-4), which were developed by ANDUAT, Kumarganj, Uttar Pradesh (Fig. 13.1). 'Goma Aishwarya' (Clonal selection) was developed at CHES, Godhra, Gujarat. 'Laxmi-52', which has bold fruit, was released at CISH, Lucknow. In addition, other promising cultivars of anola are

NA 6

NA 7

Fig. 13.1 Plant growth habit of different aonla cultivars, including NA-6, NA-7, NA-10, and Chakaiya

NA 10

Fig. 13.1 (continued)

Chakaiya

Fig. 13.1 (continued)

'Francis', 'Chakaiya' (Fig. 13.1), 'BSR-1', and 'Banarasi'. Among these cultivars, the highest yield and high-quality fruit is produced by 'Kanchan' (NA-4), while 'NA-7' is the most promising cultivar that has maximum area coverage in the country (Maholiya et al. 2015; Singh et al. 2014).

13.4.5 Varietal Description

Name of the variety	Varietal description
Early maturing group	
NA-26	The growth habit of the plant is spreading. Fruits are flattened and greenish yellow in colour. Fruit flesh is soft but fibrous. It is an ideal variety for preparation of pickle and other processed products. It is recommended for commercial cultivation under rainfed condition of arid as well as semi-arid ecosystem of India.
NA-25	The growth habit of the plant is upright and spreading. Fruits are flattened, medium in size and creamy yellow in colour. Fruits possess good keeping quality and are suitable for preserve making. It is recommended for commercial cultivation under rainfed condition of semi-arid ecosystem of India.
NA-10	It is also a chance seedling selection of Banarasi. It bears profusely. Fruit is attractive, medium to large in size. Fruit shape is flattened, round, skin has pinkish undertone and is rough and yellowish green in hue.
NA-9	It is a chance seedling selection from 'Banarasi'. It bears moderately and produces one to two female flowers per branchlet. However, the fruit size is big, flattened, and has smooth, yellowish skin. It contains very high vitamin C (881 mg/100 g). This variety is most suitable for making preserve and candy.
Krishna (NA-5)	It is a chance seedling selection of cultivar 'Banarasi'. It bears only one female flower per branchlet and produce medium to large-sized fruits (approx. 46 g) which contain around 760–790 mg vitamin C content per 100 g with around 1.4% fibre.
Banarasi	It is shy-bearing cultivar due to very less female flower count (0.51/branch). Tree growth of 'Banarasi' cultivar is upright, fruit is large (40–46 g), oblong to conical in shape with smooth and straw yellow skin, contains very high concentration of vitamin C with moderate fibre in flesh. It is most widely used for making preserves.
Mid-season varieties	
Neelum (NA-7)	It is a selection from Francis, which produces large number of female flowers (5 per branch). Fruit is medium to large in size, flattened, and oblong in shape with conical apex. Fruit skin is smooth, yellowish green in colour. Fruit flesh contains negligible amount of fruit fibre. Furthermore, the fruit flesh is soft with moderate keeping quality. Juice is highly astringent. It is free from the disorder necrosis disorder.
Amrit (NA-6)	It is also a selection from 'Chakaiya'. Its fruit is bright, medium to large, and has six thin, distinct strips. Fruit has low fibre content. It also bears large number of female flowers (profuse bearing). It is considered as the excellent variety for candy and preserves preparation.
Kanchan (NA-4)	A chance seedling selection of 'Chakaiya'. It bears profusely due to the production of high number of female flowers (4.6 branchlet^{-1}). Fruits are small to medium in size, flattened, oblong in shape, smooth-skinned, yellow in colour. Fruit possesses 6 strips. Flesh is fibrous and hard. Fruits contain high vitamin C and are highly suitable for processing.

Name of the variety	Varietal description
Francis	It is also called as 'Hathijhool' (tree has drooping habit) and native to Pratapgarh district of Uttar Pradesh, India. Its fruit is flattened, oblong, medium to large and yellowish green in hue. Fruit skin is smooth with 6 stripes. Its flesh is fibrous with low in ascorbic acid.
Late maturing groups	
Bhawanisagar	It is a selection from a local cultivar 'Thimban' which is grown in Tamil Nadu, India. Its fruits have more flesh with very low phenolic. The vitamin C content of the fruit is also very high.
Chakaiya	The fruits are flattened, medium to large in size, and have straw-greenish skins. Its fibrous flesh has high vitamin C content. It has good keeping quality and is suitable for preparation of pickles and other processed products.

13.4.6 Propagation

13.4.6.1 Seed Propagation

Aonla trees that are often grown from seeds have a long juvenile growth phase with inferior fruit yield and quality. Trees developed from naturally dispersed seeds in forests are not true to type and exhibit a wide range of fruit sizes, shapes, and yields (both number and quality) as well as irregular vegetative development patterns. Furthermore, the trees that are vegetatively propagated bear late as they need longer duration to reach to the reproductive maturity. In small polybags, aonla seeds are typically sown 5 cm deep between April and June (Gantait et al. 2021) for their use as rootstock for further budding and grafting operation.

13.4.6.2 Budding and Grafting

Among all the vegetative multiplication techniques, budding is most widely used for commercial multiplication of aonla (Javed et al. 2023). Among different budding methods, patch budding is employed for commercial multiplication of aonla. In an irrigated hot arid environment, 90% success rate was achieved in patch budding in perforated polythene bags (Saroj et al. 2001; Singh et al. 2014). Furthermore, softwood grafting has around 70% success rate, particularly in arid regions. Additionally, good results have also been reported with veneer and cleft grafting (Gantait et al. 2021; Wali et al. 2015).

13.4.7 Establishment of Aonla Orchards

It is always advisable to establish any fruit orchards during monsoon so that ample amount of water may be available. Therefore, the planning and layout of the orchard should be done during the month of May. After proper planning and layout, pit digging should also be done in the same month with the dimension of 1 × 1 × 1 m size at a distance of 7–8 m and left the soil of the pit under open conditions for about a month. For a better orchard establishment, each pit is then need to fill with soil mixture containing 20–25 kg farm yard manure per pit and drenched with chlorpyriphos (Singh et al. 2019). Healthy budded or grafted saplings are then planted at the centre of the already prepared the pit, shortly after the first rain. In semi-arid regions, rootstocks are recommended to be raised in situ at 8 × 8 m spacing to achieve the best results for successful budding as well as to avoid establishment shock. For this, rootstocks need to be planted in the already prepared pit in the month of June and after 30–45 days, budding operation can be done on the rootstock under in situ condition (Singh et al. 2014). It is better to plant more than two varieties in the same field to avoid the problem of self-incompatibility and for proper pollination and fruit setting (Singh et al. 2019; Gantait et al. 2021).

13.4.8 Nutrient Management

Combined application of organic and inorganic fertilizers promotes fruit quality and yield, while vermicompost application greatly improves quality of fruit. These nutrient sources affect the physical, biological, and chemical attributes of soil. However, the amount of manure and fertilizer used depends on the fertility of the soil, the age of the plants, and the frequency of fruiting (Mishra et al. 2023). For 1-year-old aonla plant, 10 kg FYM, 100 g nitrogen (N), 50 g phosphorus (P) along with 100 g potassium (K) need to apply. Annual increment up to 10 years should be guaranteed, afterwards, a stable dose of 1 kg N, 500 g P and 1 kg K need to apply annually (Singh et al. 2019). Full doses of FYM and P along with half of the N and K should be placed around tree basin during December–January, while remaining N and K should be applied during the month of August. Besides, 100–500 g boron, zinc sulphate, and copper sulphate are recommended to apply to micronutrient-deficient areas.

Plant growth-promoting substances also played vital role for improving the overall growth and development of aonla plants, which ultimately improves the fruit yield and quality significantly (Singh et al. 2022). The yield loss due to heavy fruit drop under sodic soils (with production constraints due to the restricted availability of specific mineral nutrients) can be reduced by two applications of thiourea in combination with GA_3 and NAA during mid-May and mid-July. Furthermore, two foliar sprays of NAA (30 ppm) during May and July are also beneficial for enhancing the fruit quality and retention significantly (Yadav et al. 2010).

13.4.9 Irrigation Management

Well-developed aonla orchards usually give optimum yield even under rainfed condition, hence, do not require extra irrigation. Furthermore, it is advisable to avoid any water application during flowering period (mid-March to mid-April) (Wali et al. 2015). However, under deficient moisture conditions in the soil, irrigation is necessary after applying fertilizers and manures. Plants that are irrigated with drip irrigation with 60% wetting had the best growth, yield, and quality of aonla, although it may be cultivated successfully in semi-arid rainfed regions without irrigation (Singh et al. 2019). Application of a total of nine irrigations along with mulching during a normal dry period can result in notable stock girth and plant height. This recommendation is beneficial to conserve about 20 cm irrigation water (four irrigations) (Gantait et al. 2021). The highest reproductive growth was obtained using drip fertigation with 125% of recommended dose of fertilizer (RDF) though water-soluble fertilizers (WSF) (Suresh and Kumar 2014; Suresh et al. 2019). According to research conducted at NDUAT, Faizabad, establishment of aonla orchard in sodic soil has been shown to be successful when drip irrigation system is installed at 0.6 CPE along with paddy straw mulch (Pathak et al. 2006).

13.4.10 Cropping System

The sparsely leafed tree canopy of aonla allows for plenty of natural light to penetrate and helps the intercrops to grow successfully at the interspace of the aonla orchard (Gantait et al. 2021). Thus, in the first 3–4 years of planting, this type of intercropping provides a unique opportunity to utilize the vacant interspaces of the orchard. When turmeric, ginger, colocasia are cultivated in the interspace of aonla orchard, the yield attributes of the crop along with the nutrient status of the soil viz. available carbon, nitrogen, phosphorus, etc. and overall farm economics showed positive effect (Das et al. 2011). In addition, another shade-loving plant i.e. Elephant Foot Yam can also be profitably grown in aonla orchards. It has been discovered that plants, including bottle gourd, coriander, cauliflower, okra, peas, and turmeric, as well as flowers like marigold and gladiolus, are excellent for intercropping in the aonla orchard. However, to improve the physico-chemical characteristics of marginal or salt-affected soils spiny sesbania can be grown as intercrop in the aonla orchards for initial few years. Even in orchards with a lot of shade, tuber crops can be cultivated commercially (Singh and Singh 2015). Furthermore, winter season crops like chickpea, beans, cumin, fenugreek, mustard, etc. and rainy season crops like moth bean are most suited for the arid region (Gantait et al. 2021). Furthermore, the deciduous, deeply rooted aonla tree with sparse leaves is also suitable for two, three, or multi-tier cropping techniques (Wali et al. 2015). Guava and karonda are two fruit crops that perform well as filler crops in multi-tier cropping techniques in aonla orchard (Pathak et al. 2006).

13.4.11 Fruit Maturity, Harvesting, and Yield

After planting, seedling trees begin to produce fruit within 6–8 years, but budded or grafted trees require 3 years to complete the juvenile phase and continue to bear fruit until they are 60–75 years old (Pareek and Kitinoja 2011). The characteristic features for determining the maturity indices of aonla include days from anthesis to maturity, total heat units, changes in skin colour of the fruit and total soluble solids to acid ratio, seed colour from creamy white to brown, etc. (Bakshi et al. 2018). Generally, during the months of November–December, fruits become ready for the harvesting. Harvesting is mainly done by hand picking. In order to prevent fruit dropping, fully mature fruits are harvested as soon as possible (either in the morning or late at night), especially for the cultivar Banarasi and Francis. Harvesting is mainly done by hand picking. A vegetatively propagated aonla tree of 10 years old can bear around 150–200 kg fruits $tree^{-1}$ with an average productivity of around 15–20 tons ha^{-1}.

13.5 Plant Protection

13.5.1 Important Pests and Their Management

Initially, aonla were not subject to pest or disease attacks or infestations. However, as the area under cultivation has grown over time, a large number of insects have begun to feed on and infect aonla plants at different stages of their growth and development (Sawant et al. 2022).

13.5.1.1 Bark-Eating Caterpillars (*Indarbela quadrinotata*)

They create tunnels within the bark and cause damage to the main trunk. In a severe infestation, the affected stem may perish but not the main trunk. Severe attack by *I. quadrinotata* makes the crop unhealthy, which ultimately reduces the yield potential of the crop.

This pest can be managed by clean cultivation and removing the overcrowded branches. Furadan spray @ 0.03% concentration during the month of February to March or September to October has shown effective results to control the pest under severe infestation (Gantait et al. 2021).

13.5.1.2 Shoot-Gall Maker (*Betousa stylophora*)

The young caterpillar of this insect burrows into fragile shoots and started to feed from the centre of pith during the month of August to September. As a result of its feeding to the plant, a gall-like structure started to develop around the wounded area. Each variety of aonla is vulnerable to the infestation of shoot gall maker (Gantait et al. 2021).

Pruning and burning of the infested plant part is helpful to reduce the infestation of this pest. In addition, the larvae present within the shoot can be killed by inserting iron or spokes or by injecting chlorpyrifo @ 0.05% in the holes. Furthermore, application of 0.05% Monocrotophos during July to August is quite beneficial (Wali et al. 2015) to control the shoot gall maker infestation in aonla.

13.5.1.3 Fruit Borer (*Deudorix isocrates*)

In India, this is one of the most dangerous pests. Newly hatched caterpillars consume growing seeds and bore into the fruits. Furthermore, fungal and bacterial infections are also accelerated in the infested fruit, which ultimately results in fruit rotting. As a result, before the maturity of the fruit, they started to deteriorate with significant decay and finally drop down. In some areas, during the rainy season, the borer has been known to destroy 40–70% of aonla fruits in cases of severe infestations (Sawant et al. 2022).

Collection and destruction of damaged fruits may reduce the population of the fruit borer in the aonla orchard. However, it can effectively be controlled during July–August through chlorpyrifo (0.05%) spray. Furthermore, it is recommended to avoid the cultivation of pomegranate in the vicinity of aonla orchards (Gantait et al. 2021).

13.5.1.4 Aonla Aphids (*Schoutedenia emblica*)

New flushes are seriously damaged by the infestation of aonla aphid. The actual feeding parts of this pest are delicate stems, leaves, reproductive buds, and the fruits. Infested leaves become dry and yellow, while the infected shoots became twisted and bent at growing points. Another indicator of an aphid infestation is the attraction of ants to the plant. The nymphs and the adult females of the aphid feed the sap from the infested plant part which ultimately affect the overall vigour and growth of the plant and eventually compromise flowering and fruiting potential of the plant. September is the peak month for the infestation of this pest, which prevailed in the aonla orchard from July to October (Devi and Rajasekaran 2011). It can be effectively controlled with a single application of neem seed kernel extract (2%) or dimethoate (0.03%) during the emergence of new flush on the plant.

13.5.2 Diseases and Its Managements

13.5.2.1 Rust (*Ravenaliia emblicae*)

Rust causes the most significant economic loss in aonla orchards. Fruits produce black pustules, which are followed by a ring pattern. Infestation of this disease can be reduced by clean cultivation and by the removal of infested plant parts through proper pruning operation. Furthermore, it is observed that rust can be managed with the of spray of Zineb (0.2%) or sulphur (0.5%) three times starting at monthly interval starting from July.

13.5.2.2 Anthracnose (*Glomerella cingulata*)

It affects fruits and leaflets in the month of August and September. It is advised to spray 0.2% Difolatan or 0.1% Carbendazim to eradicate the problem of anthracnose from aonla orchard (Nallathambi et al. 2007; Wali et al. 2015).

13.5.2.3 Penicillium Fruit Rot (*Penicillium funiculosum*)

It is a serious disease of aonla orchard causing heavy losses (Singh et al. 2019). Under severe infestation by this microorganism, yellow liquid started to release drop by drop from the damaged fruit, and the fruits eventually turn into a beaded appearance with bluish-green colour and emit an unpleasant smell.

Pre-harvest treatment with Blitox, Bavistin, or KH_2PO_4 1 week prior to harvest and postharvest application of 1.5% $CaCl_2$ and GA_3 (200 ppm) can significantly control the caused by penecillium fruit rot (Singh et al. 2014, 2019).

13.5.2.4 Fruit Rot (*Phoma putaminum*)

During the winter month (January), pre-harvest fruit rot is a commonly observed disease in aonla orchard. This kind of decay begins as tiny, lemon-coloured lesions that get larger and cover the entire skin of the fruit with funicles of conidiophores bearing spore (Wali et al. 2015). Usually, the disease caused an early fruit drop that eventually mummified on the ground. Necrotic patches, which form an eye-shaped patch on both ends of the fruit, are a defining feature of the disease. When fruits are extensively infected, several lesions combine to form larger pustules. Vitamin C content of the fruit decreases at a faster rate as a result of Phoma infection (Gantait et al. 2021). Removal of the infested plant part and application of dithane M 45 at 0.2% or carbendazim at 0.1% at the maturity stage of the fruit is highly recommended to control this disease.

13.5.3 Physiological Disorders

13.5.3.1 Chilling Injury

Aonla fruits are highly susceptible to chilling injury. Under severe cases, the fruit skin started to split which ultimately causes rotting of the fruits. Therefore, it is recommended to harvest the fruit at optimum maturity and avoid delayed harvesting. Furthermore, after harvesting, fruit should be stored at 12 °C temperature to prevent such damage by chilling injury (Gantait et al. 2021).

13.5.3.2 Necrosis

The fruits have been found to exhibit fruit necrosis, which is linked with boron deficiency. The symptom of necrosis is stated with the browning of mesocarp, which later on spreads to the epicarp and gives the flesh a brownish-black colour. It starts to appear from early September. Hence, it is recommended to spray 0.6% borax three times at 15 days interval (Singh et al. 2014) starting from first week of September. Furthermore, it is also recommended to cultivate necrosis-resistant cultivars such as NA-7, NA-6, Chakaiya, etc. (Pathak et al. 2006).

13.5.3.3 Pink Spots

Random pink patches that finally degraded the fruit quality developed on aonla fruits as a result of a boron deficiency. It is recommended to spray borax (0.6%) three times at 15 days intervals during September–October to combat the problem.

13.5.3.4 White Specks

It makes the fruits poor in appearance with spongy texture particularly during curing and pickling. By preserving the fruit segments in a solution of 0.04% $K_2S_2O_5$ along with 10% NaCl, followed by salting and dehydration with same solution mixture after 4 weeks of storage has significant effect to reduce the frequency of white specks of aonla fruit (Gantait et al. 2021).

13.6 Improvement

13.6.1 Genetic Resources of Aonla

India has significant genetic diversity in aonla, both in wild and cultivated forms, notably in the northern as well as eastern part (Singh et al. 2019), which is mainly due to the prevalence of seed propagation. This genetic diversity can be used in the breeding programs of the crop to give the offspring resistance to harsh environmental conditions (Singh et al. 2014). Bold size fruits, low fibre content in the flesh, and resistance to rust, frost and internal fruit necrosis are some desirable characters for which germplasm must be gathered (Pathak 2003).

13.6.2 Breeding Programme

Meticulous efforts have been made with respect to aonla improvement programme through selection (Pathak et al. 2006; Singh et al. 2019). The 'BSR-1' genotype showed greater yield potential (43.05 t ha^{-1}) with improved fruit quality. Under the rainfed and irrigated conditions of Tamil Nadu, India, this genotype is recommended for commercial cultivation. Other varieties such as 'Anand 1', 'Anand 2' and 'Anand 3' from Anand Agricultural University, Gujarat, India and 'NA-4' (Kanchan), 'NA-5' (Krishna), 'NA-6 (Amrit)', 'NA-7' (Neelam), 'NA-9' 'NA-10', 'NA-25' and 'NA-26' from NDUAT, Faizabad, Uttar Pradesh, India showed promising results with respect to yield and fruit quality and released for commercial cultivation in semi-arid and rainfed ecosystem (Pathak 2003; Singh et al. 2019). Furthermore, 'Laxmi-52' (bold fruit) and 'Goma Aishwarya' were developed through selection from CISH, Lucknow and CHES, Godhra, respectively for higher yield potential of aonla in the country.

13.7 Primary Processing

Fruits and vegetable crops are highly perishable in nature because physiological activities within the produce are continued even after harvest in these two categories of crops (Saroj et al. 2023a). Therefore, when these food commodities are handled and distributed after harvest, significant postharvest (quantitative and qualitative) losses occur. To reduce this, post-harvest suitable methods must be developed (Saroj et al. 2023b; Prasad et al. 2024). At ambient storage conditions, shelf life of ripe aonla fruit is only 5–6 days (Pathak et al. 2009). Furthermore, the mechanical injury and damage caused by several postharvest pathogens such as black and blue mould are very common in aonla. These ultimately reduce the marketability, fruit quality and cause about 50% annual yield losses (Sengupta et al. 2019). Hence, it is very important to curtail the post-harvest losses of aonla fruit (Muzaffar et al. 2022) by adopting different post-harvest management practices which are as follows.

13.7.1 Blanching and Drying

Aonla fruits are traditionally processed through blanching using hot water or steam. Then the blanched fruits are dried under the sun. Enzyme which alters colour and develop off flavours is inactivated by blanching treatment; thereby increasing the shelf life (Bouhile et al. 2025; Lalita et al. 2024). Furthermore, it also helps to retain the nutritional content of the fruit. Aonla treated with hot water (at 60 °C temperature for 2 minutes) can prevent the fruits from the attach of *Aspergillus niger* (Muzaffar et al. 2022). But the major concern about the sun drying is the increasing chances of dust contamination of the fruits. As an alternative to conventional drying, osmo-air drying can be used, which causes the least nutrient degradation from the fruit (Gautam et al. 2023). Presently, this method is commonly used to prolong the shelf-life and also to preserve the fruits. Through this technique, fruit is dried in two steps: first, the water is removed by submerging it for 12 hours in 40% hypertonic sucrose solution, and then it is dried in a drier until required level of moisture is achieved (Sengupta et al. 2019). As compared to other traditional drying techniques, osmo-air drying improves the nutrient retention specifically ascorbic acid and sugars (Muzaffar et al. 2022).

13.7.2 Chemical Treatments

The postharvest treatment of aonla fruits with chemicals (fungicides, calcium chloride, and calcium nitrate) has shown significant results to extend the shelf-life of the fruits. Aonla showed resistance against fungal pathogens (*Penicillium funiculosum*) when treated with 500 and 1000 ppm carbendazim (Yadav et al. 2012). 1% calcium nitrate dip treatment after harvest was proven to be an effective way to preserve the qualitative characteristics of aonla and extend the shelf life of the fruits significantly. Reduction in physiological loss of weight as well as pathological loss and retention in total sugar and vitamin C content of the fruit was reported on post-harvest treatment with calcium nitrate (1%) which also helps to increase of shelf life of the fruit by 15 days (Gangwar et al. 2012; Brar et al. 2023). Furthermore, it was also found to be improved by treating with calcium hydroxide (2%) followed by modified atmospheric packaging. Treatment of aonla fruits with waxol (an edible wax) also has significant effect for the improvements in shelf life and retention of qualitative characteristics (Shukla 2021).

13.8 Post-harvest Technology

13.8.1 Storage of the Fruit

It is very important to store the fruits at optimum temperature to prolong their shelf life, thereby increasing their availability and maintaining the market price under control (Saroj et al. 2023a). At ambient temperature, shelf life of aonla fruits is very limited (Tomar and Pradhan 2024). However, keeping the fruits at 0–2 °C with 85–90% relative humidity is optimum for improving the shelf life by 7–8 days (Muzaffar et al. 2022). Furthermore, it can also be stored under zero-energy cool chamber in which the principle of evaporative cooling is followed for lowering the temperature by water evaporation (Lal Basediya et al. 2013).

13.8.2 Packaging

Effective packaging techniques and materials are crucial for reducing the post-harvest losses of aonla fruits (Prasad et al. 2018). Packing of aonla fruits in high-density polyethylene bags effectively extended their shelf life by 15 days and also maintained their ascorbic acid and total sugar concentration significantly (Vishwakarma et al. 2012). Aonla fruits can also be stored in boxes made of corrugated fibre board (CFB). These kinds of boxes are recyclable, offer proper ventilation, and printable information. The shelf life of aonla fruits stored in CFB boxes with newspaper liners is around 11 days (Muzaffar et al. 2022).

13.9 Value-Added Products

Aonla fruit has a short shelf life of 5–6 days only. Furthermore, it is highly prone to bruises, browning, and necrosis, which cause huge post-harvest losses and decrease its market potential (Tewari et al. 2023). Furthermore, due to strong acidity and astringent taste, it is not much preferable for direct consumption (Tomar and Pradhan 2024). In addition, due to low total soluble solid content of the fruit along with poor colour and lack of flavour, consumers did not widely accept it as a table fruit. Hence, the production of different processed products from aonla is the useful way to increase its value. Thus, it is processed into a wide range of products for preservation as well as to utilize its natural medicinal and nutritional values (Rathinasamy et al. 2022). There is a large number of aonla-based products available and their demand is also increasing day by day due to growing health concerns among the peoples.

Products like candy, murabba, jam, pickle, sauce, shreds, powder, and juice are commonly prepared from aonla (Khedkar 2022; Tewari et al. 2023). A popular processed product, aonla preserve (murabba) is prepared from completely matured fruit by cooking the whole fruit or bigger-sized fruit pieces in heavy sugar syrup solution (Shukla 2021). Smaller aonla fruits that are not ideal for preparing murabba can be used to prepare pickles (Tewari et al. 2023). Furthermore, aonla candy is also a popular processed product due to its high nutritional composition and prolonged shelf life (Tandon et al. 2003). In addition, chutney is also made from aonla fruit (Pareek and Kitinoja 2011). Shred (dried aonla), a valuable product prepared by blanching in potassium metabisulphite (KMS) solution followed by drying on solar dryer with 3% common salt (Prajapati et al. 2021).

Ayurvedic medicine, i.e. Triphala is prepared by mixing *Terminalia chebula*, *Phyllanthus emblica*, and *Terminalia belerica* in equal amounts. Therapeutic properties of triphala are related to the presence of polyphenols, vitamin C along with flavonoid content of the aonla fruit. Another useful product obtained through processing of aonla fruit pulp is the aonla juice, which is also very popular for its bioactive substances and found effective to combat different health-related issues, including diabetes, gastric problems, skin disorders, high blood pressure, and ageing. Furthermore, aonla juice can be mixed with other fruit juices to enhance their nutritional value (Kumari et al. 2019). In addition, ready-to-serve (RTS) drink can be made by combining aonla juice with bitter gourd (to remove the astringency) in 75:25 ratio (Gajanana et al. 2007).

Fermentation of aonla using *Saccharomyces cerevisiae* is done for the preparation of fruit wine. It contains a total of 9.29% alcohol in the form of n-butanol, isobutanol and n-propanol. Furthermore, gram flour, aonla pulp, and spices are used for the preparation of aonla bhujia. Organoleptic assessment of aonla bhujia had high acceptable scores. In addition, mouth freshener is also prepared using aonla fruit as an alternative to pan masala, gutka and tobacco (Sohshangrit et al. 2022). Healthy and tasty aonla mouth freshener can be prepared by mixing aonla fruit pulp (from 'Desi' and 'Banarasi' cultivars), arecanut, carboxymethyl cellulose, cardamom, milk powder, and sugar (Tewari et al. 2023).

13.10 Marketing

Domestic consumers provide a major market of aonla both for fresh as well as processed products. The demand for this fruit is rising both in domestic and international markets, as a result of people's increased health consciousness and the growing acceptance of herbal products. Nearly 42 countries of the world are importing aonla from India, as per data by Department of Commerce, Ministry of Commerce and Industry, Government of India (SYSTEM ON INDIA's TRADE) (commerce.gov.in). USA (0.22 USD Million) is the biggest trade partner for India's aonla exports followed by Bangladesh (0.04 USD Million), Canada (0.01 USD Million), Sri Lanka (0.01 USD Million), and Germany (0.01 USD Million) with annual earnings from aonla export is around 0.29 USD million among which 87.88% of export earning comes from these top five nations (Sawant et al. 2022).

Aonla is a beneficial fruit to ensure human health since it is a rich source of phenolics and antioxidants. The regular intake of aonla products can be helpful for the treatment of several human ailments. Thus, in order to maximize the benefits of this fruit crop, it needs to be processed to different value-added products. Furthermore, due to its potent antimicrobials and antioxidants properties, it can also be used as natural preservative to preserve the food materials instead of using synthetic chemicals. However, during the processing of anola to different value-added products, especially during drying and blanching, the bioactive compounds of the fruit started to decrease. Therefore, extensive research into minimal processing techniques is necessary to counteract this loss. Furthermore, it is very important to develop low-cost production and processing technology related to anola so that the aonla growers get benefited.

References

Almatroodi SA, Alsahli MA, Almatroudi A, Dev K, Rafat S, Verma AK, Rahmani AH (2020) Amla (*Emblica officinalis*): role in health management via controlling various biological activities. Gene Rep 21:100820

Avinash PG, Hamid S, Dash KK, Shaikh AM, Ungai D, Kovács B (2024) Recent insights into the morphological, nutritional, and phytochemical properties of Indian gooseberry (*Phyllanthus emblica*) for the development of functional foods. Plants 13(5):574

Bakshi B, Kumar R, Kour G, Vali VK, Bhat D, Hazarika TK (2018) Maturity indices of aonla cultivars under rainfed conditions of Shivalik foothills of Himalayas. J Agric Sci Technol 20(7):1517–1524

Bouhile Y, Guo Y, Wu B, Dai J, Song C, Pan Z, Ma H (2025) Research progress in the application of infrared blanching in fruit and vegetable drying process. Compr Rev Food Sci Food Saf 24(1):e70103

Brar RS, Topno SE, Bahadur V, Sharma A (2023) Effect of different packaging materials and chemicals on shelf life and quality of amla (*Emblica officinalis* L.). Int J Plant Soil Sci 35(15):78–88

Bulo U, Nimbolkar PK, Singh S, Sahu GD, Wangchu L, Das S, Pandey DK (2024) Nutrient profiling of wild amla (*Emblica officinalis* Gaertn.) populations in Northeast India: assessing the potential of this fruit tree for ecological and human health restoration. J Food Compos Anal 125:105814

Das DK, Chaturvedi OP, Jha RK, Kumar R (2011) Yield, soil health and economics of aonla (*Emblica officinalis* Gaertn.) based agri-horticultural systems in eastern India. Curr Sci 100(6):786–790

Devi SJ, Rajasekaran B (2011) Biology and morphology of aonla aphid. Madras Agric J 98:277–278

Gajanana K, Rokhade AK, Patil PB, Kulkarni MS (2007) Standardisation of recipe for preparation of amla squash. Beverage Food World 34:55–56

Gangwar S, Shukla HS, Katiyar D, Pandey V (2012) Effect of calcium nitrate on physico-chemical changes and shelf-life of Amla (*Emblica officinalis* Gaertn.) fruits. HortFlora Res Spectrum 1(3):253–258

Gantait S, Mahanta M, Bera S, Verma SK (2021) Advances in biotechnology of *Emblicaofficinalis* Gaertn. syn. *Phyllanthus emblica* L.: a nutraceutical-rich fruit tree with multifaceted ethnomedicinal uses. Biotech 11:1–25

Gautam A, Kumar A, Bhad R, Garg M, Sharma NK, Yadav JK (2023) Sustainable preservation: exploring modern solar drying technologies for fruits and vegetables. Int J Stat Appl Math SP-8(5):801–807

Golechha M, Sarangal V, Ojha S, Bhatia J, Arya DS (2014) Anti-inflammatory effect of Emblica officinalis in rodent models of acute and chronic inflammation: involvement of possible mechanisms. Int J Inflamm 2014:178408. https://doi.org/10.1155/2014/178408

Gul M, Liu ZW, Rabail R, Faheem F, Walayat N, Nawaz A, Aadil RM (2022) Functional and nutraceutical significance of Amla (*Phyllanthus emblica* L.): a review. Antioxidants 11(5):816

Huang JI, Zhong ZG (2011) Study of gallic acid extracted from the leaves of *Phyllanthus emblica* on apoptotic mechanism of human hepatocellular carcinoma cells BEL-7404. J Chin Med Mater 34:246–249

Javed S, Nasim T, Zia-Ul-Haq M (2023) Amla. In: Essentials of medicinal and aromatic crops. Springer International Publishing, Cham, pp 53–81

Khan MS, Qais FA, Ahmad I (2019) Indian berries and their active compounds: therapeutic potential in cancer prevention. In: Khan MS, Ahmad I, Chattopadhyay D (eds) New look to phytomedicine: advancements in herbal products as novel drug leads. Apple Academic Press, pp 179–201

Khedkar R (2022) Traditional food adjuncts: sustainable and healthy option for functional foods. In: Bioactive components: a sustainable system for good health and well-being. Springer Nature Singapore, pp 503–515

Kumar A, Tantry BA, Rahiman S, Gupta U (2011) Comparative study of antimicrobial activity and phytochemical analysis of methanolic and aqueous extracts of the fruit of *Emblica officinalis* against pathogenic bacteria. J Tradit Chin Med 31(3):246–250

Kumari P, Khatkar BS, Duhan A (2019) Amla phytochemicals: extraction, identification, and quantification. J Food Sci Technol 56(4):2278

Lal Basediya A, Samuel DVK, Beera V (2013) Evaporative cooling system for storage of fruits and vegetables—a review. J Food Sci Technol 50(3):429–442

Lalita G, Giri SK, Mohapatra D, Tripathi MK, Kate A, Wahid A (2024) Optimization of process parameters of refractance window drying for amla slices and comparison with other drying techniques. J Sci Food Agric 104(13):7854–7863

Li W, Zhang X, Chen R, Li Y, Miao J, Liu G, Lan Y, Chen Y, Cao Y (2020) HPLC fingerprint analysis of *Phyllanthus emblica* ethanol extract and their antioxidant and anti-inflammatory properties. J Ethnopharmacol 254:112740

Lorenzo JM, Munekata PE, Putnik P, Kovačević DB, Muchenje V, Barba FJ (2018) Sources, chemistry, and biological potential of ellagitannins and ellagic acid derivatives. Stud Nat Prod Chem 60:189–221

Maholiya BK, Prasad H, Thakur M, Gupta AK, Solanki SPS (2015) Comparative performance of aonla (*Emblicaoffcinalis* G.) cultivars under parbhani condition. Int J Bio-resour Stress Manag 6:778–780

Mehere N, Kaushik S, Upadhyay MK, Singh D, Yadav P, Galib R (2024) Development of Amla Nutri-bar: proximate analysis, gallic acid quantification and its validation through HPTLC, and spectroscopy. J Drug Res Ayurvedic Sci 9(Suppl 1):S92–S103

Mehjabin S, Akanda MKM, Akhter N, Bosri M, Ali S (2024) Medicinal and nutritional importance of *Phyllanthus emblica* in human health. In: Medicinal plants and their bioactive compounds in human health, vol 1. Springer, Singapore, pp 143–164

Mishra DN, Kumar A, Maurya AK, Pratap R, Sonkar A, Singh S (2023) Impact of nutrient management through bio-organic manure on bio-chemical attributes of aonla (*Emblica officinalis* Gaertn.) cv. NA-10. Int J Plant Soil Sci 35(19):146–151

Muzaffar K, Sofi SA, Makroo HA, Majid D, Dar BN (2022) Insight about the biochemical composition, postharvest processing, therapeutic potential of Indian gooseberry (amla), and its utilization in development of functional foods—a comprehensive review. J Food Biochem 46(11):e14132

Nallathambi P, Umamaheswari C, Nagaraja A, Dhandar DG (2007) Aonla diseases and management. Technical Bulletin, Central Institute for Arid Horticulture, Bikaner

Nampoothiri SV, Prathapan A, Cherian OL, Raghu KG, Venugopalan VV, Sundaresan A (2011) In vitro antioxidant and inhibitory potential of Terminalia bellerica and *Emblica officinalis* fruits against LDL oxidation and key enzymes linked to type 2 diabetes. Food Chem Toxicol 49:125–131

NHB Database (2021) National Horticulture Board Gurgaon, India

Pareek S, Kitinoja L (2011) Amla (*Emblica officinalis* Gaertn.). In: Yahia EM (ed) Postharvest biology and technology of tropical and subtropical fruits. Woodhead Publishing, pp 65–99e

Pathak RK (2003) Status report on genetic resources of Indian gooseberry-amla (*Emblica officinalis* Gaertn.) in South and Southeast Asia. IPGRI Office for South Asia, National Agriculture Science Centre (NASC), DPS Marg, Pusa Campus, New Delhi, pp 1–96

Pathak RK, Singh S, Saroj PL (2006) Amla (*Emblica officinalis*). In: Saroj PL, Awasthi OP (eds) Advances in arid horticulture, vol 2. IBD Co, Lucknow, pp 1–20

Pathak PK, Preeti D, Sunil K (2009) Effect of post-harvest treatments on shelf-life of amla (*Emblica officinalis*) fruits damaged during harvesting. J Food Sci Technol Mysore 46(3):283–285

Prajapati RK, Pandey CS, Gupta PK, Singh VK, Singh SR (2021) Recent advances in diseases management of aonla (*Emblica officinalis*). In: Singh RK, Gopata (eds) Innovative approaches in diagnosis and management of crop diseases, vol 2. Apple Academic Press, New York, p 43

Prasad K, Jacob S, Siddiqui MW (2018) Fruit maturity, harvesting, and quality standards. In: Preharvest modulation of postharvest fruit and vegetable quality. Academic Press, New York, pp 41–69

Prasad K, Singh SK, Kumar S, Maurya P, Saroj N, Maurya S, Agnihotri R, Lal MK, Ahanger MA (2024) Rootstock mediates postharvest fruit quality, phytochemicals, and plant stress of fruit crops. In: Ahanger MA, Bhat JA, Ahmad P, John R (eds) Improving stress resilience in plants. Academic Press, pp 425–446

Purena R, Seth R, Bhatt R (2018) Protective role of *Emblica officinalis* hydro-ethanolic leaf extract in cisplatin-induced nephrotoxicity in rats. Toxicol Rep 5:270–277

Rathinasamy M, Ayyasamy S, Velusamy S, Suresh A (2022) Natural fruits based ready to serve (RTS) beverages: a review. J Food Sci Technol 59(12):4563–4569

Saini R, Sharma N, Oladeji OS, Sourirajan A, Dev K, Zengin G, Kumar V (2022) Traditional uses, bioactive composition, pharmacology, and toxicology of *Phyllanthus emblica* fruits: a comprehensive review. J Ethnopharmacol 282:114570

Saroj PL, Nath V, Vashishtha VV (2001) Effect of poly containers on germination, seedling vigour, root characters and budding success in aonla. Indian J Hortic 57(4):300–304

Saroj N, Prasad K, Singh SK, Kumar U, Lal MK, Maurya S, Nana LP (2023a) Regulatory role of melatonin in flowering, fruit setting, and ripening. In: Melatonin in plants: a regulator for plant growth and development. Springer Nature, Singapore, pp 67–79

Saroj N, Prasad K, Singh SK, Maurya S, Maurya P, Kumar S, Dhongabanti B (2023b) Diverse functional role of melatonin in postharvest biology. In: Melatonin in plants: a regulator for plant growth and development. Springer Nature, Singapore, pp 203–217

Sawant SS, Lee B, Song J, Seo HJ (2022) The Indian gooseberry (*Emblica officinalis*) industry and cultivation in India. J Korean Soc Int Agric 34:199–204

Sengupta P, Sen S, Mukherjee K, Acharya K (2019) Postharvest diseases of Indian gooseberry and their management: a review. Int J Fruit Sci 20(2):178–190. https://doi.org/10.1080/15538362.2019.1608889

Shukla A (2021) Value addition, packaging, and storage of Amla in India. Int J Adv Agric Sci Technol 8(10):128–141

Singh SK, Singh PK (2015) Intercropping elephant foot yam is an economical cultivation practice for Indian gooseberry (*Phyllanthus emblica*) orchard management. New Agric 26:357–363

Singh AK, Singh S, Hiwale SS, Appa Rao VV, Joshi HK (2014) Production technology of amla under rainfed conditions of western India. Technical Bulletin, Pub. CHES (ICAR-CIAH), pp 1–36

Singh AK, Singh S, Saroj PL, Mishra DS, Singh PP, Singh RK (2019) Amla (*Emblica officinalis*) in India: review of its improvement, production, and diversified uses. Indian J Agric Sci 89:1773–1781

Singh MK, Dwivedi S, Yadav SS et al (2020) Anti-diabetic effect of *Emblica officinalis* (Amla) against arsenic-induced metabolic disorder in mice. Indian J Clin Biochem 35:179–187

Singh S, Mishra DS, Singh AK, Singh AK, Verma S, Mishra P (2022) Plant growth regulators in aonla. In: Plant growth regulators in tropical and sub-tropical fruit crops. CRC Press, pp 52–63

Sohshangrit S, Prasad VM, Nura CS (2022) Standardization of amla candy (*Emblica officinalis* L.) cv. Kanchan. Int J Plant Sci 34(20):608–614

Suresh PM, Kumar S (2014) Impact of drip fertigation levels on flower characters in aonla (*Emblica officinalis* Gaertn.) cv. NA-7. Trends Biosci 7:1723–1725

Suresh PM, Kumar S, Swaminathan V, Gurusamy A, Shivakumar T (2019) Influence of drip fertigation on soil NPK status of aonla (*Emblica officinalis* Gaertn.) cv. NA-7. Chem Sci Rev Lett 8:29–34

Tandon DK, Yadav RC, Sood S, Kumar S, Dikshit A (2003) Effect of blanching and lye peeling on the quality of aonla candy. Indian Food Packag 57(6):147–149

Tewari R, Kumar V, Sharma HK (2019) Physical and chemical characteristics of different cultivars of Indian gooseberry (*Emblica officinalis*). J Food Sci Technol 56(3):1641–1648

Tewari R, Kumar V, Sharma HK (2023) Thermal and nonthermal processing of an underutilized fruit *Emblica officinalis* (Amla): a sustainable approach. Sustain Food Technol 1(5):658–680

Tomar MS, Pradhan RC (2024) Effect of storage temperatures on physicochemical, textural, bioactive, and microstructure changes in amla fruit. J Food Meas Charact 18(11):9658–9668

Variya BC, Bakrania AK, Patel SS (2016) *Emblica officinalis* (Amla): a review for its phytochemistry, ethnomedicinal uses, and medicinal potentials with respect to molecular mechanisms. Pharmacol Res 111:180–200

Vishwakarma SK, Shukla HK, Singh BK, Singh AK (2012) Effect of different gauge polythene and newspaper on storage life of Kanchan cultivar of aonla (*Emblica officinalis* Gaertn. Sy. Phyllanthus emblica L.). Environ Ecol 30(2):424–426

Wali VK, Bakshi P, Jasrotia A, Bhushan B, Bakshi M (2015) Aonla. Directorate of Extension, SKUAST-Jammu, pp 1–30

Yadav S, Shukla HS, Ram RA (2010) Studies on foliar application of NAA, GA_3, boric acid and Ca $(NO_3)_2$ on fruit retention, growth, yield and quality of aonla (*Emblica officinalis* Gaertn.) cv. Banarasi. Hortic J 23:64–67

Yadav SM, Patil RK, Yadav DL (2012) Biological control of Penicillium fruit rot (*Penicillium funiculosum* Thom.) of aonla. Green Farming 3(1):78–79

Yang CJ, Wang CS, Hung JY, Huang HW, Chia YC, Wang PH, Weng CF, Huang MS (2009) Pyrogallol induces G2-M arrest in human lung cancer cells and inhibits tumor growth in an animal model. Lung Cancer 66:162–168

MAHUA (*Madhuca indica*): The Multifunctional Tree of Tradition and Sustenance

14

Shalini Purwar, Vinita Bisht, and Ena Gupta

Abstract

It is often known that many rural households in India rely on natural resources for their livelihood. Natural resources are therefore beneficial to the economy and give a community access to a variety of resources, including food and revenue. Trees and green areas are associated with mortality, income, and climate change in arid settlements. Farmers are compelled to implement tree-based systems in order to guarantee their income and livelihood due to the challenging climate. Humans can obtain fresh water, food, and lumber goods from natural resources. One of the most important natural resources in urban and woodland settings is *Madhuca longifolia*, also known as Mahua, and belonging to the Sapotaceae family. Rich in sugar, protein, vitamins, and minerals, flowers and fruits are utilized as food and to make value-added products like laddu, jam, and pickles that are sold in the local market. They are also used to support their livelihoods. Oil is made from the seed for use in biodiesel, laundry soaps, detergents, cooking, and lighting, as well as in the medical industry. Biodiesel is a nontoxic, renewable, and biodegradable fuel. The high-yielding mass-propagated clones using vegetative propagation, grafting, and seedlings as part of an afforestation effort in Eastern Ghats village are areas based on our collection and vegetative propagation. Trees with a high flower/fruit yield were chosen based on the weight, length, and quantity of blooms produced in a branch. From those chosen

S. Purwar (✉)
Department of Basic and Social Science, Banda University of Agriculture and Technology, Banda, Uttar Pradesh, India

V. Bisht
Department of Silviculture and Agroforestry, Banda University of Agriculture and Technology, Banda, Uttar Pradesh, India

E. Gupta
Department of Community Science, University of Allahabad, Prayagraj, Uttar Pradesh, India

A. Kumar et al. (eds.), *Fasting Superfoods*,
https://doi.org/10.1007/978-981-95-2762-5_14

trees, stem cuttings were taken, and they were then multiplied in the nursery. Farmers who live in arid regions will gain from this in terms of marketing, value addition, and sustainable harvesting.

Keywords

Livelihood · *Madhuca longifolia* · Natural resources · Sapotaceae

14.1 Introduction

The Mahua tree (*Madhuca indica* or *Madhuca longifolia*), belonging to the family Sapotaceae, is a deciduous species indigenous to the Indian subcontinent. Its origin can be traced back to tropical regions of South Asia, where it has thrived in arid and semi-arid ecosystems for centuries. This tree is deeply rooted in the cultural and ecological landscape of Central and Southern India and holds immense traditional, economic, and ecological importance (Bisht et al. 2018).

Mahua (*Madhuca longifolia* (Roxb.)) is a plant species of significant ecological and economic value for local communities. Due to its importance, it experiences considerable anthropogenic pressure, particularly through the collection of its flowers and seeds. This continued exploitation could potentially alter the natural vegetation structure associated with the species (Nayak and Sahoo 2020a, b). Mahua is extensively found throughout the Indian subcontinent, with a prominent presence especially in the states of Madhya Pradesh and Chhattisgarh, Odisha, Jharkhand, Maharashtra, Uttar Pradesh, and Andhra Pradesh. It is also found in neighboring countries such as Nepal, Sri Lanka, and Bangladesh. The tree is well-adapted to dry, tropical climates and is often a dominant species in mixed deciduous forests and agroforestry systems. Its remarkable drought resistance enables it to flourish in poor soils and withstand challenging environmental conditions, making it an invaluable resource for rural and tribal communities. In India, the two prominent species belonging to the genus *Mahua* are *Madhuca latifolia* and *Madhuca longifolia*. *Madhuca longifolia*, commonly known as Mahua, is a versatile tropical tree primarily grown or collected from the wild in South Asia for its edible flowers and oil-bearing seeds (Fern 2014). This species is predominantly found across the central, southern, and northern plains and forested regions of India, as well as in neighboring countries like Nepal, Myanmar, and Sri Lanka. It is a rapidly growing tree that can reach a height of around 20 m, characterized by its evergreen or semi-evergreen foliage, and belongs to the family *Sapotaceae*. Mahua is native to regions including India, Sri Lanka, Nepal, and Myanmar (Fern 2014). This species is capable of withstanding frost and thrives in dry tropical and subtropical forest zones, particularly in marginal areas at elevations ranging from 1200 to 1800 m. It is commonly observed growing in scattered patterns across pastures, agricultural fields in central India, and along riverbanks within semi-evergreen forests (Trifed.nic.in 2022; Orwa et al. 2009). The tree performs optimally in areas receiving annual rainfall between 500 and 1500 mm and can tolerate temperatures ranging from 2 to 46 °C. Although it grows best in deep, well-drained loamy or sandy-loam soils, Mahua can also adapt to more challenging conditions, such as shallow, rocky, clay-rich, and calcareous soils (Orwa et al. 2009). Mahua is one of the most important food items of the tribal community in India. For rural tribal people living in the remote forest areas of Central and Western India, the Mahua tree is highly significant as a source of employment and food (Caritasindia n.d.). Various dishes such as Rasukutka, Mahua Kholli, Mahua Biscuit, Dobhri, Halwa, Ladoo, Makhani, Jam, etc. are made from Mahua flowers. Mahua is rich in carbohydrates, fats, and proteins, as well as in calcium, phosphorus, iron, carotene, and vitamin C, which boost immunity. The seeds of Mahua contain fatty content, which is quite beneficial. Boiling it in milk and giving it to undernourished children provides nutrition. Besides eradicating malnutrition, Mahua cookies will also help reduce unemployment (Ishan Dubey et al. 2023). Its flowers, seeds, and bark have been integral to traditional practices, serving as food, medicine, and raw material for various industries. The tree's role in supporting tribal economies, maintaining biodiversity, and offering climate-resilient solutions underscores its unparalleled importance (Dwivedi et al. 2022). As a source of income, nutrition, and ecological stability, the Mahua tree continues to be a lifeline for millions, embodying a harmonious relationship between nature and humanity.

14.2 History of Mahua (*Madhuca longifolia*)

The Mahua tree (*Madhuca longifolia*) is often called the "Tree of Life" or "Kalpavriksha" in many tribal communities of India. This tree holds deep traditional, cultural, and economic significance, especially in Central and Eastern India. The term Mahua holds significance across various fields including Hinduism, Sanskrit, Jainism, Prakrit, Hindi, and biological sciences. The history of the Mahua tree (*Madhuca longifolia*) is deeply rooted in the cultural, medicinal, economic, and ecological fabric of India. Its significance can be traced back to ancient times, continuing through the medieval period, colonial rule, and into the present day (Lal et al. 2014). More than just a botanical species, Mahua symbolizes sustenance, resilience, and spiritual importance for many communities across the Indian subcontinent (Kuthadi 2024). Reverence for the Mahua tree reflects the tribal communities' harmonious relationship with nature and their enduring commitment to environmental stewardship. Furthermore, this plant is associated with the Revati star, one of the constellations in the traditional almanac, and its presiding deity is Pusha (Gogate and Mittal 1983).

14.2.1 Ancient and Classical References

Mahua, known as Madhuka in Sanskrit, finds early mention in the Atharvaveda, where it is associated with mystical and ritualistic applications, including its use in love-spells and the preparation of intoxicating beverages (Sinha et al. 2017). This early reference indicates the tree's significant spiritual and medicinal value in Vedic society. The Charaka Samhita, a seminal Ayurvedic text, also highlights Mahua for its therapeutic potential. Different parts of the Mahua tree—such as its flowers, fruits, seeds, and bark—have long been utilized in traditional medicinal systems for addressing various health conditions, highlighting its extensive role in the ancient healthcare traditions of India (Kuthadi 2024).

14.2.2 Medieval Period and Mughal References

During the medieval period, Mahua continued to be part of the Indian landscape, both physically and culturally. The Baburnama, the memoirs of Babur, the founder of the Mughal Empire, includes references to the Mahua tree (Chatterjee et al. 2021; Kuthadi 2024). This attests to its widespread presence and utility during the Mughal period, where it was likely valued for both its economic and culinary contributions (Kuthadi 2024).

14.2.3 Colonial Era and British Policies

The British colonial period marked a significant shift in the perception and regulation of Mahua. What was once a free gift of nature became a commodity under state control. British authorities saw Mahua as an important resource for producing country liquor and feeding cattle. This led to the enforcement of colonial laws to control its collection, trade, and processing. One such measure was the Bombay Abkari Act of 1878, which was one of the earliest legislations in British India aimed at regulating the production and sale of alcohol (Saldanha 1995; Anon 1898; Kuthadi 2024). The Act introduced a licensing system, imposed taxes and duties, and attempted to curb illicit trade in Mahua-based liquor. This regulation disproportionately affected tribal and rural communities, who had traditionally collected and used Mahua flowers freely. Kuthadi (2024) remarked that such legislation led to a state monopoly over Mahua flower procurement, disrupting traditional livelihoods.

14.2.4 Mahua During Famines and Scarcity

Historical accounts reveal Mahua's crucial role as a lifeline during periods of famine and scarcity. In her comparative study of the 1770 Bengal famine and the 1897 Chotanagpur famine, historian Damodaran (1995, 2006) highlights how Mahua became an essential dietary staple for the poor. During droughts, epidemics, and food shortages under colonial rule, Mahua provided a vital source of nutrition and calories. This is supported by records from Sir Charles Watt, who quoted Mr. Lockwood, a magistrate of Monghyr (now Munger), stating that the Mahua crop saved thousands from starvation during the famine of 1873–74 (Kuthadi 2024). Even folk songs from Chhattisgarh celebrate Mahua's role in sustaining communities during hard times (Kuthadi 2024), further cementing its place in the collective memory of the people.

14.2.5 Cultural and Culinary Significance

Mahua's significance extends far beyond sustenance, serving as an essential element of festivals and communal life in tribal societies. It is consumed in multiple forms—as a fermented beverage, an ingredient in traditional sweets, or simply chewed fresh—reflecting its versatility and cultural embeddedness. A Tamil proverb, "Alaaiillaathaoorililuppai poo sarkarai" ("In a place where there is no sugar factory, Mahua flowers are sugar"), underscores the flower's high sugar content and its role as a vital local sweetener (Jayasree et al. 1998). The Mahua tree (*Madhuca longifolia*) holds deep cultural and spiritual value, especially in regions like Parghat, where it is venerated as the *Kalpavriksha* or wish-fulfilling tree. Symbolizing fertility, prosperity, and ecological harmony, its flowers and seeds are gathered with ritualistic reverence. Central to tribal rituals and ceremonies, Mahua-based products are offered as expressions of hospitality and abundance. The tree is also deeply rooted in oral traditions, songs, and folklore, symbolizing endurance and sustenance. Its seasonal harvesting is a collective effort, especially among women. Promoting social unity, safeguarding indigenous wisdom, and strengthening cultural heritage are key aspects associated with this practice (Fig. 14.1).

14.2.6 Rasashastra (Alchemy and Herbo-Mineral Preparations)

Madhukataila (also spelled *Madhūkataila* or *Madhuka-taila*) refers to Mahua oil, traditionally derived from the Madhuka tree (*Madhuca indica* or *Madhuca longifolia*), and also holds significant value in both Ayurveda and Rasashastra (Indian alchemy). In Ayurvedic literature, particularly in the fifteenth-century Keralite text *Yogasārasaṅgraha*, Madhukataila is mentioned in the *Tailakhaṇḍa* section (verse 3.159) as a medicated oil used in treating conditions such as fever, cough, asthma, and dyspnoea. Its formulation is part of a broader system of therapeutic oils known for their curative and preventive properties. Beyond its medicinal use, Madhukataila also appears in Rasashastra, in the thirteenth-century text *Rasaratnākara* by Nityanātha. It is used in the *Vādakhaṇḍa* section in the process of creating artificial rubies. Heated stones, when treated with Mahua oil, are said to attain a divine radiance, underscoring the oil's role in early Indian alchemical and gem-making practices. The wide-ranging applications of Madhukataila—from healing ailments to enhancing gemstones—reflect the interdisciplinary knowledge systems of ancient India. Its relevance continues today through modern Ayurvedic research, including studies on its repeated processing (avartana) and efficacy in managing disorders like Vātarakta (akin to gout), as well as its role in nasya (nasal) therapies (https://www.wisdomlib.org/definition/madhukataila).

Fig. 14.1 Cultural and culinary significance of Mahua

14.3 Production: Cultivation Practices, Climatic and Soil Requirements, and Global Production Statistics

14.3.1 Cultivation Practices

The Mahua tree (*Madhuca indica* or *Madhuca longifolia*) is predominantly grown in tropical and subtropical regions. While the tree is primarily propagated naturally, plantation efforts are increasingly adopted to meet growing demand. The best time to move Mahua plants is usually during the rainy season, specifically in July and August, when there is plenty of rain and the ground is wet. Still, the specific dates can change depending on the age of the seedlings and the area where they are growing (Shedage et al. 2024).

14.3.1.1 Vegetative Propagation

- *By Seeds*

- Mahua is commonly propagated through seeds. Mahua seeds are recalcitrant, meaning they are highly sensitive to desiccation (drying) and cannot be stored for long periods in a conventional manner, and there germination rates are moderate to high. Recalcitrant seeds typically have a higher moisture content at maturity and are easily damaged by drying or freezing. In this reasons, the seeds are shown in the nursery that is collected in June to July month of current year because of their short viability period. The Mahua seeds are sown directly in the field or in nurseries used as a raised bed for germination (Kulkarni et al. 2013).

Karunarathna et al. (2023) results showed that mechanical scarification was the most effective method for enhancing germination, with up to 90% germination under light conditions and 85.45% in darkness within 28 days. Interestingly, seeds in dark conditions germinated more quickly during the first week. While chemical treatments such as GA_3, KNO_3, and water improved shoot growth (about 8.3–8.6 cm), scarified seeds developed stronger root systems, indicating a trade-off between shoot elongation and root robustness. According to Shedage et al. (2024), the rate of seed germination and survival percentages range from 73.6% to 46%, and 88.6% to 71%, respectively. Mahua trees typically begin fruiting about 10 years after planting and can continue to produce fruits for as long as 60 years. Older trees tend to be more fruitful compared to younger ones. While irrigation can be beneficial during the fruit development stage, it is generally discouraged during the periods of flowering and leaf fall. In India, the fruit harvesting season usually falls between May and July. On average, a single tree can produce between 60 and 80 kg of seeds annually. In addition to fruit, Mahua trees are also valued for their flowers, with each tree capable of yielding approximately 100–150 kg of dry flower matter per year (Singh et al. 2020). Nayak and Sahoo (2020b) studied significant variation in germination rates among different seed sources of *Madhuca latifolia* in Odisha, with germination percentages ranging from 53% to 93.8%. The Nuapada provenance (P8) showed the highest germination and early seedling growth, making it the most promising seed source. Phenotypic variation was higher than genotypic variation, indicating environmental factors had a greater influence on germination and growth. Collar diameter exhibited higher heritability (50%) compared to seedling length and biomass, suggesting it is a more reliable trait for selection in breeding programs. Overall, the results emphasize the importance of selecting seed sources based on both genetic potential and environmental adaptability for successful reforestation efforts.

- *Macropropagation*
- Fresh Mahua seeds generally exhibit high fertility; however, their viability diminishes rapidly during storage, making them highly vulnerable to insect infestations and fungal infections. To overcome these challenges, various vegetative propagation techniques have been developed for Mahua multiplication, including softwood grafting, wedge grafting, veneer grafting, and air layering, each showing different levels of success. Notably, softwood grafting has recently shown a success rate of 70–80% (Singh and Ravishankar 2010; Singh and Singh 2015), while veneer grafting has demonstrated up to 90% success (Singh et al. 1999). Among the commonly used methods, cleft grafting is widely practiced, wherein scions taken from elite or superior "plus trees" are grafted onto rootstocks of a particular age to ensure genetic uniformity and desirable traits in the propagated trees. This process helps in passing on preferred genetic qualities and is effective for establishing orchards in arid and tougher soil conditions. The study on macropropagation of *Madhuca latifolia*, by Dobriyal et al. (2002), demonstrated successful vegetative propagation techniques for this important tree species. They found that stem cuttings treated with specific concentrations of

growth hormones, particularly auxins like indole-3-butyric acid (IBA), significantly improved root initiation and development. Rooting percentages varied depending on the hormone concentration and cutting type, with semi-hardwood cuttings responding best to treatment. The study also highlighted that environmental factors such as humidity and temperature played a crucial role in enhancing rooting success. Overall, the findings provided an effective propagation method that can support conservation and large-scale multiplication of *Madhuca latifolia*, helping to meet reforestation and commercial planting demands. The study by Suryawanshi and Mokat (2021) analyzed seed traits of *Madhuca longifolia* across different locations in the Western Ghats. The results revealed significant morphophysiological variability in seed size, weight, oil content, and germination percentage among the selected mahua trees. This variability has a direct impact on the traditional livelihoods of local tribal communities, who rely on mahua seeds for oil extraction, food, and cultural practices. The findings underscore the importance of conserving genetically diverse populations and selecting superior seed sources to enhance productivity and support sustainable use of this valuable tree species in tribal regions.

- *Grafting*
- This technique involves grafting on one-year-old rootstock seedlings that have begun new vegetative growth, characterized by light green leaves. Scion material is collected from 3 to 4-month-old shoots that display a healthy and prominent apical bud. Prior to grafting, these shoots are defoliated 8–12 days in advance. Naturally shed shoots collected during March–April can also be used for improved grafting success. Scions are carefully detached using a sharp grafting knife or secateurs and joined to the rootstock using the cleft grafting method. When the compatibility between the rootstock and scion is optimal, grafting success rates can reach between 60% and 70%, especially when performed during the months of March, April, July, and August (Chadha et al. 2001). A study by Singh and Singh (2015) explored the most favorable period for carrying out softwood grafting in Mahua and Khirni under the semi-arid climatic conditions of western India. Conducted at the Central Horticultural Experiment Station in Vejalpur, Gujarat, the research focused on identifying the most effective months for grafting in order to enhance success rates and promote better plant development. The results revealed that July was the most effective month for grafting in mahua, with a success rate of 77.50%, while August proved best for khirni, achieving an 85.83% success rate. Grafts performed during these months also exhibited superior growth in terms of shoot length, number of leaves, and sprouting percentage. In contrast, grafting conducted during the extreme summer or winter months showed poor outcomes due to unfavorable temperature and humidity conditions. The study concluded that the timing of grafting plays a critical role in its success and recommended July for mahua and August for khirni as the ideal periods for softwood grafting in the semi-arid regions of western India. This finding is valuable for improving nursery practices and promoting the successful propagation and conservation of these important arid-zone fruit species.

- Pohare et al. (2016) evaluated the feasibility and success of using khirni as a rootstock for softwood grafting of mahua. Conducted at the Department of Forestry, Dr. Panjab Rao Deshmukh Krishi Vidyapeeth, Akola, the experiment aimed to enhance graft survival and compatibility between the two species. The results indicated a notable survival percentage of grafts, demonstrating that khirni can serve as a compatible and effective rootstock for mahua. Among different grafting dates and conditions tested, the timing and physiological condition of both scion and rootstock significantly influenced graft success. The study concluded that khirni is a promising rootstock option for improving propagation success in mahua, offering potential benefits for large-scale cultivation and conservation. In situ grafting is also successful in Mahua plantations in rocky or dry soils, where nursery-transplanted grafts often show high mortality. Seeds are freshly extracted and germinated in polybags. Once the seedlings reach a height of 5–10 cm, they are transplanted directly into the field at the desired spacing. Softwood grafting is then performed in situ using scions from elite varieties. These plants tend to grow vigorously and develop rapidly into healthy trees (Chadha et al. 2001). Veneer grafting is carried out on 1-year-old rootstocks using shoots from the spring flush. The procedure is ideally performed in July and August, with July being the most favorable period for successful graft union. Air layering is another effective propagation technique for Mahua. The success rate improves significantly with the application of rooting hormone Indole-3-butyric acid (IBA) at a concentration of 1% (Chadha et al. 2001).
- *Micro Propagation*

- This technique involves growing plant tissue in a clean setting, which can yield many genetically identical plants. Under in vitro conditions, the highest germination percentage (77.71%) was observed on half-strength Murashige and Skoog (½ MS) medium supplemented with 0.5 mg/L benzylaminopurine (BAP), suggesting that cytokinin plays a key role in promoting germination in controlled environments (Karunarathna et al. 2023). The study by Akkenapally et al. (2024) established an effective protocol for in vitro propagation of Madhuca longifolia using leaf explants. The highest callus induction was achieved using Woody Plant Medium (WPM) supplemented with 2 mg/L thidiazuron (TDZ). For somatic embryogenesis, Murashige and Skoog (MS) medium containing 6 mg/L benzyl amino purine (BAP) and 0.5 mg/L naphthalene acetic acid (NAA) was most effective. Additionally, the study explored the development of synthetic seeds from somatic embryos has enabled the successful regeneration of whole plants under both controlled laboratory (in vitro) and natural (ex vitro) conditions, even outside of typical growing seasons. This approach presents a viable and scalable method for the mass propagation and extraction of valuable bioactive compounds, which are essential for use in the pharmaceutical and nutraceutical industries. In addition, this technique provides a foundational platform for advanced research focused on genetic enhancement, resilience to environmental stresses, and the metabolic pathways involved in the production of secondary metabolites in *Madhuca longifolia*. The study by Rout and Das (1993) on micropropagation of *Madhuca longifolia* var. latifolia successfully developed an in vitro propagation protocol using nodal explants. The researchers found that Murashige and Skoog (MS) medium supplemented with 1.0 mg/L benzylaminopurine (BAP) was most effective for multiple shoot induction. The shoots were rooted efficiently on half-strength MS medium containing 0.5 mg/L indole-3-butyric acid (IBA). The micropropagated plantlets showed a high survival rate when transferred to soil, demonstrating the protocol's potential for rapid clonal multiplication and conservation of this valuable tree species.

- The study successfully established an in vitro propagation protocol for *Madhuca longifolia* using nodal segments. Axillary bud break was achieved at 64.44% on Murashige and Skoog (MS) medium supplemented with 3 mg/L benzyladenine (BA), particularly from explants collected during the rainy season (July-September). Shoot multiplication was maximized on the same medium and hormone concentration with a 30-day subculture cycle, producing the highest number of shoots, leaves, and longest shoots. Rooting of excised shoots was achieved at 55.55% on MS medium supplemented with 2 mg/L indole-3-butyric acid (IBA) after 40 days. A two-step hardening process involving transfer to half-strength MS liquid medium followed by acclimatization in soilrite with inorganic salts ensured successful plantlet survival. This protocol offers an efficient method for in vitro regeneration and large-scale multiplication of Madhuca longifolia, aiding conservation and commercial use of this valuable tropical tree (Shirin et al. 2020).

14.3.2 Climatic and Soil Requirements

Mahua (*Madhuca longifolia*) prefers a tropical climate and is highly drought-tolerant but does not survive in waterlogged conditions. It can grow in degraded rocky areas, including salt-affected soils and crevices between barren rocks. However, for optimal growth and productivity, well-drained, deep loamy soil is ideal. Mahua is a medium to large deciduous tree, reaching 12–15 m in height and found across forested regions of India up to 1200 m altitude. It has a short trunk, thick dark cracked bark, and a broad, rounded crown. The leaves are thick, leathery, and elliptic-oblong (7.5–23 × 3.8–11.5 cm), clustered at the ends of branches. Flowering occurs annually from March to April. The small, fleshy flowers are pale white to yellow, aromatic, and arranged in fascicles. Fruits are fleshy, greenish, and 2–6 cm long (Bisht et al. 2018). Mahua thrives in regions receiving annual rainfall between 500 and 1500 mm and can withstand a wide temperature range from 2 to 46 °C. While the species shows optimal growth in deep, well-drained loamy or sandy-loam soils, it is also capable of adapting to less ideal conditions, including shallow, rocky, clay-rich, and calcareous soils (Orwa et al. 2009).

Garai et al. (2021) assessed the Mahua tree habitat under current and changing climatic conditions. It found that approximately 29,550.29 km^2 in eastern India is suitable for M. longifolia, with Jharkhand covering about 72% of this area and

Chhattisgarh around 17%. The study highlighted concerns over declining populations and poor regeneration due to overexploitation and climate change. Given the tree's ecological importance and its critical role in the livelihood of tribal communities—providing food, medicine, and economic benefits—the authors emphasized the need for targeted conservation strategies to ensure its survival and continued socio-economic contribution. Though not a legume, mahua develops mycorrhizal associations able to fix atmospheric N. Its extensive superficial root system binds the soil and limits erosion. In India, Mahua has been utilized for reclaiming degraded lands, particularly in areas with hard lateritic soils. Its fallen leaves contribute to soil enrichment by adding organic matter, although they decompose more slowly than the leaves of species like teak or *Shorea robusta* (Siddique 2019). Additionally, the seed cake derived from Mahua is commonly applied as an organic fertilizer and has also been found effective in managing root-knot nematode infestations (Orwa et al. 2009; Yadav et al. 2005). Pradhan et al. (2024) investigate the potential distribution of Madhuca longifolia (mahua) under current and future climate scenarios across India. Utilizing the maximum entropy (MaxEnt) modeling approach, the researchers assessed habitat suitability for the period 2041–2060 under four Representative Concentration Pathways (RCPs): 2.6, 4.5, 6.0, and 8.5. The model incorporated ten bioclimatic variables selected through multicollinearity testing. Findings indicate that mean diurnal temperature range (Bio_2) is the most influential factor under current conditions, while rainfall during the warmest quarter (Bio_18) and annual rainfall (Bio_12) are significant under future scenarios. Presently, *M. longifolia* shows high habitat suitability in eastern India, particularly in Jharkhand. However, projections suggest a southward shift in suitable habitats under all RCP scenarios. Specifically, the study predicts a 4.5% increase in suitable area under RCP 2.6 but declines of 7.37%, 7.78%, and 3.54% under RCPs 4.5, 6.0, and 8.5, respectively. Despite these shifts, the species demonstrates resilience to climatic changes. The authors emphasize the importance of these insights for conservation planning and the sustainable management of this vital non-timber forest product, which holds significant socio-economic value for local communities.

14.3.3 Global Production Statistics

Mahua (*Madhuca longifolia*), predominantly cultivated and utilized in India, holds immense economic and cultural significance, particularly among tribal communities. India, the largest producer, yields approximately 45,000 metric tons of mahua flowers annually, primarily from states like Madhya Pradesh, Odisha, Jharkhand, and Chhattisgarh. The tree thrives in arid, tropical mixed deciduous forests and is deeply integrated into non-timber forest product (NTFP) economies, supporting rural livelihoods through the collection of flowers and seeds for liquor, food, and traditional medicine (Panda et al. 2010; Rao and Singh 1996). While its presence in Nepal, Sri Lanka, and Bangladesh is modest, the species is gaining attention in parts of Africa and Southeast Asia for its drought tolerance and utility. The commercial value of mahua is expanding globally, especially its seed oil, which is used in cooking, biofuels, and cosmetics, alongside growing interest in mahua-based alcoholic beverages and nutraceuticals (Fern 2014). Enhancing cultivation techniques and ensuring sustainable harvesting can further strengthen Mahua's position as a crucial ecological and economic asset worldwide.

14.4 Protection: Major Pests, Diseases, and Management Strategies

Although the Mahua tree (Madhuca indica or Madhuca longifolia) is a hardy species, it is susceptible to several pests and diseases that can reduce its productivity and vitality. Effective management practices are crucial for promoting robust growth and ensuring consistent and sustainable yields. In forestry nurseries, seedlings are the primary source of planting material, making it important to assess and implement preventive measures either before sowing the seeds or during the seedling phase to protect plant health (Verma and Verma 2015). Both nursery beds and polythene containers are susceptible to various plant diseases (Sehgal 1983), which can result in seedling mortality and negatively impact plant vigor, ultimately hampering reforestation and plantation efforts (Harsh et al. 1992). Root rot fungi are known to infect the root system, leading to decay and death of roots (Omukhua and Godwin-Egein 2011). Affected plants often exhibit symptoms such as yellowing leaves, dull appearance, and a general decline in vitality, frequently culminating in plant death. In newly infected seedlings, early signs include dieback at the tips, wilting, and stunted growth as the infection worsens. The root systems of diseased plants typically show poor development of fine roots and extensive degradation of the cortical tissues, making it easy to peel away the outer layers from the inner root core. These root rot pathogens can rapidly spread between neighboring plants, resulting in large-scale seedling losses within nursery settings (Harsh and Gupta 1993; Torgeson et al. 1954).

14.4.1 Major Pests and Management Strategies

- *Leaf vein gall symptoms*: Kumar and Singh (2018) reported the first observation of leaf vein gall symptoms on the multipurpose tree *Madhuca longifolia*, which were caused by an infestation from an unidentified insect. This phenomenon was initially recorded in 2012 in Ranchi, Jharkhand, and subsequently spread to several other districts within the state. By 2015, similar vein gall formations were also documented at the Forest Research Institute (FRI) Campus in Dehradun, Uttarakhand, indicating a broader geographic distribution of the infestation and raising concerns regarding its impact on this ecologically and economically significant species.
- *Polychorosis cellifera*: *Polychorosis cellifera* causes leaf roller in Mahua. The larvae web the leaves by folding the tip downwards on both margins parallel to the mid-rib and feed inside. In a severe attack, 1/4th of the lamina is eaten up.

Regular clipping and burning of affected leaves can keep the population under control (Chadha et al. 2001).

- *Scopella echinulata*: In the parasitic rust fungus *Scopella echinulata* was identified infecting the lower surface of leaves of the host plant *Madhuca latifolia* (Roxb.) Chev. The rust infection manifested as dark chocolate-colored pustules on the underside of the leaves. Microscopic examination revealed the presence of teliospores, which appeared spiny and exhibited an irregularly round shape. These dark brown teliospores were predominantly located on the lower leaf surface (Pathak et al. 2015).
- *Bark-eating caterpillars* (*Indarbela* spp.): It has been observed to be the most serious pest of Mahua. Larvae of this moth feed on the bark of the stem, interrupting translocation of sap, and thus the tree becomes weak and unproductive. The pest is very active from February to March (Chadha et al. 2001).
- *Miresaalbipuncta*: During a survey of insect pests in forest areas of India, larvae of Miresaalbipuncta were found infesting Madhuca latifolia(*M. longifolia*) in nurseries and young plantations in Jabalpur, Madhya Pradesh, India. The limacodid was of minor importance and was found only on the older leaves (Meshram et al. 1991).
- *Termites*: Arumugam and Rahman (2015) reported the termite assemblage in Mahua is primarily composed of higher termite species, particularly those from the Termitinae and Nasutitermitinae subfamilies. The Termitinae subfamily consists mostly of soil-feeding termites, which play a crucial role in the ecosystem as soil engineers and in the decomposition process, and thus require conservation. However, natural factors may contribute to the decline of these species.

14.5 Improvement (Genetics and Genomics): Breeding Efforts, Genetic Diversity, and Recent Advancements

Mahua (*Madhuca longifolia*) is a highly valued multipurpose tree, and its improvement has become a priority to enhance its productivity, resilience, and utility. Breeding efforts in Mahua focus on the selection of superior germplasm for traits such as high flower yield, oil content, and stress tolerance. Mass selection and clonal propagation have been employed to identify and multiply elite phenotypes, while vegetative propagation techniques like grafting and tissue culture have been standardized for preserving and rapidly propagating superior genotypes (Vir and Kumar 2021). Although hybridization efforts are limited due to the complex reproductive biology of Mahua, advancements in molecular tools and biotechnology are offering new opportunities. It is highly heterozygous, cross-pollinated fruit crop and as such seedlings exhibit a wide range of variations, which aids in the selection of the superior desirable genotypes. Owing to natural cross-pollination and the dominance of seed-based propagation over generations, there exists a vast potential for identifying superior trees with desirable horticultural characteristics. In regions such as Uttar Pradesh and Gujarat, substantial variability has been documented in traits like sweetness, acidity, fruit size and shape, as well as fruiting patterns in Mahua (Singh et al. 1999; Anonymous 2002; Singh and Singh 2005). Similar patterns of notable genetic diversity have also been observed across different agro-climatic zones in other underutilized fruit species, including tamarind and jamun (Keskar et al. 1989a, b; Karale et al. 1999; Devi et al. 2002; Rai and Mishra 2005; Singh and Singh 2005; Singh et al. 2020). In the study by Yadav et al. (2011), a comprehensive assessment of genetic variability and divergence among *Madhuca longifolia* (mahua) accessions was conducted by evaluating key seed and oil traits. The parameters analyzed included seed weight, kernel weight, kernel-to-seed ratio, and oil content—traits of high economic relevance due to the species' use in biofuel production, traditional medicine, and tribal livelihoods. The analysis revealed a significant range of variability across accessions, particularly in oil content (ranging from 30.9% to 47.6%), indicating the existence of genotypes with superior oil-yielding potential. Multivariate analysis, including D^2 statistics and cluster analysis, further confirmed the genetic divergence among the accessions, grouping them into distinct clusters based on their trait performance. This divergence suggests the presence of unique genotypes suitable for selection and breeding efforts. The study concluded that this natural variability can be effectively utilized to improve seed and oil yield in *M. longifolia*, reinforcing its potential as a sustainable resource for energy and rural economic development.

Studies on genetic diversity using molecular markers like RAPD, SSR, and AFLP have revealed high variability in natural populations, driven by the species' wide distribution and outcrossing nature. The study by Nimbalkar et al. (2018) focused on assessing the genetic diversity of *Madhuca indica* (Mahua) candidate trees using inter-simple sequence repeat (ISSR) molecular markers. The research involved analyzing 32 candidate trees collected from different geographical locations to understand the extent of genetic variation within the species. The ISSR markers revealed a high level of polymorphism, indicating substantial genetic diversity among the sampled trees. Cluster analysis grouped the trees into distinct genetic clusters, reflecting their genetic relationships and geographic origins. This diversity is crucial for the species' adaptability, conservation, and breeding programs. The study emphasized that maintaining this genetic variability is essential for the sustainable use and improvement of *Madhuca indica*, especially given its socio-economic importance for rural communities and its potential in forestry and agroforestry systems. Overall, the findings provide valuable insights into the genetic makeup of mahua populations, aiding in the development of strategies for conservation and genetic improvement. However, challenges such as habitat loss and population fragmentation threaten this diversity, underscoring the importance of conservation efforts. Ex situ approaches, including seed and gene banks, complement in situ measures like protecting natural stands to preserve the species' genetic resources.

Several improved varieties of Mahua (*Madhuca longifolia*) have been developed to enhance its yield and quality traits. Among them, NM-2, developed by the N.D. University of Agriculture and Technology, Faizabad (Uttar Pradesh), is a notable selection. This variety flowers early and its fruits ripen by the second week of June. The fresh flowers of NM-2 contain 25.50% total sugar, 19.60% reducing sugar, 5.6% nonreducing sugar, 60.90 mg/100 g vitamin C, 650.70 IU vitamin A, and 0.60% mineral content. The average fruit weighs about 16.80 g, with a husk content of 58.80%. The fruit contains 13.60% total sugar, 7.8% reducing sugar, 5.80% nonreducing sugar, 51.30 mg/100 g vitamin C, 586.70 IU vitamin A, and 1.90% mineral content. The average seed and kernel weights are 7.2 g and 5.50 g, respectively, with an oil content of 47.40%. Another promising variety is NM-4, developed at the Central Horticultural Experiment Station, Godhra. This variety begins flowering in the 4th week of April, and fruits mature by the third week of June. NM-4 flowers contain 24.50% total sugar, 19.60% reducing sugar, 4.80% nonreducing sugar, 49.20 mg/100 g vitamin C, 528.20 IU vitamin A, and 0.67% mineral content. The fruits have an average weight of 19.50 g with 65.30% husk content. They contain 11.20% total sugar, 8.20% reducing sugar, 3.20% nonreducing sugar, 55.80 mg/100 g vitamin C, 860 IU vitamin A, and 2.10% mineral content. The average seed and kernel weights are 6.00 g and 4.70 g, respectively, with an oil content of 45.20%. These improved varieties not only offer better flower and fruit composition but also show potential for higher oil yield and nutrient content, making them valuable for cultivation and commercial use (Chadha et al. 2001).

Recent genomic advancements, including transcriptome sequencing and pathway analyses, have provided insights into genes associated with oil biosynthesis, secondary metabolites, and stress responses. Marker-assisted selection (MAS) and genome editing tools like CRISPR/Cas9 hold promise for improving traits such as oil yield, flower sweetness, and drought tolerance. The study by Wang et al. (2024) presents a high-quality chromosome-level genome assembly of *Madhuca longifolia*, providing significant insights into its molecular adaptation to low-temperature stress in the southern subtropical regions of China. The assembled genome spans approximately 885 Mb, organized into 15 pseudochromosomes with high completeness and accuracy. A total of over 37,000 protein-coding genes were annotated, many of which are involved in stress responses and metabolic regulation. Comparative genomic and transcriptomic analyses revealed the expansion and positive selection of gene families associated with cold tolerance, including key transcription factors such as CBF/DREB, late embryogenesis abundant (LEA) proteins, and antioxidant enzymes like superoxide dismutase (SOD) and catalase (CAT). These genes contribute to enhanced antioxidant defense, osmotic regulation, and signal transduction under low-temperature conditions. Phylogenomic analysis further demonstrated evolutionary adaptations through gene family expansion, distinguishing Madhuca longifolia from closely related tropical species. The findings provide a foundational genomic resource for future research and breeding programs aimed at improving cold tolerance, thus supporting the wider cultivation and conservation of this valuable multipurpose tree species.

14.6 Primary Processing: Postharvest Handling and Processing Techniques

Postharvest handling and processing of Mahua are crucial to maintaining the quality of its flowers, seeds, and oil, which have diverse applications in food, medicine, and industry. Proper postharvest practices ensure minimal loss of valuable nutrients and bioactive compounds, enhancing the market value of Mahua products (Lungade and Karadbhajne 2023). The primary processing of Mahua begins with the collection of flowers, typically carried out manually during their peak flowering season. To maintain quality, flowers are collected early in the morning to prevent spoilage due to high temperatures (Mishra and Poonia 2020). Once collected, they are spread thinly on a clean surface in a shaded area or under solar dryers to prevent fermentation and microbial growth. Proper drying is essential to reduce the moisture content to an optimal level for storage, typically below 10%. Traditional sun drying remains the most common method, but modern solar dryers are increasingly used for better efficiency and hygiene (Mishra and Poonia 2020). After drying, the flowers are sieved to remove impurities like dust, stones, and foreign materials. They are then stored in airtight containers or jute bags in a cool, dry environment to prevent moisture absorption and fungal contamination. Proper storage extends their shelf life and ensures their suitability for various uses, including liquor production and sweetener preparation (Mishra and Poonia 2020; Behera et al. 2016; Kulkarni and Mehta 2020).

The seeds of Mahua, which are rich in oil, are processed separately. After collection, seeds are sun-dried to reduce their moisture content, which facilitates better oil extraction. Traditional methods of oil extraction involve crushing the seeds using wooden or metal presses. However, modern techniques such as mechanical expellers and solvent extraction methods are now used for higher efficiency and yield. The crude oil extracted is filtered and refined, depending on its intended use, whether as edible oil, in cosmetics, or for biodiesel production (Hegde et al. 2019).

Efficient pre- and post-harvest technologies are essential for maintaining the quality and economic value of Mahua (*Madhuca indica* or *Madhuca longifolia*) flowers, seeds, and oil. These technologies focus on minimizing losses, preserving bioactive compounds, and ensuring optimal conditions for storage, drying, and transportation. Preharvest practices directly impact the postharvest quality of Mahua. Trees should be maintained through regular pruning, pest management, and soil fertility enhancement to ensure high-quality produce. Avoiding the use of harmful chemicals close to the harvesting period is crucial to ensure the safety of flowers and seeds for consumption or processing. Lungade P and Karadbhajne SV study the *rationalization of Post-Harvesting Unit Operation of Mahua* by investigating the impact of improved postharvest practices on the collection and processing of Mahua flowers, a vital non-timber forest product for tribal communities. Traditionally, Mahua flowers are collected by burning the undergrowth beneath the trees, a method that poses environmental risks such as forest fires and loss of biodiversity. The study introduces a net collection technique as a sustainable alternative. This method not only mitigates environmental hazards but also enhances economic outcomes for tribal collectors. Specifically, flowers collected using nets fetched significantly higher market prices (₹110–₹120 per kg) compared to those gathered through traditional means (₹25–₹30 per kg). Additionally, the net collection approach reduced labor and time requirements by approximately 80%. The research also highlights the potential for value addition through the development of various Mahua-based food products, including syrups, cookies, jams, spreads, jellies, gummies, energy drinks, and hard-boiled candies, utilizing 10–60% Mahua content. These interventions not only improve the livelihoods of tribal communities by increasing income and reducing labor but also promote sustainable harvesting practices and product diversification. Behera et al. (2016) explore the multifaceted role of the Mahua tree in the lives of the Santhal tribe residing in Chandbill village, located on the northern border of the Similipal Biosphere Reserve in Odisha. Mahua (*Madhuca latifolia*) is integral to the Santhal community, serving as a source of food, beverage, and medicine. The edible flowers are traditionally fermented to produce a distilled alcoholic beverage known locally as "mahuli," which contains an alcohol content of 30–40%. This beverage is commonly consumed alongside rice, forming a staple part of their diet. Beyond its use as a drink, various parts of the Mahua tree—including flowers, seeds, and bark—are utilized for their medicinal properties, addressing a range of health issues within the community. The research emphasizes the deep-rooted traditional knowledge possessed by the Santhal tribe regarding the Mahua tree. Elders in the community act as custodians of this knowledge, ensuring its transmission to younger generations. The study highlights the importance of preserving this ethnobotanical wisdom, which is crucial for the sustainability of tribal livelihoods and the conservation of biodiversity in the region. Kulkarni and Mehta (2020) explore an innovative approach to address the challenges associated with traditional Mahua flower collection methods in Central India. Traditionally, collectors burn the undergrowth beneath Mahua trees to facilitate flower collection, a practice that often leads to uncontrolled forest fires, posing significant threats to biodiversity and the environment. To mitigate these risks, the study introduced the use of Mahua collection sheets—simple nets made from repurposed materials like old saris—deployed beneath Mahua trees to catch falling flowers. Trials conducted in the Melghat Tiger Reserve demonstrated that this method significantly reduced the time required for flower collection from an average of 6.78 h to 1.58 h per day. Additionally, the use of collection sheets minimized the need for manpower, reduced encounters with wild animals, and eliminated the necessity of burning undergrowth, thereby preventing forest fires. The intervention not only enhanced the efficiency and safety of Mahua flower collection but also contributed to environmental conservation and improved livelihoods for forest-dwelling communities.

Other parts of the Mahua tree, such as leaves and bark, may also be processed for medicinal purposes. Leaves are typically dried and ground into powder for herbal formulations, while bark is sun-dried and used in traditional medicine. Throughout the processing chain, maintaining hygiene and adhering to quality standards is vital to ensure the safety and efficacy of Mahua products (Dave et al. 2018; Roat et al. 2023).

14.7 Value Addition and Products: Potential for Creating High-Value Products

The Mahua tree (*Madhuca indica*) has been a cornerstone of traditional livelihoods and cultural practices for centuries. Its historical significance is deeply woven into the fabric of rural and tribal communities in India, where it is revered as a sacred and life-sustaining tree. Mahua (Madhuca indica or Madhuca longifolia) offers immense potential for value addition, thanks to its multifaceted uses across food, medicinal, and industrial sectors. By leveraging advanced processing techniques, modern research, and sustainable practices, various high-value products can be developed, improving both the economic output and social impact of this tree. Singh et al. emphasize Mahua's role as a vital non-timber forest product (NTFP) that contributes to the livelihoods of tribal communities. The study discusses traditional uses of Mahua flowers in food, beverages, and medicine, highlighting their nutritional and therapeutic properties (Dave et al. 2018). It also addresses the economic potential of Mahua through value-added products and the importance of sustainable harvesting practices to ensure ecological balance. Kumari et al. focus on developing and accessing a nutritious, value-added product using dried Mahua (*Madhuca longifolia*) flower powder. Recognizing the underutilization of Mahua flowers beyond traditional liquor production, the researchers aimed to explore alternative food applications to enhance their value and promote broader consumption (Hegde et al. 2019).

Mishra and Poonia (2020) and Hegde et al. (2019) explained that Mahua flowers, with their natural sweetness, can serve as sugar replacers in the development of functional food products. The fruits, when minimally processed, offer a nutritious vegetable option, while the seed oil holds promise as a butter substitute or a component in blended oils. To truly empower tribal livelihoods and integrate this versatile tree into mainstream industry, it is essential to recognize and promote the Mahua tree as a vital economic resource across India's forest regions (Fig. 14.2), Ahirwar et al. (2023) developed a sweet snack by combining dried Mahua flower powder with roasted peanuts and jaggery. The preparation involved collecting and cleaning Mahua flowers, drying and grinding them into a fine powder, roasting peanuts and grinding them separately, preparing a jaggery syrup, and mixing these components to form the final product. This confectionery item was then subjected to sensory evaluation by a panel of trained evaluators using a 9-point hedonic scale, assessing attributes such as taste, appearance, texture, and overall acceptability. The sensory evaluation results indicated that the product was well-received, particularly in terms of taste and texture, suggesting its potential appeal to consumers. The study concludes that Mahua flowers can be effectively utilized to create palatable and nutritious food products beyond their traditional use in liquor production. Developing such value-added products can offer new avenues for income generation among tribal communities and promote the sustainable use of Mahua tree.

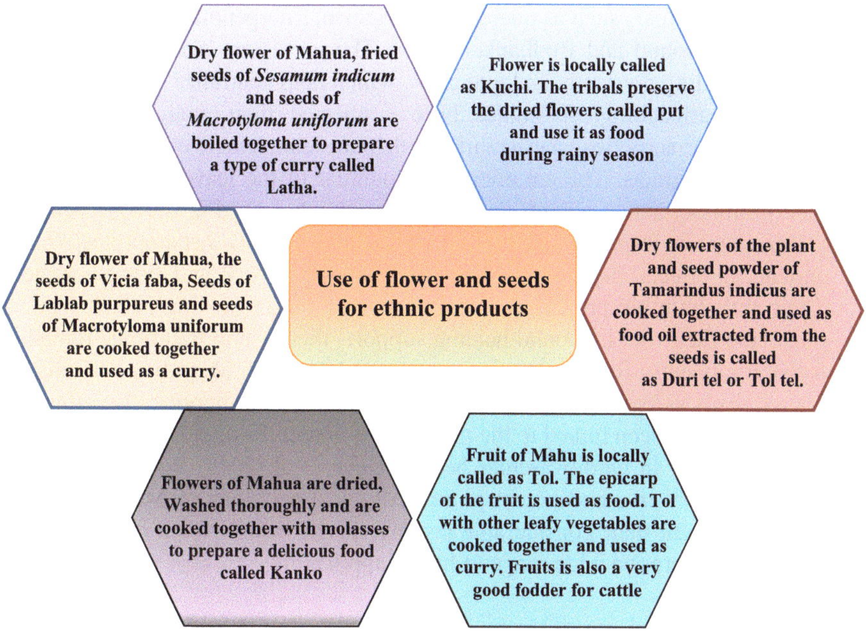

Fig. 14.2 Different parts of Mahua have specific commercial value

- *Mahua Leaves*
- Mahua leaves are not only used in traditional medicine but also as fodder for livestock. Nutritionally, they offer a source of crude fiber and have been reported to contain trace amounts of proteins, flavonoids, and micronutrients such as zinc and manganese. In medicine, leaves are used for their anti-inflammatory and wound-healing properties, indicating the presence of bioactive compounds beneficial to health (Dave et al. 2018; Roat et al. 2023).
- The leaves of Madhuca indica exhibit a broad range of pharmacological properties, underscoring their significant therapeutic potential. One of the most prominent activities is their antifertility effect, demonstrated by a marked reduction in fertility in rats following a 20-day oral administration of methanolic leaf extract (Shivabasavaiah et al. 2011). These leaves also possess notable antifungal and antibacterial activities, with both aqueous and methanolic extracts showing strong efficacy against various pathogenic microorganisms (Nimbekar et al. 2012). In the context of metabolic disorders, the leaves have been shown to exert potent antihyperglycemic effects, indicating their potential role in diabetes management (Khan et al. 2011). Their antiulcer activity is well documented, with studies demonstrating that aqueous and methanolic extracts, as well as flavonoids like 3,5,7,3′,4′-pentahydroxy flavone, contribute to ulcer healing by reducing ulcerated areas, decreasing pepsin content, and increasing gastric pH (Nimbekar et al. 2012; Mohod et al. 2016). Additionally, M. indica leaves offer protection against naproxen-induced gastric antral ulcers, likely due to the pres-

ence of compounds such as quercetin, myricitrin, triterpenoids, β-sitosterol, and quercitrin (Mohod and Bodhankar 2013). The leaves are also rich in antioxidants, including myricetin and quercetin, which help neutralize oxidative stress.

- Further pharmacological studies have confirmed their analgesic and anti-inflammatory activities, particularly with petroleum ether, ethyl acetate, and methanolic extracts. The antimicrobial potential of the leaves is supported by their inhibitory action against microbial growth when used in aqueous, acetone, and methanolic forms. Beyond these, the leaves also show hepatoprotective and anticancer effects, especially when used alongside bark and flower extracts. Their anthelminthic properties make them valuable in treating parasitic infections, and their role in wound healing supports their use in traditional remedies for skin ailments. Toxicity assessments have shown that M. indica leaf extracts are relatively safe at therapeutic doses. Moreover, the cardioprotective effects of the leaves have been linked to the presence of potent flavonoids, further enhancing their value in cardiovascular health (Dave et al. 2018; Roat et al. 2023). Overall, the diverse and potent pharmacological activities of Madhuca indica leaves highlight their considerable promise in the development of plant-based therapeutics across a wide range of health applications.
- *Mahua Flowers*
- Mahua flowers are one of the most nutritionally rich and widely used parts of the tree. When fresh or dried, they serve as a source of food and fermented beverage. They are rich in sugars (up to 60–70%), particularly glucose, fructose, and sucrose, which make them a valuable energy source. Besides sugars, the flowers also contain moderate amounts of proteins, vitamins (particularly vitamin C and B-complex), and essential minerals such as calcium, phosphorus, and iron. Due to their high carbohydrate content, Mahua flowers are also used in alcohol production through natural fermentation, especially in tribal communities. Tribal communities have relied on Mahua flowers as a vital food source, especially during periods of food scarcity. Mahua is a versatile tree species that plays a vital role in the livelihood and local economy of tribal communities in Odisha, offering a wide array of food products, traditional medicines, and other useful resources. It is frequently collected from natural habitats and is also commonly grown in tropical regions. There exists a widespread misconception that Mahua is solely associated with the production of liquor and thus considered harmful to health; however, this association arises only after fermentation. The present review focuses on the overall characteristics and chemical composition of *Madhuca longifolia*, highlighting its broad economic value, including its therapeutic and other practical applications, while drawing attention to its most significant biological activities (Patel et al. 2012). The flowers, which are sweet and fleshy, are consumed in multiple ways—either eaten fresh, dried, ground into powder, or mixed with flour for culinary preparations, used as a sweetener or fermented to make alcohol (Fern 2014; Fig. 14.3).

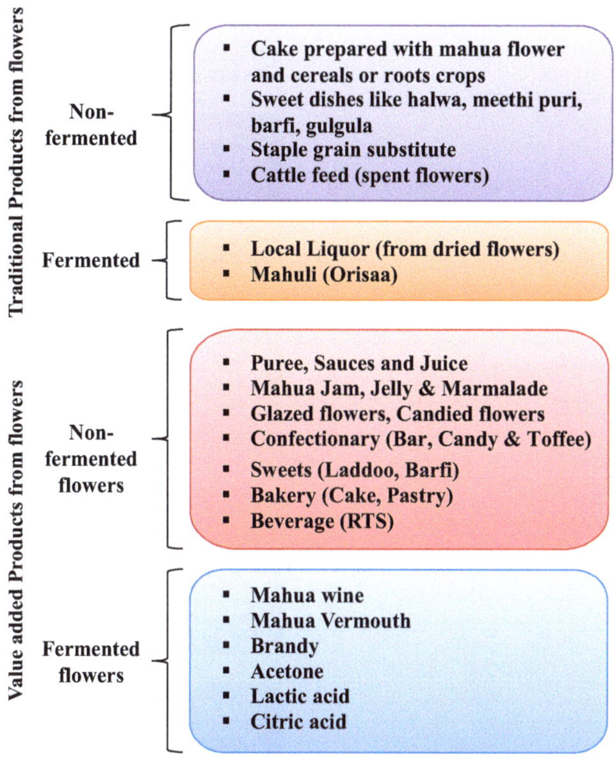

Fig. 14.3 Different value-added product prepared from Mahua

1. *Alcoholic Beverages*

The fermentation of Mahua flowers produces a distinct alcoholic beverage that has cultural and economic importance. This traditional drink is often central to social gatherings, festivals, and rituals in tribal societies. At the stage when Mahua flowers reach full maturity and are about to fall, they contain the highest concentration of total sugars. A significant portion of these dried flowers is traditionally utilized in the preparation of indigenous alcoholic beverages or country liquor (Bisht et al. 2018; Suryawanshi and Digambar 2020).

Medicinal Applications

Various parts of the Mahua tree have been used in traditional medicine. The flowers are known for their cooling and tonic properties, while the seeds yield oil with applications in treating skin ailments and rheumatism. The bark is used for its astringent properties to address wounds and infections. Mahua, a plant native to India, possesses immense therapeutic value and diverse potential applications; however, its full benefits remain underutilized due to a general lack of awareness among the population. Various parts of the Mahua tree exhibit significant pharmacological properties and are traditionally employed in the management of numerous health conditions. These include ailments such as tuberculosis, rheu-

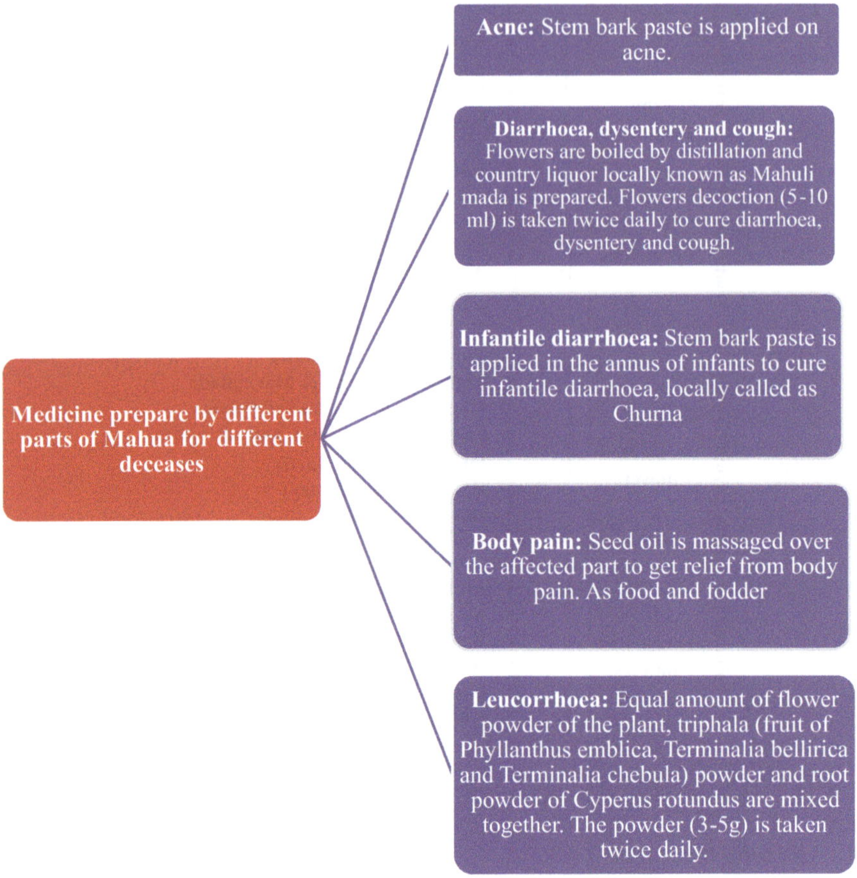

Fig. 14.4 Medicine prepared by different parts of Mahua for different diseases

matoid arthritis, cholera, paralysis, snake bites, general weakness, tonsillitis, influenza, hemorrhoids, joint pain, worm infestations, headaches, flatulence, and various infections. Additionally, Mahua is valued for its role in blood purification and is also traditionally used as a remedy against poisoning. The Madhuca longifolia commonly known as Mahua is an important economic plant growing throughout the subtropical region of the Indo-Pakistan subcontinent (Siddiqui et al. 2010; Dave et al. 2018; Fig. 14.4).

2. *Oil Extraction and Honey*

   Mahua seeds are a rich source of oil, traditionally extracted and used for cooking, lighting lamps, and as a base for making soaps and ointments. This oil is also valued in Ayurvedic medicine. Mahua is recognized as an oil-producing plant, with its seeds containing an oil content ranging from 35% to 47% (Ratnabhargavi 2013). The extracted Mahua oil is traditionally utilized in the production of soaps and candles, and it also serves as a natural preservative to protect seeds from pest infestation (Orwa et al. 2009). Oil produced by small-scale or marginal farmers is often of inferior quality and is commonly used as a substitute or adulterant for ghee (Fern 2014). In India, the projected potential for Mahua oil production has been estimated to reach as high as 60 million tons per year (CJP 2007). The by-product of oil extraction, known as oil cake, is frequently applied as an organic fertilizer. Due to its high saponin content, it also exhibits potential in managing root-knot nematodes and fungal diseases by inhibiting phytopathogenic organisms (Orwa et al. 2009). Additionally, Mahua is valued for its robust, durable, and dense reddish-colored wood, making it suitable for timber purposes (Orwa et al. 2009). The nectar-rich flowers of Mahua are also an important resource for honey bees, especially during times when alternative floral sources are scarce (Singh and Upadhyay 2008).

3. *Fodder, Animal Feed, and Fuel*

   The leaves of the Mahua tree are used as fodder for livestock, and its wood serves as a sustainable source of fuel for rural households. The fleshy outer covering of Mahua fruit is traditionally consumed as a vegetable by tribal communities. In times of food scarcity in India, a mixture of Mahua flowers and Sal seeds (*Shorea robusta*) is boiled and used as a substitute for staple grains (Sunita and Sarojini 2013). Mahua has been widely recognized for its diverse uses in traditional healing systems and its contribution to environmental sustainability. Various parts of the tree, such as leaves, flowers, and fruits, are commonly fed to livestock, particularly goats and sheep, through regular lopping practices (Singh and Upadhyay 2008). The oil cake derived from Mahua seeds—commonly referred to as Mahua seed cake in India—is also used as animal feed, especially in regions where oil extraction is prevalent. However, due to its high saponin content, this seed cake is generally considered to have adverse effects on livestock health and productivity (Singhal et al. 1986). Residual flowers left over from alcohol production are occasionally repurposed as animal feed. Indigenous communities have also used Mahua oil for lighting lamps and as a natural hair setting agent (Trifed.nic.in 2022). Additionally, Nayak and Pattanaik (2014) notes that Mahua oil has been employed as a biodiesel source. In several areas of Bihar, especially villages within the Siwan district, Mahua flowers are sun-dried, ground into flour, and used in the preparation of different types of bread. The species is cultivated in warm and humid climates for its seeds—which yield anywhere from 20 to 200 kg annually per tree depending on age—along with its flowers and timber. The fat extracted from the seeds, which solidifies at room temperature, is applied in skin care and is used in the production of soaps, detergents, and as a form of vegetable butter. The seed cake left after oil extraction also serves as an effective organic fertilizer. Medicinal properties are attributed to various parts of the Mahua tree, including its bark. The tree holds sacred significance for many tribal communities due to its wide-ranging benefits. Interestingly, alkaloids found in the seed press cake have been traditionally used to stun or kill fish in aquaculture ponds in certain regions of India. The practice not only aids in harvesting fish but also enriches the pond soil; the pond is subsequently dried, refilled, and restocked with fish fingerlings (Pillay and Kutty 2005). While Mahua seed cake is often regarded as a suboptimal feed component for ruminants, negatively affecting protein digestion and growth performance (Singhal et al. 1986), some studies have reported acceptable tolerance levels. In particular, trials have shown that when the seed cake is water-washed, it can be incorporated into ruminant diets at levels up to 20% without adverse effects (EFSA 2009). The palatability of unprocessed Mahua seed cake was found to be very poor but improved notably after a period of adaptation (Singhal et al. 1986). In lactating Murrah buffaloes, processed Mahua seed cake was incorporated into the concentrate mixture to replace groundnut cake at 0%, 25%, 50%, 75%, and 100%. DM intake and nutrient digestibility did not differ between the five treatments. Milk fat, solids-corrected milk and solids-not-fat yields, percentage of fat, lactose, ash and total solids, and gross energy values of milk did not differ between the five groups (Sharma et al. 2019). Despite its availability, several experimental studies have indicated that livestock fed with Mahua seed cake tend to show poorer performance compared to animals maintained on standard diets (Singhal et al. 1986). In one study, crossbred calves aged between 12 and 14 months that were provided a complete diet incorporating 20% Mahua seed cake demonstrated improved weight gain; however, their crude protein digestibility was found to be lower than that of animals on a conventional diet (EFSA 2009). Interestingly, other research has shown that the inclusion of Mahua seed cake did not significantly influence nutrient intake, digestibility, or overall energy utilization in animals (Inamdar et al. 2015).

Biodiesel Production and Industrial Use of Mahua Oil

In recent years, biodiesel has garnered significant interest due to its environmentally friendly characteristics and its status as a renewable energy source, in contrast to conventional diesel, which is derived from finite fossil fuels and poses the risk of eventual depletion. Among the various feedstocks available, Mahua oil—an inedible oil abundantly found in tropical regions such as India—has shown promising potential for biodiesel production. Classified as a nonedible oil crop, Mahua presents two key advantages over conventional edible oil sources used for biodiesel. Firstly, due to the presence of naturally occurring toxic compounds, these nonedible oil crops are not fit for human consumption, thereby avoiding conflicts with food security that often arise from diverting edible oils for fuel purposes. Secondly, such crops are capable of thriving and yielding satisfactory outputs even in suboptimal agricultural conditions, including nutrient-poor soils, limited rainfall or water availability, minimal fertilizer input, and extreme temperature fluctuations. These traits make them highly adaptable and cost-effective for biodiesel production, requiring fewer intensive cultivation practices compared to edible oil crops (Bankovic-Ilic et al. 2012).

14.8 Market Opportunities and Economic Impact

The value addition to Mahua products can lead to enhanced economic benefits, particularly in rural areas where the tree is abundantly available. By creating high-value products, communities can generate employment opportunities, boost local economies, and improve livelihoods. Sustainable value chains, from cultivation to processing, ensure that both small-scale producers and larger industries benefit from Mahua's diverse applications. Additionally, integrating traditional knowledge with modern technologies can position Mahua as a key ingredient in the global market for natural, sustainable, and health-focused products. Hegde et al. (2019) discuss the challenges faced in the Mahua trade, including low market prices and lack of developed market infrastructure. To address these issues, the authors advocate for the establishment of financially viable enterprises based on ecological sustainability and community empowerment. They suggest that with proper training in systematic collection, grading, primary processing, and marketing of Mahua products, tribal communities can significantly benefit economically. In conclusion, the underscores the importance of recognizing and promoting the diverse applications of Mahua resources. By doing so, it is possible to enhance the livelihoods of tribal communities, conserve biodiversity, and contribute to the rural economic growth of the country.

14.9 Conclusion

Plants are vital economic resources, serving as the basis for many well-established medicines. Among them, *Madhuca longifolia* stands out for its wide-ranging potential and multifaceted uses. Given its adaptability and value, cultivating Mahua on a large scale—especially on unproductive and wastelands—could significantly support the livelihoods of poor and landless families. Traditionally recognized primarily for its use in liquor production, the Mahua tree's extensive medicinal properties have remained largely unrecognized by the general public. To harness its full potential, there is an urgent need to shift perceptions and promote the propagation of high-quality Mahua trees through plant tissue culture techniques such as micropropagation. Collaborative efforts between researchers and rural communities can facilitate knowledge exchange and enhance conservation. The coming generations are likely to place increased importance on this versatile tree due to its therapeutic efficacy, accessibility, affordability, and low toxicity. Historically, the Mahua tree has traversed centuries of Indian tradition—revered in Vedic texts, used as a famine food during colonial times, and cherished as a symbol of tribal resilience and ecological wisdom. Today, its significance as both a cultural emblem and a sustainable resource underscores the need for renewed attention, conservation, and responsible utilization in the face of environmental and societal change.

Acknowledgments The authors extend their appreciation to the Researchers Supporting Project No. (BUAT/DR/2020/282/9) at Bonda University of Agriculture and Technology, Bonda.

References

Ahirwar SD, Singh N, Sagar P (2023) Development and sensory evaluation of value added product from Mahua (*Madhuca longifolia*) flower. Int J Adv Res Innov Ideas Educ 9(3):3878–3881

Akkenapally S, Mudalkar S, Bodiga S, Sundaram R, Bokam AK (2024) In vitro callus generation and somatic embryogenesis from leaf explant of *Madhuca longifolia* var. latifolia (Roxb.) A. Chev. Vegetos. https://doi.org/10.1007/s42535-024-01044-6

Anon (1898) The staple crops of Khasi and Jaintia hills. Agric Ser 3:18

Anonymous (2002) Annual report. Central Institute for Arid Horticlture, Bikaner, p 31

Arumugam N, Rahman H (2015) Termite fauna (Blattodea: Termitoidae) of Mahua, Crocker Range Park, Sabah, Malaysia. Malay Nat J 67(4):403–411

Bankovic-ILic IB, Olivera S, Stamenkovic, Vlada BV (2012) Biodiesel production from non-edible plant oils. Renew Sust Energ Rev 16(6):3621–3647

Behera S, Ray RC, Swain MR, Mohanty RC, Biswal AK (2016) Traditional and current knowledge on the utilization of Mahua (*Madhuca latifolia* L.) flowers by the Santhal tribe in Similipal biosphere reserve, Odisha, India. Ann Trop Res 38(1):94–104. https://doi.org/10.32945/atr3818.2016

Bisht V, Neeraj, Solanki VK, Dalal N (2018) Mahua an important Indian species: a review. J Pharmacogn Phytochem 7(2):3414–3417

Caritasindia (n.d.). https://www.caritasindia.org/mahua-a-reliable-source-of-tribal-livelihood/

Chadha KL, Singh SK, Kumar AT (2001) Handbook of horticulture, vol 1. Published by Indian Council of Agriculture Research. ISBN 978-81-7164-187-1

Chatterjee R, Mukherjee SK, Paul B, Chattopadhyaya S (2021) Comparative spectroscopic analysis, performance and emissions evaluation of Madhuca longifolia and Jatrophacurcas produced biodiesel. Environ Sci Pollut Res 28:62444–62460

CJP (2007) Madhuca indica the new source of biodiesel. Centre Jatropha Promotion Biodiesel, Churu, Rajasthan

Damodaran V (1995) Famine in a forest tract: ecological change and the causes of the 1897 famine in Chotanagpur, Northern India. Environ Hist 1(2):129–158

Damodaran V (2006) Famine in Bengal: a comparison of the 1770 Famine in Bengal and the 1897 Famine in Chotanagpur. Medieval Hist J 10(1–2):143–181

Dave JP, Kumar V, Kumar A, Gat Y, Suri S, Sharma K (2018) Mahua: a boon for pharmacy and food industry. Curr Res Nutr Food Sci J 6(2):498–507. https://doi.org/10.12944/CRNFSJ.6.2.12

Devi PS, Thangam M, Desai AR, Adsule PG (2002) Studies on variability in physico-chemical characters of different jamun (Syzygium cumini Skeels) accessions from Goa. Indian J Hortic 59:153–156

Dobriyal MJ, Surendran C, Parthiban KT, Sashikumar K (2002) Macropropagation of *Madhuca latifolia*. Indian J For 25(2):180–184. https://doi.org/10.54207/bsmps1000-2002-V5Y4Y0

Dwivedi A, Priyadarshini A, Induar S (2022) Mahua (Madhuca longifolia) flower and its application in food industry: a review. Int J Chem Stud 10(1):80–84

EFSA (2009) Saponins in Madhuca longifolia L, as undesirable substances in animal feed. EFSA J 979:1–36. https://doi.org/10.2903/j.efsa.2009.979

Fern K (2014) Madhuca longifolia. Useful Tropical Plants Database

Garai S, Ranjan A, Rahaman SK, Khatun M, Kumar R, Kumar B, Tiwari S (2021) Climate change and habitat adaptability of Madhuca longifolia (J. Koenig ex L.) J.F. Macbr.: an endangered tree species of high socio-economic significance to tribal community of eastern India. Indian J Ecol 48(4):1062–1067

Gogate MG, Mittal RC (1983) Tree worship based on astrology-a tool for agro-forestry. In: Mathur RS, Gogate MG (eds) Agroforestry in India. National Workshop on Agroforestry, Karnal, Haryana, pp 463–473

Harsh NSK, Gupta BN (1993) Root disease in forest nurseries of Madhya Pradesh and their control. TFRI, Information Series No 1, p 31

Harsh NSK, Dadwal VS, Jamaluddin (1992) A new post emergence damping off disease of Eucalyptus seedling. Indian Forester 118:279–283

Hegde HT, Gunaga RP, Thakur NS (2019) Current trends and future prospects for utilization of Mahua resources. Van Sangyan 6(2):7–11. Published by Tropical Forest Research Institute, Jabalpur, Madhya Pradesh, India. Available from: https://www.researchgate.net/publication/332751536_Current_trends_and_future_prospects_for_utilization_of_mahua_resources. Accessed 20 May 2025

Inamdar AI, Chaudhary LC, Agarwal N, Kamra DN (2015) Effect of Madhuca longifolia and Terminalia chebula on methane production and nutrient utilization in buffaloes. Anim Feed Sci Technol 201:38–45

Ishan Dubey I, Chhajed M, Chourasiya R, Chouhan PS, Walvekar A, Bhandari Y, Verma K (2023) A review of the botanical, conventional applications, phytochemical constituents, and pharmacology of Madhuca longifolia (Koenig) J.F. Macb. Pharmacogn Rev 17(34):392–405

Jayasree B, Harishankar N, Rukmini C (1998) Chemical composition and biological evaluation of Mahua flowers. J Oil Technol Assoc 30:170–172

Karale AR, Wagh AP, Pawar BG, More TA (1999) Association of fruit characters in tamarind. J Maharashtra Agric Univ 24(3):319–320

Karunarathna HGMK, Eeswara JP, Iqbal MCM (2023) Effect of seed treatments on seed germination and seedling growth of *Madhuca longifolia var. longifolia* ("Mee"). Trop Agric Res 34(3):188–199

Keskar BG, Karale AR, Dhawale BC, Chaudhary KG (1989a) Improvement of tamarind by selection. Maharashtra J Hort 4:121–124

Keskar BG, Karale AR, Ohawale BC, Chaudhary KG (1989b) Improvement of jamun by selection. Maharashtra J Hort 4:117–120

Khan S, Zahan D, Das R, Nasrin D, Ahsan S, Ahmed R, Rahmatullah M (2011) Antihyperglycemic activity studies with methanol extract of Madhuca indica JF Gmel. leaves and Paederia foetida L. stems in mice. Adv Nat Appl Sci 5(2):122–126

Kulkarni J, Mehta PC (2020) Mahua flower collection sheets: a simple intervention for facilitating collection of Mahua flowers and reducing forest fire. Indian Forester 146(9):843–847. https://doi.org/10.36808/if/2020/v146i9/154535

Kulkarni PS, Sharanappa G, Ramesh MR (2013) Mahua (Madhuca indica) as a source of biodiesel in India. Int J Sci Eng Res 4(7):2319–2329

Kumar A, Singh S (2018) A new report of leaf vein gallon madhucalongifolia (sapotaceae) from India. Indian Forester 144(4):390–393. http://www.indianforester.co.i3

Kuthadi RA (2024) Mahua flower through the ages. Sampratyaya 1(Spl. Issue, No-3):52–58. https://doi.org/10.21276/smprt.202409.1s3.a6

Lal HS, Singh S, Mishra PK (2014) Trees in Indian mythology. Discovery 12(29):16–23

Lungade P, Karadbhajne SV (2023) Rationalization of post-harvesting unit operation of Mahua flowers (Madhuca longifolia, Madhuca indica): systematic interventions and benefits for tribals. Progress Plants Agric Sci 6(1):Article 2224. https://doi.org/10.18686/ppas.v6i1.2224

Meshram PB, Jamaluddin, Pathak SC (1991) A new report of slug caterpillar, Miresaalbipuncta Herr-Schaff (Lepidoptera: Limacodidae) as a pest of Mahua, Madhuca latifolia. Indian J Appl Pure Biol 6(1):79

Mishra A, Poonia A (2020) Mahua (*Madhuca longifolia*) flowers: review on processing and biological properties. Centre of Food Science and Technology, Banaras Hindu University, Varanasi. Retrieved from https://www.researchgate.net/publication/338682193_Mahua_Madhuca_longifolia_flowers_review_on_processing_and_biological_properties. Accessed 20 May 2025

Mohod SM, Bodhankar SL (2013) Antiulcer activity of aqueous extract of leaves of Madhuca indica JF Gmel against naproxen induced gastric mucosal injury in rats. J Acute Dis 2(2):127–133

Mohod SM, Kandhare AD, Bodhankar SL (2016) Gastroprotective potential of Pentahydroxy flavone isolated from Madhuca indica JF Gmel. leaves against acetic acid-induced ulcer in rats: the role of oxido-inflammatory and prostaglandins markers. J Ethnopharmacol 182:150–159

Nayak SK, Pattanaik BP (2014) Experimental investigation on performance and emission characteristics of a diesel engine fuelled with mahua biodiesel using additive. Energy Procedia 54:569–579

Nayak S, Sahoo UK (2020a) Tree diversity and ecological status of Madhuca latifolia (Roxb.) J.F. Macbr in forests of Odisha. Indian J Ecol 47(1):138–149

Nayak S, Sahoo UK (2020b) Effect of seed source on germination and early growth in *Madhuca latifolia* in Odisha. Int J Ecol Environ Sci 46(2):133–140

Nimbalkar SD, Chavan LN, Agale SJ, Kauthale VK (2018) Madhuca indica JF Gmel. (Mahua) candidate tree selection based on the perception of tribal community and scientific validation. Environ Conserv J 19(1&2):161–165

Nimbekar T, Bais Y, Katolkar P, Wanjari B, Chaudhari S (2012) Antibacterial activity of the dried inner bark of Madhuca indica JF GMEL. Bull Env Pharmacol Life Sci 1(2):26–29

Omukhua GE, Godwin-Egein MI (2011) Root rot disease of five fruit tree seedlings in the nursery. J Agric Soc Res 11(1):191–195

Orwa C, Mutua A, Kindt R, Jamnadass R, Simons A (2009) Agroforestree Database: a tree reference and selection guide version 4.0. World Agroforestry Centre, Kenya. AgroforesTree: http://apps.worldagroforestry.org/treedb2/

Panda BB, Mishra PK, Thakur R (2010) Report of the study on Mahua sub sector. The Livelihood School, Bhopal

Patel PK, Prajapati NK, Dubey BK (2012) Madhuca indica: a review of its medicinal property. Int J Pharma Sci Res 3:1285–1293

Pathak H, Maru S, Satya HN, Silawat SC (2015) Fungal diseases of trees in forest nurseries of Indore, India. Plant Pathol Microb 6:8. https://doi.org/10.4172/2157-7471.1000297

Pillay TVR, Kutty MN (2005) Aquaculture: principles and practices, 2nd edn. Blackwell Publishing Ltd., Hoboken, p 623

Pohare CV, Choudhari SW, Mane VP, Shambharkar VB, Raut UA, Deshmukh HK, Taide YB (2016) Studies on survival percentage of softwood grafting in Mahua (*Madhuca indica* Gmel.J.F.) by using khirni (*Manilkara hexandra* Roxb.) as a root stock. Int J For Crop Improv 7(2):167–171

Pradhan M, Malakar A, Sinha A (2024) Appraisal of the potential habitat distribution of *Madhuca longifolia* manifested remarkable resilience under various socio-climatic scenarios pan-India. Model Earth Syst Environ 10:2435–2446. https://doi.org/10.1007/s40808-023-01913-0

Rai D, Mishra KK (2005) Studies on genetic divergence in Bael (Aeglemarmelos Correa). Indian J Hort 62:152–154

Rao AR, Singh BP (1996) Non-wood forest products contribution in tribal economy (A case study in south Bihar and southWest Bengal). Indian For 122:337–342

Ratnabhargavi AR (2013) Setting up of a Mahua oil extraction unit. Vidarbha Livelihood Forum, Oxfam India

Roat P, Hada S, Chechani B, Yadav DK, Kumar S, Kumari N (2023) Madhuca indica: a review on the phytochemical and pharmacological aspects. Pharm Chem J 57(2):284–295

Rout GR, Das P (1993) Micropropagation of Madhuca longifolia (Koenig) MacBride var. latifolia Roxb. Plant Cell Rep 12:513–516. https://doi.org/10.1007/BF00236098

Saldanha IM (1995) On drinking and 'drunkenness': history of liquor in colonial India. Econ Polit Wkly:2323–2331

Sehgal HS (1983) Disease problem of Eucalyptus in India. Indian Forester 109:909–816

Sharma P, Singh DK, Nayak S, Sharma DP (2019) Performance of lactating buffaloes fed on De-oiled Mahua (Madhuca indica) seed cake in Chhattisgarh Hills Region of Central India. Buffalo Bull 38(4):653–657

Shedage S, Aayate D, Dobriyal MJ (2024) Variability in seed morphology, germination, and seedling growth of Madhuca indica: implications for seed source selection. Agric Res 13(4):684–691

Shirin F, Bhadrawale D, Singh S, Panika S, Gupta T (2020) Effect of environmental factors on successful in vitro morphogenesis of Madhuca longifolia. Adv Res 21(11):9–19

Shivabasavaiah KR, Pavana T, Ramyashree MJ (2011) Anti-fertility Effect of Madhuca indica Leaves in Male Swiss Albino Rats. Pharm Res 4:323–326

Siddique MRH (2019) Nutrient dynamics in soil associated with leaf litter of Shorea robusta Gaertn. f. in Madhupur Sal forest of Bangladesh. Khulna Univ Stud 16:1–8

Siddiqui BS, Khan S, Kardar MN (2010) A new isoflavone from the Madhuca latifolia. Nat Prod Res 24:76–80

Singh AK, Ravishankar H (2010) Prospects of production of quality planting material of underutilized fruits. In: Abstract national conference on production of quality seeds and planting material health management in horticulture crops, New Delhi, pp 149–150, 11–14 Mar 2010

Singh S, Singh AK (2005) Genetic variability in jamun (Syzygium cumini Skeels) in Gujarat. Prog Hort 37:144–148

Singh S, Singh AK (2015) Standardization of time of softwood grafting in Mahua (*Bassia latifolia*) and khirni (*Manilkara hexandra*) under semi-arid environment of western India. Indian J Agric Sci 85(2):166–170. https://doi.org/10.56093/ijas.v85i2.46467

Singh RP, Upadhyay SK (2008) The beneficial effects of feeding Mahua (Bassia latifolia Roxb.) flower syrup to honey bee (Apis mellifera) colonies during periods of dearth. J Apic Res 47(4):261–264

Singh IS, Srivastava AK, Singh V (1999) Improvement of some under-utilized fruits through selection. J Appl Hort 1:34–37

Singh V, Singh J, Kushwaha R, Singh M, Kumar S, Rai AK (2020) Assessment of antioxidant activity, minerals and chemical constituents of edible Mahua (Madhuca longifolia) flower and fruit of using principal component analysis. Nutr Food Sci 51(2):387–411

Singhal KK, Chahil SM, Sharma DD (1986) Versatile Mahua (Madhuca indica) seed cake-a review. Agric Rev 7(2):105–123

Sinha J, Singh V, Singh J, Rai AK (2017) Phytochemistry, ethnomedical uses and future prospects of Mahua (Madhuca longifolia) as a food: a review. J Nutr Food Sci 7(573):2

Sunita M, Sarojini P (2013) Madhuca longifolia (Sapotaceae): a review of its traditional uses and nutritional properties. Int J Humanit Soc Sci Invent 2(5):30–36

Suryawanshi Y, Digambar M (2020) Variability studies in Madhuca longifolia var. latifolia flowers from Northern Western Ghats of India. Indian J Hill Farming 33(2):261–266

Suryawanshi YC, Mokat DN (2021) Morphophysiological seed variability in Mahua trees from Western Ghats and its impact on tribal life. Proc Natl Acad Sci India Sect B Biol Sci 91:227–239. https://doi.org/10.1007/s40011-020-01223-w

Torgeson DC, Young RA, Milbrath JA (1954) Phytophthora root rot diseases of Lawson cypress and other ornamentals. Agricultural Experiment Station Oregon State College Corvallis, Station Bulletin 537, pp 1–19

Trifed.nic.in (2022) Retrieved on Trees India Madhuca longifolia latifolia (Roxb.) A. Chev. India Biodiversity Portal

Verma P, Verma RK (2015) A new wilt disease of maida chhal (Litsea glutinosa) in central India. World J Pharm Res 4(3):1483–1491

Vir R, Kumar V (2021) Role of plant tissue culture in improvement of horticultural crops. Ann Hortic 14(1):1–6

Wang S, Lin H, Ye S et al (2024) High-quality chromosome-level genomic insights into molecular adaptation to low-temperature stress in Madhuca longifolia in southern subtropical China. BMC Genomics 25:877. https://doi.org/10.1186/s12864-024-10769-2

Yadav YS, Siddiqui AU, Parihar A (2005) Management of root-knot nematode *Meloidogyne incognita* infesting gram through oil cakes. J Phytol Res 18(2):263–264. http://www.cabi.org/cabdirect/FullTextPDF/2006/20063145825.pdf

Yadav S, Suneja A, Hussain Z, Abraham Z, Mishra SK (2011) Genetic variability and divergence studies in seed and oil parameters of Mahua (Madhuca longifolia Koenig) J.F. Macribide accessions. Biomass Bioenergy 35(5):1773–1778

Part V

Herbs and Functional Crops

Poppy Seeds (Khus Khus) (*Papaver somniferum*): The Tiny Seed of Flavor and Nutrient Density

15

Sivendra Joshi, K. T. Venkatesha, Anil Kumar Gupta, Rajendra Chandra Padalia, Rakesh Kumar Upadhyay, Dipender Kumar, Amit Chauhan, Amit Kumar Tiwari, and Sonveer Singh

Abstract

Poppy seeds (*Papaver somniferum* L.), also known as Khus Khus, are prized for their delicious flavor, high nutritional density, and versatility. Poppy is prized for their alkaloids (morphine and codeine). This plant was domesticated in Mediterranean, its cultivation dates back to 6500 BCE and has long been utilized for food, medicine, and hallucinogenic uses throughout Europe. Afghanistan is a major producer of opium, as well as European nations. Cultivation varies according to climate and farming practices, affecting alkaloid content. It is grown to produce alkaloids (morphine, codeine, and thebaine) used in pharmaceuticals. In India, it is predominantly cultivated in Uttar Pradesh, Madhya Pradesh, and Rajasthan states, and its cultivation is monitored by the Central Bureau of Narcotics (CBN) under Ministry of Finance, Government of India. Poppy seeds have high value potential for culinary, nutritional, and medicinal purposes. Poppy seeds contain protein (11.60%), fat (29.40%), carbohydrate (1.4%), and fiber (10.05%). They are high in vital fatty acids and antioxidants and can be used to make food, health supplements, and skincare products. Genetics, molecular breeding, CRISPR, and genomics advances have improved its characteristics.

S. Joshi · K. T. Venkatesha (✉) · R. C. Padalia · R. K. Upadhyay · D. Kumar · A. Chauhan
A. K. Tiwari · S. Singh
Division of Plant Breeding and Genetic Resources Conservation, CSIR-Central Institute of Medicinal and Aromatic Plants (CIMAP), Research Centre, Pantnagar, Uttarakhand, India
e-mail: venkatesha@cimap.res.in

A. K. Gupta
Division of Plant Breeding and Genetic Resources Conservation, CSIR-Central Institute of Medicinal and Aromatic Plants (CIMAP), Research Centre, Pantnagar, Uttarakhand, India

Division of Plant Breeding and Genetic Resources Conservation, CSIR-Central Institute of Medicinal and Aromatic Plants (CIMAP), Near Kukrail Picnic Spot, Lucknow, India

A. Kumar et al. (eds.), *Fasting Superfoods*,
https://doi.org/10.1007/978-981-95-2762-5_15

Sustainable farming, regulatory compliance, and market expansion can help India establish itself as major producer in global poppy seed sector.

Keywords

Poppy seed · Alkaloids · Morphine · Codeine · Fatty acids · Culinary uses

15.1 Introduction

Papaver somniferum, also known as opium poppy, belongs to Papaveraceae family and both ethnobotanically and agriculturally significant due to its seeds, alkaloidal compounds, particularly morphine and codeine (Agarwal et al. 2016; Celik et al. 2016). Morphological characteristics of *P. somniferum* include its erect, hairy, or glaucous stem, with an average height ranging from 80 to 120 cm; the leaves are large, alternate, and can vary from unlobed to pinnately lobed, typically showcasing a glaucous sheen (Ngernsaengsaruay et al. 2023; Kumar and Patra 2011). The flowers of the opium poppy exhibit a vast color range—white, yellow, violet, and blue—with each flower usually having four to six conspicuous petals and a predominant central capsule known for its unique structure, which contains seeds (Ngernsaengsaruay et al. 2023; Salavert et al. 2018). Figure 15.1 depicts morphological characteristics of plant.

The plant karyotype comprises a total of 22 chromosomes, categorized into two sets of 11, which display considerable variation in length and centromere positioning (Rasekh and Karimzadeh 2023). This chromosomal variability contributes to the diverse morphological traits in *P. somniferum*, leading some scholars to categorize various forms as subspecies or distinct morphotypes (Celik et al. 2016; Saunders et al. 2001). Previous findings indicate that differentiating between cultivated *P. somniferum* and its wild relative *P. setigerum* is complicated due to their considerable morphological and genetic similarities, which can sometimes lead to taxonomic ambiguities (Hong et al. 2021; Jesus et al. 2023).

This plant typically flowers from late spring to early summer, culminating in the formation of seed capsules (Ngernsaengsaruay et al. 2023). The capsules are bulbous, with an opening at the apex that allows for the dispersal of tiny seeds, commonly used for culinary purposes due to their oil content, which is rich in unsaturated fatty acids (Celik et al. 2016). During the growth phase, *P. somniferum* exhibits phenological variances based on environmental conditions and genetic backgrounds, suggesting that breeding practices may be influenced by climate and market demands for alkaloids and seed production (Ngernsaengsaruay et al. 2023; Hong et al. 2022).

Research has demonstrated that the genetic diversity of *Papaver somniferum* is extensive, with various accessions exhibiting different levels of alkaloid production and morphological traits (Hong et al. 2022; Celik et al. 2016). This species is believed to have been domesticated in the Western Mediterranean region, with its wild ancestor, *Papaver setigerum*, being indigenous to this area (Salavert et al.

Fig. 15.1 Opium plant: key parts and structure

2018; Jesus et al. 2021). Mediterranean region is considered as center of origin for opium poppy, with evidence suggesting that it was domesticated by Neolithic farmers. Archaeological evidence suggests that the earliest cultivation of opium poppy dates back to around 6500 BCE, with findings of wild poppy seeds in locations such as the submerged village of Atlit-Yam in Israel (Presley and Lindsley 2018). Additionally, charred seeds dated to 5270–4990 BC were discovered in Taï cave in South France, indicating that the cultivation of poppy was likely associated with early agricultural practices involving cereal grains (Jesus et al. 2023). The dispersal of *Papaver somniferum* across Europe is well-documented, with significant archaeological evidence from sites associated with the Linear Pottery Culture (5300–4900 BC) indicating its spread beyond the Mediterranean (Salavert et al. 2018). This suggests that the poppy was not only valued for its seeds but also for its psychoactive properties, which were recognized in ancient cultures. The Sumerians, for instance, referred as "Hul Gil," or "joy plant," around 3400 BC, highlighting its significance in early medicinal and recreational use (Mella-Raipán et al. 2020). The geographical distribution of poppy reflects a complex interplay of historical cultivation practices and modern agricultural trends. Afghanistan is recognized as the largest producer of opium poppy globally, primarily for the illicit opiate market. However, poppy seeds, which are distinct from opium production, are cultivated in several countries, including Czech Republic, Spain, Germany, France, Turkey and Croatia, where they are harvested and exported for culinary uses (Schuppener and Corliss 2018; Lošák et al. 2022). The cultivation practices and the geographical variations in the harvesting of poppy seeds have a profound impact on the alkaloid content of the seeds, particularly morphine and codeine, which can vary significantly based on factors such as method of harvest and specific variety (Sproll et al. 2006; Yamaguchi et al. 2011). This variability is crucial, as it affects not only the culinary uses of poppy seeds but also their implications in drug testing and potential for abuse (Chen et al. 2014; Kauppila and Eagen 2023). The seeds are often contaminated with latex from unripe capsules during harvesting, which can lead to the presence of these narcotic compounds in food products (Yamaguchi et al. 2011). Archaeological findings indicate that the cultivation of poppy may have begun in the western Mediterranean, where the wild progenitor of the opium poppy is indigenous (Jesus et al. 2023; Hong et al. 2022). The spread of poppy cultivation from this region to central and northern Europe has been documented, with various cultivars adapted to local climates and agricultural practices (Jesus et al. 2021). For instance, Czech Republic has a seed production tradition, which has positioned it as a significant exporter in Europe (Lošák et al. 2022). The cultivation of poppy seeds has several factors, including climatic conditions, soil quality, and agricultural techniques. Studies have shown that variations in these factors can significantly affect yield and poppy seed quality. For example, timing of sowing and density of planting can lead to substantial differences in seed yield, as demonstrated in research conducted in Pannonian climate conditions (Neugschwandtner et al. 2023). Additionally, the alkaloid content in poppy seeds, which can vary widely depending on the cultivar and harvesting methods, has implications for both food safety and culinary applications (Sproll et al. 2006; Samano et al. 2015).

In India, *Papaver somniferum* is mainly grown in Madhya Pradesh, Uttar Pradesh, and Rajasthan. Seeds are used in curries, breads, sweets, and confectionery (Srinivas and Narasinga Rao 1981). Poppy (*Papaver somniferum*) is highly significant medicinal plant widely recognized for pain-relieving properties (Prochazka and Smutka 2012). Poppy seed has around protein—11.60%, fat—29.40%, carbohydrate—1.4%, and fiber—10.05% (Kukade et al. 2015).

15.2 History

The plant is believed to have originated in western Mediterranean region, with its wild ancestor likely being *Papaver setigerum*, native to coastal Mediterranean areas (Pushpangadan et al. 2012; Machado et al. 2025; Jesus et al. 2023). Archaeological evidence shows poppy seeds and pods were present in Neolithic Europe as early as the 6th millennium BC, indicating early cultivation and use. By around 5000 BC, the opium poppy was already being cultivated in Sumer (ancient Mesopotamia, modern-day Iraq and Kuwait), where it was valued for its psychoactive properties and medicinal uses (Administration DE 1992; Carlin et al. 2020). The spread of the plant continued through the Balkans into Asia Minor and further across Europe, becoming one of the earliest domesticated crops outside the Fertile Crescent (Jesus et al. 2021; Salavert et al. 2020). The opium poppy has been cultivated for multiple purposes throughout history. Its most notable products are opium and poppy seeds.

Opium, derived from latex of poppy capsule, has been used since antiquity for its analgesic and sedative properties. Ancient Sumerians referred to poppy as "joy plant," and both Egyptians and Greeks utilized opium for medicinal and ritualistic purposes (Administration DE 1992). Poppy seeds, in contrast, are nonnarcotic and have been used as a food source since ancient times. They were eaten raw, used to make porridge, pressed for edible oil, and incorporated into cakes and breads. In the modern era, poppy seeds are a staple in many cuisines, valued for their flavor and nutritional content.

Cultural significance of *Papaver somniferum* is profound and multifaceted. In addition to its role as a food and medicine, the opium poppy has deep symbolic and ritual associations. In many ancient cultures, the poppy was linked to sleep, dreams, and death, owing to its sedative effects—reflected in its Latin name "somniferum," meaning "sleep-bringing" (Administration DE 1992).

The plant's psychoactive properties made it a subject of both reverence and controversy. It played a central role in religious ceremonies, healing rituals, and even as a symbol of fertility and rebirth in some societies. Over time, the cultivation of opium poppy also became intertwined with global trade, colonial expansion, and conflict, most notably during the Opium Wars of the nineteenth century (Carlin et al. 2020). Today, Papaver somniferum continues to be cultivated for three main purposes: as a source of edible seeds, for ornamental gardening, and for the extraction of alkaloids such as morphine, codeine, and thebaine, which are essential to the pharmaceutical industry (Carlin et al. 2020). The duality of its uses—nutritious seeds versus narcotic latex—has led to complex legal and cultural attitudes toward its cultivation and trade.

15.3 Utilization Pattern

Papaver somniferum has been cultivated for millennia, primarily for seeds and latex that contains various alkaloids. In India, the seeds of this plant are utilized in a myriad of applications spanning culinary, medicinal, and industrial domains. The traditional and modern uses of *Papaver somniferum* seeds reflect a rich cultural heritage intertwined with contemporary practices.

In culinary applications, the seeds of *Papaver somniferum* are widely recognized for their nutritional value and flavor. A detailed description of fatty acid profile and protein content of poppy and other important crops is given in Tables 15.1 and 15.2. They are commonly incorporated into various food products such as bakery items, including bread, buns, and pastries, where they serve as toppings or ingredients (Casado-Hidalgo et al. 2021, 2022). The seeds are also used in traditional Indian sweets and snacks, contributing to their unique taste and texture. The oil extracted from these seeds is utilized in cooking and food preparation, further enhancing the culinary versatility of *Papaver somniferum* (Knutsen et al. 2018). The seeds are rich in essential fatty acids, proteins, and dietary fiber, making them valuable addition to a balanced diet (Avula et al. 2023).

Table 15.1 Comparison of fatty acid profile in poppy seed and important oil crops

Crop	Oil content %	Unsaturated fatty acid profile in oil (%)			
		Oleic acid (18:1)	Linoleic acid (18:2)	Linolenic acid (18:3)	Total
Rapseed	45.0	24.0	15.0	8.0	47.0
Mustard	40.0	10.7	16.7	18.5	45.9
Sunflower	45.0	19.0	68.0	Nil	87.0
Soyabean	20.0	23.0	54.2	8.0	85.0
Poppy	45.0–58.0	15.8	66.3	Nil	82.1

Table 15.2 Comparison of protein content in poppy seed and important oil and cereal crops

Crops	Seed protein content (%)	Crops	Seed protein content (%)
Rice	7–8	Soybean	38–42
Wheat	12–15	Sunflower	15–22
Pulse	15–20	Poppy	23–27

From a medicinal perspective, *Papaver somniferum* seeds have been historically employed in traditional medicine systems, particularly in Ayurveda. The seeds are believed to possess various therapeutic properties, including analgesic, antitussive, and antispasmodic effects, attributed to the presence of opium alkaloids (morphine and codeine) (Krošlák et al. 2017; Skopikova et al. 2020). Recent studies have identified specific compounds within the seeds that exhibit antiepileptic properties, suggesting a potential for treating neurological disorders (Tittarelli et al. 2022). The integration of modern pharmacological techniques, such as network pharmacology and molecular docking, has provided insights into the mechanisms through which these compounds exert their effects, thereby validating traditional uses (Tittarelli et al. 2022).

In addition to their culinary and medicinal uses, *Papaver somniferum* seeds have found applications in various industrial sectors. The seeds can be processed to produce biodiesel, showcasing their potential as a sustainable energy source (Shahzadi et al. 2024). The oil extracted from the seeds undergoes transesterification to create biodiesel, which can be utilized in various applications, including transportation and heating (Shahzadi et al. 2024). This industrial application not only highlights the versatility of *Papaver somniferum* but also aligns with global efforts to seek renewable energy sources.

Opium alkaloids in *Papaver somniferum* raised concerns for food safety and regulatory standards. Studies have shown that the concentration of these alkaloids can vary significantly based on processing methods, such as thermal treatments and washing (Aksoy and Sözbilir 2015; Shetge et al. 2020). Understanding levels of these compounds in food is crucial for ensuring consumer safety, especially in regions where poppy seeds are a staple ingredient (Knutsen et al. 2018). Regulatory authorities have set guidelines to track and regulate presence of opium alkaloids in food products, ensuring that their levels stay within safe thresholds for human consumption.

The cultivation of *Papaver somniferum* in India is subject to strict regulations due to its association with narcotic substances. Farmers are required to adhere to government guidelines that dictate the cultivation practices and the intended use of the crop (Vera-Baquero et al. 2022). This regulatory framework aims to prevent the illicit production of narcotics while allowing for the legitimate use of the seeds in food and medicinal applications. The balance between traditional practices and modern regulatory measures is essential for the sustainable cultivation of this economically significant crop.

15.4 Production

15.4.1 Climatic Requirements

Papaver somniferum grows best in a moderate climate. For successful germination, temperatures range is 15–25 °C. During the growing season, the plant thrives when daytime temperatures range from 18 to 24 °C, ensuring optimal development and productivity (Shetge et al. 2020). The species is sensitive to frost, waterlogging, particularly during the flowering stage, which can adversely affect seed set and quality. Therefore, regions with a stable climate and minimal frost risk are preferred for cultivation (Shetge et al. 2020; Butnariu et al. 2022).

Additionally, poppies require a certain amount of sunlight, with full sun exposure being optimal for growth. Insufficient light can lead to leggy plants and reduced alkaloid production (Shetge et al. 2020). Rainfall is also a consideration; while poppies can tolerate dry conditions, they benefit from regular rainfall during the early growth stages, with a total annual precipitation of about 400–600 mm being ideal (Shetge et al. 2020).

15.4.2 Soil Requirements

The soil type is a crucial factor in the successful cultivation of *Papaver somniferum*. Well-drained loamy soils rich in organic matter are preferred, as they provide the necessary nutrients and support healthy root development (Shetge et al. 2020; Butnariu et al. 2022). Heavy clay soils can lead to waterlogging, which is detrimental to the plant's health. Soil drainage can be improved through the addition of organic matter or by implementing raised beds (Shetge et al. 2020). Soil fertility is also important; the incorporation of well-rotted manure or compost can enhance nutrient availability. Soil testing is recommended for nutrient levels and pH, allowing for timely amendments to optimize growth conditions (Shetge et al. 2020; Butnariu et al. 2022). The presence of essential nutrients, particularly nitrogen, phosphorus, and potassium, is vital for the growth and alkaloid production of the opium poppy (Shetge et al. 2020).

15.4.3 Cultivation Practices

Poppy cultivation typically involves selecting appropriate varieties based on the intended use, whether for culinary or medicinal purposes. The seeds are usually sown in well-prepared soil, with a preference for a pH range of 6.0–7.5 (Shetge et al. 2020; Butnariu et al. 2022). For optimal growth, *Papaver somniferum* should be cultivated at a density of approximately 70 plants per square meter, with a row spacing of 15 cm × 10 cm. A recommended seed rate of 750 g/ha ensures uniform distribution. Seeds should be sown at a depth of 5–10 mm in the soil. Nutrient management involves applying 20–30 tons of FYM/ha through broadcasting. Additionally, a fertilizer application of 60–80 kg (N_2) and 40–50 kg of phosphorus (P_2O_5) per hectare is advised. At the time of sowing, half of the nitrogen and the entire phosphorus should be applied through placement, while the remaining nitrogen should be incorporated at rosette stage. Proper irrigation management is crucial, as water stress during early growth stages can lead to stunted plant development.

Maintaining a weed-free environment for first 60–75 days is essential, as this period is critical for minimizing competition. Preplanting application of Simazine (1.5 kg/ha), Diuron (1 kg/ha), or Oxyfluorfen (0.25 kg/ha) is recommended for effective weed management and to promote healthy crop growth. Generally, two weeding, one at 20–25 and another at 40–45 days after planting, should be done. Proper irrigation practice is essential to maintain optimal soil moisture levels. Rainfall plays a significant role in cultivation of this species. *Papaver somniferum* prefers well-drained soils and does not tolerate excessive moisture, which can lead to root rot and fungal infections (Voglmayr et al. 2014).

15.5 Protection

15.5.1 Major Pest of *Papaver somniferum*

The primary pests affecting *P. somniferum* in India include aphids, cutworms, and various beetles. Aphids, particularly the cotton aphid (*Aphis gossypii*), are notorious for their ability to reproduce rapidly and can cause significant damage by feeding on the sap of the plants. Aphids and thrips are sap-sucking insects that dine on the juices of the poppy.

Aphids and thrips are common sap-feeding pests that can severely affect poppy plants by extracting essential nutrients. Aphids, which are small and typically green or yellowish, often cluster on the undersides of leaves and stem joints. A severe infestation may cause the leaves to curl, become deformed, and eventually weaken the plant. Thrips, on the other hand, are tiny, elongated insects that range in color from black and brown to yellow. Their feeding activity damages both leaves and flowers, resulting in a speckled or stippled appearance, distorted blooms, and unopened buds. In addition to direct damage, these pests can stunt plant growth, cause leaf yellowing, and spread viral infections (Kundrátová et al. 2021). Another significant threat comes from cutworms, the larvae of nocturnal moths, which target young plants by severing them at the soil surface, often leading to plant death (Muhammad et al. 2021). Another significant pest is the poppy seed weevil (*Rhinocyllus conicus*), which infests the seed capsules. The larvae of this weevil feed on the seeds, leading to reduced seed viability and yield losses (Celik et al. 2016). Effective management of these pests is crucial, as they can lead to economic losses for farmers and affect the overall production of opium poppy in the region.

15.5.2 Diseases Affecting *Papaver somniferum*

Downy mildew—This fungal disease appears as reddish spots on the upper leaf surfaces, with gray, tan, or purple mold-like growth on the undersides. Downy mildew (*Plasmopara*) thrives in cool temperatures between 40 and 60 °F and spreads rapidly in high humidity (above 90%). Spores can infect wet foliage within 12 h, making prevention crucial. Managing this disease involves improving air circulation, watering plants at the base to keep leaves dry, and clearing plant debris. Severe outbreaks may require fungicidal treatments as per label instructions.

Gray mold—Most prevalent in cool, damp conditions, gray mold (*Botrytis cinerea*) primarily affects aging or damaged plant tissue, causing brown lesions that decay into fuzzy gray spore masses. The disease is most active between 69 and 77 °F and spreads when moisture appears on plant surfaces. To reduce infection risks, water poppies at the base in the morning to allow leaves to dry quickly, remove affected leaves and flowers, and apply fungicides like captain or chlorothalonil if needed.

Powdery mildew—This fungal infection is characterized by a powdery white coating on poppy leaves, stems, and flowers. It flourishes in dry, shaded conditions at temperatures between 60 and 80 °F but cannot survive intense heat above 95 °F. Preventative measures include ensuring full sun exposure and occasionally washing leaves to remove spores. If an outbreak occurs, treating the plants with horticultural oil in cooler conditions (below 90 °F) can help manage the disease.

Tomato spotted wilt virus—This incurable viral disease causes dark streaks on stems and distinct ringed or spotted patterns on flower petals, often delaying bloom development and distorting flowers. The virus spreads through thrips—tiny insects that feed on young leaves and floral tissues. To control the spread, dislodge thrips with a strong stream of water, apply neem or horticultural oil at the first sign of infestation, and remove infected plants along with nearby weeds. Avoid excessive nitrogen fertilizers, as they promote tender growth that attracts thrips.

Root rot disease (caused by Fusarium and Phytophthora) can lead to wilting, yellowing, and eventual death of the plants. The management strategies for root rot include the use of well-drained soils, crop rotation, and the application of biological control agents (Kundrátová et al. 2021; Landa et al. 2007).

15.6 Improvement (Genetics and Genomics)

Historically, *Papaver somniferum* has been cultivated for its alkaloids, which are vital for pharmaceutical applications. Primary alkaloids include morphine, codeine, and thebaine, which are used in pain management and other medical treatments (Celik et al. 2016). In India, specific varieties are grown under strict government oversight to ensure compliance with international narcotic control agreements. Indian Council of Agricultural Research (ICAR) and CSIR-Central Institute of Medicinal and Aromatic Plants (CIMAP) have been successful in developing varieties through breeding methods, which are given in Table 15.3.

Genetic diversity is a key factor for successful breeding of *P. somniferum*. Studies utilizing genotyping-by-sequencing have revealed significant genetic variation among different cultivars and wild relatives, such as *Papaver setigerum*, which is considered a progenitor of cultivated opium poppy (Hong et al. 2021, 2022). Genetic diversity is crucial for breeding programs as it offers a wider genetic pool from which desirable traits, such as increased alkaloid content, disease resistance, and better adaptability to different climatic conditions, can be selected (Celik et al. 2016). The use of molecular markers, such as EST-SSR markers, has facilitated the discrimination of different *P. somniferum* varieties, enabling breeders to make informed decisions during the selection process (Kaňuková et al. 2024; Vašek et al. 2020).

Biotechnological advancements have also played a significant role in enhancing genetic potential *P. somniferum*. Techniques such as CRISPR/Cas9 gene editing and RNA interference have been explored to target specific genes involved in alkaloid biosynthesis and stress responses (Boke et al. 2015). These technologies allow for precise modifications that can lead to improved traits without the introduction of

Table 15.3 Varieties of *Papaver somniferum* released by ICAR and CSIR-CIMAP

Variety	Breeding method	Year of release	morphine content
Shyama	Half-sib selection	1984	12–15% in latex
Shweta	Half-sib selection.	1984	22% in latex
Vivek	Induced mutation breeding	1984	0.9% in straw
Rakshit	Selected from intra-specific hybrids for enhanced resistance to various stress factors.	2001	0.5–0.87% in straw
Shubhra	Mass selection	2001	0.86% in straw
Sujata	Mutation breeding	2002	52–59% in seeds
Sampada	Induced mutation	2002	18% in latex.
CIM-Ajay	Hybridization followed by selection	2010	1.036% in straw
Chetak Aphim	–	–	11.9%
Jawahar opium 539	–	1997	14.85%

foreign DNA, which is particularly important in a crop with stringent regulatory oversight.

Moreover, the integration of omics technologies, including transcriptomics and proteomics, has provided insights into the metabolic pathways involved in alkaloid production (Desgagné-Penix et al. 2010). For example, the identification of key enzymes and regulatory factors in the biosynthetic pathways of benzylisoquinoline alkaloids has paved the way for metabolic engineering approaches aimed at increasing the yield of valuable compounds like morphine and codeine (Beaudoin and Facchini 2014; Menéndez-Perdomo et al. 2021).

15.7 Primary Processing

Postharvest handling and processing techniques for *Papaver somniferum* seeds are critical to ensuring their quality, safety, and usability in various applications, particularly in the food industry. Seeds of *Papaver somniferum*, commonly known as opium poppy, are utilized not only for culinary purposes but also in medicinal applications due to their association with alkaloids such as morphine and codeine. The postharvest phase encompasses several stages, including harvesting, drying, cleaning, storage, and processing, each of which is crucial in maintaining seed quality and preventing contamination.

Harvesting of *Papaver somniferum* seeds typically occurs approximately 110 days after sowing, following the natural maturation of the plant. This timing is crucial, as harvesting too early or too late can adversely affect seed quality and yield (Fig. 15.2). The latex, which contains various alkaloids, is collected shortly after the flowering petals have fallen, usually between 88 and 93 days after sowing (Ngernsaengsaruay et al. 2023). This latex collection process must be carefully managed to avoid contamination of the seeds, which can occur if the latex comes into contact with them during harvesting.

Once harvested with moisture content (7–9%) is essential for preventing mold growth and deterioration. Proper drying techniques include using controlled temperature and airflow to ensure even drying without damaging the seeds. Studies indicate that inadequate drying can lead to significant losses in seed viability and quality (Muui et al. 2020). Following drying, seeds should be cleaned to remove any foreign materials, including plant debris and potential contaminants. This cleaning process is vital for ensuring that the seeds are safe for consumption and meet food safety standards (Casado-Hidalgo et al. 2021).

Storage conditions are another critical aspect of postharvest handling. Seeds should be stored in a cool, dry environment to minimize the risk of spoilage and pest infestation. The use of airtight containers can help maintain optimal humidity levels and prevent contamination from external sources (Spitzer et al. 2014). Furthermore, the implementation of modified atmosphere packaging has increased shelf life of seeds by reducing the incidence of internal rot and other spoilage factors (Frans et al. 2021).

Fig. 15.2 Different phenological stages of opium (*Papaver somniferum*)

There are some specific processing techniques that can further enhance the safety and quality of *Papaver somniferum* seeds. For instance, thermal treatments such as dry heat or steam have been investigated for their effectiveness in reducing alkaloid levels in seeds. Research indicates that these treatments can significantly lower the concentrations of morphine and codeine, making seeds safer for consumption (Shetge et al. 2020). Moreover, washing the seeds with water or acidic solutions has also been shown to reduce alkaloid content, although the effectiveness can vary based on the method used (Thangavel et al. 2020).

The potential for contamination with pathogenic fungi during postharvest handling is another concern that must be addressed. Studies have identified various fungi that can infect poppy seeds, leading to quality degradation and health risks (O'Neill et al. 2000). Implementing seed treatment methods, such as washing with hypochlorous acid or sodium hypochlorite, has been shown to effectively reduce transmission of seed-borne diseases (Thangavel et al. 2020). These treatments not only enhance seed health but also contribute to overall crop yield and quality.

In addition, understanding the chemical composition of *Papaver somniferum* seeds is vital for their safe utilization. Although the seeds themselves have minimal alkaloid content, they may become contaminated with these compounds during harvesting if appropriate precautions are not taken (Garcia et al. 2020). Therefore, continuous monitoring and analysis of alkaloid levels in seeds are necessary to ensure compliance with food safety regulations (Knutsen et al. 2018). Advanced analytical techniques such as HPLC are employed to quantify alkaloid concentrations in seeds and processed products (Casado-Hidalgo et al. 2022).

15.8 Value Addition and Products

The potential for creating high-value products from *Papaver somniferum* seeds in India is multifaceted, encompassing culinary, nutritional, and medicinal aspects. The seeds of the opium poppy, traditionally valued for their culinary applications, are gaining attention for their nutritional advantages and health-enhancing qualities. They are rich in essential fatty acids, especially linoleic acid, and are rich in tocopherols, which are crucial for maintaining human health (Ozkaya et al. 2013). Incorporating poppy seeds into various food products, such as bakery items, has gained traction due to their appealing flavor and texture, as well as their health benefits, including antioxidant properties and contributions to dietary fiber intake (Casado-Hidalgo et al. 2021, 2022; Vera-Baquero et al. 2022).

In India, the culinary application of *Papaver somniferum* seeds can be expanded significantly. The seeds are traditionally used in various regional cuisines, but their potential as a health food ingredient remains underexplored. The seeds can be added to bread, pastries, and salads, enhancing not only the flavor but also the nutritional profile of these products (Casado-Hidalgo et al. 2021, 2022). The increasing consumer awareness regarding health and nutrition presents an opportunity for food manufacturers to innovate and create high-value products that cater to this demand. The seeds can also be processed into oil, which is rich in unsaturated fats and can serve as a healthier alternative to traditional cooking oils (Gök 2015).

Moreover, the medicinal properties of *Papaver somniferum* seeds are noteworthy. The seeds contain polyphenols and alkaloids, which have been shown to possess antioxidant, anti-inflammatory, and analgesic properties (Butnariu et al. 2022; Kamal et al. 2017). Traditional medicinal uses of poppy seeds, particularly in Unani and Ayurvedic practices, highlight their potential as a natural remedy for various ailments (Shah et al. 2025). This aspect could be leveraged to develop health supplements or functional foods that appeal to health-conscious consumers. Growing trend toward natural and organic products further supports potential for high-value product development in this area.

The regulatory landscape surrounding the cultivation and processing of *Papaver somniferum* in India is also a critical factor. Cultivation of opium poppy strictly regulated due to its association with narcotic alkaloids; the seeds themselves are generally considered safe for consumption when sourced appropriately (Knutsen et al. 2018). This regulatory framework can provide a pathway for the development of a sustainable supply chain for poppy seeds, ensuring that they are cultivated and processed in compliance with health and safety standards. Furthermore, advancements in agricultural practices and biotechnology could enhance the yield and quality of poppy seeds, making them more viable for commercial production (Hong et al. 2022; Özcan et al. 2025).

Research into genetic diversity of *Papaver somniferum* also holds promise for improving the quality and nutritional content of the seeds. By utilizing modern breeding techniques and biotechnological approaches, it is possible to develop new cultivars that are higher in beneficial compounds and more resilient to environmental stresses (Hong et al. 2022). This genetic enhancement can lead to the production of seeds that not only meet consumer demands but also contribute to the sustainability of agricultural practices in India.

In addition to culinary and medicinal applications, the potential for using *Papaver somniferum* seeds in the cosmetic and personal care industry should not be overlooked. The antioxidant properties of the seeds make them suitable for incorporation into skincare products, where they can provide benefits such as skin hydration and protection against oxidative stress (Krošlák et al. 2017).

Furthermore, the environmental sustainability of cultivating *Papaver somniferum* should be considered. The cultivation of this crop can be incorporated into sustainable farming practices like crop rotation and intercropping, which help improve soil health and minimize the reliance on chemical fertilizers (Muntaha et al. 2021). By promoting sustainable farming practices, India can position itself as a leader in the production of high-quality, ethically sourced poppy seeds.

15.9 Marketing

The global demand for *Papaver somniferum* seeds, commonly known as poppy seeds, has been influenced by various factors, including their culinary applications, nutritional benefits, and the presence of pharmacologically active alkaloids. The seeds are primarily utilized in bakery products, where they serve as toppings for bread, pastries, and other baked goods, contributing to their increasing popularity in the food industry (Casado-Hidalgo et al. 2022; Kanukova et al. 2020). The culinary use of poppy seeds is not limited to traditional recipes; they are also incorporated into modern health-conscious diets due to their rich nutritional profile, which includes essential fatty acids, vitamins, and trace elements (Butnariu et al. 2022; Ozkaya et al. 2013).

In terms of market trends, the demand for poppy seeds has seen a significant rise, particularly in regions where they are cultivated legally, such as Turkey and the Czech Republic. These countries are among the leading exporters of seeds in Europe, capitalizing on the growing interest in health foods and the culinary versatility of poppy seeds (Lošák et al. 2022; Butnariu et al. 2022). Czech Republic is also a big poppy seed producer, which has positioned it as a key player in the European market (Lošák et al. 2022). Seeds are not only valued for their flavor but also for antioxidant properties and potential contributions to cardiovascular health (Krošlák et al. 2017; Vera-Baquero et al. 2022).

The trade of *Papaver somniferum* seeds is also impacted by regulations related to alkaloids, which naturally found in seeds. Alkaloids morphine and codeine has raised concerns regarding food safety and public health, leading to stringent regulations in various countries (Knutsen et al. 2018). European Food Safety Authority (EFSA) has issued opinions on risks associated with opium alkaloids in food products, which have prompted manufacturers to adopt measures to minimize alkaloid contamination in poppy seeds (Knutsen et al. 2018). This regulatory scrutiny has, in some cases, affected market dynamics, as producers must ensure compliance with safety standards while meeting consumer demand.

Furthermore, the cultivation practices of *Papaver somniferum* have evolved to enhance seed yield and quality. Research indicates that factors such as potassium and magnesium significantly impact the seed yield and overall quality of poppy seeds (Lošák et al. 2022). These advancements in agricultural practices are essential for meeting rising global demand while maintaining the quality of the seeds, which is crucial for both culinary and medicinal applications.

In addition to their culinary uses, poppy seeds are recognized for medicinal properties. Seeds contain various alkaloids that have analgesic and therapeutic effects (Skopikova et al. 2020; Ashrafi et al. 2023). The pharmacological potential of these alkaloids has led to increased interest in the seeds not only as a food source but also as a component in pharmaceuticals (Skopikova et al. 2020; Ashrafi et al. 2023). This duality of use—culinary and medicinal—has further fueled the demand for *Papaver somniferum* seeds in global markets.

The nutritional composition of poppy seeds is another factor contributing to their popularity. Studies have shown that poppy seeds are rich in tocopherols, which are important antioxidants, as well as essential fatty acids that are beneficial for heart health (Butnariu et al. 2022; Ozkaya et al. 2013). Adding poppy seeds to a variety of food products boosts their nutritional value (Krošlák et al. 2017; Vera-Baquero et al. 2022).

Additionally, the global trade of poppy seeds involves a multifaceted supply chain encompassing cultivation, processing, and distribution. Countries that produce poppy seeds often export them to regions with high demand, such as North America and parts of Asia (Butnariu et al. 2022; Kanukova et al. 2020). The global market for poppy seeds is also influenced by consumer preferences, which have shifted toward organic and sustainably sourced products. This shift has prompted producers to adopt more environmentally friendly cultivation practices to meet the expectations of discerning consumers (Butnariu et al. 2022; Kanukova et al. 2020).

The importance of technology in the production and analysis of poppy seeds is undeniable. Innovations in analytical methods, such as HPLC and mass spectrometry, have allowed researchers to gain a deeper understanding of the chemical composition of poppy seeds and how processing impacts alkaloid concentrations (Casado-Hidalgo et al. 2021, 2022). These technologies play a vital role in ensuring safety and quality of poppy seeds in market, especially given the regulatory challenges related to opium alkaloids (Knutsen et al. 2018).

References

Administration DE (1992) Opium poppy cultivation and heroin processing in Southeast Asia. Office of Intelligence, Washington, DC

Agarwal P, Pathak S, Lakhwani D et al (2016) Comparative analysis of transcription factor gene families from Papaver somniferum: identification of regulatory factors involved in benzylisoquinoline alkaloid biosynthesis. Protoplasma 253:857–871. https://doi.org/10.1007/S00709-015-0848-8

Aksoy L, Sözbilir NB (2015) Trace and major element levels in rats after oral administration of diesel and biodiesel derived from opium poppy (Papaver somniferum L.) seeds. Toxicol Ind Health 31:890–897. https://doi.org/10.1177/0748233713484656

Ashrafi S, Alam S, Sultana A et al (2023) Papaverine: a miraculous alkaloid from opium and its multimedicinal application. Molecules 28:3149

Avula B, Katragunta K, Adams SJ et al (2023) Applicability of LC-QToF and microscopical tools in combating the sophisticated, economically motivated adulteration of poppy seeds. Foods 12. https://doi.org/10.3390/FOODS12071510

Beaudoin GAW, Facchini PJ (2014) Benzylisoquinoline alkaloid biosynthesis in opium poppy. Planta 240:19–32. https://doi.org/10.1007/S00425-014-2056-8

Boke H, Ozhuner E, Turktas M et al (2015) Regulation of the alkaloid biosynthesis by miRNA in opium poppy. Plant Biotechnol J 13:409–420. https://doi.org/10.1111/PBI.12346

Butnariu M, Quispe C, Herrera-Bravo J et al (2022) Papaver plants: current insights on phytochemical and nutritional composition along with biotechnological applications. Oxidative Med Cell Longev 2022. https://doi.org/10.1155/2022/2041769

Carlin MG, Dean JR, Ames JM (2020) Opium alkaloids in harvested and thermally processed poppy seeds. Front Chem 8:539741. https://doi.org/10.3389/FCHEM.2020.00737/BIBTEX

Casado-Hidalgo G, Pérez-Quintanilla D, Morante-Zarcero S, Sierra I (2021) Mesostructured silica-coated magnetic nanoparticles to extract six opium alkaloids in poppy seeds prior to ultra-high-performance liquid chromatography-tandem mass spectrometry analysis. Foods 10. https://doi.org/10.3390/FOODS10071587

Casado-Hidalgo G, Martínez-García G, Morante-Zarcero S et al (2022) New validated method for the determination of six opium alkaloids in poppy seed-containing bakery products by high-performance liquid chromatography-tandem mass spectrometry after magnetic solid-phase extraction. J Agric Food Chem 70:7594–7606. https://doi.org/10.1021/ACS.JAFC.2C01664

Celik I, Camci H, Kose A et al (2016) Molecular genetic diversity and association mapping of morphine content and agronomic traits in Turkish opium poppy (Papaver somniferum) germplasm. Mol Breed 36. https://doi.org/10.1007/S11032-016-0469-8

Chen P, Braithwaite RA, George C et al (2014) The poppy seed defense: a novel solution. Drug Test Anal 6:194–201. https://doi.org/10.1002/DTA.1590

Desgagné-Penix I, Khan MF, Schriemer DC et al (2010) Integration of deep transcriptome and proteome analyses reveals the components of alkaloid metabolism in opium poppy cell cultures. BMC Plant Biol 10:252. https://doi.org/10.1186/1471-2229-10-252

Frans M, Aerts R, Ceusters N et al (2021) Possibilities of modified atmosphere packaging to prevent the occurrence of internal fruit rot in bell pepper fruit (Capsicum annuum) caused by Fusarium spp. Postharvest Biol Technol 178. https://doi.org/10.1016/J.POSTHARVBIO.2021.111545

Garcia MR, Swortwood MJ, Aune CN, Ahmad KA (2020) Maternal poppy seed tea ingestion and ensuing neonatal abstinence syndrome. Neonatology 117:529–531. https://doi.org/10.1159/000508446

Gök V (2015) Effect of replacing beef fat with poppy seed oil on quality of Turkish sucuk. Korean J Food Sci Anim Resour 35:240–247. https://doi.org/10.5851/KOSFA.2015.35.2.240

Hong UVT, Tamiru-Oli M, Hurgobin B et al (2021) Insights into opium poppy (*Papaver* spp.) genetic diversity from genotyping-by-sequencing analysis. https://doi.org/10.1101/2021.09.28.462245

Hong UVT, Tamiru-Oli M, Hurgobin B et al (2022) Insights into opium poppy (Papaver spp.) genetic diversity from genotyping-by-sequencing analysis. Sci Rep 12. https://doi.org/10.1038/S41598-021-04056-3

Jesus A, Bonhomme V, Evin A et al (2021) A morphometric approach to track opium poppy domestication. Sci Rep 11(1). https://doi.org/10.1038/S41598-021-88964-4

Jesus A, Bonhomme V, Evin A et al (2023) Morphometrics of waterlogged archaeological seeds give new insights into the domestication and spread of Papaver somniferum L. in Western Europe. PLoS One 18:e0286190. https://doi.org/10.1371/JOURNAL.PONE.0286190

Kamal SB, Kumar A, Tanwar T (2017) Physico-chemical, proximate, sensory and storage quality attributes analysis of Papaver somniferum (poppy) fortified chevon nuggets. J Appl Nat Sci 9:114–120. https://doi.org/10.31018/JANS.V9I1.1158

Kanukova S, Mrkvova M, Mihalik D, Kraic J (2020) Procedures for DNA extraction from opium poppy (papaver somniferum l.) and poppy seed-containing products. Foods 9. https://doi.org/10.3390/FOODS9101429

Kaňuková Š, Ondreičková K, Mihálik D, Kraic J (2024) New set of EST-STR markers for discrimination of related Papaver somniferum L. varieties. Life 14. https://doi.org/10.3390/LIFE14010072

Kauppila GR, Eagen KV (2023) Opioid use disorder from poppy seed tea use: a case report. Am J Case Rep 24. https://doi.org/10.12659/AJCR.938675

Knutsen HK, Alexander J, Barregård L et al (2018) Update of the Scientific Opinion on opium alkaloids in poppy seeds. EFSA J 16. https://doi.org/10.2903/J.EFSA.2018.5243

Krošlák E, Maliar T, Nemeček P et al (2017) Antioxidant and proteinase inhibitory activities of selected poppy (Papaver somniferum L.) genotypes. Chem Biodivers 14. https://doi.org/10.1002/CBDV.201700176

Kukade DS, Talib MI, Parate VR, Shrirame V (2015) Development of lemongrass flavored high nutrient cookies by fortification of flaxseed and poppy seed flour. Int J Eng Res Technol 4. https://doi.org/10.17577/IJERTV4IS100248

Kumar B, Patra NK (2011) Combining ability for inheritance of economic traits in opium poppy (Papaver somniferum L.). Bangladesh J Plant Breed Genet 24:01–08. https://doi.org/10.3329/BJPBG.V24I2.17000

Kundrátová K, Bartas M, Pečinka P et al (2021) Transcriptomic and proteomic analysis of drought stress response in opium poppy plants during the first week of germination. Plants 10. https://doi.org/10.3390/PLANTS10091878

Landa BB, Montes-Borrego M, Muñoz-Ledesma FJ, Jiménez-Díaz RM (2007) Phylogenetic analysis of downy mildew pathogens of opium poppy and PCR-based in planta and seed detection of Peronospora arborescens. Phytopathology 97:1380–1390. https://doi.org/10.1094/PHYTO-97-11-1380

Lošák T, Elbl J, Kintl A et al (2022) Effect of poppy (Papaver somniferum L.) fertilization with potassium and magnesium on the seed yield and its quality. Acta Fytotech Zootech 25:242–246. https://doi.org/10.15414/AFZ.2022.25.03.242-246

Machado RSR, Bonhomme V, Soteras R et al (2025) The origins and spread of the opium poppy (Papaver somniferum L.) revealed by genomics and seed morphometrics. Philos Trans B 380:20240198. https://doi.org/10.1098/RSTB.2024.0198

Mella-Raipán J, Romero-Parra J, Recabarren-Gajardo G (2020) DARK classics in chemical neuroscience: heroin and desomorphine. ACS Chem Neurosci 11:3905–3927. https://doi.org/10.1021/ACSCHEMNEURO.0C00262

Menéndez-Perdomo IM, Hagel JM, Facchini PJ (2021) Benzylisoquinoline alkaloid analysis using high-resolution Orbitrap LC-MSn. J Mass Spectrom 56. https://doi.org/10.1002/JMS.4683

Muhammad A, Akhtar A, Aslam S et al (2021) Review on physicochemical, medicinal and nutraceutical properties of poppy seeds: a potential functional food ingredient. Funct Foods Health Dis 11:522–547. https://doi.org/10.31989/FFHD.V11I10.836

Muntaha S-T, Mubarak SS, Sherwani FK, Azim N (2021) Allelopathic potential of Papaver somniferum L. on seed germination against three different varieties of Zea mays. Int J Sci Res Publ 11:310–317. https://doi.org/10.29322/IJSRP.11.04.2021.P11241

Muui CW, Muasya RM, Nguluu S, Kambura S (2020) Evaluation of seed quality attributes of sorghum germplasm accessions from eastern coastal and Nyanza regions Kenya. Sustain Agric Res 9(3):9–20. https://doi.org/10.5539/sar.v9n3p9

Neugschwandtner RW, Dobos G, Wagentristl H et al (2023) Yield and yield components of winter poppy (Papaver somniferum L.) are affected by sowing date and sowing rate under Pannonian climate conditions. Agriculture 13. https://doi.org/10.3390/AGRICULTURE13050997

Ngernsaengsaruay C, Leksungnoen N, Chanton P et al (2023) Morphology, taxonomy, anatomy, and palynology of the opium poppy (Papaver somniferum L.) cultivation in Northern Thailand. Plants 12. https://doi.org/10.3390/PLANTS12112105

O'Neill NR, Jennings JC, Bailey BA, Farr DF (2000) Dendryphion penicillatum and Pleospora papaveracea, destructive seedborne pathogens and potential mycoherbicides for Papaver somniferum. Phytopathology 90:691–698. https://doi.org/10.1094/PHYTO.2000.90.7.691
Özcan DE, Köm D, Doğan O et al (2025) Modification of gene expression involved in alkaloid production in opium poppy by VIGS combined with pretreatment of macerozyme enzyme. Plant Direct 9. https://doi.org/10.1002/PLD3.70034
Ozkaya A, Ciftci H, Yilmaz O et al (2013) Vitamin, trace element, and fatty acid levels of Vitex agnus-castus L., Juniperus oxycedrus L., and Papaver somniferum L. plant seeds. J Chem. https://doi.org/10.1155/2013/845743
Presley CC, Lindsley CW (2018) DARK classics in chemical neuroscience: opium, a historical perspective. ACS Chem Neurosci 9:2503–2518. https://doi.org/10.1021/ACSCHEMNEURO.8B00459
Prochazka P, Smutka L (2012) Czech Republic as an important producer of poppy seed. AGRIS On-Line Pap Econ Inform 04:1–13. https://doi.org/10.22004/AG.ECON.131361
Pushpangadan P, George V, Singh SP (2012) Poppy. In: Handbook of herbs and spices, 2nd edn. Elsevier Inc., pp 437–448. Woodhead Publishing Limited, Cambridge, UK
Rasekh SZ, Karimzadeh G (2023) Chromosomal and genome size variations in Opium poppy (Papaver somniferum L.) from Afghanistan. Caryologia 76:15–22. https://doi.org/10.36253/CARYOLOGIA-1955
Salavert A, Martin L, Antolín F, Zazzo A (2018) The opium poppy in Europe: exploring its origin and dispersal during the Neolithic. Antiquity 92. https://doi.org/10.15184/AQY.2018.154
Salavert A, Zazzo A, Martin L et al (2020) Direct dating reveals the early history of opium poppy in western Europe. Sci Rep 10(1):1–10. https://doi.org/10.1038/s41598-020-76924-3
Samano KL, Clouette RE, Rowland BJ, Barry Sample RH (2015) Concentrations of morphine and codeine in paired oral fluid and urine specimens following ingestion of a poppy seed roll and raw poppy seeds. J Anal Toxicol 39:655–661. https://doi.org/10.1093/JAT/BKV081
Saunders JA, Pedroni MJ, Penrose LDJ, Fist AJ (2001) AFLP analysis of opium poppy. Crop Sci 41:1596–1601. https://doi.org/10.2135/CROPSCI2001.4151596X
Schuppener LM, Corliss RF (2018) Death due to complications of bowel obstruction following raw poppy seed ingestion. J Forensic Sci 63:614–618. https://doi.org/10.1111/1556-4029.13562
Shah HA, Khan S, Ahmed MM (2025) Unravelling the medicinal secrets of Khashkhaash (Papaver somniferum L.) seeds: a powerful blend of Unani wisdom and modern science. J Drug Deliv Ther 15:155–158. https://doi.org/10.22270/JDDT.V15I1.6957
Shahzadi K, Rasool D, Irshad F et al (2024) Identification and characterization of anti-epileptic compounds from Papaver somniferum using quantification techniques (GC-MS, FTIR), integrated network pharmacology, molecular docking, and molecular dynamics simulations. https://doi.org/10.1101/2024.07.29.605558
Shetge SA, Dzakovich MP, Cooperstone JL et al (2020) Concentrations of the opium alkaloids morphine, codeine, and thebaine in poppy seeds are reduced after thermal and washing treatments but are not affected when incorporated in a model baked product. J Agric Food Chem 68:5241–5248. https://doi.org/10.1021/ACS.JAFC.0C01681
Skopikova M, Hashimoto M, Richomme P, Schinkovitz A (2020) Matrix-free laser desorption ionization mass spectrometry as an efficient tool for the rapid detection of opiates in crude extracts of Papaver somniferum. J Agric Food Chem 68:884–891. https://doi.org/10.1021/ACS.JAFC.9B05153
Spitzer T, Spitzerová D, Matušinský P, Kazda J (2014) Possibility of using seed treatment to suppress seed-borne diseases in poppy. Plant Prot Sci 50:78–83. https://doi.org/10.17221/76/2012-PPS
Sproll C, Perz RC, Lachenmeier DW (2006) Optimized LC/MS/MS analysis of morphine and codeine in poppy seed and evaluation of their fate during food processing as a basis for risk analysis. J Agric Food Chem 54:5292–5298. https://doi.org/10.1021/JF0608975
Srinivas H, Narasinga Rao MS (1981) Studies on the proteins of poppy seed (Papaver somniferum L.). J Agric Food Chem 29:1232–1235. https://doi.org/10.1021/JF00108A033

Thangavel T, Scott J, Jones S et al (2020) Effect of physio-chemical seed treatments on opium poppy downy mildews caused by Peronospora meconopsidis and P. somniferi. PLoS One 15. https://doi.org/10.1371/JOURNAL.PONE.0230801

Tittarelli R, Gismondi A, Di Marco G et al (2022) Forensic application of genetic and toxicological analyses for the identification and characterization of the opium poppy (Papaver somniferum L.). Biology (Basel) 11. https://doi.org/10.3390/BIOLOGY11050672

Vašek J, Čílová D, Melounová M et al (2020) New EST-SSR markers for individual genotyping of opium poppy cultivars (Papaver somniferum L.). Plants 9. https://doi.org/10.3390/PLANTS9010010

Vera-Baquero FL, Morante-Zarcero S, Sierra I (2022) Evaluation of thermal degradation of tropane and opium alkaloids in gluten-free corn breadsticks samples contaminated with stramonium seeds and baked with poppy seeds under different conditions. Foods 11. https://doi.org/10.3390/FOODS11152196

Voglmayr H, Montes-Borrego M, Landa BB (2014) Disentangling Peronospora on Papaver: phylogenetics, taxonomy, nomenclature and host range of downy mildew of opium poppy (Papaver somniferum) and related species. PLoS One 9. https://doi.org/10.1371/JOURNAL.PONE.0096838

Yamaguchi K, Hayashida M, Hayakawa H et al (2011) Urinary morphine and codeine concentrations after ingestion of bean-jam buns decorated with poppy seeds. Forensic Toxicol 29:69–71. https://doi.org/10.1007/S11419-010-0099-5

Saffron (*Crocus sativus* L.): Gold of the Spices

16

Mudasir Hafiz Khan, Niyaz Ahmad Dar, Sher Ahmad Dar, Ghulam Hassan Mir, Firdous Ahmad Nehvi, Bashir Ahmad Alie, Mehraj Din Sofi, Uzma Fayaz, Ulfat Jan, and Tawkeer Nabi

Abstract

This chapter explores saffron (*Crocus sativus* L.), a valuable perennial spice prized for its vibrant color, unique flavor, and wide-ranging medicinal properties. Although Iran leads global production, the saffron cultivated in Jammu and Kashmir, India, is noted for its exceptional quality. Efforts to enhance India's presence in international markets emphasize improved cultivation techniques, streamlined supply chains, and the adoption of modern technologies. A member of the Iridaceae family, *C. sativus* is propagated through corms and is distinguished by its bioactive constituents: crocin (color), picrocrocin (bitterness), and safranal (aroma). Due to its high market value, saffron is frequently targeted for adulteration, making robust quality control essential. Historically used in traditional healing systems like Ayurveda, saffron remains relevant in treating various health conditions and finds roles in perfumery, cosmetics, and textiles. Its pharmacological effects are largely attributed to phytochemicals such as crocetin, crocin, safranal, picrocrocin, and flavonols like kaempferol and quercetin. These contribute to its antioxidant, antimicrobial, anti-inflammatory, and anticancer potential.

Keywords

Saffron · Agrotechnology · Crocin · Picrocrocin · Safranal · Uses

M. H. Khan (✉) · F. A. Nehvi · M. D. Sofi
Division of Genetics and Plant Breeding, Faculty of Agriculture, SKUAST-Kashmir, Wadura, Sopore, J&K, India
e-mail: drmhkhan@skuastkashmir.ac.in

N. A. Dar · S. A. Dar · G. H. Mir · B. A. Alie · U. Fayaz · U. Jan · T. Nabi
Advanced Research Station for Saffron and Temperate Seed Spices, SKUAST-Kashmir, Srinagar, India

A. Kumar et al. (eds.), *Fasting Superfoods*,
https://doi.org/10.1007/978-981-95-2762-5_16

16.1 Introduction

Saffron (*Crocus sativus* L.) is recognized as one of the most valuable spices globally and is typically grown as a perennial crop. Known for its deep crimson stigmas, commonly referred to as saffron threads, it is widely used in culinary applications, as a natural dye, in fragrances, and in traditional medicine. The spice is derived by hand-harvesting of these delicate stigmas. Global saffron production is estimated at around 418 metric tons annually, with Iran accounting for nearly 90% of this output (Cardone et al. 2020). In India, saffron cultivation covers approximately 3715 hectares yielding about 18 metric tons annually with an average productivity of 4.8 kg per hectare.

Despite its historical significance and superior quality (Table 16.1), Indian saffron represents only a small fraction of global production. In comparison, countries like Iran dominate the market, producing nearly 90% of the world's saffron supply. To address this gap and strengthen India's presence in the global saffron market, several strategic initiatives have been introduced. These initiatives aim to enhance production through improved agricultural practices, optimize supply chain logistics, establish robust branding and marketing strategies, and integrate modern technology to boost efficiency and productivity.

Saffron is propagated asexually through underground daughter corms that emerge from the primary mother corm. The unique sensory attributes of saffron, its vibrant color, distinct taste, and aromatic quality are attributed to three principal bioactive compounds. Crocin imparts the characteristic deep yellow hue, picrocrocin contributes to its notable bitter flavor, and safranal is responsible for aroma and fragrance (Cagliani et al. 2015). Given its high market value and limited yield, saffron is frequently targeted for adulteration. Common fraudulent practices include the addition of inexpensive substitutes such as maize stigmas or safflower parts to increase volume and mimic the appearance of genuine saffron. Such practices not only degrade the quality of the product but also harm the market value of genuine saffron (Babaei et al. 2014). Consequently, ensuring the purity and authenticity of saffron remains a significant challenge for producers and traders alike.

Table 16.1 Quality Comparison of Indian and Iranian Saffron (IIKSTC 2022)

Component	Indian saffron	Iranian saffron
Crocin (440 nm)	>200	<200
Safranal (330 nm)	>50	<50
Picro-crocin (257 nm)	>70	<70
Filament length	Long and thick head	Short and thin head
Color	Deep red	Red

16.2 Origin and Distribution

The exact origin of saffron (*Crocus sativus* L.) remains an area of scientific uncertainty, with two primary regions proposed as its ancestral homeland: the eastern Mediterranean, particularly Greece and the broader region spanning Turkey, Iran, and India. Archaeobotanical and historical records suggest that saffron was domesticated as early as 2000 to 1500 BCE (Negbi 1999). Over the centuries, the cultivation of this valuable spice expanded across various regions, and today it is extensively grown for commercial purposes in several countries including Iran, India, Spain, Greece, Morocco, Italy, and Turkey. Other nations such as France, Switzerland, Israel, Pakistan, Azerbaijan, China, Egypt, the UAE, and Japan also contribute to global production. In recent years, nontraditional saffron-growing nations like Australia, New Zealand, the United States, Argentina, and Chile have initiated commercial cultivation (Sharifi et al. 2010).

In the Indian subcontinent, saffron is believed to have been introduced to the Kashmir Valley by Persian settlers around 500 BCE (Singh and Dhar 1976; McGee 2004). Today, the spice is cultivated primarily in the union territory of Jammu and Kashmir, particularly in the districts of Pulwama, Budgam, Srinagar and Kishtwar.

Geographically, the natural distribution of the genus *Crocus* spans longitudes from 10°W to 80°E and latitudes from 30°N to 50°S (Fig. 16.1). Saffron thrives at varying elevations, growing well at around 650 meters above sea level in Greece and between 1200 and 1300 meters in the Kashmir region of India. From

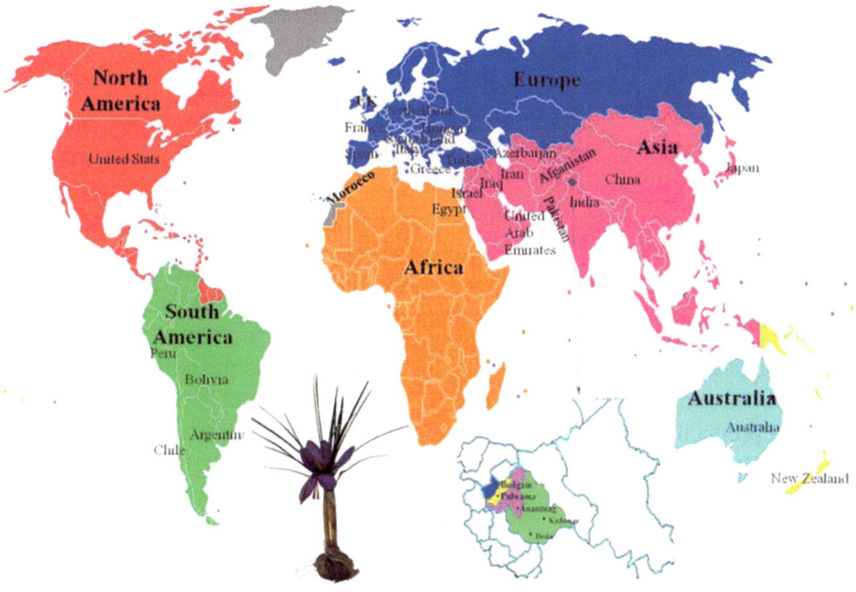

Fig. 16.1 Saffron-growing countries in the world

a genetic standpoint, *Crocus sativus* is a sterile triploid species, and its most likely ancestors are *Crocus cartwrightianus* and *Crocus thomasii*, both of which are diploid species (Mathew 1982). Among them, *C. cartwrightianus* shares greater DNA homology and karyotypic similarity with *C. sativus*. Research in plant cytogenetics, including studies on reproductive biology and DNA content (Grilli Caiola and Canini 2010), supports the hypothesis that *C. sativus* emerged from natural hybridization events. These may have involved either the union of a reduced female gamete with an unreduced male gamete from the same species or a cross-involving gametes from different but closely related species. The absence of strong reproductive barriers among these *Crocus* species likely facilitated the formation of this unique triploid, sterile lineage that we now cultivate as saffron.

16.3 Biology of Saffron

The genus *Crocus*, part of the Iridaceae family, represents a diverse group of perennial plants primarily distributed across a broad swath from Western Europe and North Africa to parts of Western Asia. With over 160 recognized species, the genus reaches its highest concentration of biodiversity in the Balkans and Turkey, areas known for their complex topography and variable climates. These conditions likely contributed to the rapid speciation and ecological specialization seen across the genus. Geographically, *Crocus* species are confined within the longitudinal boundaries of 10°W to 80°E and latitudes 30°N to 50°N. The majority inhabit the Mediterranean phytogeographical region, although some extend into the more arid Irano-Turanian zone. Most *Crocus* species are adapted to climates with pronounced seasonal fluctuations, entering dormancy during dry summer periods and resuming growth after the first autumn rains or snowmelt.

Taxonomic classification within *Crocus* has long been a subject of debate. While morphology and chromosomal characteristics remain central to species delineation, high variability within traits such as flower morphology, leaf structure, and chromosome number continues to challenge researchers. The genus is currently divided into two main subgenera: *Crociris*, which includes the atypical *C. banaticus*, and *Crocus*, which encompasses the majority of species. This second subgenus is further split into sections *Crocus* and *Nudiscapus*, though the distinction is often blurred due to overlapping traits. Efforts to establish clear species boundaries are complicated by the genus's wide ecological amplitude and potential for hybridization. Chromosomal studies suggest that variation within and between species may reflect reproductive isolation mechanisms, but without thorough investigation into mating systems and fertility, such conclusions remain speculative. Some of the species belonging to the genus *Crocus* showcasing the diversity within this group are illustrated in Fig. 16.2.

Fig. 16.2 Species of genus *Crocus*

16.3.1 Botany

Saffron scientifically known as *Crocus sativus*, is a sterile, triploid species cultivated for its valuable stigmas. As a geophyte, it persists through underground corms that regenerate annually. The corm is rounded, covered in a fibrous tunic, and produces both contractile and feeder roots. Leaves are narrow, grass-like, and typically emerge alongside or just after floral development.

Each corm usually gives rise to one or two flowers during the growing season. These flowers are composed of six violet tepals arranged in two concentric whorls. Internally, three stamens bear pollen-laden anthers, though these lack any economic significance. The most important floral structure is the style, which is divided into three vivid red stigmas, the part harvested for spice production. The stigmas contain high concentrations of apocarotenoids such as crocin, picrocrocin, and safranal which contribute to the characteristic hue, flavor and fragrance of saffron. The flower's morphology is highly specialized. Flowers emerge directly from the corm without a true above-ground stem, supported instead by a floral tube. This adaptation protects reproductive organs from climatic extremes and enables flowering even in cooler autumn temperatures.

16.3.2 Cytogenetics and Chromosomal Characteristics

One of the defining features of *C. sativus* is its triploid chromosomal makeup ($2n = 3x = 24$), a configuration that renders it sterile and incapable of producing seeds. The base number of chromosomes ($x = 8$) is shared with several wild relatives, notably *C. cartwrightianus*, a species long suspected to be involved in saffron's evolutionary origin.

Early hypotheses suggested that saffron emerged through autotriploidy, where chromosome duplication occurred within a single species. However, more recent molecular and cytological data lean toward an allopolyploid origin likely involving hybridization between related species followed by chromosome doubling. Structural differences in chromosome morphology and the presence of homeologous chromosomes support this theory. Because of its sterile nature, saffron relies entirely on human-assisted vegetative propagation. This has ensured the genetic uniformity of cultivated material but also limits opportunities for natural adaptation and improvement.

16.3.3 Reproductive Biology and Sterility Mechanisms

The reproductive biology of *C. sativus* is marked by a cascade of anomalies linked to its triploid status. During meiosis, chromosomes fail to align properly, resulting in irregular division and malformed gametes. Male gametogenesis, in particular, is severely compromised. Pollen grains are often misshapen or nonviable, and even viable pollen shows poor germination rates under in vitro conditions. Female reproductive structures are somewhat less affected. Ovule development occurs early in the sprouting phase, and embryo sacs can be detected while flowers are still enclosed within protective cataphylls. Experiments have shown that these ovules can survive beyond the flowering period and, under controlled conditions, can form parthenocarpic fruits.

Attempts to cross *C. sativus* with other species have yielded mixed results. When used as the female parent and pollinated with species like *C. cartwrightianus* or *C. thomasii*, saffron can produce fruits and seeds. However, reciprocal crosses where *C. sativus* serves as the pollen donor typically fail, suggesting that the primary barrier to sexual reproduction lies in male sterility. These findings underscore the importance of exploring alternative breeding strategies, including biotechnological approaches, for crop improvement.

16.3.4 Morphology

Saffron is an herbaceous perennial plant that propagates from an underground corm. It bears narrow leaves, violet-colored flowers, and crimson red stigmas. Flowering of saffron takes place in the month of October (Dhar and Mir 1997). Each corm of saffron produces 3–5 flowers from which dark red to reddish-brown stigmas are produced. The stigmas are trifid in shape and 25 mm in length. The styles are about 10 mm long, cylindrical in shape and are yellowish brown to yellowish orange with strong odor, aroma with bitter taste (Srivastava et al. 2010).

16.4 History

16.4.1 Traditional Use of Saffron

Saffron has been used for centuries in traditional medicine for its diverse therapeutic applications particularly in Ayurveda, where it is incorporated into various herbal formulations and home remedies. It is recognized for its multifunctional therapeutic roles, including its use as a stimulant, liver tonic, lactation enhancer, nerve tonic, aphrodisiac, detoxifying agent, digestive aid, immunity booster, fever reducer, diuretic, sedative, heart tonic, menstrual regulator, and as a relaxant with antianxiety and stress-relief properties. Historically, saffron has been used in managing ailments such as smallpox, measles, scarlet fever, respiratory conditions, and gastrointestinal problems (Mousavi and Bathaie 2011). Research also highlights the anticancer potential of saffron's bioactive components, showing promise in treating cancers of the breast, pancreas, lungs, and blood (Samarghandian et al. 2013). Further, it has been employed in addressing general fatigue, alcohol-related health issues, metabolic disorders like diabetes, inflammation, pediatric illnesses of unknown origin, insect bites, and fluid retention.

16.4.2 General Uses of Saffron

Saffron has been utilized for over three millennia, initially as a culinary spice and natural food colorant. Over time, its applications expanded into both traditional and contemporary medicine, where it is valued for its potential anticancer, antiseptic, antidepressant, antispasmodic, and carminative properties. Furthermore, it has been used in managing respiratory ailments such as colds, coughs, scarlet fever, and asthma. Beyond its medicinal significance, saffron also finds widespread use in religious rituals, herbal fand cosmetic formulations, perfumery and as a natural dye in the textile industry.

16.4.3 Therapeutic Properties

Crocin, a key bioactive compound found in saffron, has demonstrated notable therapeutic potential, particularly in regulating gene expression and inducing apoptosis in cancerous cells. Research suggests that crocetin, another active constituent, plays a role in hindering the synthesis of essential macromolecules like proteins, DNA and RNA within tumor cells, thereby restricting their growth. Furthermore, saffron extracts exhibit promising antitumor effects, largely attributed to their ability to disrupt nucleic acid formation and neutralize harmful free radicals. These antioxidant properties contribute significantly to saffron's potential in cancer prevention and therapy. The collective action of its compounds not only interferes with cancer cell proliferation but also enhances cellular protection making saffron a subject of increasing interest in oncology-related studies.

16.5 Utilization Pattern

Saffron (*Crocus sativus* L.), often hailed as the "golden spice," has been valued for centuries due to its vibrant color, unique aroma, and therapeutic properties. Its uses have evolved from ancient traditions to contemporary innovations across various domains.

16.5.1 Traditional and Modern Food Applications

Historically, saffron has played a central role in traditional cuisines, particularly in regions like Persia, India, and the Mediterranean. It was commonly used to flavor and color rice-based dishes, desserts, and herbal teas. In Kashmir, saffron-infused *kahwa* remains a cultural staple. The spice was also used in religious offerings and royal banquets, symbolizing purity and opulence. Today, saffron is used globally in gourmet foods, beverages, and luxury confectioneries. Its application extends to processed foods such as saffron milk, herbal drinks, and energy supplements, where it is appreciated not only for flavor but also for its bioactive health benefits.

16.5.2 Traditional Medicinal Uses

In traditional medicine systems like Ayurveda, Unani, and Chinese medicine, saffron was used for treating ailments such as cough, asthma, digestive issues, and menstrual irregularities. It was believed to balance bodily humors and was applied topically for skin disorders and to enhance complexion.

16.5.3 Modern Medicinal and Pharmaceutical Applications

Nowadays, increase in use of traditional medicine leads to extensive research on application and use of medicinal plants for treatment purposes. Research conducted on saffron confirms the vast scope of saffron and related Crocus species for medicinal purposes.

16.5.3.1 Antimicrobial Effect

Worldwide, antibiotic resistance to various microorganisms is rapidly increasing. Studies depicted the application of medicine plant derivatives and natural products as antimicrobial agents without imposing any harm. Similarly, research on saffron highlights the antimicrobial activity of stamen and petal of *Crocus sativus* (Vahidi et al. 2002). Comprehensive investigations into the bioactive properties of *Crocus sativus* have demonstrated its significant antibacterial potential against a variety of microbial strains (Muzaffar et al. 2016). Studies have shown that different solvent-based extracts, namely, ethanolic, aqueous, and methanolic from the flower petals, possess notable antimicrobial effects. These extracts have been particularly effective against several foodborne pathogens and disease-causing bacteria, indicating their possible application in food safety and pharmaceutical formulations (Gandomi et al. 2012). Such findings support the traditional medicinal uses of saffron and highlight its potential as a natural antimicrobial agent.

16.5.3.2 Antioxidant Effect

Different compounds in saffron as crocin phenols, safranal, and Crocetin extracted from the stigma of *C. sativus* possess potent intracellular free radical scavenging activity that helps in protecting cells and tissues against oxidation. The antioxidant activity of crocin has been a subject of study in differentiated pheochromocytoma neuronal cells. Cells were deprived of glucose that led to decreased intercellular superoxide dismutase activity and peroxidation of cell membrane lipids. The findings revealed an increased antioxidant capacity in cells subjected to glucose deprivation, with crocin successfully ameliorating the observed consequences. This indicates that crocin functions as a uniquely effective antioxidant, diminishing oxidative stress within neuronal cells (Maqbool et al. 2022).

16.5.3.3 Anti-inflammatory Effect

To investigate the anti-inflammatory and antinociceptive properties of saffron, researchers administered both ethanolic and aqueous extracts derived from its flower petals to mice. Separate investigations have shown that *Crocus sativus* and its constituent safranal exert protective effects on serum inflammatory markers in sensitized guinea pigs. Furthermore, ethanolic extracts have been linked to a reduction in chronic inflammation. These observed benefits are likely attributable to the rich presence of bioactive compounds, including flavonoids, tannins, anthocyanins, alkaloids and saponins (Bukhari et al. 2018).

16.5.3.4 Antidiabetic Effect

The prominent antidiabetic properties of saffron are attributed to its principal active constituents: crocin, crocetin, picrocrocin, and safranal. Specifically, crocin therapy at a dosage of 50 mg/kg has been shown to contribute to reduced blood lipid levels and the normalization of blood sugar and cholesterol (Bukhari et al. 2018).

16.5.3.5 Antihypertensive Effect

Hypertension is the major cause of many metabolic disorders and cardiovascular diseases. Use of different drugs for the treatment of hypertension imposes various adverse health risks. Alternatively, saffron and its bioactive components proved effective in treatment of hypertension and cardiovascular damage without harming normal cells (Razavi and Hosseinzadeh 2017).

16.5.3.6 Anticancer Effect

Therapeutic value of saffron alongside with its vibrant coloration and unique fragrance is largely conferred by compounds such as crocin, crocetin, picrocrocin and safranal. Investigations have highlighted the ability of these saffron constituents to selectively induce toxicity in cancer cells while sparing normal cells. Given its capacity to suppress tumor growth via antioxidant, anti-inflammatory and pro-apoptotic actions, saffron holds considerable promise as an adjunctive treatment in oncology (Mokhtarian et al. 2024).

16.5.4 Traditional and Modern Industrial Applications

Traditionally, saffron was used as a natural dye for fabrics, particularly for garments worn by nobles and religious figures. It also featured in perfumery and cosmetics, appreciated for its deep hue and fragrance.

In present times, saffron finds application in high-end cosmetic and skincare products due to its antiaging and brightening effects. It is used in serums, facial creams, and herbal formulations. In biotechnology and pharmacological industries, saffron's bioactive compounds are being explored for drug development and therapeutic applications. It also holds potential in the development of organic dyes and aroma-based products.

16.6 Production Technologies of Saffron

16.6.1 Climatic Requirement

Saffron thrives in regions characterized by a temperate and dry climate coupled with abundant sunny days. For economically viable flower production, optimal conditions include mean daytime temperatures ranging from 15 to 20 °C and night temperatures of 6 to 8 °C (Menia et al. 2018). The timing of precipitation is also crucial: early autumn rains significantly boost flower production, while spring rains are beneficial for corm multiplication, ensuring future yields. Further research provides specific temperature insights for various growth stages. Koocheki et al. (2010) observed that maximum vegetative growth occurs at a temperature of 27 °C, whereas an ideal temperature of 17 °C is required for optimal flowering. Complementing these findings, Yasmin and Nehvi (2018) indicated that proper vegetative development of the saffron plant necessitates an average temperature of 27.5 °C, alongside a precipitation of 418.90 mm per hectare. These climatic parameters collectively underscore the specific environmental needs for successful saffron cultivation and high-quality yields.

16.6.2 Soil Requirement

Saffron (*Crocus sativus* L.) flourishes in well-drained, adequately irrigated and clay-calcareous soils that are rich in organic matter (Golmohammadi 2014). While saffron adapts to a broad spectrum of soil types, it exhibits optimal growth in well-drained, calcareous and loose clay soils which facilitate unhindered root development (Menia et al. 2018). Conversely, saffron can also grow in saline soil, though a deficiency in calcium carbonate may impede its growth (Dar et al. 2017).

16.6.3 Planting Time

The optimal period for planting saffron corms generally spans from the second half of August to the first half of September. Research by Koocheki et al. (2016) emphasizes the critical importance of saffron planting, demonstrating that delayed planting significantly hinders growth and reduces flowering indices. Studies indicate that the highest yields are typically achieved when corms are planted in June, while planting as late as October results in the lowest yields. Specifically in India, the ideal sowing period for saffron corms extends from the last week of August through mid-September (Husaini et al. 2010). This timeframe is further corroborated by Kafi et al. (2018), who identified the last week of August to mid-September as the most suitable period for sowing corms to ensure optimal saffron growth.

16.6.4 Nutrient Management

Balanced nutrient management is a critical component in sustaining saffron cultivation and enhancing crop productivity. Optimal nutrient input directly influences yield outcomes, making it a key factor for long-term sustainability (Kafi et al. 2018; Mohammad et al. 2012). Globally, a common and effective fertilization strategy involves the incorporation of 20–30 metric tons per hectare of organic manure as a basal application, which improves soil structure and nutrient availability (Koocheki 2003). In several regions, this organic amendment is supplemented with chemical nitrogen at a rate of 23 kg per hectare, a combination known to enhance soil fertility and stimulate plant growth (Mohammad et al. 2012).

Further advancements in nutrient management have been observed with the integration of bioorganic inputs. For instance, the use of vermicompost and arbuscular mycorrhizal fungi has been associated with a significant rise in flower production, with studies reporting an increase of over 28% in the second year of application (Jami et al. 2020).

16.6.5 Water Management

Water management has a crucial role in cultivation of saffron, since saffron is a highly valued crop and has special requirements. Proper irrigation practices lead to good flowering, corm development, and high yield of saffron. Areas receiving seasonal rainfall of about 600 mm water application of about 50 mm is required for attaining economic output. However, areas receiving 400 and 200 mm require additional irrigation at 24- or 15-day intervals (Sepaskhah and Kamgar 2009). Koocheki et al. (2014) investigated the impact of water availability on saffron yield, specifically highlighting its role in enhancing both flower and dried stigma production. Their research indicated that applying 50% of the saffron's total water requirement (SWR) in the initial year was beneficial. However, sustained reduced irrigation had a detrimental effect. Nehvi and Makhdoomi (2007) observed a subsequent decrease in flower count and dry stigma yields during the second year when only 50% of the SWR was provided in contrast to treatments receiving 75% or 100% of the SWR. This suggests that while moderate irrigation might suffice initially, consistent and adequate water supply becomes crucial for maintaining optimal saffron productivity in subsequent years.

16.6.6 Weed Management

Effective weed management is crucial for maintaining healthy crop development in saffron cultivation. In regions where saffron is cultivated as an annual crop, manual weeding remains the primary method. Conversely, perennial saffron fields are typically managed using chemical herbicides such as Atrazine (Gesaprim 50) and Simazine (Gesatop 50) applied at a rate of 1.0 kg per hectare (Dar et al. 2017). In the saffron-growing areas of Jammu and Kashmir, India, several weed species are commonly encountered. These include *Polygonum aviculare*, *Medicago lupulina*, *Lithospermum arvense*, *Tulipa stellata*, *Poa bulbosa*, *Papaver rhoeas*, *Lepidium virginicum*, *Euphorbia helioscopia*, *Erodium cicutarium*, *Salvia moorcroftiana*, *Galium tricorne*, *Ranunculus arvensis*, *Filago arvense*, and *Chenopodium album* (Husaini et al. 2010).

Postharvest herbicide applications, particularly at the 6–8 leaf stage, have shown effective control of these weeds. Notably, Tribenuron-methyl (18.75 g active ingredient per hectare) and Ioxynil (750 g a.i. ha^{-1}) have demonstrated strong efficacy in suppressing weed populations. Furthermore, Metribuzin, applied at 560 g a.i. ha^{-1} during spring or autumn, has proven to be effective in targeting undesirable plants while preserving saffron crop (Menia et al. 2018).

16.6.7 Harvesting

The harvesting of saffron is a meticulous, multistage process involving both flower collection and subsequent stigma separation. The onset of saffron flower collection varies geographically but generally begins as soon as flowers emerge in the fields, typically from October to November in regions like Khorasan, Iran and J&K, India (Kafi et al. 2018; Kafi and Showket 2007). This timing is highly dependent on local climatic conditions, geographical topography, and initial irrigation schedules. Flowers are carefully picked in the early morning to preserve their delicate quality. Following collection, the vital process of stigma separation takes place indoors, protecting the delicate saffron threads from environmental elements. Once separated, these precious flower parts are traditionally spread out to dry under shade for 2–3 days (Husaini et al. 2010).

Different saffron-producing regions have distinct flowering seasons and harvesting customs. In Italy, for instance, the flowering season typically spans from mid-October to November. Spain celebrates this crucial period with "*the day of the mantel*" a term reflecting the peak of anthesis when the landscape appears covered in a vibrant blanket of saffron flowers. Ultimately, the precise timing of saffron flowering is heavily influenced by prevailing weather conditions with plants generally blooming in the autumn, approximately 90 days after planting (Dar et al. 2017).

16.6.8 Crop Productivity

The production of saffron flowers and stigmas is significantly influenced by a combination of environmental conditions and genetic factors. Achieving a substantial yield, such as 1 kg of dry saffron, necessitates an enormous quantity of fresh flowers approximately 160,000 weighing around 78.5 kg (Husaini et al. 2010). A key determinant of flower production is the weight of the saffron corm. Research by Dar et al. (2017) demonstrated a positive correlation between corm weight and the number of flowers a corm produces, indicating that larger corms lead to a greater number of flowers. This finding is further supported by Emadi (2009), whose work validated that corms weighing over 45 grams can yield as many as twelve flowers per corm.

16.7 Protection Technologies of Saffron

16.7.1 Diseases and Their Management

Saffron cultivation, though economically rewarding, faces several biological challenges with plant health being a primary concern. One of the most serious threats is corm rot, a disease caused by soil-borne fungi such as *Fusarium oxysporum*, *Rhizoctonia solani*, and *Penicillium* species. Infected corms often decay, emit a foul odor and fail to germinate leading to significant crop loss. The disease is exacerbated by poor drainage and continuous cultivation in the same field. To manage this, farmers are advised to use healthy, disease-free corms, ensure well-drained soils, and treat planting material with appropriate fungicides like carbendazim or captan. Rotating crops and conducting deep ploughing during summer can further help to reduce the fungal load in the soil.

In addition to corm rot, foliar diseases such as leaf spot and blight caused by *Alternaria alternata* and *Botrytis cinerea* pose challenges during the vegetative stage. These diseases result in yellowing, necrosis, and premature drying of leaves, which adversely affects photosynthesis and flower production. Management includes removing infected plant debris, ensuring adequate spacing to promote air circulation and timely application of preventive fungicides. Maintaining field hygiene and monitoring weather conditions also play an important role in limiting the spread of these diseases.

Nematode infestation particularly by root-knot nematodes (*Meloidogyne* spp.) is another hidden yet damaging issue in saffron fields. These microscopic pests impair root function and reduce nutrient uptake resulting in stunted growth and poor corm development. Effective control strategies include the use of nematode-free corms, application of organic soil amendments like neem cake and practicing crop rotation with nonhost plants. An integrated disease and pest management approach combining cultural, chemical and biological practices is essential to safeguard saffron production and ensure long-term sustainability of this high-value crop.

16.7.2 Rodent Management

Effective rodent management presents a significant challenge due to the complex and adaptive behavior of rodent species. A comprehensive and systematic approach is essential to reduce the damage caused by these pests, especially in saffron-growing regions. Prior to initiating control strategies, it is critical to understand certain behavioral characteristics of rodents:

- Rodents typically limit their movement to a radius of 5–10 meters around their burrows.
- Rodent control measures are most effective when implemented across large, contiguous areas and repeated periodically. To achieve a meaningful reduction in population, it is necessary to eliminate more than 90% of the rodents; otherwise, their rapid breeding quickly restores their numbers.
- Rodent migration must also be factored into control programs. Isolated efforts by individual farmers, grain handlers, or officials are insufficient. Instead, coordinated community-based management with structured planning is necessary to integrate rodent control into broader agricultural protection frameworks.

In the saffron-producing zones of the Kashmir Valley, two rodent species are of primary concern:

16.7.2.1 Blyth's Vole (*Pitymys leucurus*)

This small, diurnal burrowing rodent is active throughout the vegetative phase of saffron (November to April). It poses a major threat by damaging the base of the saffron shoots where they connect to the corms, leading to detachment of radical leaves and ultimately the loss of the plant. The voles often transport and store corms in nearby fields as a food reserve for the summer dormancy period. Damage is evident from patches of dry, withered foliage that resemble corm rot symptoms.

Integrated Management Strategies

(i) *Agronomic Factors*: Traditional practices, such as longer corm planting cycles, use of raised beds, intercropping saffron with almond, and poor weed management have turned saffron fields into favorable habitats for voles. These fields remain active during October to March, while other agricultural lands lie fallow, increasing the risk of infestation due to lack of routine pest control.

(ii) *Field Sanitation*: Remove crop residues, weeds, and debris to reduce available food and nesting material for rodents.

(iii) *Bund Modification*: Minimize bund (field boundary) width to about 30 cm to discourage burrowing. This modification reduces habitat availability and exposes rodents to predators.

(iv) *Mechanical Control—Burrow Fumigator*: A specially designed burrow fumigator has been tested and found effective in saffron fields. The device suffocates rodents by releasing oxygen-depleting fumes into their burrows for 10–15 min.

- Application Procedure:

Fuel (cow dung, maize straw/pith, and local herbs) is ignited at the top of the fumigator tank. A blower mechanism directs fumes into the burrows. One operator manages the fumigator, while a second person seals the burrow entrances and prevents escape.

(v) *Chemical Control—Rodenticide Baiting Schedule*:

- *Day 1*: Identify and plug all visible burrows.
- *Day 2*: Identify live burrows and pre-bait with 10–15 g of plain bait (crushed rice + wheat flour + jaggery/sugar + mustard oil).
- *Day 3*: Apply 2.0% Zinc Phosphide poison bait (6–10 g/burrow) in the late evening.
- *Day 4*: Collect and safely dispose of dead rodents. Re-seal all burrows.
- *Day 5*: Re-check for active burrows.
- *Day 6*: Fumigate reopened burrows using Aluminum Phosphide pellets (2 pellets/burrow) or pouches (5–10 g), sealing the entrances with moist soil.

Precautions:

- Rodents may develop bait shyness after one application of Zinc Phosphide. A gap of at least 50–60 days should be maintained before reuse.
- During snowy or rainy periods, baiting should be done using bait stations.

(vi) *Trapping*: Using Sherman traps at a density of 15–20 traps/ha can help control residual rodent populations. Trapping is most effective when integrated with chemical control programs.

16.7.2.2 Indian Crested Porcupine (*Hystrix indica*)

This is the largest rodent species found in India and is recognized by its long, stiff bristles (15–30 cm) on the back and neck that form a prominent crest when alarmed. The animal is covered in quills with alternating dark brown and white bands, and the tail is equipped with short, needle-like spines. These features provide it with excellent defense and make it a difficult pest to control.

Management Strategies

(i) *Lighting*: Installing bright lights around the field perimeter can deter porcupines from entering at night.
(ii) *Fencing*: Use metallic sheets (45–60 cm in height) to enclose the field. This acts as a physical barrier to porcupine entry.
(iii) *Tree Protection*: Wrap tree trunks with 30-inch-wide aluminum sheets to prevent porcupines from damaging mature trees.
(iv) *Legal Considerations*: The Jammu and Kashmir Wildlife (Protection) Act, 1978 (amended up to 2002) currently prohibits population control measures against Indian crested porcupines due to their protected status. This necessitates nonlethal, preventive strategies only.

16.8 Saffron Improvement (SKUAST-Kashmir Achievements)

16.8.1 Geographical Indication (GI) Tagging and Its Impact

One significant step toward safeguarding the authenticity of Indian saffron was the acquisition of a Geographic Indication (GI) tag for Kashmir saffron in 2020. The GI tag is a form of intellectual property that indicates the origin of a product and guarantees its quality, thus preventing the sale of adulterated or counterfeit saffron under the "Kashmir Saffron" name.

In addition, the Saffron Heritage of Kashmir in India was recognized by the Food and Agriculture Organization (FAO) in 2011 by recognizing Pampore's Saffron Heritage Site as a *Globally Important Agricultural Heritage System* (GIAHS) and efforts are being made to preserve and promote its unique traditional farming practices, cultural significance and ecological value on an international platform. This recognition is unique among saffron-producing countries of the world. These systems make an outstanding contribution to brand protection, consumer trust, market premiums, food security, biodiversity, indigenous knowledge and cultural diversity, thereby supporting sustainable development.

16.8.2 Technological Interventions and Innovations

Technological advancements are crucial for improving the quality and quantity of saffron production in India. Several initiatives taken by SKUAST-Kashmir have been introduced to modernize cultivation, harvesting, and processing:

16.8.2.1 Varietal Development

SKUAST-Kashmir developed the first global variety in saffron, *Shalimar Saffron-1* (Fig. 16.3), through clonal selection followed by mutation breeding. Some of the salient features of the variety are:

Fig. 16.3 Corm and yield attributes of Shalimar Saffron-1

- *Adapted Ecology*: Suitable for both traditional and nontraditional areas of the Kashmir Valley, up to an altitude of 1600–1800 m above mean sea level (amsl).
- *Quality*: It shows an increase of 11.81%, 28.90%, and 12.46% in crocin, picrocrocin and safranal contents, respectively, compared to the natural population.
- *Corm Characteristics*: This variety displays a high corm size index, a low multiplication rate and a strong flower production index. It shows moderate resistance to diseases like corm rot and leaf blight, as well as pests such as mites and demonstrates good resistance to thrips.
- *Yield*: Over a 3-year cultivation period, this variety yields an average of 6.18 kg/ha, marking a 26.04% improvement over the local population benchmark (natural population).
- *Benefit-Cost (B:C) Ratio*: Under high-density planting conditions, the variety delivers a B:C ratio of 4.15:1 outperforming the 3.51:1 ratio recorded for the local check (natural population).

16.8.2.2 Genetic Characterization of Saffron Germplasm

A total of 86 saffron germplasm lines, 44 of exotic origin from the Netherlands and Spain and 42 from the Kashmir region were analyzed for their genetic variation. Genomic DNA was isolated and subjected to PCR-based amplification. The resulting DNA sequences were aligned and examined to construct a partial phylogenetic framework for the *Crocus* genus. The phylogenetic analysis revealed a distinct divergence of *Kashmiri* lines from those of European origin. In particular, *Crocus sativus* populations from Kashmir formed a separate clade, indicating early evolutionary separation. The European lines clustered into a separate subgroup, highlighting geographical differentiation. Further research by Mudasir et al. (2021) and Busconi et al. (2021) supports these findings on saffron's genetic diversity.

16.8.2.3 Biochemical Markers of Kashmir Saffron

Kashmir saffron (*Crocus sativus L.*) is recognized not only for its unique aroma and vibrant color but also for its rich nutritional profile. Each floral component holds a distinct biochemical value. The stamens are abundant in essential nutrients, including ash, proteins, lipids, phosphorus, magnesium, potassium and inositol. Tepals, often overlooked, serve as a source of various sugars such as glucose, sucrose, maltose, mannitol and reducing sugars. The pistil, which is the primary commercial part of the flower, is notably rich in carbohydrates, calcium, sodium, iron, fructose, and sorbitol. Research highlights the exceptional quality of Kashmir saffron particularly in terms of its high carotenoid content. Comprehensive chemical profiling has also been documented by several researchers underscoring the health-promoting potential of this valuable spice.

16.8.2.4 Morphological Descriptors of Kashmir Saffron

Morphological descriptors for Kashmir saffron have been systematically developed focusing on characteristics related to corms, flowers, and foliage. Research conducted in the Pampore saffron heritage zone revealed that the saffron cultivated in this region represents a stable natural subpopulation. Due to its triploid nature, the crop does not produce viable seeds, which explains the absence of new genetic recombinants. The observed variations in floral and foliar structures appear to be nonheritable, implying that such differences may arise from somatic changes during corm division rather than from genetic recombination. This variation, especially in economically important traits and corm characteristics, presents significant potential for genetic improvement through clonal selection within the existing population. Detailed morphological characterization of saffron grown in other regions has also been reported in earlier studies, including those by Saxena (2010) and Cardone et al. (2021), further supporting the diverse phenotypic expression within *Crocus sativus* across geographical locations.

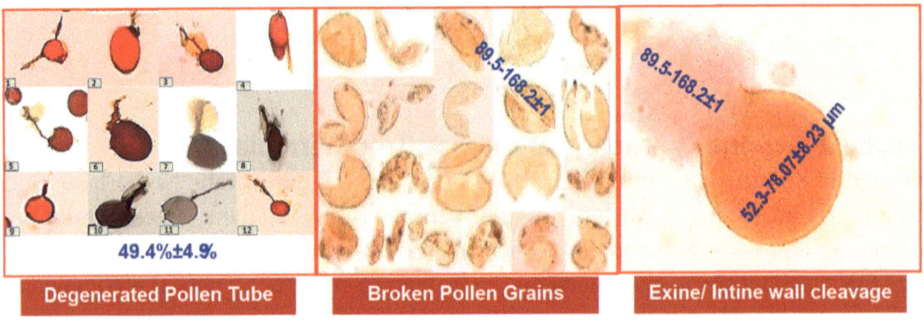

16.8.2.5 Pollen Morphology of Kashmir Saffron

An in-depth investigation into the pollen morphology of Kashmir-grown saffron (*Crocus sativus*) has revealed critical insights into its reproductive limitations. The pollen grains vary in size from approximately 80.5–161.9 μm, with an average germination rate of just 16.18%, ranging between 10.5% and 21.2%. Notably, a significant increase in the proportion of abnormal pollen grains was observed under decreasing temperature conditions, suggesting that colder climates negatively affect pollen viability. One of the key abnormalities identified was exine breakage, which is strongly linked to the high rate of pollen sterility. Furthermore, only a minimal 0.53% of pollen grains successfully developed normal pollen tubes, highlighting a major barrier to successful fertilization. These findings affirm that the reproductive sterility observed in *C. sativus* cultivated in Kashmir is primarily due to both structural deformities in pollen grains and poor pollen tube formation. These results are consistent with earlier studies conducted by Sani et al. (2013) and Chichiricco and Grilli (2014) emphasizing the sterility of saffron under temperate climatic conditions.

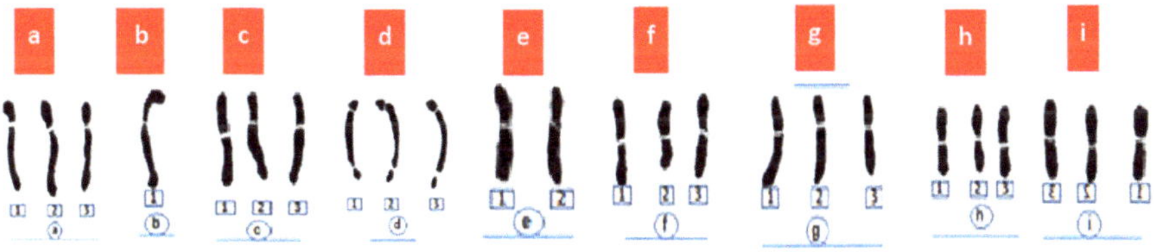

16.8.2.6 Cytological Markers

Cytogenetic analysis of Kashmir saffron (*Crocus sativus*) has confirmed that it possesses a triploid chromosome number with $2n = 3x = 24$ chromosomes in its somatic cells. Detailed karyotype assessments conducted across 30 metaphase plates (10 per subpopulation) revealed a total chromosomal length of approximately 160.56 μm, with an average chromosome size of 6.69 μm. The overall arm ratio of the karyotype was calculated to be 2.10, indicating a predominance of submetacentric chromosomes. Interestingly, the chromosome length in Kashmir saffron was found to be greater than that recorded in Italian varieties (reported at 85.36 μm) and closely aligned with Iranian saffron as reported by Aghayev (2002), which measured 164.21 μm in total length with a slightly higher average chromosome size of 6.84 μm and an arm ratio of 2.18. These findings suggest a potential regional variation in karyotype structure and size, reflecting the genetic uniqueness of Kashmir saffron. Previous studies by Aghayev (2002) and Aghayev et al. (2004) have also contributed significantly to understanding the chromosomal architecture of this valuable crop.

16.8.2.7 Standardized Cost-Effective In-vitro Protocol for Corm Production

Due to its triploid genetic structure, saffron (*Crocus sativus*) is incapable of sexual reproduction and thus relies entirely on vegetative propagation. However, the naturally low multiplication rate coupled with susceptibility to fungal infections often

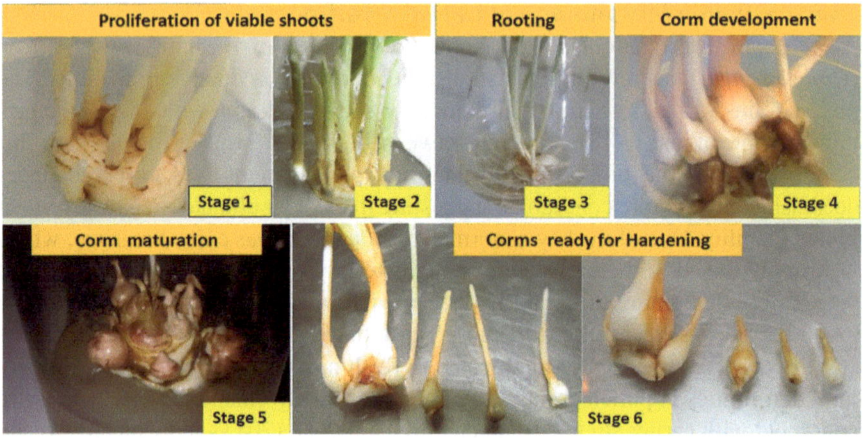

Fig. 16.4 Sequential stages of in-vitro corm multiplication in saffron (*Crocus sativus* L.)

exacerbated by erratic weather conditions in Kashmir has limited the availability of healthy quality planting material. In response to these constraints, SKUAST-Kashmir has successfully developed a reliable *in-vitro* direct regeneration protocol for producing saffron corms under controlled conditions (Fig. 16.4). This innovative approach presents a promising solution for scaling up saffron cultivation beyond traditional zones within J&K. By providing disease-free, uniform planting stock, the protocol not only supports increased productivity but also promotes long-term sustainability of saffron farming.

16.9 Postharvest Management

16.9.1 Chemistry of Saffron

To date, approximately 150 chemical constituents, both volatile and nonvolatile, have been identified from the stigmas of *Crocus sativus* (saffron), though detailed structural and functional characterization exists for fewer than 50 of these (Winterhalter and Straubinger 2000). These compounds contribute to the unique aroma, color, and therapeutic properties of saffron. Among the key chemical groups found in saffron are terpenes, terpene alcohols and their respective esters, which play important roles in its bioactivity and sensory characteristics. The nonvolatile fraction of saffron is particularly rich in crocetin, crocin, picrocrocin, and various flavonoids such as kaempferol and quercetin (Liakopoulou-Kyriakides and Kyriakidis 2002). Crocin, a glucosyl ester of crocetin, is the primary pigment responsible for saffron's vivid golden-yellow to orange hue (Shahi et al. 2016). Chemically defined as 8,8-diapocarotene-8,8-dioic acid, crocin exists in several forms, of which α-crocin (also known as crocin-1) is the most abundant. This compound is a trans-crocetin di-(β-D-gentiobiosyl) ester, which imparts the characteristic color due to its high pigment concentration and solubility.

In terms of flavor, picrocrocin is the second most significant component, contributing a distinct bitter taste and constituting 1–13% of saffron's dry weight (Melnyk et al. 2010). This compound also serves as the biochemical precursor for safranal, the dominant volatile compound responsible for saffron's aroma. Safranal is chemically identified as 2,6,6-trimethyl-1,3-cyclohexadiene-1-carboxaldehyde, and it is primarily formed through the degradation of picrocrocin. During this transformation, intermediate molecules such as HTCC (4-hydroxy-2,6,6-trimethyl-1-cyclohex en-1-carboxaldehyde) are produced, further enhancing the aromatic profile of the spice (Maggi et al. 2010). Table 16.2 shows the phytochemicals isolated from different parts of saffron. The diverse and complex phytochemical composition of saffron underlines its importance not only as a culinary ingredient but also as a valuable resource in medicinal, cosmetic, and nutraceutical industries.

16.9.2 Major Secondary Metabolites of Saffron

The distinctive flavor, color, and aroma of saffron are primarily attributed to a group of specialized secondary metabolites. These bioactive compounds, which are naturally synthesized by the plant play a crucial role in defining the sensory and medicinal qualities of saffron. The key secondary metabolites include crocin, which imparts the vibrant golden-yellow hue; picrocrocin, responsible for its characteristic bitter taste and safranal, which contributes to the rich aromatic fragrance. Each of these compounds not only enhances the culinary value of saffron but also possesses therapeutic properties making saffron a highly valued spice in both traditional and modern medicinal systems.

16.9.2.1 Carotenoids

Saffron in aqueous solution gives intense yellow color, which is mainly because of the presence of a group of chemical compounds commonly known as Carotenoids. These compounds are mostly water insoluble, which constitutes an extensive group of natural pigments as it contains 600 different chemical components (Kull and Pfander 1995). They play an important role against photo-oxidation in the presence of light and thus act as protectors (Adams et al. 1996). Furthermore, their synergistic effects not only define the spice's culinary appeal but also contribute to its recognized medicinal and antioxidant potential.

16.9.2.2 Crocetin

Crocetin is a naturally occurring lipophilic carotenoid, recognized for its multi-unsaturated molecular structure and a chemical formula of $C_{20}H_{24}O_4$ (Samarghandian and Borji 2014). As a key bioactive compound found in saffron, crocetin has been extensively studied for its diverse therapeutic properties. Research has highlighted its potential in combating cancer by inhibiting tumor growth and promoting apoptosis in cancerous cells (Bhandari 2015). Furthermore, it exhibits promising antidepressant effects by modulating neurotransmitter activity and improving mood-related

Table 16.2 Chemical constituents reported from saffron

S.no.	Compound name	Plant part	References
1.	Crocusatin I	Stigma	Li and Wu (2002)
2.	Crocusatin F	Stigma	Li and Wu (2002)
3.	Crocusatin G	Stigma	Li and Wu (2002)
4.	Crocusatin H	Stigma	Li and Wu (2002)
5.	2,2-Dimethyl cyclohexane carboxaldehyde	Stigma	Kanakis et al. (2004)
6.	Crocusatin D	Pollen	Kanakis et al. (2004)
7.	Lanierone	Stigma	Kanakis et al. (2004)
8.	Crocusatin B	Pollen	Kanakis et al. (2004)
9.	Crocusatin A	Pollen	Kanakis et al. (2004)
10.	4-Hydroxy-3,5,5- trimethyl-2-cyclohexen-1-one*O*-*β*—glucopyranoside	Stigma	Kanakis et al. (2004)
11.	4-Hydroxy-2,6,6-trimethyl-3-oxo-1,4-cyclohexadiene-1-carboxaldehyde	Stigma	Kanakis et al. (2004)
12.	7-Megastigmen-9-one	Leaf oil and stigma	Li and Wu (2002)
13.	7-Acetyl-4-methyl-1-azulenecarbox carboxylic acid		Li and Wu (2002)
14.	Safranal	Stigma	Escribano et al. (1996), Tarantilis et al. (1994), and Rezaee and Hosseinzadeh (2013)
15.	3,3,4,5-Tetramethyl cyclohexanone	Stigma	Kanakis et al. (2004)
16.	2,4,6-Trimethyl benzaldehyde	Stigma	Rodel and Petrzika (1991)
17.	2,6,6-Trimethyl-1, 4-cyclohexadiene-1-carboxaldehyde	Stigma	Tarantilis (1997)
18.	Dihydro-4-oxoisophorone	Stigma	Zarghami and Heinz (1971)
19.	*β*-Cyclocitral	Stigma	Li and Wu (2002)
20.	3,5,5-Trimethyl-2-cyclohexen-1-one	Stigma	Li and Wu (2002)
21.	,35,5-Trimethyl-3-cyclohexen-1-one	Stigma	Li and Wu (2002)
22.	3,5,5-Trimethyl-4-methylene-2-cyclohexen-1-one	Stigma	Kanakis et al. (2004)
23.	Crocin 1*b*	Stigma	Pfister et al. (1996) and Hosseinzadeh and Nassiri-Asl (2013)
24.	Crocin 2*b*	Stigma	Moradzadeh et al. (2018)
25.	Crocin 3*b*	Stigma	Xiang et al. (2017)
26.	Crocin 4*b*	Stigma	Pradhan et al. (2019)
27.	Crocetin	Stigma	Moradzadeh et al. (2018)

(continued)

Table 16.2 (continued)

S.no.	Compound name	Plant part	References
28.	Mangicrocin	Stigma	Ghosal et al. (1989)
29.	3-*O*-[6-*O*-Acetyl-β—glucopyranosyl-(1→2)-β—glucopyranoside	Petals	Montoro et al. (2008)
30.	Picrocrocin*b*	Stigma	Buchecker and Eugster (1973)

behaviors (Ohba et al. 2016). Studies have also suggested its cardioprotective role, indicating that crocetin may help improve heart function and reduce the risk of cardiovascular disorders by enhancing blood flow and reducing oxidative stress in cardiac tissues (Wang et al. 2014). These multifaceted health benefits position crocetin as a valuable compound in the field of natural medicine and pharmacology.

16.9.2.3 Crocin

Crocin, a diester derivative of crocetin, makes up nearly 80% of saffron's total bioactive constituents and plays a central role in imparting the spice's distinctive golden-yellow to reddish hue (Shahi et al. 2016). This water-soluble carotenoid has the molecular formula $C_{44}H_{64}O_{24}$ and is a primary pigment in saffron (Fig. 16.5; Gregory et al. 2005). Among its several forms, crocin 1 also known as α-crocin is the most prevalent (Melnyk et al. 2010). It appears dark red in its solid form but gives off a vibrant reddish-orange color when dissolved in water, making it widely used as a natural dye and food colorant. Crocin also exhibits potent antioxidant properties by neutralizing free radicals and protecting cellular structures from oxidative damage (Papandreou et al. 2006). Unlike safranal, which tends to degrade under harsh environmental conditions, crocin remains relatively stable. Its peak absorbance (λmax) occurs at 440 nm, reflecting its strong pigmentation and chemical stability (Shahi et al. 2016; Kabiri et al. 2017).

16.9.2.4 Picrocrocin

Picrocrocin ranks as the second most abundant compound in saffron by weight, following crocin. Although present in smaller quantities, it plays a vital role in imparting saffron's distinct bitter flavor (Alonso et al. 2001). This compound has a molecular formula of $C_{16}H_{26}O_7$ and displays a maximum absorbance (λmax) at 254 nm (Fig. 16.6; Kabiri et al. 2017). It originates from the oxidative cleavage of zeaxanthin, a carotenoid precursor (Lozano et al. 1999). Upon natural de-glycosylation typically triggered during the drying of saffron stigmas, picrocrocin is transformed into safranal, the key volatile responsible for the spice's characteristic aroma and astringent taste (Pitsikas 2016; Shahi et al. 2016). In its native form, picrocrocin is colorless and odorless but has a distinctly bitter flavor. Freshly harvested saffron flowers lack any notable scent; however, drying activates enzymatic and thermal processes that convert picrocrocin into safranal, giving rise to saffron's signature fragrance.

16.9.2.5 Safranal

Safranal, a key aromatic compound in saffron, was first discovered by Kuhn and Winterstein and later synthesized by Kuhn and Wendt. This compound exhibits a

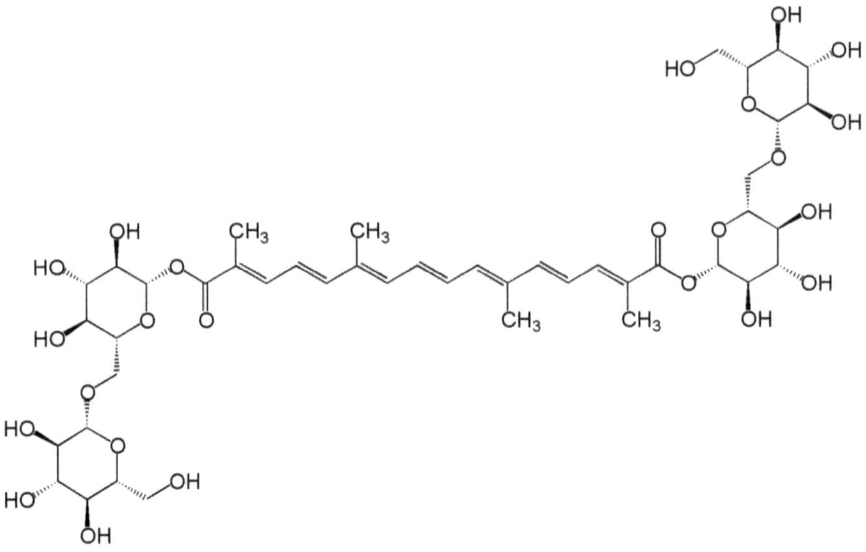

Fig. 16.5 Chemical structure of crocin

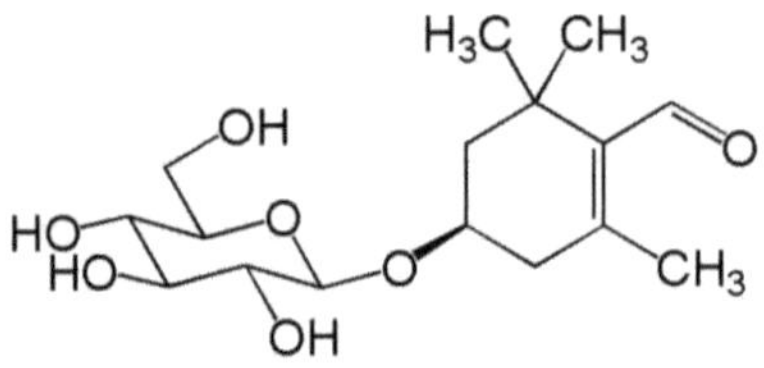

Fig. 16.6 Chemical structure of picrocrocin

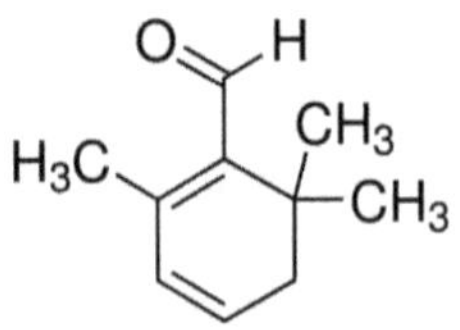

Fig. 16.7 Chemical structure of safranal

maximum absorbance (λmax) at 330 nm (Fig. 16.7; Kabiri et al. 2017). Interestingly, freshly harvested stigmas lack safranal and is generated during postharvest processing. Its formation is catalyzed by the enzyme β-glucosidase, which acts on picrocrocin during the dehydration phase of drying and storage (Maggi et al. 2010; Shahi et al. 2016). Safranal is chiefly responsible for saffron's distinct fragrance and accounts for up to 70% of its total volatile content (Aytekin and Acikgoz 2008). It comprises approximately 30–70% of the essential oil composition and between 0.001% and 0.006% of saffron's dry weight (Carmona et al. 2006; Maggi et al.

2010). Analytical techniques such as gas chromatography and spectrophotometry are commonly used to determine its concentration and purity.

16.9.2.6 Flavonols

Flavonols, a subgroup within the larger class of flavonoids, are bioactive compounds widely present in saffron. Chemical analyses of saffron extracts have confirmed the presence of glycosylated forms of two major flavonols: kaempferol and quercetin. Among these, kaempferol derivatives are predominant, making up approximately 90.3% of the total flavonol content in *Crocus sativus* L. flowers, while quercetin and its related compounds constitute about 9.7% (Kumar et al. 2009). Moreover, high levels of di- and triglucoside derivatives of these flavonols have also been observed with noticeable variability in concentration. Optimal extraction of kaempferol, measured at 1.26 μg/mg occurs at a temperature of 100 °C over a 30-min period (Emami et al. 2012). These flavonoids are known for their diverse biological activities including antioxidant, anti-inflammatory, antibacterial, and anticancer properties all of which contribute significantly to the therapeutic value and natural preservative ability of saffron.

16.9.2.7 Glycosidic Aroma Precursors

Aromatic components constitute the nonvolatile glycosidic precursors besides picocrocin (Winterhalter and Straubinger 2000). In 1997, three new glysodic aroma precursors candidates were isolated by using MLCCC chromatography in saffron as 4S, 4-hydroxy-3,5,5-trimethyl-2-cyclohexanone, 4R, 4-hydroxy-3,5,5-trimethyl-2-cyclohexanone and 4S, 4-hydroxy-methyl 3,5,5-trimethyl-2-cyclohexanone (Straubinger et al. 1997). Later on, 13 new glycoside compounds were identified by employing the same technique as (a) 4R, 4-hydroxy-2,6, 6 trimethyl 1-cyclohexenecarboxyaldehyde (b) 2-hydroxy-6-6, dimethyl-2-cyclohexen-1,4-dione, (c) 4R, 4-hydroxy-2,6, 6 trimethyl 1-cyclo hexenecarboxylic acid, (d) 6-hydroxy-3-hydroxymethy-2,4,4-trimethyl-2,5 cyclo-hexadienone, (e) 5-hydroxy-7-7, dimethyl-4,5,6,7-tetrahydro-3H-isobenzofuran-1-one, (f) Roscoide (g) 2-Phenylethanol (h) Benzenemethanol (i) 4R, 3, 5,5-trimethyl-3-cyclohexen-1-ol, (j) 4-hydroxy-2,2,6- trimethyl-1-cyclohexen-carboxaldehyde, (k) 4,4,6-trimethyl-7-oxabyciclo heptan 2-one.

16.9.2.8 Evolution of Saffron Aroma

During the evolution of experiments on treatment of various diseases by saffron, two more additional compounds were isolated due to their significant importance. The first one was identified as 4-hydroxy-β cyclocitral or HTCC (2,6,6-trimethyl-4-hydroxy-1-cyclohexen-1-carbaldehyde) having retention time of 15.35 min. Identification of the second important and complex compound with a retention time of 12.48 min appears approximately similar to compound in the spectral library as dihydroactinidiolide (DHA) compound. The presence of this compound has great importance, since it is present in several plants and plays a major role in specific enzymatic β-carotene degradation product (Kanasawud and Crouzet 1990).

16.9.3 Drying

The postharvest processing of saffron, particularly drying, plays a pivotal role in determining its final quality. The concentration of its key bioactive compounds is highly influenced by how quickly and efficiently the drying process is carried out. Prompt dehydration of *Crocus sativus* L. stigmas immediately after harvest is essential to preserve these secondary metabolites and to prevent the breakdown of crocin into crocetin, a common issue observed in Kashmir where traditional drying methods are still widely used.

In conventional practices, farmers typically dry the stigmas in the shade on cloth or under direct sunlight. These methods are slow, often taking over 50 h, especially when drying whole flowers. Prolonged exposure to sunlight and air results in the gradual degradation of biochemical compounds, leading to a decline in saffron's color, flavor, and aroma. During the drying process, saffron stigmas lose approximately 80% of their initial weight due to moisture loss. This dehydration not only contributes to the desired physical characteristics of the spice but also enhances its shelf life, making it suitable for long-distance transportation and commercial storage.

To ensure quality and consistency, it is recommended to dry saffron using controlled methods such as hot air ovens, vacuum dryers, or electric dehydrator ideally at around 60 °C. The optimal drying temperature may vary depending on whether the stigmas are harvested from opened or unopened flowers as moisture content differs between the two. Attaining a final moisture content of around 10–12%, as specified by the ISO 3632 quality standards is essential for preserving the quality of the spice. Traditional methods often fail to reach this threshold contributing to quality loss during storage and marketing. Hence, modernization of drying techniques is crucial to maintain the high value and global competitiveness of saffron. Acar et al. (2011) evaluated the impact of freezer-drying (−40 °C for 4 h) versus sun-drying (18 °C ambient temperature). Their findings indicated that freezer-dried stigmas contained the highest levels of crocin and safranal. Subsequent studies have corroborated the importance of controlled drying environments. Feili et al. (2012) reported that indirect solar drying resulted in higher quality saffron compared to traditional techniques. Consistent with these observations, Chaouqi et al. (2018) found that oven-drying at 40 °C produced better quality saffron than traditionally dried samples.

16.9.4 Packaging and Storage

Proper packaging and storage of saffron are critical to maintaining its quality, potency, and commercial value. As a delicate and high-value spice, saffron is highly sensitive to moisture, light, oxygen, and temperature fluctuations. Therefore, it must be handled with care from the point of harvesting to final consumption.

Saffron is available in various forms, including filaments (stigmas), cut filaments, and powder. Regardless of the form, it must be packed in rigid, airtight containers that are clean, nonreactive, and tamper-proof. The material used for packaging should be food-grade and must not alter the color, aroma, or chemical composition of saffron. Common materials include glass jars, aluminum foil containers, and high-barrier plastic or composite pouches. These containers protect the product from environmental factors such as humidity, oxygen, and light, which can degrade its active compounds like crocin, safranal, and picrocrocin. Furthermore, the packaging should be waterproof and easy to handle, ensuring convenience for consumers.

From a regulatory standpoint, saffron packaging must comply with national and international food safety standards, including guidelines on material safety, labeling, and environmental sustainability. For long-term storage, saffron should be kept in cool, dry, and dark conditions ideally at temperatures below 20 °C and in humidity levels under 40%. Storing it in opaque containers away from strong odors and direct sunlight helps preserve its sensory and biochemical properties. Proper storage not only extends shelf life but also ensures that saffron retains its vibrant color, strong aroma, and therapeutic qualities until the point of use.

16.10 Value Addition and Products

Saffron, renowned globally for its distinct aroma, color, and flavor, offers significant potential for value addition beyond its raw stigmas. By transforming raw saffron into a range of high-value products, producers can unlock new markets and boost profitability. This includes processing saffron into extracts, tinctures, and essential oils for use in cosmetics, pharmaceuticals, and perfumery, where its unique bioactive compounds like crocin, crocetin, and safranal are highly sought after for their antioxidant, anti-inflammatory, and other beneficial properties (Goli et al. 2019; Razavi et al. 2019).

Culinary applications extend beyond simple threads to encompass saffron-infused gourmet salts, honeys, teas and spice blends, catering to discerning palates in diverse cuisines worldwide, from Spanish paella to Indian biryani and Persian rice dishes. Furthermore, the pigment crocin, responsible for saffron's vibrant color, can be extracted to create natural dyes for textiles and food colorants, offering a safer and more sustainable alternative to synthetic options (Deveoglu 2024).

Innovative products like saffron-based dietary supplements and functional foods also hold promise, leveraging its purported health benefits such as mood improvement, anticancer properties, and positive effects on cardiovascular and cognitive health. This diversification not only increases the economic returns for saffron growers but also establishes a robust, multifaceted industry built on the inherent value of this precious spice, further supported by the valorization of saffron by-products, including tepals and stamens, which also contain valuable compounds (Lakka et al. 2019).

16.11 Marketing: Global Demand, Trade, and Market Trends

Saffron continues to command significant global demand due to its unique culinary, medicinal, and cosmetic applications. The market is primarily driven by its indispensable role in traditional and gourmet cuisines across the Middle East, India and Mediterranean regions with a rising global appreciation for exotic and high-quality ingredients (EssFeed 2024). Beyond the kitchen, its bioactive compounds, such as crocin and safranal are increasingly recognized for their antioxidant, anti-inflammatory, and mood-enhancing properties fueling demand in the pharmaceutical and nutraceutical sectors. The cosmetic industry also contributes to this surge, incorporating saffron into skincare products and perfumes as consumers seek natural and effective ingredients (Straits Research 2025).

The global saffron market was valued at approximately USD 631.48 million in 2024 and is projected to reach USD 1160.95 million by 2033, growing at a CAGR of 7.0% (Straits Research 2025). This growth is, however, tempered by the inherently high cost of saffron, largely attributable to its labor-intensive cultivation and hand-harvesting process, which requires thousands of delicate *Crocus* flowers for a small yield (Straits Research 2025; MarkNtel Advisors 2024).

In terms of global trade, Iran remains the dominant producer accounting for over 90% of the world's saffron supply, while Spain is a significant exporter known for its high-quality products (Straits Research 2025). Major importing countries include Spain, India, mainland China, the United States, and Italy collectively representing a substantial portion of global saffron imports by value (WITS 2023).

16.11.1 India's Saffron Export Landscape

Despite being the second-largest saffron producer globally by land area, India's domestic production falls significantly short of its estimated annual demand of 100 tonnes leading to substantial imports, primarily from Iran and Afghanistan (StatLedger 2024; ExportImportData.in 2025). Nevertheless, India maintains an export presence for its high-quality Kashmiri saffron which is highly valued for its distinct aroma, color, and flavor.

In 2023, India's exports of Saffron were $628.13 K (approximately ₹5.37 Crores INR) with a quantity of 15,089 Kg (WITS 2023). India exported Saffron to five major countries demonstrating specific market strongholds:

- United Arab Emirates: $128.83 K (501 Kg)
- United States: $122.39 K (694 Kg)
- Australia: $79.98 K (1,162 Kg)
- Saudi Arabia: $56.91 K (30 Kg)
- United Kingdom: $33.94 K (289 Kg)

These figures indicate that while India's overall export volume is modest compared to its import needs, it caters to niche markets for its premium produce. The demand for Indian saffron is sustained by the rising popularity of Indian cuisine globally and the recognized quality of its Kashmiri variety (ExportImportData.in 2025). However, challenges such as declining production in Kashmir due to environmental factors, climate change and pollution as well as the need for improved infrastructure and quality control continue to impact India's saffron industry (Drishti IAS 2024; PMC 2015). Addressing these issues will be crucial for India to enhance its position in the competitive global saffron market.

16.12 Conclusion

Worldwide, saffron is regarded as the most expensive spices in the world and is referred to as the "golden herb" due to its high versatility and value. Since the production is labor-intensive, as small quantity of stigmas was yielded from each flower making their collection time-consuming process, but it revealed high materialistic value as it explores the health benefits of spices. Beyond its traditional value, research has extended its beneficial effects on treatment of various biological disorders. A detailed exploration into the chemical composition of saffron provides invaluable understanding directly correlating with its multifaceted benefits. This knowledge opens significant avenues for its application, creating value in diverse fields such as culinary arts, nutraceuticals and the broader pharmaceutical landscape.

References

Acar B, Sadikoglu H, Ozkaymak M (2011) Freeze drying of saffron (*Crocus sativus* L.). Dry Technol 29:1622–1627

Adams DB, Gilmore AM, Adams WW-III (1996) In vivo functions of carotenoids in higher plants. Faseb J 10(4):403-412

Aghayev Y (2002) New features in karyotype structure and origin of saffron, Crocus sativus L. Cytologia 67(3):245–252

Aghayev Y, Shakib AM, Fathi M (2004) Study of the karyotypes of saffron (*C. sativus* L.) and its close wild relatives with application of routine and banding technique. Agricultural Biotechnology Research Institute of Iran – ABRII

Alonso GL, Salinas MR, Garijo J, Sanchez-Fernandez MA (2001) Composition of crocin and picrocrocin from Spanish saffron (*Crocus sativus* L.). J Food Qual 24:219–233

Aytekin A, Acikgoz AO (2008) Hormone and microorganism treatments in the cultivation of saffron (*Crocus Sativus* L.) plants. Molecules 13:1135–1147

Babaei S, Talebi M, Bahar M (2014) Developing an SCAR and ITS reliable multiplex PCR based assay for safflower adulterant detection in saffron samples. Food Control 35:323–328

Bhandari PR (2015) *Crocus sativus* L. (saffron) for cancer chemoprevention: a mini review. J Tradit Complement Med 5:81–87

Buchecker R, Eugster CH (1973) Absolute configuration von Picrocrocin. Helv Chim Acta 56:1121–1124

Bukhari SI, Manzoor M, Dhar MK (2018) A comprehensive review of the pharmacological potential of *Crocus sativus* L. and its bioactive apocarotenoids. Biomed Pharmacother 98:733–745

Busconi M, Wischnitzki E, Del Corvo M, Colli L, Soffritti G, Stagnati L, Fernández JA (2021) Epigenetic variability among saffron crocus (*Crocus sativus* L.) accessions characterized by different phenotypes. Front Plant Sci 12:349

Cagliani LR, Culeddu N, Chessa M, Consonni R (2015) 1H NMR metabolite fingerprinting of saffron samples. Molecules 20(5):10001–10010

Cardone L, Castronuovo D, Perniola M, Cicco N, Candido V (2020) Saffron (*Crocus sativus* L.), the king of spices: an overview. Sci Hortic 272:109560

Cardone L, Castronuovo D, Perniola M, Cicco N, Molina RV, Renau-Morata B, Candido V (2021) *Crocus sativus* L. ecotypes from mediterranean countries: phenological, morpho-productive, qualitative and genetic traits. Agronomy 11(3):551

Carmona M, Zalacain A, Sanchez AM, Novella JL, Alonso GL (2006) Crocetin esters, picrocrocin and its related compounds present in *Crocus sativus* stigmas and Gardenia jasminoides fruits. Tentative identification of seven new compounds by LCESI- MS. J Agric Food Chem 54:973–979

Chaouqi S, Moratalla-Lopez N, Lage M, Lorenzo C, Alonso GL, Guedira T (2018) Effect of drying and storage process on Moroccan saffron quality. Food Biosci 22:146–153

Chichiricco G, Grilli CM (2014) *Crocus sativus* pollen tube growth in intra- and interspecific pollinations. Caryologia 37(1–2):115–125

Dar MH, Groach R, Razvi SM, Singh N (2017) Saffron crop (golden crop) in modern sustainable agricultural system. Int J Res Appl Sci Eng Technol 5:247–259

Deveoglu O (2024) Saffron in natural dyeing. J Text Color Polym Sci 21(3):259–267

Dhar AK, Mir GM (1997) Saffron in Kashmir-VI: a review of distribution and production. Int J Geogr Inf Syst 4:83–90

Drishti IAS (2024) Decline of saffron production in Kashmir. Retrieved from. https://www.drishtiias.com/daily-updates/daily-news-analysis/decline-of-saffron-production-in-kashmir

Emadi B (2009) Separating saffron flower parts using vertical air column. World Acad Sci Eng Technol 37:25–28

Emami SA, Nadjafi F, Amine GH, Amiri MS, Khosravi MT, Nasseri M (2012) Les espèces de plantes médicinales utilisées par les guérisseurs traditionnels dans la province de Khorasan, nord-est de l'Ira. J Ethnopharmacol 48:48–59

Escribano J, Alonso GL, Coca-Prados M, Fernandez JA (1996) Crocin, safranal and picrocrocin from saffron (*Crocus sativus* L.) inhibit the growth of human cancer cells *in vitro*. Cancer Lett 100(1–2):23–30

EssFeed (2024) Saffron production trade and consumption a 360° industry report. Retrieved from https://essfeed.com/saffron-production-trade-and-consumption-a-360-industry-report-saffronproduction-trade-and-consumption-a-360-industry-report/

ExportImportData.in (2025) Saffron export from India: a strategic guide for success in 2024–25. https://www.exportimportdata.in/blogs/saffron-export-from-india.aspx?utm

Feili HR, Aghaee EM, Taslimi A (2012) Study of chemical and microbiological properties of saffron dehydrated by using solar drying system. Int J Renew Energy Res 2:627–630

Gandomi H, Misaghi A, Abbaszadeh S, Azami L, Shariatifar N, Tayyar N (2012) Antibacterial effect of aqueous and alcoholic extracts from petal of saffron (*Crocus sativus* L.) on some foodborne bacterial pathogens. Med Plants 11:189–196

Ghosal S, Singh SK, Battacharya SK (1989) Mangicrocin, an adaptogenic xanthone-carotenoid glycosidic conjugate from saffron. J Chem Res 20:70–71

Goli M, Razavi BM, Hosseinzadeh H (2019) Review of the effects of *Crocus sativus* L. and its constituents on the cardiovascular system. J Pharm 22(4):199–208

Golmohammadi F (2014) Saffron and its farming, economic importance, export, medicinal characteristics and various uses in South Khorasan Province-east of Iran. Int J Farm Allied Sci 3:566–596

Gregory MJ, Menary RC, Davies NW (2005) Effect of drying temperature and air flow on the production and retention of secondary metabolites in saffron. J Agric Food Chem 53:5969–5975

Grilli Caiola M, Canini A (2010) Looking for saffron's (Crocus sativus L.) parents. Func Plant Sci Biotech 4:1–14

Hosseinzadeh H, Nassiri-Asl M (2013) Avicenna's (Ibn Sina) the canon of medicine and saffron (*Crocus sativus*): a review. Phytother Res 27(4):475–483

Husaini AM, Hassan B, Ghani MY, Teixeira da Silva JA, Kirmani NA (2010) Saffron (*Crocus sativus* Kashmirianus) cultivation in Kashmir: practices and problems. Funct Plant Sci Biotechnol 4:108–115

IIKSTC (2022) Adding value to Kashmir's 'red gold'. India International Kashmir Saffron Trading Centre. Rising Kashmir

Jami N, Rahimi A, Naghizadeh M, Sedaghati E (2020) Investigating the use of different levels of mycorrhiza and vermicompost on quantitative and qualitative yield of saffron (*Crocus sativus* L.). Sci Hortic 262:109027

Kabiri M, Rezadoost H, Ghassempour A (2017) A comparative quality study of saffron constituents through HPLC and HPTLC methods followed by isolation of crocins and picrocrocin. LWT 84:1–9

Kafi M, Showket T (2007) A comparative study of saffron agronomy and production systems of Khorasan (Iran) and Kashmir (India). Acta Hortic 739:123–132

Kafi M, Kamili AN, Husaini AM, Ozturk M, Altay V (2018) An expensive spice saffron (*Crocus sativus* L.): a case study from Kashmir, Iran, and Turkey. In: Ozturk M, Hakeem K, Ashraf M, Ahmad M (eds) Global perspectives on underutilized crops. Springer, Cham, pp 109–149

Kanakis CD, Daferera DJ, Tarantilis PA, Polissiou MG (2004) Qualitative determination of volatile compounds and quantitative evaluation of safranal and 4-hydroxy-2,6,6-trimethyl-1-cycloheene- 1-carboxaldehyde (HTCC) in Greek saffron. J Agric Food Chem 52(14):4515–4521

Kanasawud P, Crouzet JC (1990) Mechanism of formation of volatile compounds by thermal degradation of carotenoids in aqueous medium, 1. β-Carotene degradation. J Agric Food Chem 38:237–243

Koocheki AA (2003) Indigenous knowledge in agriculture with particular reference to saffron production in Iran. In: International symposium on saffron biology and biotechnology, vol 650, pp 175–182

Koocheki A, Alizadeh A, Ganjali A (2010) The effect of increased temperature on flowering behaviour of saffron (*Crocus sativus* L.). Iran J Food Crop Res 8:324–335

Koocheki A, Seyyedi SM, Eyni MJ (2014) Irrigation levels and dense planting affect flower yield and phosphorus concentration of saffron corms under semi-arid region of Mashhad, Northeast Iran. Sci Hortic 2014(180):147–155

Koocheki A, Rezvani MP, Fallahi HR (2016) Effects of planting dates, irrigation management and cover crops on growth and yield of saffron (*Crocus sativus* L.). Agroecology 8:435–451

Kull D, Pfander H (1995) Appendix: list of new carotenoids, in Carotenoids, Vol. 1A (G. Brittou, S. Liaaen-Jensen & H. Pfander, eds) pp. 295–317, Birkhäuser Verlag, Basel

Kumar M, Paul YK, Anand V (2009) An ethnobotanical study of medicinal plants used by the locals in Kishtwar, Jammu and Kashmir, India. Ethnobot Leaflets 13:1240–1256

Lakka A, Grigorakis S, Karageorgou I, Batra G, Kaltsa O, Bozinou E, Makris DP (2019) Saffron processing wastes as a bioresource of high-value added compounds: development of a green extraction process for polyphenol recovery using a natural deep eutectic solvent. Molecules 24(24):4508

Li CY, Wu TS (2002) Constituents of the stigmas of *Crocus sativus* and their tyrosinase inhibitory activity. J Nat Prod 65(10):1452–1456

Liakopoulou-Kyriakides M, Kyriakidis DA (2002) *Crocus sativus*-biological active constituent's. In: Rahman AU (ed) Studies in natural products chemistry. Elsevier, pp 293–312

Lozano P, Castellar MJ, Simancas MJ, Iborra JL (1999) Quantitative high-performance liquid chromatographic method to analyze commercial saffron (*Crocus sativus* L.) products. J Chromatogr A 830:477–483

Maggi L, Carmona M, Zalacain A, Kanakis CD, Anastasaki E, Tarantilis PA, Polissiou MG, Alonso GL (2010) Changes in saffron volatile profile according to its storage time. Food Res Int 2010(43):1329–1334

Maqbool Z, Arshad MS, Ali A, Aziz A, Khalid W, Afzal MF, Lorenzo JM (2022) Potential role of phytochemical extract from saffron in development of functional foods and protection of brain-related disorders. Med Cell Longevity 2022:6480590

MarkNtel Advisors (2024) Global saffron market research report: forecast (2024–2030) https://www.marknteladvisors.com/infographics/saffron-market.html?utm

Mathew B (1982) The Crocus: a revision of the genus crocus (Iridaceae). Timber Press. Inc, Portland

McGee HJ (ed) (2004) On food and cooking: the science and lore of the kitchen, vol 1982. Scribner Press, New York, p 896

Melnyk JP, Wang S, Marcone MF (2010) Chemical and biological properties of the world's most expensive spice: saffron. Food Res Int 43:1981–1989

Menia M, Iqbal S, Zahida R, Tahir S (2018) Production technology of saffron for enhancing productivity. J Pharmacogn Phytochem 7:1033–1039

Mohammad M, Amiri ME, Sharghi Y (2012) Respond of saffron (*Crocus sativus* L.) to animal manure application. J Med Plants Res 6:1323–1326

Mokhtarian R, Rajabi S, Zahedian S, Jafarinejad-Farsangi S, Hadizadeh M, Sadeghinejad M (2024) The effect of saffron and its extracts on the treatment of breast cancer: a narrative review. Ann Pharm Fr 82(4):629–640

Montoro P, Tuberoso CIG, Maldini M, Cabras P, Pizza C (2008) Qualitative profile and quantitative determination of flavonoids from *Crocus sativus* L. petals by LC-MS/MS. Nat Prod Commun 3:2013–2016

Moradzadeh M, Sadeghnia HR, Tabarraei A, Sahebkar A (2018) Anti-tumor effects of crocetin and related molecular targets. J Cell Physiol 233(3):2170–2182

Mousavi SZ, Bathaie SZ (2011) Historical uses of saffron: identifying potential new avenues for modern research. Avicenna J Phytomed 1:57–66

Mudasir AM, Sheikh M, Sugapriya M, Mohammed N, Leonard W, Parvaiz A (2021) Deciphering genetic diversity analysis of saffron (*Crocus sativus* L.) using RAPD and ISSR markers. Saudi J Biol Sci 28:1308–1317

Muzaffar S, Rather SA, Khan KZ (2016) *In vitro* bactericidal and fungicidal activities of various extracts of saffron (*Crocus sativus L.*) stigmas from Jammu and Kashmir, India. Cogent Food Agric 2(1):1158999

Negbi M (1999) Saffron cultivation: past, present and future prospects. In: Negbi M (ed) Saffron: Crocus sativus L. Harwood Academic Publishers, Amsterdam, pp 1–18

Nehvi FA, Makhdoomi MI (2007) Importance of irrigation in saffron production. Indian Farm 59:15–16

Ohba T, Ishisaka M, Tsujii S, Tsuruma K, Shimazawa M, Kubo K, Umigai N, Iwawaki T, Hara H (2016) Crocetin protects ultraviolet A-induced oxidative stress and cell death in skin *in-vitro* and *in-vivo*. Eur J Pharmacol 2016(789):244–253

Papandreou MA, Kanakis CD, Polissiou MG, Efthimiopoulos S, Cordopatis P, Margarity M, Lamari FN (2006) Inhibitory activity on amyloid-β aggregation and antioxidant properties of *Crocus sativus* stigmas extract and its crocin constituents. J Agric Food Chem 54:8762–8768

Pfister S, Meyer P, Steck A, Pfander H (1996) Isolation and structure elucidation of carotenoid-glycosyl esters in gardenia fruits (*Gardenia jasminoides* Ellis) and saffron (*Crocus sativus* L.). J Agric Food Chem 44:2612–2615

Pitsikas N (2016) Constituents of saffron (*Crocus sativus* L.) as potential candidates for the treatment of anxiety disorders and schizophrenia. Molecules 21:303–314

PMC (2015) Challenges of climate change: omics-based biology of saffron plants and organic agricultural biotechnology for sustainable saffron production. GM Crops Food 5(2):97–105

Pradhan J, Mohanty C, Sahoo SK (2019) Protective efficacy of crocetin and its nanoformulation against cyclosporine A-mediated toxicity in human embryonic kidney cells. Life Sci 216:39–48

Razavi BM, Hosseinzadeh H (2017) Saffron: a promising natural medicine in the treatment of metabolic syndrome. Sci Food Agric 97(6):1679–1685

Razavi BM, Hosseinzadeh H, Ghorbani A (2019) Saffron (*Crocus sativus* L.) and its constituents as potential therapeutic agents in neurological disorders: a review. Curr Neuropharmacol 17(7):633–645

Rodel W, Petrzika M (1991) Analysis of the Volatile Components of Saffron. J High-Resolution Chromatogr 14:771–774

Rezaee R, Hosseinzadeh H (2013) Safranal: from an aromatic natural product to a rewarding pharmacological agent. Iran J Basic Med Sci 16(1):12–26

Samarghandian S, Borji A (2014) Anticarcinogenic effect of saffron (*Crocus sativus* L.) and its ingredients. Pharmacogn Rev 6:99–107

Samarghandian S, Borji A, Farahmand SK, Afshari R, Davoodi S (2013) Crocus *sativus* L. (saffron) stigma aqueous extract induces apoptosis in alveolar human lung cancer cells through caspase-dependent pathways activation. Biomed Res Int 2013:1–12

Sani AM, Abbas HK, Elahe M (2013) Chemical composition and nutritional value of saffron's pollen (*Crocus sativus* L.). Nutr Food Sci 43(5):490–495

Saxena RB (2010) Botany, taxonomy and cytology of Crocus sativus series. AYU 31(3):374–381

Sepaskhah ALIR, Kamgar HAA (2009) Saffron irrigation regime. Int J Plant Prod 3:1–16

Shahi T, Assadpour E, Jafari SM (2016) Main chemical compounds and pharmacological activities of stigmas and tepals of 'red gold'; saffron. Trends Food Sci Technol 58:69–78

Sharifi G, Ebrahimzadeh H, Ghareyazie B, Karimi M (2010) Globular embryolike structures and highly efficient thidiazuron-induced multiple shoot formation in saffron (Crocus sativus L.). In Vitro Cell Dev Biol Plant 46:274–280

Singh G, Dhar U (1976) Origin of Kashmir saffron- a possible clue from weeds. Sci Cult 42:485–487

Srivastava R, Ahmed H, Dixit RK, Dharamveer SSA (2010) *Crocus sativus* L.: comprehensive review. Pharmacogn Rev 4:200–208

StatLedger (2024) India saffron market- trend analysis & forecast to 2030. https://statledger.com/products/iran-saffron-market?utm_source=chatgpt.com&variant=45494914908398

Straits Research (2025) Saffron market size, growth & demand analysis report by 2033. https://www.imarcgroup.com/saffron-market?utm

Straubinger M, Knapp H, Oka N, Watanabe N, Winterhalter P (1997) Isolation of a Glucosidic β-Damascenone Precursor from Rose Petals. J Agr Food Chem 45(10):4053-4056

Tarantilis PA (1997) Isolation and Identification of the Aroma Components from Saffron (Crocus sativus). J Agri Food Chem 45(2):459- 462

Tarantilis PA, Polissiou M, Manfait M (1994) Separation of picrocrocin, cis-trans-crocins and safranal of saffron using high-performance liquid chromatography with photodiode-array detection. J Chromatogr A 664(1):55–61

Vahidi H, Kamalinejad M, Sedaghati N (2002) Antimicrobial properties of *Crocus sativus* L. Iran J Pharm Res 2002(1):33–35

Wang Y, Sun J, Liu C, Fang C (2014) Protective effects of crocetin pretreatment on myocardial injury in an ischemia/reperfusion rat model. Eur J Pharmacol 741:290–296

Winterhalter P, Straubinger M (2000) Saffron-renewed interest in an ancient spice. Food Rev Int 16:39–59

World Integrated Trade Solution (WITS) (2023) India Saffron exports by country | 2023 | Data. https://wits.worldbank.org/trade/comtrade/en/country/IND/year/2023/tradeflow/Exports/partner/ALL/product/091020?utm

Xiang M, Yang R, Zhang Y, Wu P, Wang L, Gao Z, Wang J (2017) Effect of crocetin on vascular smooth muscle cells migration induced by advanced glycosylation end products. Microvasc Res 112:30–36

Yasmin S, Nehvi F (2018) Phenological growth stages of saffron (*Crocus sativus* L.) under temperate conditions of Jammu and Kashmir-India. Int J Curr Microbiol App Sci 7:3797–3814

Zarghami NS, Heinz DE (1971) Monoterpene aldehydes and isophorone-related compounds of saffron. Phytochemistry 10(11):2755–2761

Tulsi (Holy Basil) (*Ocimum tenuiflorum*): The Sacred Herb of Wellness

17

Channayya Hiremath and B. S. Chandana

Abstract

Tulsi, or *Ocimum*, is a large genus of aromatic medicinal herbs with a long history of traditional use that belongs to the Lamiaceae family. It has approximately 64 species. The Asian continent is where tulsi originated. In addition to their many medical uses, the fragrant plants of the genus Ocimum have long been employed as flavorings. In addition to being used as a culinary herb, *Ocimum tenuiflorum* has several pharmacological properties that can be used to prevent or treat cancer, diabetes, menstrual cramps, digestive issues, cardiovascular diseases, and neurodegenerative diseases. Furthermore, reports of its larvicidal, antibacterial, and antioxidant properties have been made. Assam, Bihar, Jammu, Maharashtra, Madhya Pradesh, West Bengal, Punjab, Haryana, and Uttar Pradesh are the states with the highest amounts of tulsi farming. In Karnataka, 603 hectares are used to grow tulsi, which produces 2343 metric tons per hectare. In India, two types of Ocimum sanctum are cultivated: Sri tulsi, sometimes called Rama tulsi, which has green leaves, and Krishna tulsi, which has purple foliage. Most traditional Hindu families cultivate Sri tulsi, the most popular kind. Krishna tulsi is preferred in the industry because of its higher level of medicinal efficacy. The aforementioned activities are mostly caused by chemical elements like methyl eugenol, 1, 8-cineone, eugenol, and anthocyanins. CIM-Ayu, CIM-Kanchan, and CIM-Angana are high-yielding, eugenol-rich varieties of O. sanctum that were developed by CSIR-CIMAP, Lucknow, using the traditional breeding method. The genetic diversity is confirmed by molecular characterization employing AP-PCR, RAPD, and RFLP markers of several *Ocimum* genotypes. The total genome size of Krishna tulsi, as determined by whole-genome

C. Hiremath (✉) · B. S. Chandana
Divisional Unit, Plant Breeding and Genetic Resource Conservation, CSIR-Central Institute of Medicinal and Aromatic Plants, Research Centre, Bangalore, India
e-mail: channayyahiremath@cimap.res.in

A. Kumar et al. (eds.), *Fasting Superfoods*,
https://doi.org/10.1007/978-981-95-2762-5_17

sequencing, is 374 Mb, with a 61% genome coverage (612 Mb projected genome size). Even with its remarkable collection of bioactive substances and significant significance in traditional Indian medicine, the crop still requires further research and conservation in it to be used in the future.

Keywords

Tulsi · *Ocimum tenuiflorum* · Medicinal herb · Traditional medicine

17.1 Introduction

The genus *Ocimum* (Lamiaceae), commonly referred to as basil, is thought to have originated in tropical and subtropical regions of Asia, Africa, and Central and South America. It includes around 160 annual and perennial shrubs and herbs (Darrah 1980). It is used as a culinary herb, a source of essential oils and fragrance compounds (Simon et al. 1990), and a fragrant, visually appealing ornamental plant (Morales et al. 1993; Morales and Simon 1996). Ocimum sanctum L., also known as holy basil, has long been considered a cornerstone of traditional herbal medicine. The tropics and subtropical regions of Asia, Africa, and Central and South America are its native habitats, according to Singh and Chaudhuri (2018). Tulsi has been cultivated in these regions for over 3000 years, primarily for its medicinal and religious significance. The herb is erect, periannual, annual, and heavily branched. It can reach a height of 30 to 60 cm. The leaves have whole or subserrate-oriented margins and can be simple, opposite, elliptic, oblong, obtuse, or acute. The green leaves have a slightly spicy and aromatic flavor and can grow up to 5 cm. The flowers are scarlet to purple and grouped in tight clusters on a single cylindrical spike. The dried stems are upright, herbaceous, woody, branching, hairy, and subquadrangular in shape. The root is thin, wiry, branching, hairy, soft, greenish-brown outside, and pale blackish inside. Corrado et al. (2020) describe the seeds as spherical to oval, 0.1 cm long, odorless, with a pungent taste and brownish appearance.

A wide range of ailments, including anxiety, cough, asthma, diarrhea, fever, dysentery, arthritis, eye conditions, otalgia, indigestion, hiccups, vomiting, gastric, cardiac, and genitourinary disorders, back pain, skin conditions, ringworm, insect, snake, and scorpion bites, and malaria, have traditionally been treated with tulsi's leaves, stem, flower, root, seeds, and even the entire plant (Cohen 2014; Jamshidi and Cohen 2017). Numerous tulsi extracts have been demonstrated to possess various antispasmodic (Mondal et al. 2009), antifertility (Sethi et al. 2010), anti-inflammatory (Manaharan et al. 2014), antidiabetic (Parasuraman et al. 2015), antimicrobial (Yamani et al. 2016), hepatoprotective (Satapathy et al. 2017), cardioprotective (Shah et al. 2019), and anticancer (Harsha et al. 2020) qualities.

Tulsi grows natively in wild habitats in tropical and subtropical locations around the world. It can reach a height of 2000 meters and grow throughout India. It grows in homes, temples, and gardens. The most diverse populations of this species occur in Africa, South America (Brazil), and Asia. Tulsi grows in Italy, Bulgaria, France,

South America, Egypt, Hungary, the Comoro Islands, Thailand, Guatemala, and Haiti. Holy basil is extensively cultivated in India, where it holds a sacred place in both religion and medicine. It is considered one of the most widely cultivated herbs in India, growing in households, temples, and gardens. In India, holy basil is grown in most parts, from the Himalayan foothills in the north to the southern states. It is especially prevalent in Uttarakhand, Uttar Pradesh, Bihar, and Tamil Nadu, Assam, Bihar, Jammu, Maharashtra, Madhya Pradesh, West Bengal, Punjab, and Haryana (Yadava et al. 2018). Tulsi is grown on 603 hectares in Karnataka, yielding 2343 metric tons per hectare. In India, there are two varieties of *Ocimum sanctum.* Two varieties of *Ocimum sanctum* are grown in India: Krishna tulsi, which has purple leaves (Fig. 17.1) and Sri tulsi, also known as Rama tulsi, which has green leaves value (Fig. 17.2). The most popular variety, Sri tulsi, is cultivated in pots in the majority of traditional Hindu households. Because of its greater medical effectiveness, Krishna tulsi is favored in the trade (Barickman et al. 2021).

Holy basil has long been regarded as a medicinal herb in Ayurvedic and traditional medicine. Its wide range of therapeutic uses stems from its bioactive compounds, including eugenol, rosmarinic acid, ursolic acid, and flavonoids (Ezeorba et al. 2024). Economically, holy basil has gained importance as a high-value herb in agriculture. Its demand in global markets for health supplements, herbal teas, essential oils, and cosmetic products has made it a highly traded commodity. Its essential oil, dried leaves, and extracts are used in a variety of consumer goods, including dietary supplements, skin-care products, herbal medicines, and functional foods. These products have increased their commercial importance globally (Abhay

Fig. 17.1 Krishna tulsi

Fig. 17.2 Sri tulsi or Rama tulsi

Kumar et al. 2014). In addition to its traditional agricultural applications, holy basil is also a key player in the eco-tourism and agro-processing sectors. With an emphasis on organic farming, holy basil cultivation offers economic opportunities for small-scale farmers and communities (Sharma et al. 2021). It also plays a role in sustainable agriculture. It is known to have pest-repellent properties, helping to protect crops from various pests, thus promoting sustainable farming practices. Moreover, its use in organic farming is growing, as it can be grown without the need for synthetic pesticides. Holy basil is not only a sacred plant with profound cultural and spiritual significance but also a valuable agricultural crop with broad medicinal, economic, and industrial applications. Its global distribution has expanded beyond its native India to other regions, making it a widely recognized herb in both traditional and modern wellness sectors. As the demand for natural and herbal remedies continues to rise, holy basil's medicinal and economic importance is expected to grow, establishing it as a key plant in the global market.

17.2 History: Traditional Uses and Cultural Significance

It has a rich and deep history rooted in both traditional medicine and cultural practices, particularly within India and other parts of South Asia. Revered for its sacred and medicinal properties, holy basil has been an integral part of Ayurvedic medicine for thousands of years, with cultural significance that transcends its medicinal uses. Holy basil has a long-standing history of use in traditional medicine, particularly in Ayurveda (the traditional Indian system of medicine), where it has been prescribed for a variety of ailments and health concerns. Its use spans over 3000 years, and it is considered one of the most important herbs in Ayurvedic healing practices. In Ayurveda, holy basil is often referred to as "The Queen of Herbs" and is recognized for its adaptogenic properties, which help the body adapt to stress. It is known to promote physical, mental, and spiritual well-being. Some of the key traditional medicinal uses of Tulsi include Respiratory Health: Holy basil has been widely used for treating respiratory conditions such as coughs, bronchitis, asthma, and sinusitis. The herb is believed to have antibacterial, antiviral, and anti-inflammatory properties, making it effective in clearing up respiratory infections. Due to its adaptogenic nature, tulsi is commonly used to reduce stress, anxiety, and fatigue (Saurabh and Komal 2018). It is often consumed in the form of herbal tea or tinctures to enhance mental clarity and reduce the effects of stress on the body.

Tulsi is widely recognized for its immune-boosting effects. It has been traditionally used to strengthen the body's immune system, particularly during the change of seasons, to help prevent colds and fevers (Yadava et al. 2018). It is used to treat digestive disorders such as indigestion, constipation, and gastritis. It is known to promote digestive enzyme secretion and has antioxidant and anti-inflammatory effects that soothe the stomach (Sharma et al. 2021). Holy basil has been used traditionally to regulate blood sugar levels, making it a valuable herb for managing diabetes. Studies have shown that it can help maintain blood glucose levels (Saurabh and Komal 2018). It is incorporated into various Ayurvedic preparations, including tinctures, oils, powders, teas, and pastes. Ayurvedic Formulations: it is a key ingredient like Chyawanprash (a herbal tonic), Tulsi Rasayana (a rejuvenating medicine), and Tulsi Ghee (ghee infused with Tulsi). While its most well-known uses stem from Ayurveda, holy basil also features in other traditional medicine systems across South and Southeast Asia. In Traditional Chinese Medicine (TCM), tulsi is used to treat conditions like cold, cough, and digestive issues. It is also utilized in various folk remedies across Southeast Asia for its calming and health-promoting properties.

Holy basil is considered a sacred plant in Hinduism and holds great spiritual and cultural significance, particularly in India. Sacred Plant: Holy basil is often planted in courtyards or near the temples of Hindu homes. A specific type of holy basil, known as Shri Tulsi or Vrinda, is revered as the sacred Tulsi plant that represents the goddess Lakshmi. The leaves of the plant are considered holy and are used in rituals and offerings to honor the deities. Ritualistic Uses: Tulsi leaves are traditionally offered to the gods in pujas (religious ceremonies) and are used in the preparation of prasad (holy offerings). Devotees also place tulsi leaves on idols or statues of deities, especially Vishnu and Krishna, to invite blessings of prosperity and spiritual purification. The water used to irrigate the holy basil plant is often considered sacred and is used in purification rituals. In some regions, tulsi water is sprinkled around the house or used in bathing rituals to cleanse the body and soul.

Beyond its religious importance, tulsi plays an integral role in daily life in many Indian households. Its use in daily medicinal preparations, such as tulsi tea for digestion or as a remedy for colds, is commonplace. The practice of drinking tulsi water in the morning is a traditional habit that is believed to improve immunity and purify the body (Yadava et al. 2018). In traditional farming communities, holy basil is planted not only for its medicinal and religious importance but also for its role in eco-protection. It is believed to have pest-repellent properties, making it a natural and sustainable method for pest control in household gardens. Moreover, it is often grown alongside other crops as a companion plant to improve soil quality and prevent insect infestations.

17.3 History of Basil Cultivation

It is widely grown and naturalized in locations all over the world, including the Caribbean, Pacific islands, and portions of Africa. It is thought to be native to the Indian subcontinent, which includes the Himalayas, Malesia, and other tropical and subtropical regions of Asia (Govaerts 2014; Wagner and Lorence 2014). Since the species has been utilized for at least 3000 years for culinary and medicinal purposes, its introduction outside of Asia has mostly been for these purposes. Tulsi crop has a long history that dates back to ancient India. It is mentioned in the Vedas, which are the oldest scriptures of Hinduism. The Vedas describe tulsi as a sacred plant that has healing properties and is used in various religious rituals. The plant is also mentioned in the Puranas, which are ancient Hindu texts that describe the history of the universe and the mythology associated with it. This herb was utilized in many ways by Ayurvedic and Unani doctors to treat ringworm, rashes, and other skin conditions. Since 1930, Ocimum has been used as a cooking herb and decorative plant in addition to being used to investigate the chemical components of basil oil (Chang et al. 1977). In addition to shrubs, the genus Ocimum includes more than 150 species of herbs (Lachowicz et al. 1997; Khazaeia et al. 1997). Because of its flavor, it is widely utilized in the culinary, pharmaceutical, cosmetic, aromatherapy, and perfumery sectors (Zheljazkov 2008).

For millennia, people have utilized *O. tenuiflorum*, an annual or short-lived perennial plant, as a culinary and medicinal herb. According to reports, it is invasive in Cuba, and Puerto Rico, India, Guam, Malaysia, and Palawan (Philippines) all have it as a weed (Stone 1970; Holm et al. 1979; Globinmed 2014). The species spreads by its many tiny seeds, grows quickly, and can be found in a range of soil types (Makri and Kintzios 2008; NMPB 2014). Although it is easy to remove, it can become more troublesome in areas where it is weedy after being pruned. Humans are the main source of *O. tenuiflorum's* global proliferation as a cultivated plant used for culinary and medicinal purposes. Unconfirmed causes for the species' alleged weediness and even invasiveness in some introduced locations, like Cuba, including cultivation escape, deliberate release, or accidental introduction through garden debris and human waste. The species spreads through its seeds, which are very tiny, abundant, and covered in nutlets (Stone 1970; Wagner et al. 2014). Animals that feed on the edible leaves may also transport these seeds.

17.4 Production

Soil and Climatic Conditions for Tulsi Production

Soil condition: It is grown in vast array of soil. Soil ranging from pH 5.0–7.5 is more suitable. Soil with loam, poor laterite, saline, and alkaline soil.

Climate: It thrives in humid environments with a considerable amount of rainfall. Plant growth and oil production have been demonstrated to benefit from long days and high temperatures. It can reach a height of 900 meters. The plant can withstand frost and drought to a moderate extent. Although it has a low oil content, the plant can be grown in partially shaded settings.

Propagation: Tulsi is propagated by means of seeds. Because of the high level of cross-pollination, seeds will degrade over generations. Therefore, the growers must use fresh seeds from the pedigree stock for new crops.

Nursery Preparation: A hectare of land requires around 200–300 g of seeds to produce seedlings for transplantation. Since the seeds are small, they should be mixed with sand and sown 2 centimeters deep. Apply a thin layer of soil and farm yard manure to the seeds after placing them in the nursery, then water them with a sprinkler hose. After 8–12 days, the seeds sprout, and after 4–5 leaf stages, the seedlings are ready for transplantation in roughly 6 weeks. Healthy seedlings for transplanting can be obtained by spraying nursery plants with a 2% urea solution 15–20 days before transplanting.

Varieties: Three varieties, namely, CIM-Ayu, CIM-Angana and CIM-Kanchan released by the CSIR-CIMAP, Lucknow. Their characteristics features are presented in Table 17.1.

Transplanting: To obtain high herbage and oil yield, seedlings with four to five leaves that are 6 weeks old are transplanted at 40 by 40 cm, 40 by 50 cm, and 50 by 30 cm intervals. After transplantation, the field is immediately irrigated. By the time of the second irrigation, the seedlings will have established properly. In order to achieve a homogeneous plant stand, gap filling and the replacement of inferior plants are also carried out at this stage.

Manure and Fertilizers: The required dose of fertilizer includes about 120 kg of Nitrogen, 60 kg of Phosphorous and 60 kg of Potash per ha during various phases of cultivation, 6 tons of FYM and 1–2 tons of green manure per hectare along with 25–30 kg of Neem cake powder as will enhance the yield.

Table 17.1 Characteristics of varieties released by CSIR-CIMAP, Lucknow

Type of tulsi	Varieties	Characteristics of varieties
Krishna Tulsi	CIM- Ayu	Maturity:100 days, 200 ql /ha herb, 0.6-0.7% oil, 110 kg/ha oil, Eugenol 83%, β- elemene 7.47%
Shyama Tulasi	CIM-Angna	Maturity:90 days, 180 ql /ha herb, 0.5-0.6% oil, 80–90 kg/ha oil yield, Eugenol 40%
	CIM-Kanchan	Maturity:120 days, 180 ql /ha herb, 0.5-0.6% oil, 100 kg/ha oil yield, Methyl eugenol rich 70%, β-caryophyllene 15.7%, β- elemene 7.6%

Irrigation: The season of year and soil moisture level determine how much tulsi has to be watered. Three irrigations per month are required in the summer, but in the other seasons—aside from the rainy season, when no irrigation is needed—it should be done as needed.

17.5 Plant Protection

Pest Management

Holy basil faces several pests, which can damage the plants and reduce yield quality. Integrated pest management (IPM) strategies help minimize the impact of pests without relying heavily on chemical pesticides.

Aphids: These pests suck sap from leaves and can transmit plant viruses. Aphids can be controlled using insecticidal soaps, neem oil, or natural predators like ladybugs.

Whiteflies: These pests can cause leaf chlorosis and reduce photosynthesis. Control methods include using yellow sticky traps and applying neem oil or insecticidal soap.

Mealybugs: These insects form cotton-like masses on the leaves and stems. They can be controlled with neem oil or insecticidal soaps.

Spider Mites: They can cause leaf discoloration and webbing. Spider mites are controlled through the use of predatory mites or miticides and by maintaining good plant hygiene.

Biological Control: The use of natural predators and parasitoids, such as lady beetles, lacewings, and predatory mites, can significantly reduce pest populations.

Botanical Pesticides: Neem oil, garlic extract, and chili-based sprays are effective, eco-friendly options for controlling various pests on holy basil.

Trap Crops: Some farmers use trap crops like marigolds to attract pests away from holy basil, thereby reducing pest pressure on the primary crop.

Disease Management

Holy basil is susceptible to several fungal, bacterial, and viral diseases. Early detection and proper disease management techniques are essential for maintaining healthy crops.

Powdery Mildew: A fungal disease that causes white fungal growth on the surface of leaves. To prevent powdery mildew, avoid overcrowding, and ensure good air circulation. Fungicides like 0.1% sulfur or potassium bicarbonate can be used for control.

Fusarium Wilt: This soil-borne fungal disease causes yellowing and wilting of the leaves. To manage Fusarium wilt, use disease-resistant varieties, practice crop rotation, and apply biological controls like Trichoderma species.

Downy Mildew: Downy mildew affects basil leaves, causing yellow spots and a fuzzy growth. Fungicides like copper-based treatments or organic solutions like neem oil can help reduce infection.

Bacterial Leaf Spot: This bacterial infection causes black or brown spots on basil leaves. Good sanitation practices and crop rotation can help minimize the spread of this disease.

Weed Management

Mulching: Applying organic mulch around holy basil plants can suppress weed growth, conserve moisture, and add organic matter to the soil.

Manual Weeding: Regular manual weeding, especially during the early growth stages of holy basil, ensures that weeds do not compete with the plants for nutrients and water.

17.6 Crop Improvement

In order to improve qualities including increased essential oil yield, enhanced disease resistance, and improved tolerance to a variety of climatic situations, tulsi has been the focus of intensive breeding efforts. The different basic chromosomal numbers $x = 8$, 10, 12, or 16 indicate that holy basil is a cross-pollinated crop (Morton 1962). The chromosomal number is found to be $2n = 36$, although individuals with $2n = 32$, $2n = 36$, and $2n = 76$ have been documented (Carović-Stanko et al. 2009). The variations in *Ocimum* species' genome sizes and chromosome counts suggest that both sequence amplification and deletion, chromosomal rearrangements, and polyploidization occurred during the genomes' evolution (Carović-Stanko et al. 2009).

Essential Oil Yield

The focus of breeding has been to increase the content of bioactive compounds, especially eugenol, which is the key compound in Tulsi essential oil. *Ocimum sanctum* varieties with higher essential oil yields are in demand for both medicinal and commercial purposes (Sharma et al. 2013). The Central Institute of Medicinal and Aromatic Plants (CSIR) in India developed multiple varieties through traditional breeding methods. The high-yielding, dark purple-pigmented varieties CIM-Angana of Shyam tulsi and CIM-Kanchan were also developed by Lal et al. (2003), CIM-Ayu and CIM-Angana of Shyam tulsi (Lal et al. 2003).

Genetics, Genomics, and Recent Advancements in Tulsi

Recent advancements in the genetics and genomics of holy basil (*Ocimum sanctum*) have greatly enhanced the understanding of its biological processes, medicinal properties, and cultivation practices. These advancements have opened doors for improving crop yield, disease resistance, and the production of bioactive compounds. Here's an overview of the latest developments in holy basil genetics and genomics:

Genome Sequencing of Holy Basil

The genome of holy basil has been sequenced, providing valuable insights into its genetic makeup and potential for improving its cultivation and medicinal uses. In a major breakthrough, the complete genome of *Ocimum sanctum* was sequenced. Upadhyay et al. (2015) sequenced the draft genome of *Ocimum tenuiflurum* L. (subtype Krishna Tulsi). The Illumina HiSeq 1000 was used to sequence the whole genome, producing paired-end and mate-pair sequencing libraries. The assembled genome was 374 Mb in size, with a 61% genome coverage (612 Mb estimated genome size). The transcriptomes (RNA-Seq) of two subtypes of *O. tenuiflorum*, Krishna and Rama Tulsi, have also been examined, and the relative gene expression in each variety has been reported.

Transcriptome and Metabolome Studies

Rastogi et al. (2014) published the first comparative transcriptome analysis of Ocimum species. Comparative genomics has been employed to study the differences between *Ocimum sanctum* and other closely related species like *Ocimum basilicum* (sweet basil). These studies help identify genetic traits that can be leveraged to improve the agronomic and medicinal properties of holy basil. A comparative analysis of *Ocimum sanctum* and *Ocimum basilicum* genomes revealed genetic variations related to the biosynthesis of essential oils and other secondary metabolites.

Genetic Diversity and Population Studies

The genetic diversity studies of *Ocimum sanctum* have helped to understand the different varieties of holy basil and how they adapt to various climatic and soil conditions. Studies on the genetic variability of holy basil populations help in identifying superior genotypes that possess traits like high essential oil content, disease resistance, and better adaptation to specific growing conditions. Such findings are crucial for breeding programs aimed at developing high-yielding and climate-resilient varieties. An allele-specific PCR (AS-PCR) technique was developed by Balaji and Parani (2024) to differentiate between green and purple tulsi. Two subtypes of *O. tenuiflorum*, green and purple tulsi, share the same therapeutic qualities. Using green Tulsi as a reference, they discovered ten insertions/deletions (InDels) and twelve single-nucleotide polymorphisms (SNPs) in the purple Tulsi's chloroplast genome.

Future Directions

As more genomic data is collected, we can expect the following advancements in holy basil research. The potential to use synthetic biology to create modified strains of *Ocimum sanctum* that produce higher quantities of essential oils and other medicinal compounds could revolutionize the production of holy basil-based products. The Genomic Selection for Precision Agriculture, combining genomics with machine learning and big data approaches, could lead to precision agriculture techniques that tailor farming practices based on the genetic makeup of holy basil cultivars.

17.7 Harvesting, Primary Processing, Pre- and Post-harvest Technology

Preharvest Technologies

Cultivation Practices: Preharvest practices play a vital role in ensuring the plant's optimal growth and quality. These include the use of good agricultural practices (GAP), pest management, and nutrient management.

Soil and Fertilization: Holy basil grows well in well-drained, fertile soils. Organic compost or farmyard manure is often used to improve soil quality. The proper application of fertilizers and nutrients is critical for enhancing the aromatic and medicinal compounds in the plant.

Pest and Disease Control

Integrated pest management (IPM) practices are used to manage pest issues in holy basil. This includes biological control, use of organic pesticides, and manual removal of infected parts. Chemical pesticides should be used cautiously due to their potential to affect the plant's medicinal properties.

Water Management

Regular irrigation is essential, especially in regions with low rainfall. However, overwatering should be avoided, as it can lead to root diseases. Proper water management practices help maintain plant health and prevent contamination.

Harvesting

The process of steam distillation or hydro-distillation is used to extract the oil from the entire herb or from the immature inflorescence. The most popular method for processing leaves is steam distillation, which takes less time.

- The crop is ready for the first harvest after 80–90 days of transplanting. Furthermore, harvests can be made after every 85–90 days interval.
- Tropical and subtropical conditions of South India, 3–4 harvests per year.
- Crop should be harvested on bright sunny days to get good quality and higher quantity oil.
- The crop should be cut at 15–20 cm above ground level in the first harvest, 25–30 cm in the second harvest and 35–45 cm high in the third harvest.
- The eugenol content in the oil is maximum at the pre-flowering stage so the crop should be harvested before flowering.

Yield and Economics

Fresh herbage is steam distilled for a period of 3–4 h to obtain normal yield. Oil recovery is 0.3%

- Tulsi Herb Yield/ha/year (in 3–4 Harvests) 20–25 Tonnes
- Oil Yield/ha/year 60–75 Kgs
- Oil rate/Kg Rs. 3000/–
- Cost of Cultivation /ha/year Rs. 45,000/–
- Income from oil/ha/year Rs. 1.8–2.25 Lakhs
- Net Profit/ha/year Rs. 1.35–1.8 Lakhs

Harvesting Techniques

Harvesting Time: The optimal time for harvesting holy basil is just before flowering, when the leaves contain the highest concentration of essential oils and bioactive compounds (José Luiz et al. 2006). Leaves should be harvested early in the morning, after the dew has dried but before the heat of midday.

Manual Harvesting: The most common method of harvesting holy basil is manual plucking. This ensures minimal damage to the leaves and the plant, preserving the quality of the product. A clean cutting tool should be used to avoid bruising and contamination.

Postharvest Handling and Processing Technologies of Holy Basil (Ocimum sanctum)

Postharvest handling and processing of holy basil (*Ocimum sanctum*) play a critical role in preserving the quality, medicinal properties, and aromatic oils of the herb. Proper techniques are required to ensure the quality of leaves, oils, and extracts while extending shelf life.

Postharvest Handling

Sorting and Grading: After harvesting, basil leaves are carefully sorted and graded according to quality. Damaged, yellowed, or diseased leaves are removed to ensure that only the highest-quality leaves are used for further processing. The grading process involves classifying leaves based on size, color, and overall freshness. Grading ensures that high-value products like essential oils or medicinal extracts come from the healthiest plants.

Washing and Cleaning: Holy basil leaves are washed with clean water to remove dirt, dust, and any pesticide residues. Washing should be gentle to avoid damaging the fragile leaves. After washing, the leaves should be carefully dried with clean towels or air-dried to reduce moisture content before further processing.

Drying: Drying is one of the most common methods used to preserve basil leaves for long-term use. Proper drying ensures the retention of essential oils and preserves the medicinal and aromatic properties of the leaves.

Air Drying

The traditional method involves air-drying holy basil in a well-ventilated, shaded area. Basil leaves should be spread out in a thin layer on drying racks, avoiding direct sunlight to prevent the degradation of volatile compounds such as eugenol (Rudrakshi et al. 2018). Air drying allows the leaves to retain their color and aroma, making it a preferred method for both small-scale and large-scale producers.

Oven Drying: Oven drying is commonly used in commercial operations for faster results. The basil leaves are dried at temperatures between 30 °C and 40 °C (86 °F to 104 °F). Care should be taken to monitor the temperature carefully, as excessive heat can lead to the evaporation of essential oils.

Freeze Drying: Freeze-drying (lyophilization) is an advanced method that involves freezing the basil leaves and then removing moisture through sublimation. This method helps to preserve the chemical composition of the leaves and their nutrients (Rudrakshi et al. 2018). Freeze-dried basil retains much of the flavor, color, and medicinal value, making it ideal for high-quality products.

Powdering and Grinding

After drying, holy basil leaves can be ground into a fine powder. The powder is used in teas, supplements, or cosmetics. Mechanical grinding is done in a clean environment to avoid contamination and preserve the leaf's bioactive properties.

Herbal Extracts and Tinctures

Alcoholic Extracts: Holy basil leaves are often soaked in ethanol or a mixture of ethanol and water to extract the bioactive compounds. The extract is left for several weeks, during which it absorbs compounds like flavonoids, alkaloids, and phenolic acids. This extract is then filtered and concentrated (Romano et al. 2022).

Water-Based Extracts: Water-based extracts are also made by boiling holy basil leaves to create teas or concentrated solutions, which are then dried into powders or liquid herbal extracts.

Packaging and Storage

Packaging: After processing, holy basil in dried, powdered, or oil form is packed in airtight containers, such as dark-colored glass jars or vacuum-sealed pouches, to prevent exposure to light, moisture, and oxygen, which could degrade its quality (Brindisi and Simon 2023).

Storage: The packaged products are stored in cool, dry, and dark conditions to prevent spoilage and preserve the potency of the bioactive compounds.

Short-Term Storage

- *Cool Storage*: Freshly harvested basil leaves are perishable and should be stored in cool, shaded conditions. Temperature control is crucial, especially in tropical climates where the plant can quickly wilt or spoil. A temperature range of 5–10 °C (41–50 °F) is ideal for short-term storage to preserve freshness (Brindisi and Simon 2023).
- *Humidity Control*: The relative humidity of the storage environment should be around 60–70% to prevent wilting while avoiding mold growth. Humidity and temperature should be carefully controlled during storage to avoid loss of essential oils and degradation of the leaves.

Long-Term Storage

Drying: To preserve basil for extended periods, the leaves can be dried. Proper drying helps maintain the plant's medicinal and aromatic properties. The dried leaves can be stored for months or even longer, provided they are kept in airtight containers in cool, dry, and dark conditions.

17.8 Utilization Pattern and Value Addition of Product and Marketing

Utilization Patterns

Holy basil (*Ocimum sanctum*) is one of the most revered plants in traditional medicine and culinary practices, particularly in Indian culture. It has been utilized for centuries due to its extensive medicinal, aromatic, and nutritional properties. The applications of holy basil are diverse, and it has both traditional and modern uses across various sectors, including food, medicine, and industry. Below is an exploration of its traditional and modern uses.

Traditional Uses

Holy basil has been a key component of Ayurvedic medicine for over 3000 years. It is regarded as a sacred herb and is often referred to as the "queen of herbs" in traditional healing systems, particularly for its potent medicinal properties.

- Adaptogen and Immunity Booster: Tulsi is traditionally used to enhance the body's ability to cope with stress and illness. It has been used to boost immunity, reduce inflammation, and fight infections. It is believed to support the respiratory, digestive, and cardiovascular systems.

- Treatment for Fever and Respiratory Disorders: The leaves are often used as a remedy for fever, colds, coughs, and respiratory issues. They are typically consumed as tea or in the form of decoctions (Yadava et al. 2018).
- Antioxidant and Anti-inflammatory: In traditional practices, Tulsi is used to treat a wide range of conditions such as acne, arthritis, and skin disorders due to its antioxidant and anti-inflammatory properties (Sharma et al. 2013).
- Antimicrobial Properties: Traditional uses also include topical applications for skin conditions, as the plant is believed to have antimicrobial and antiseptic effects.

Culinary Uses

In traditional culinary applications, holy basil leaves are used to enhance the flavor and aroma of various dishes. It is particularly prevalent in South and Southeast Asian cuisines.

- Herbal Teas: One of the most common traditional uses of holy basil is in the preparation of herbal teas. Tulsi tea is brewed from fresh or dried leaves and is commonly consumed for its health benefits, especially for digestive and respiratory health.
- Flavoring Agent: The aromatic leaves are used in the preparation of soups, curries, and sauces. It is commonly added to dishes in Indian, Thai, and other Southeast Asian cuisines for its spicy, slightly sweet, and peppery flavor.
- Herbal Medicine Preparations: Traditional recipes also involve combining Tulsi leaves with other herbs to make pastes or syrups for treating ailments such as headaches, sore throats, and digestive issues.

Modern Medicinal and Nutritional Applications

In contemporary medicine, holy basil is increasingly being used as a dietary supplement and in pharmaceutical formulations due to its proven health benefits.

- Anti-Stress and Anxiety: Recent studies have shown that holy basil is an adaptogen, helping to balance cortisol levels and reduce stress and anxiety. It is commonly used in modern supplements formulated for mental well-being (Saurabh and Komal 2018).
- Antioxidant and Anticancer Properties: Tulsi's rich phytochemical composition, including flavonoids and essential oils, has led to its use in modern nutraceuticals. It is promoted for its antioxidant, anticancer, and anti-aging benefits, often in capsule or tincture form (Ifeoma et al. 2023).
- Diabetes Management: Modern research has demonstrated the potential of Tulsi in controlling blood sugar levels. It is being used in the form of extracts and powders for managing type 2 diabetes and improving insulin sensitivity (Ifeoma et al. 2023).

- Cardiovascular Health: Studies have suggested that holy basil has cardioprotective properties, and modern supplements containing Tulsi extract are marketed to support heart health (Ifeoma et al. 2023).
- Aromatherapy and Cosmetics: Holy basil essential oil is widely used in aromatherapy due to its calming and antimicrobial effects. It is also used in cosmetic products, including soaps, lotions, and skin creams, to promote skin health and provide a natural fragrance (Pandey et al. 2014).
- Pharmaceutical and Perfume Industries: The high demand for Tulsi essential oil is driven by its widespread use in pharmaceutical preparations for stress, immunity, and skin health. It is also used as a natural fragrance in the perfume industry.

Food Products

In addition to traditional uses as a flavoring herb, holy basil has modern applications in the food and beverage industry, driven by its health benefits.

- Herbal Beverages and Juices: Holy basil extracts are used in the production of herbal beverages and health drinks, which are marketed for their detoxifying and immunity-boosting properties. Tulsi is also used in functional beverages, such as health shots and energy drinks.
- Functional Foods: As a rich source of vitamins, minerals, and antioxidants, holy basil is incorporated into energy bars, snack foods, and dietary supplements. Its antimicrobial and anti-inflammatory properties make it an attractive ingredient for functional foods that promote health (Sharma et al. 2013).
- Tulsi-Infused Products: The use of holy basil in infusions such as oils, vinegar, and honey is gaining popularity. These products not only add flavor but also promote health benefits, particularly in the areas of immunity and stress management.

Agricultural Products and Natural Pesticides

Holy basil is also gaining traction in the agricultural sector due to its bioactive compounds, which have pesticidal properties.

- Natural Pesticides: Tulsi's essential oils, particularly eugenol, are used as natural pesticides. It has been shown to repel a wide range of pests and is being incorporated into organic farming practices as a safer alternative to synthetic chemicals (Ghosh et al. 2025).
- Fertilizers and Bio-fertilizers: The plant's ability to enhance soil health has led to its use in organic farming as a natural fertilizer. Tulsi leaves can be composted or used in liquid manure formulations to enrich the soil and improve crop yields.

Marketing: Global Demand, Trade, and Market Trends

Holy basil has become increasingly popular on the global market, driven by rising consumer demand for natural health products, herbal medicine, and sustainable goods. Its use in pharmaceuticals, cosmetics, food, and wellness sectors has helped create a growing industry for holy basil, with significant potential for trade and expansion. Below, we examine the global demand, trade trends, and market trends related to holy basil, supported by citations.

Global Demand for Holy Basil

The demand for holy basil is driven primarily by its health benefits, which include immunity-boosting, stress-reducing, anti-inflammatory, and antioxidant properties. Several factors contribute to the increasing global demand for holy basil:

Rising Popularity of Herbal and Natural Products

Consumers are increasingly shifting away from synthetic chemicals and pharmaceuticals, opting for natural, organic, and herbal products. This trend is particularly strong in developed markets, where wellness and personal care products are increasingly incorporating holy basil as a key ingredient.

- Herbal Medicine: Tulsi is used in a variety of products, including herbal teas, supplements, and topical creams, to address common health concerns such as stress, anxiety, skin disorders, and digestive issues (Cohen 2014).
- Functional Foods and Beverages: The growth of the health-conscious consumer base has expanded the use of holy basil in beverages (e.g., herbal teas, juices), energy bars, and other functional food products aimed at improving immunity, enhancing mental clarity, and supporting overall wellness.

Increased Health Awareness

There is a growing global focus on wellness and disease prevention. Holy basil, with its documented health benefits, is gaining recognition as a natural alternative for disease management and as a preventive measure. As consumers increasingly seek out adaptogens (natural substances that help the body cope with stress), holy basil has become a popular ingredient in supplements and functional foods designed to alleviate stress and improve mood (Saurabh and Komal 2018).

Export Markets and Key Producers

India is the world's largest producer of holy basil, and its cultivation and export have grown significantly. Other countries that grow and trade holy basil include Thailand, Indonesia, and Vietnam. These regions not only cater to domestic demand but also export the herb to countries in North America, Europe, and Asia-Pacific.

- India as a Key Exporter: India is the leading exporter of holy basil products, primarily dried leaves, essential oils, and extracts. The Indian herbal product industry is estimated to grow rapidly due to the increasing use of herbs like holy basil in wellness and alternative medicine (Ghosh et al. 2025).
- Export Markets: The key markets for Tulsi include the United States, Germany, United Kingdom, and other parts of Europe, where consumer demand for natural

remedies and wellness products is high. These markets are primarily focused on herbal teas, dietary supplements, and essential oils (Yamani et al. 2016).

Acknowledgments The author expresses gratitude to the Director of CSIR-CIMAP, Lucknow, for providing facilities.

References

Abhay Kumar P, Pooja S, Nijendra NT (2014) Chemistry and bioactivities of essential oils of some *Ocimum species*: an overview. Asian Pac J Trop Biomed 4(9):682–694

Balaji R, Parani M (2024) Development of an allele-specific PCR (AS-PCR) method for identifying high-methyl eugenol-containing purple Tulsi (*Ocimum tenuiflorum* L.) in market samples. Mol Biol Rep 51:439. https://doi.org/10.1007/s11033-024-09365-0

Brindisi LJ, Simon JE (2023) Preharvest and postharvest techniques that optimize the shelf life of fresh basil (*Ocimum basilicum* L.): a review. Front Plant Sci 14:1237577. https://doi.org/10.3389/fpls.2023.1237577

Carović-Stanko K, Liber Z, Besendorfer V, Javornik B, Bohanec B, Kolak I (2009) Genetic relations among basil taxa (*Ocimum* L.) based on molecular markers, nuclear DNA content, and chromosome number. Plant Syst Evol 285:13–22

Chang SS, Ostric-Matijasevic B, Hsieh OAL, Huang CL (1977) Natural antioxidant from rosemary and sage. J Food Sci 42:1102–1107. https://doi.org/10.1111/j.1365-2621.1977.tb12676.x

Cohen MM (2014) Tulsi- Ocimum sanctum: a herb for all reasons. J Ayurveda Integr Med 5:251–259

Corrado G, Formisano L, De Micco V, Pannico A, Giordano M, El-Nakhel C, Chiaiese P, Sacchi R, Rouphael Y (2020) Understanding the morpho-anatomical, physiological, and functional response of sweet basil to isosmotic nitrate to chloride ratios. Biology 9(7):158

Darrah HH (1980) The cultivated basils. Buckeye Printing Co., University of Wisconsin, Madison

Ezeorba TPC, Chukwuma IF, Asomadu RO, Ezeorba WFC, Uchendu NO (2024) Health and therapeutic potentials of Ocimum essential oils: a review on isolation, phytochemistry, biological activities, and future directions. J Essent Oil Res 36(3):271–290. https://doi.org/10.1080/10412905.2024.2338117

Ghosh A, Majumdar D, Biswas H (2025) Nano-biopesticide formulation comprising of silver nanoparticles anchored to *Ocimum sanctum*: a sustainable approach to pest control in jute farming. Sci Rep 15:3414. https://doi.org/10.1038/s41598-025-87727-9

Globinmed (2014) Medicinal herbs and plants online database. Global Information hub on Integrative Medicine (Globinmed), Kuala Lumpur

Govaerts R (2014) World checklist of Lamiaceae. Royal Botanic Gardens, Kew, Richmond, London. http://apps.kew.org/wcsp/

Harsha M, Mohan Kumar KP, Kagathur S, Amberkar VS (2020) Effect of Ocimum sanctum extract on leukemic cell lines: a preliminary in-vitro study. J Oral Maxillofac Pathol 24:93–98

Holm LG, Pancho JV, Herberger JP, Plucknett DL (1979) A geographical atlas of world weeds. Wiley, New York

Ifeoma FC, Nene OU, Rita OA, Wisdom FCE, Timothy PCE (2023) African and holy basil-a review of ethnobotany, phytochemistry, and toxicity of their essential oil: current trends and prospects for antimicrobial/anti-parasitic pharmacology. Arab J Chem 16(7):104870. https://doi.org/10.1016/j.arabjc.2023.104870

Jamshidi N, Cohen MM (2017) The clinical efficacy and safety of tulsi in humans: a systematic review of the literature. Evid Based Complement Alternat Med 2017:9217567

José Luiz S, Carvalho F, Arie FB, Péricles BA, Polyana ADE, Alberto SM, Sócrates CHC, Maria de Fátima A, Renata SM (2006) Influence of the harvesting time, temperature and drying period on basil (*Ocimum basilicum* L.) essential oil. Rev Bras Farmacogn 16(1):24–30

Khazaeia N, Esmaiilia M, Djomehb ZE, Ghasemlouc M, Joukid M (1997) Characterization of new biodegradable edible film made from basil seed (*Ocimum basilicum* L.) gum. Carbohydr Polym 102:199–206. https://doi.org/10.1016/j.carbpol.2013.10.062

Lachowicz KJ, Jones JP, Briggs DR, Bienvenu FE, Palmer MV, Mishra V, Hunter MM (1997) Characteristics of plants and plant extracts from five varieties of basil (*Ocimum basilicum* L.) grown in Australia. J Agric Food Chem 45(7):2660–2665. https://doi.org/10.1021/jf960791h

Lal RK, Khanuja SPS, Agnihotri AK, Misra HO, Shasany AK, Naqvi AA, Dhawan OP, Kalra A, Bahl JR, Darokar MP (2003) High yielding eugenol rich oil producing variety of Ocimum sanctum 'CIM-Ayu'. J Med Aromat Plant Sci 25:746–747

Makri O, Kintzios S (2008) Ocimum sp. (basil): botany, cultivation, pharmaceutical properties, and biotechnology. Int J Geogr Inf Syst 13(3):123–150. https://doi.org/10.1300/J044v13n03_10

Manaharan T, Thirugnanasampandan R, Jayakumar R, Ramya G, Ramnath G, Kanthimathi MS (2014) Antimetastatic and anti-inflammatory potentials of essential oil from edible Ocimum sanctum leaves. Sci World J 2014:239508

Mondal S, Mirdha BR, Mahapatra SC (2009) The science behind sacredness of Tulsi (*Ocimum sanctum* Linn.). Indian J Physiol Pharmacol 53:291–306

Morales MR, Simon JE (1996) New basil selections with compacts inflorescence for the ornamental market. In: Janick J (ed) Progress in new crops. ASHS, Alexandria, pp 543–546

Morales MR, Charles DJ, Simon JE (1993) New aromatic lemon basil germplasm. In: Janick J, Simon JE (eds) New crops. Wiley, New York, pp 632–635

Morton JK (1962) Cytotaxonomic studies on the West African Labiatae. J Linn Soc Bot 58:231–283

NMPB (2014) Cultivation practices of some commercially important medicinal plants (online datasheets): Ocimum sanctum. Government of India National Medicinal Plants Board (NMPB), Delhi. http://www.nmpb.nic.in/WriteReadData/links/615264999tulsi.pdf

Pandey AK, Singh P, Tripathi NN (2014) Chemistry and bioactivities of essential oils of some *Ocimum species*: an overview. Asian Pac J Trop Biomed 4(9):682–694. https://doi.org/10.12980/apjtb.4.2014c77

Parasuraman S, Balamurugan S, Christapher PV, Petchi RR, Yeng WY, Sujithra J, Vijaya C (2015) Evaluation of antidiabetic and antihyperlipidemic effects of hydroalcoholic extract of leaves of *ocimum tenuiflorum* (Lamiaceae) and prediction of biological activity of its phytoconstituents. Pharm Res 7:156–165

Rastogi S, Meena S, Bhattacharya A, Ghosh S, Shukla RK, Sangwan NS (2014) De novo sequencing and comparative analysis of holy and sweet basil transcriptomes. BMC Genomics 15:588

Romano R, De Luca L, Aiello A, Pagano R, Di Pierro P, Pizzolongo F, Masi P (2022) Basil (*Ocimum basilicum* L.) Leaves as a source of bioactive compounds. Foods 11(20):3212. https://doi.org/10.3390/foods11203212

Rudrakshi S, Surekha B, Preetinder K (2018) Effect of drying methods on biochemical quality of basil leaf. Agric Res J 5(2):331–335. https://doi.org/10.5958/2395-146X.2018.00059.5

Satapathy S, Das N, Bandyopadhyay D, Mahapatra SC, Sahu DS, Meda M (2017) Effect of tulsi (*Ocimum sanctum* Linn.) supplementation on metabolic parameters and liver enzymes in young overweight and obese subjects. Indian J Clin Biochem 32:357–363

Saurabh G, Komal K (2018) Comparative characterization for antimicrobial activity and bioactive compounds present in leaf extract of *ocimum sanctum*. J Food Ind Microbiol 3:1–5

Sethi J, Yadav M, Sood S, Dahiya K, Singh V (2010) Effect of tulsi (*Ocimum Sanctum* Linn.) on sperm count and reproductive hormones in male albino rabbits. Int J Ayurveda Res 1:208–210

Shah SMA, Akram M, Riaz M, Munir N, Rasool G (2019) Cardioprotective potential of plant-derived molecules: a scientific and medicinal approach. Dose Response 17:1559325819852243

Sharma A, Kumar N, Kumar D, Kumari V, Saraswati S, Chandel KA (2013) Review paper on antimicrobial activity of medicinal plant Tulsi (*Ocimum* spp.) and pudina (Mentha spp.). Int J Curr Res 5:487–489

Sharma G, Regmi S, Lamichhane R, Bhetwal H, Subedi SS, Thapa S (2021) Phytoconstituents, conventional and chemical uses of Tulsi: a review. Asian J Pharmacogn 4(1):17–23

Simon JE, Quinn J, Murray RG (1990) Basil: a source of essential oils. In: Janick J, Simon JE (eds) Advances in new crops. Timber, Portland, pp 484–498

Singh D, Chaudhuri PK (2018) A review on phytochemical and pharmacological properties of Holy basil (*Ocimum sanctum* L.). Ind Crop Prod 118:367–382

Stone BC (1970) The flora of Guam. Micronesica 6:218

Upadhyay AK, Chacko AR, Gandhimathi A (2015) Genome sequencing of herb Tulsi (*Ocimum tenuiflorum*) unravels key genes behind its strong medicinal properties. BMC Plant Biol 15:212. https://doi.org/10.1186/s12870-015-0562-x

Wagner WL, Lorence DH (2014) Flora of the Marquesas Islands website. Smithsonian Institution, Washington

Wagner WL, Herbst DR, Tornabene MW, Weitzman A, Lorence DH (2014) Flora of Micronesia website. Smithsonian Institution, Washington. http://botany.edu/pacificislandbiodiversity/micronesia/index.htm

Yadava DK, Hossain F, Mohapatra T (2018) Nutritional security through crop biofortification in India: status & future prospects. Indian J Med Res 148(5):621–631. https://doi.org/10.4103/ijmr.IJMR_1893_18. PMID: 30666987.

Yamani HA, Pang EC, Mantri N, Deighton MA (2016) Antimicrobial activity of Tulsi (*Ocimum tenuiflorum*) essential oil and their major constituents against three species of bacteria. Front Microbiol 17(7):681. https://doi.org/10.3389/fmicb.2016.00681

Chia Seeds (*Salvia hispanica*): The Tiny Powerhouse of Omega-3 and Fiber

18

Ashish Kumar Tiwari, Sonal Tiwari, Jitendra Kumar Tiwari, and Dinesh Chandra Joshi

Abstract

Chia seeds (*Salvia hispanica* L.), native to Central America and historically revered by ancient civilizations such as the Aztecs and Mayans, have reemerged in the modern era as a globally celebrated superfood. Popular for their exceptional nutritional profile—including higher dietary fiber, omega-3 fatty acids (ALA), protein, antioxidants, and essential minerals. Seeds offer multiple health benefits, mostly for cardiovascular health, metabolic function, and good digestion. Unique hydrophilic properties allow them to take up multiple times their weight in water, making them particularly valuable in fasting diets by promoting satiety, hydration, and controlling blood sugar levels. This flexibility has improved their addition in various dietary regimens such as vegan, ketogenic, and gluten-free lifestyles. In India, the production of chia seeds is gaining momentum across numerous states due to favorable agro-climatic situation and increasing domestic and export demand. With a probable yield of up to 2.5 tons/acre and comparatively low input necessities, chia seeds present a capable crop for sustainable agriculture.

A. K. Tiwari (✉)
Indira Gandhi Krishi Vishwavidyalaya, Raipur, Chhattisgarh, India

S. Tiwari
Mahatma Gandhi University of Horticulture and Forestry, Durg, Chhattisgarh, India

J. K. Tiwari
Rani Lakshmi Bai Central Agricultural University, Jhansi, Uttar Pradesh, India

D. C. Joshi
Crop Improvement Division, Vivekananda Parvatiya Krishi Anusandhan Sansthan, Almora, Uttarakhand, India

A. Kumar et al. (eds.), *Fasting Superfoods*,
https://doi.org/10.1007/978-981-95-2762-5_18

Keywords

Chia seed · Health benefits · Nutritional profile · Consumption process · Seed production

18.1 Introduction

Chia seeds (*Salvia hispanica*), still tiny in size, are categorized among the most nutritive food known to humanity. Native to Central America and belonging to the mint ancestors (Lamiaceae), chia seeds have been cultivated for centuries, with chronological records tracing their importance back to the prehistoric Aztec and Mayan civilizations (Ayerza and Coates 2005). Chia seeds are used not only as a staple food but moreover their traditional and medicinal uses but were also believed to present stamina and vitality—attribute that earn them the name "warrior food" (Ullah et al. 2016). In recent years, chia seeds garnered global credit as a superfood due to their extraordinary nutritional properties. They are mainly loaded with dietary fiber, omega-3 fatty acids (alpha-linolenic acid, or ALA), protein, antioxidants, and essential minerals like calcium (Ca), magnesium (Mg), and phosphorus (P) (Mohd Ali et al. 2012). These components have a wide range of health benefits, together with enhanced cardiovascular function, improved digestive health, and support metabolic activities (Coelho and Salas-Mellado 2014).

Their purposeful property makes chia seeds particularly suitable for fasting diets, which have seen reappearance in popularity for spiritual and health-related purpose. Chia seed can take up more amount of water up to 12 times their mass, form a gel-like material that holdup gastric empty, promote satiety, and control blood glucose level (Ulbricht et al. 2009). This hydrophilic feature also aids in maintaining hydration—vital during extended fasting period—while sustaining intellectual clearness and continuous power. Agriculturally, chia seed cultivation is no longer restricted to its inhabitant regions. In India, states such as Uttar Pradesh (UP), Rajasthan (RJ), Andhra Pradesh (AP), and Karnataka have emerged as promise new hubs for chia seeds farming, fueled by mounting awareness of its health profit and increasing demand in both domestic and abroad areas (Kumar et al. 2021). Under most favorable agronomic situation, yield can achieve up to 2.5tons/acre, position of chia seeds as a high-potential, low-input crop suitable for sustainable farming systems.

As worldwide consumers gradually more prioritize hygienic consumption, plant-based nutrition, and purposeful foods, chia seeds are flattering essential to developing a dietary pattern. Their addition in fast responsive foodstuffs not only meets necessary dietary requirements but also resonates with broader standard of living trends toward sustainability and high appearance wellness food (Poudel et al. 2020). This chapter delves into the expedition of chia seeds from their prehistoric roots in Mesoamerican civilization to their vibrant responsibility in present agriculture, nutrition discipline, and worldwide food system.

18.2 Historical Significance of Chia Seeds

18.2.1 Origins and Early Uses

Chia seed, derived from the plant of *Salvia hispanica*, is native to Central and Southern Mexico and Guatemala. These seeds are traced back to approximately *3500 BCE*, where archaeological and botanical confirmations suggest they were extensively cultivated by prehistoric Mesoamerican civilizations (Ayerza and Coates 2005). Flourishing in semi-arid and arid region, chia seed becomes a most important crop due to its flexibility and very low maintenance necessities; making it appropriate for various environmental situations.

18.2.2 Role in Aztec and Mayan Civilizations

Chia seed was a very important part of the diet and culture of the Aztec and Mayan civilization. The Mayan words for chia, chiabaan, translate to "strength," representing its standard role in energy and stamina. Aztec warriors are believed to consume a single tablespoon of chia seeds to maintain them for 24 h of fight (Coates 2011). Chia seed was also used for:

- *Religious offerings* to deity like *Tlaloc*, the god of rain.
- *Tribute expenses* to rulers and elites.
- *Medicinal uses*, including healing treatment of skin irritation and gastrointestinal issue.

18.2.3 Cultural Importance and Ritualistic Uses

Ahead of nutrition, chia seeds symbolize productiveness, life, and abundance. In Aztec ritual, they were decorated, presented to gods, and used in religious ceremony. The gel formed from saturated seeds was used to treat dehydration during famine and as a cooling agent for fever. This multifunctional property made chia seed intensely embedded in both daily life and holy rituals (MexicoLore 2023).

18.2.4 Decline During the Colonial Era

With the influx of the Spanish Conquistadors in the sixteenth century, chia seed agriculture declined sharply. The colonizers replaced resident crops with European crops such as barley and wheat, and dynamically concealed native religious practices, many of which incorporated chia seed contributions (BackThenHistory 2023). As a consequence, chia seed cultivation almost disappeared, existing only in remote communities in Mexico and parts of Central America.

18.2.5 Rediscovery in the Twentieth Century

Chia seeds experienced resurgence in the 1980s thanks to researchers Dr. Wayne Coates, who promoted chia outstanding nutritional profile. Chia seeds are at present extensively acknowledged as a super food, habitually appreciated for their: Higher omega-3 fatty acid, extra Dietary fiber, loaded Antioxidants, plenty of Proteins, and micronutrients.

By 2000, chia seeds had entered the worldwide health food market. Major producers nowadays include *Mexico, Argentina, Bolivia, and Australia.*

18.3 Botanical Overview

Salvia hispanica is a flowering plant in the mint relatives (Lamiaceae). This yearly herb grows between 1.0 and 1.2 meters in stature, forming clusters of bluish-purple or white inflorescence that ultimately give up oval-shaped seeds (Figs. 18.1 and 18.2). These seeds, colored black, gray, or white, are extremely adaptive and resistant to pests, making them appropriate for sustainable agriculture in insensitive environments.

Botanical explanation of chia seed

- *Scientific name*: *Salvia hispanica L.*

Family: Lamiaceae (Mint family)
Plant nature: Annual herb
Seed appearance: Small (≈1–2 mm), oval shaped, black, white, or gray spotted.

Fig. 18.1 Chia plant during flowering stage, B- Chia

Fig. 18.2 Chia seeds

Peripheral structure of seed

- *Shape*: Small, oval (≈ 1–2 mm long).

Color: Varies as of black, gray, white, or spotted.

Seed cover (outer Layer): Very skinny but tough, Smooth and glossy outside.

Hydrophilic—when exposed to water, the seed cover absorbs water quickly and forms a mucilaginous gel (a slimy, jelly-like layer).

Endosperm: wealthy in lipids (oil), especially alpha-linolenic acid (ALA)—a type of omega-3 fatty acid. More dietary fiber content (~34–40% of seed weight) and they contain antioxidants that defend the fats from spoiling.

Embryo: Stores proteins, necessary fats, and enzymes desirable for germination.

18.4 Utilization Pattern of Chia Seeds

18.4.1 Traditional Utilization of Chia Seeds as Food

Chia seeds were a staple in the diets of prehistoric Aztec and Mayan civilizations. The traditional uses are presented in Table 18.1.

18.4.2 Modern Utilization of Chia Seed as Food

Chia seeds are gained worldwide popularity in health-conscious diets and are now extensively used in many culinary and industrial food applications (Table 18.2).

Table 18.1 Traditional uses of chia seeds

Traditional uses	Description
Chia Fresca (Iskiate)	A traditional energy drink made by mixing together of chia seeds, natural sweetener like honey, lemon juice, and water. It was commonly used for hydration and stamina by Aztec messengers and warriors.
Chia seed flour	Seeds were grind into flour and added to maize-based dishes like tortillas and tamales to boost nutritional substance.
Porridge and gruels	Soaked seeds were used in porridge-like arrangements for breakfast or medicinal purposes.
Medicinal tonics	Mutually combined with herbs to treat gastrointestinal issues or applied as poultices.
Thickening agent	Used to thicken stews, sauces, or soups because of their mucilage-forming property when soaked with water.
Cultural and ritual use	Seeds were offered in religious rituals, believed to bring more strength and vitality.

Table 18.2 Modern utilization of chia seed as food

Modern uses	Description
Smoothies and juices	Add whole seed or flour to drinks for texture and nutritional enhancement (omega-3, fiber, protein).
Chia pudding	A popular dish where seeds are saturated in milk or plant-based milk to form a gelatinous pudding without cooking.
Baked goods	Used in bread, crackers, and cookies either as a flour alternative or as an egg alternate of chia gel.
Egg alternate	Use 1 tablespoon chia and 4 tablespoons water to make 1 egg substitute for vegan recipes.
Salads and yogurt toppings	Use as toppings to sprinkled raw seeds over the salads and yogurt for crunch and nutrition.
Energy bars and snacks	Included into protein bars and granola for continuous energy release.
Chia seed oil	Oil is extracted for use in health supplements, loaded with ALA (omega-3 fatty acid) and gourmet oils.
Functional foods	Incorporated in fortified foods for health benefits like heart health, weight management, and improve digestion.
Packaged foodstuffs	Used in ready-to-eat foodstuffs, beverages, and dairy alternatives.

18.4.3 Health Benefits of Chia Seeds

Chia seeds present various health benefits, mainly due to their rich omega-3 fatty acid, which can give to control heart health, reduce inflammation, and improve blood sugar. They are also a very good source of fiber, protein, and antioxidants, additional enhancing their nutritional value.

Omega-3 Fatty Acids

Chia seeds are great source of healthy omega-3 fatty acids and polyunsaturated fat, which are not prepared by the body and necessity be obtained from foods. The three main omega-3 fatty acids which we can easily get are: Eicosa pentaenoic acid (EPA) and Docosa hexaenoic acid (DHA). Alpha linolenic acid (ALA) comes from plants, vegetable oils, nuts, and seeds. Eicosapentaenoic acid (EPA) and Docosahexaenoic acid (DHA) come from fish and fish oils. Our body changes ALA to EPA 14 Backyards Beyond and DHA, which have the most effective health benefits like reducing risk factors for heart disease, as well as elevated cholesterol and blood pressure, and significant roles in cognition and vision. Chia seeds contain about 26–35% oil by mass and have one of the uppermost known contents of alpha linolenic acid (ALA). A tablespoon of chia seeds and walnuts contains about 2.5 g of alpha linolenic acid, compared to ½ cup of tofu containing 0.4–0.7 g alpha linolenic acid.

Fiber

Chia seeds have more amounted of fiber, one tablespoon of seeds presents a total of 5 grams of fiber, which contributes to the healthy heart personality of chia seeds. A soluble fiber, found in chia seeds, dissolves in water and passes through the digestive system extra slowly than insoluble fiber. Soluble fiber with a healthy diet may possibly assist lower LDL (bad) cholesterol level without lowering HDL (good) cholesterol. Chia seeds also present 2 grams of protein per tablespoon, which is 4% of the everyday recommended value (based on a 2000 calorie diet). Protein helps in building and preserves body muscle and tissues.

Minerals

Chia seeds are loaded with several minerals, contributing extensively to the daily-recommended value of numerous of them. Containing 15% of the daily-recommended value of magnesium (Mg) and phosphorus (P), chia seeds are a good quality source of electrolytes, which control fluid and are significant for muscle function. Calcium (Ca), characteristically obtained in the diet through foods like milk and cheese, is relatively high in chia seeds. A milk-based smoothie with a small amount of chia seeds mixed in would have nearly 500 milligrams of calcium (Ca), about half of a distinctive adult's calcium (Ca) needs (1 small amount of chia seeds provides 180 mg of calcium and 1 cup of every type of milk provides 300 mg of calcium). Cook with some calcium-rich vegetables, such as broccoli, bok choy (leafy Chinese cabbage), and a sprinkle of chia seeds at the finish of cooking could build half of a daily aim of 1000 mg of calcium (Ca) in adults. Moderate amount of Zinc (Zn) is also found in chia seeds, making them an enormous source for vegetarians or vegans. Zinc (Zn) plays a role in immune function, protein synthesis, and injury healing.

Last but not least, iron (Fe) is found in reasonable amounts in chia seeds; however, another nutrient typically found in meats. Current studies have shown that chia seeds are loaded with antioxidants (phenolic compounds). Antioxidants are the substances found in foods, which slow down ("anti-") a process called oxidation ("-oxidant") in the body. Oxidation is chemical responses that produce free radicals. Free radicals cause damage of our body's cells and antioxidants discontinue them. Studies have shown that antioxidants decrease the risk of heart disease and cancer.

The phenolic compounds found in chia seeds can reduce the invasiveness of cancer cells and get better the clinical outcomes.

Heart Health

Chia seeds have higher amount of Omega-3 fatty acids that can help decrease blood pressure and triglyceride levels, which are danger factors for heart disease. Studies recommend that they be able to also decrease the risk of heart attack and stroke.

Antioxidant and Anti-inflammatory Behavior of Chia Seed Protein

Chia seed protein has gained attention growing concentration in nutritional science due to its distinguished *antioxidant* and *anti-inflammatory* properties. Upon digestion, chia protein is easily broken down into different bioactive peptides and amino acids that play an essential responsibility in modulating oxidative pressure and inflammatory responses in the human body.

The *antioxidant activity* of these digestion resulting peptides is particularly significant in counteracting oxidative pressure—a foremost contributor to aging and the pathogenesis of chronic diseases such as cardiovascular disorders, cancer, and neurodegenerative situation. These bioactive compounds have the capability to hunt reactive oxygen species (ROS), by this means preventing cellular injury and maintaining the reliability of biological systems.

In similar, the *anti-inflammatory property* of chia protein hydrolysates further enhances its strength-promoting potential. Whereas inflammation is a protective response to wound or disease, chronic inflammation is an input driver of diseases like arthritis, diabetes, and atherosclerosis. Peptides generated from chia seed protein absorption have been revealed to adjust inflammatory pathways by inhibiting the construction of pro-inflammatory cytokines and mediators.

Blood Sugar Control

Chia seeds contain Omega-3 and fiber may assist in improving blood sugar control, which is useful for individuals with diabetes or persons at risk of developing the situation. High fiber slows down the digestion of carbohydrates and prevents blood sugar spike and helps in supervision of diabetes.

Digestive Health

Chia seeds contain very higher quantity of fiber content (10–11 g per 2 tablespoons); the high fiber content in chia seeds is able to aid digestion by promoting healthy bowel movements and providing bulk to the stool.

Weight Management

Chia seeds play a vital role in contributing to weight management by rising satiety and potentially reducing appetite, chia seeds get bigger in water and form a gel, manufacture you feel fuller for longer.

Excellent Source of Plant-Based Protein

Chia seeds contain good amount of protein, about 20% of chia seed weight is protein. Protein content of chia seeds is almost 17%, greater than the protein content in all other cereals like corn has 9.4%, rice has 6.5%, quinoa has 14.1%, and wheat has 12.6% protein content. They contain all 9 essential amino acids—uncommon for plant foods and excellent for muscle repair and vegetarian/vegan diets. The U.S. Department of Agriculture has confirmed that chia seeds contain a number of exogenous amino acids (arginine, leucine, phenylalanine, valine, and lysine) and a number of endogenous amino acids (glutamic and aspartic acid, alanine, serine, and glycine).

Strengthens Bones

Chia seeds are fully loaded with excellent amount of calcium (Ca), phosphorus (P), magnesium (Mg), and protein. One small amount (about 28 g) of chia seeds has 18% of the daily calcium (Ca) requirement—gram for gram, that's more calcium (Ca) than most dairy items and products.

Naturally Gluten-Free: Chia seeds are naturally gluten free which means they don't have gluten in their seeds that's way there utilization is very secure for people with gluten intolerance or celiac disease.

Supports Hydration:

Chia seeds help maintain *hydration levels*, particularly for athletes or throughout fasting periods, because they absorb water so well and have Gel-forming capability that helps to keep moisture, reducing dehydration risk.

18.4.4 Nutritional Value (per 28 g, ~2 tbsp) Presented in Table 18.3

18.5 How to Consume Chia Seeds

Chia seeds can be easily consumed raw, soaked in liquid, or included into various foods and beverages. Saturating them in water or milk allows them to absorb liquid and create a gel-like structure, making them easier to digest and enhancing nutrient absorption.

Here Are some Ways to Consume Chia Seeds

Chia Pudding:

Chia seeds are combined with cow milk/coconut milk (dairy or nondairy), mix them in a 1:10 ratio, and allow them to saturate for at least 10–15 min, or overnight for a thicker consistency (Fig. 18.3).

Table 18.3 Nutritional value of chia seeds

Nutrient	Amount
Calories	138 kcal
Fiber	10–11 g
Protein	4.7 g
Fat	8.7 g (including 5 g Omega-3 ALA)
Calcium (Ca)	177 mg
Magnesium (Mg)	95 mg
Phosphorus (P)	244 mg
Antioxidants	High levels

Fig. 18.3 Chia pudding

Chia Seeds Water with Lemon:

Chia seeds are soaked with water and add some amount of lemon juice to prepare a popular and easy beverage that can be a stimulating and potentially healthy addition to your diet. The key is to soak chia seeds in water for 10–20 min, awaiting them form a gel, and after that add lemon juice for flavor and additional benefits (Fig. 18.4).

Chia Smoothies:

Chia seeds can add to your favorite smoothie for additional nutrients and a slightly thicker consistency. Place cherries, peaches, strawberries, almond milk, yogurt, dates, and chia seeds in a blender and mix them all to form healthy smoothies (Fig. 18.5).

Sprinkle on:

Chia seeds are used as a topping ingredient sprinkled on oatmeal, yogurt, cereal, or salad for a nutritional enhancement and a delicate crunch.

Fig. 18.4 Chia seeds water with lemon

Fig. 18.5 Chia smoothies

18.6 Production Practices of Chia Seeds

Climate and Temperature

Chia seeds cultivation practices thrive in tropical and subtropical regions, ideally between 400 to 2500 meters above sea point. They prefer light to medium sandy loam or clay soils that are well drained and fairly fertile, and they can tolerate acid soils and moderate drought conditions. Most favorable growth occurs within a temperature range of 16–26 °C, with minimum and maximum tolerance levels of temperature, respectively, 11 and 36 °C.

Rainfall

Chia seeds can easily grow in areas with varying rainfall, from 300 to 1000 mm per year. However, for germination it requires moist soil and good rainfall.

Propagating Chia Seeds

Chia seeds are propagated from seeds, rising chia plants from seeds can be the best job. Soil must be well drained and carry it to well tilth for better germination, spread the seeds over it and cover up them with soil. Watering must be done at regular intervals. Chia seeds initiate sprouting within 3–5 days. After germination, thinning is to be done when the seedling grows up at least 20–30 cm tall with 6–7 pairs of leaves (Arya and Kumar 2021).

Soil Preparation for Chia Cultivation

Preparation of soil plays a very significant role in the successful cultivation of chia seeds. They grow well in sandy loam to clay soil with good drainage. For the proper growth and development pH level of the soil should be between 6.0 and 8.5. Before the planting of seedlings, farmers should clear the field of unwanted weeds and plough the soil thoroughly to improve aeration.

Well-decomposed organic compost or farmyard manure that has been enhancing the fertility of the soil. As the crop chia seeds do not need much amount of fertilizers, using organic nutrients will enhance proper growth along with good fertility of soil.

Sowing and Seed Rate

The cultivation of chia seed requires light to medium sandy loam to clay soils. The crop can provide higher yield in well drained and moderately fertile soils. For sowing of chia seeds and seedlings, needed fully moistures soil for proper establishment, while the maturing chia plant cannot defend against wet soils during enlargement. Conventional cultivation techniques of chia include soil preparation by disruption and loosening followed by seed broadcasting. In modern profitable production, a typical seed rate of 1.0 kg ha^{-1} and spacing of 60 cm × 15 cm will be most favorable.

Land Preparation: Systematic land preparation, including plowing and planking, is important.

Sowing Time

Chia seeds naturally require a temperature of 20–25 °C to germinate properly; as a result, sowing should take place in delayed autumn or near the beginning winter when temperatures are appropriate.

Fertilization and Irrigation

Chia seeds can be easily cultivated under low inputs of fertilizer requirement, using 40 N:20P:20 K kg ha^{-1} or in several cases, no fertilizer is used for cultivation. Irrigation frequency in chia seed cultivation field may vary from none to eight irrigations per rising seasons, depending on climatic situation and rainfall conditions.

The plants require very modest water once established. However, at the germination time, they must have adequate moisture to grow up. Light irrigation at intervals is suitable for the early weeks. After the plants have reached a constant height, they can tolerate arid spells and need fewer watering. Drip irrigation is an excellent practice for chia cultivation because it provides water to plants with no much wastage. In adding up, farmers have to prevent overwatering since the plants are vulnerable to root rot under waterlogged soil.

18.7 Plant Protection of Chia Seeds from Biotic Stress

18.7.1 Weed and Pest Management

The plants move toward close up to one another, thereby stopping the weeds from growing. However, when in the growth stage, weeds have to be physically eliminated to keep away from struggle for nutrients. Mulching using dried straw or vegetation retard soil moisture defeat and prevents weeds growth. The crop is habitually resistant to pests but can be attacked by aphids and whiteflies as a field pest and confused flour beetle, Indian meal moth and weevils which can be infest the stored seeds and cause damage if seeds are stored inappropriately. Natural predator like ladybugs, or organic pesticides, such as neem oil, may aid in pest control without destroying the crop.

18.7.2 Disease Management

Crop is not affected by major diseases. Essential oils in Chia foliage make them more resistant to diseases, making them appropriate for organic cultivation. Virus infections, sometimes transmitted by whiteflies, may arise. Weeds can produce a major problem during the premature growing period of the chia crop awaiting its canopy closes, but because Chia is responsive to the most frequently used herbicides, mechanical weed control is favored. Some viruses that contaminate the genus Salvia are Cucumber mosaic virus, the extensive bean wilt virus, Mung bean Yellow Mosaic Virus, Tomato yellow leaf curl virus, and other putative golden mosaics. These viruses can influence crop production to a greater degree. One has to take essential measures as soon as the crop is contaminated with these viruses.

18.8 Harvesting and Postharvest Handling

Harvesting of Chia seeds depends on the days necessary to dry a plant. Farmers are harvested manually and doing mound until it dries fully. The seeds are separated from the capsules by thrashing with sticks or machine can also be used. Seed yield and composition of Chia yield of chia seeds vary on cultivars, mode of cultivation, and growing conditions by geographic area. For example, in commercial fields, the yield ranges from 1000 to 1250 kg ha^{-1}. The higher yields can be predictable when there is positive growing environment and cultivar interaction.

18.8.1 Harvesting Chia Seeds

The plants are ready to be harvest when the flowers dry, and the seed heads turn brown in 90–120 days. If this process is late, seed loss can happen because mature seeds are simply spread when wind or rain disturbs them. Farmers can harvest the plants at the base and dry them in a shaded area for a few days. After drying, the seeds can be easily removed from the plant by threshing or winnowing. Drying appropriately makes them last longer and is free from mold or moisture injury.

18.8.2 Postharvest Processing and Storage

After that, the seeds have to be appropriately cleaned to remove any contaminants. They have to be sun-dried earlier than they are stored in air-tight containers. Storage is essential to protect the quality and shelf life. They contain a high percentage of oil in their seeds that is a reason making them spoilage prone when they are not properly stored in storage. By storing the seeds in a cold and dry condition, their nutritional value is maintained for a longer time. Farmers also packed them in vacuum packs to prevent oxidation.

18.9 Future Global Market Potential and Commercialization of Chia Seeds

The global demand for chia seeds is increasing quickly day by day due to growing consumer awareness about health and nutrition. Chia seeds are an excellent source of omega-3 fatty acids, rich in antioxidants, and are extensively recognized for their health benefits. Increasing health consciousness and a move toward organic and chemical-free diets have significantly contributed to the expansion of the Global Packaged Chia Seed Market during the predicted period of 2021–2026 (IndustryARC 2021).

Chia seed production is geographically partial, making it a significant export product. The initial COVID-19 lockdowns in 2020 disrupt logistics and worldwide supply chains, leading to a provisional turn down in the chia seed market. However, as the pandemic sensitive health concerns, demand for super foods, including chia seeds, rebound and is predictable to continue rising through 2026 (Mordor Intelligence 2022).

Unlike a lot of agricultural commodities, chia seeds have not so far followed the path of full commoditization. With rising awareness of their benefits, the market is predictable to experience robust growth worldwide. The United States and Australia lead in chia seed consumption and utilization, whereas emerging markets such as the United Kingdom, Brazil, Chile, and Spain have shown extraordinary growth (Mordor Intelligence 2022). The rush in popularity of prehistoric grains and purposeful foods is also fueling this requirement.

Chia seeds are gluten-free and higher in protein content, attractive especially to consumers with celiac disease and those following gluten-free or plant-based diets (Mordor Intelligence 2022).

18.9.1 Market Segmentation by Product Category

Based on *The Mysore Journal of Agricultural Sciences* (Anand et al. 2024), the Global Packaged Chia Seeds Market is segmented by product category as follows:

Black Chia Seeds: Accounted for the largest profits share in 2020. Because of their higher protein content and small size construct they are preferred by health-aware consumers, particularly in the food and beverage division.

White Chia Seeds: Predictable to register the greatest compound annual growth rate (CAGR) of 7.5% from 2021 to 2026 due to their exclusive texture, mild flavor, and high fiber content that supports metabolism.

Brown Chia Seeds: Although a lesser segment, they grasp potential in niche applications.

18.9.2 Market Segmentation by Application/Request

Based on consumer's request in worldwide, the chia seeds market is segmented into:

Food and Beverages: This segment detained the largest market share in 2020, ambitious by increasing use in bakery products, peanut butter, cereals, and gluten-free products (IndustryARC 2021).

Pharmaceuticals: Predictable to grow at a CAGR of 8.1% during 2021–2026. Chia seeds are included into supplements for cardiovascular health (Anand et al. 2024). Based on higher Omega-3 fatty acid percentage in the chia seeds, they should be used for making Omega-3-rich capsules for health benefits.

Personal Care and Cosmetics: Chia seed oil is gaining use in skin and hair care due to its Omega-3 fatty acid content, moisturizing, and antioxidant properties.

18.9.3 Regional Insights of Chia Market

North America leads the global market with a 45% share in 2020 due to the high acceptance of gluten-free diets and the increasing occurrence of lifestyle diseases such as cardiovascular issues. The American Heart Association's suggestions to include Omega-3 fatty acid-rich foods like chia seeds in the diet have supported market growth (IndustryARC 2021).

Asia Pacific (APAC) is predictable to register the uppermost CAGR of 3.05% from 2021 to 2026, ambitious by increasing health consciousness and lifestyle-associated diseases such as diabetes, obesity, and hypertension, particularly in countries like India (Anand et al. 2024).

18.10 Commercialization Pathways and Value-Added Products

The commercial potential of chia seeds extends beyond raw utilization. Various forms are being developed to meet market demands:

Whole Chia Seeds: Used for culinary applications.
Ground Chia Seeds: Added to smoothies, bakery items, and dietary supplements.
Chia Oil: Used for its high omega-3 content in health supplements and culinary products.
Chia-Based Foods: Includes energy bars, puddings, and ready-to-drink beverages.

Certifications and Quality Standards

For successful international commercialization and increasing worldwide demand of chia seeds, seed products often necessitate:

Organic Certification: Enhance worldwide demand trust and market value of chia seeds.
Non-GMO and Gluten-Free Labels: Provide to specific dietary requirements.
Safety Standards: Regular testing for microbial safety, pesticide residues, and heavy metals is essential for export compliance (Market Research Future 2022).

Opportunities in India and International Trade

Chia cultivation is still rising in India. With rising domestic demand and favorable climatic conditions, successful large-scale farming might enhance both national nutrition levels and economic development. Collaboration with international institutions is recommended to access better germplasm lines and improve crop productivity (Anand et al. 2024).

Chia imports into Europe are primarily sourced from Paraguay, Bolivia, Argentina, Peru, Mexico, Uganda, Chile, and Australia (Mordor Intelligence 2022). This trend highlights the potential for new players, including India, to enter the international supply chain.

References

Anand SR et al (2024) Chia (*Salvia hispanica* L): a nutri-rich potential crop, its present scenario and future perspectives in India. Mysore J Agric Sci 58(2):1–14

Arya RK, Kumar P (2021) Shamtawan faslon ki kheti avam upyog. Anand Kalan Manch Publication, Bhiwani, Haryana, pp. 17–20

Ayerza R, Coates W (2005) Chia: rediscovering a forgotten crop of the Aztecs. University of Arizona Press

Back Then History (2023) The history of chia seeds. Retrieved from https://www.backthenhistory.com/articles/the-history-of-chia-seeds

Coates W (2011) The nutritional composition of chia seeds and their role in pre-Columbian diets. J Ethnobiol Ethnomed 7(1):12. https://doi.org/10.1186/1746-4269-7-12

Coelho MS, Salas-Mellado M (2014) Chemical characterization of chia (Salvia hispanica L.) for use in food products. J Food Sci Technol 51(3):547–552

Industry ARC (2021) Global packaged chia seed market report: Forecast (2021–2026). Industry ARC. https://www.industryarc.com/Research/Global-Packaged-Chia-Seed-Market-Research-509520

Kumar A, Singh R, Rathi A (2021) Cultivation potential and commercialization scope of chia seeds (Salvia hispanica) in India. J Agric Biotech S Dev 3(2):22–30

Market Research Future (2022) Chia seeds market research report – forecast to 2027. Market Research Future. https://www.marketresearchfuture.com/reports/chia-seeds-market-4299

MexicoLore (2023) Aztec advances in health and medicine – Chia. Retrieved from https://www.mexicolore.co.uk/aztecs/health/aztec-advances-17-chia

Mohd Ali N, Yeap SK, Ho WY, Beh BK, Tan SW, Long K (2012) The promising future of chia, Salvia hispanica L. J Biomed Biotechnol 2012:1

Mordor Intelligence (2022) Chia seeds market – growth, trends, COVID-19 impact, and forecasts. https://www.asdreports.com/market-research-report-615324/chia-seed-market-growth-trends-covid-impact-forecasts

Poudel P, Nandurkar NS, Koche D (2020) Functional properties and health benefits of chia seeds: a review. Plant Arch 20(Supplement 1):1021–1026

Ulbricht C, Chao W, Nummy K et al (2009) Chia (Salvia hispanica): an evidence-based systematic review by the natural standard research collaboration. J Diet Suppl 6(1):54–78

Ullah R, Nadeem M, Imran M (2016) Nutritional and therapeutic perspectives of chia (Salvia hispanica L.): a review. J Food Sci Technol 53(4):1750–1758

Halim Seeds (Garden Cress): The Ancient Seed of Iron and Immunity

19

Sarita Agrawal, Bhavana Tiwari, and Aprajita Tiwari

Abstract

The garden cress seed (*Lepidium sativum* L.) is an underutilized plant that plays a vital role as nutritional and medicinal plant in India. It belongs to family cruciferae/brassicaceae. It is originated from highland region of Ethiopia and Eritrea. It is fast growing annual herbaceous plant, mainly cultivated for seeds. Water-soaked seed is mucilage. In addition to vitamins like carotene, riboflavin, thiamine, niacin, dietary fiber, calcium, and protein, these are abundant sources of folate, iron, vitamin A, vitamin C, and vitamin E. Numerous health benefits, including anticancer, antimicrobial, antioxidant, anti-inflammatory, antianemic, antidiabetic, hepatoprotective, antipyretic, and analgesic qualities, are found in halim seeds. It helps in the prevention of many diseases such as anemia, menstrual cycle disorders, diabetes, heart disease, high blood pressure, cancer, bone diseases, and many skin diseases. It is beneficial for memory enhancement, boosting milk production in nursing moms, etc. It is used in various industrial sectors like food industry, cosmetic industry, and pharmaceutical industry. It is cool-season crop. It grows in all types of soils, but well-draining soil with pH of 6.0–8.5 is ideal for its cultivation. It is propagated by seeds (8–10 kg/ha). It is mature after 90 days of sowing. The average seed yield is 10–12 quintals per hectare in un-irrigated areas and 14–16 quintals per hectare in irrigated areas. The market price of its seeds generally ranges from Rs. 7000 to 8000 per kg. Dahiwala bread, brioche bread, biscuits high in omega-3 fatty acids and iron,

S. Agrawal (✉)
Indira Gandhi Agriculture University, Raipur, Chhattisgarh, India

B. Tiwari
Rani Lakshmi Bai Central Agricultural University, Jhansi, Uttar Pradesh, India

A. Tiwari
Ayurvedic Medical Practitioner, Jhansi, Uttar Pradesh, India

A. Kumar et al. (eds.), *Fasting Superfoods*,
https://doi.org/10.1007/978-981-95-2762-5_19

vegetable oils mixed with garden cress oil rich in alpha-linoleic acid, fortified burfi, fortified chikki, ladoo, noodles, cookies, flakes, and other foods and beverages are all fortified with garden cress seeds to improve their nutritional and functional qualities.

Keywords

Garden cress · *Lepidium sativum* · Chandrasur · Halim seed · Medicinal · Iron · Immunity

Traditional medicine practitioners widely use garden cress, a fairly well-known plant. There are several colloquial terms used to describe the garden cress plant in India, including chand shura (Sanskrit), chansur (Hindi), halim (Urdu), holan (Punjabi), alian (Kashmiri), haliv (Marathi), aseliyo (Gujarati), candriki (Assamese), chandasura (Oriya), halim (Bengali), allibija (Kannada), aadalu (Telugu), allivirai (Tamil), and asali (Malayalam). Garden cress is known as chansur in Hindi and raktabija in Sanskrit. It is commonly known in English as pepper wort, garden pepper grass, pepper cress, or common cress (Vaishnavi et al. 2020).

According to Patil et al. (2015), garden cress seed has been considered an important nutritional and medicinal herb in India since the Vedic era (c. 500–1700 B.C.). Garden cress is another potential new oilseed crop (Angelini et al. 1997).

19.1 Origin and Distribution

The highland areas of Ethiopia and Eritrea are where garden cress first appeared. According to Stchenkova, Europe and western Asia are considered secondary places of origin. It comes from Ethiopia and is disseminated in different places. The temperate and sub-temperate regions are home to over 150 species (Wadhwa et al. 2012). It is grown on over 5000 hectares of land in spots throughout the Indian states of Madhya Pradesh, Gujarat, Uttar Pradesh, Rajasthan, and Maharashtra (Choudhari et al. 2010). It is grown in North America and several regions of Europe in addition to India (Gokavi et al. 2004). Due to its tolerance in many temperate zones, garden cress is currently grown on a limited scale in a number of countries, most notably India, Pakistan, Europe, and much of Africa (El-Salam et al. 2019; Wadhwa et al. 2012).

19.2 Nutritional Value

Garden Cress is a rich source of folate, iron, vitamin A, vitamin C, vitamin E, and vitamins like carotene, riboflavin, thiamine, niacin, dietary fiber, calcium, and protein (Table 19.1). It also contains acids like rachidic acid, folic acid, ascorbic acid, and beta carotene, which all help to strengthen our immune systems (Balasubramanian

Table 19.1 Nutritional composition of garden cress seeds

Nutrient	Nutrient amount per 100 gram of garden cress seeds	References
Energy(kcal)	474 ± 1.06	Gokavi et al. (2004) Zia-Ul-Haq et al. (2012)
Carbohydrate(g)	30.7 ± 1.2	
Moisture (g)	2.9 ± 0.1	
Protein (g)	24.2 ± 0.5	
Fibre (g)	11.9 ± 0.4	
Lipids (g)	23.2 ± 0.2	
Ash(g)	7.1 ± 0.1	
Vitamin(dry weight basis)		
Vitamin A (μg)	90	Yadav et al. (2018) Zia-Ul-Haq et al. (2012)
Thiamine (mg)	0.59	
Riboflavin (mg)	0.61	
Niacin (mg)	14.30	
Vitamin C (mg)	10.62	
Vitamin E (ppm)	258.74	
Mineral(dry weight basis)		
Calcium(mg)	480.72	Gokavi et al. (2004) Yadav et al. (2018)
Cooper (mg)	12.51	
Iron (mg)	28.82	
Manganese(mg)	46.07	
Magnesium(mg)	631.06	
Phosphorus(mg)	637.25	
Potassium(mg)	1635.62	
Sodium(mg)	36.25	
Zinc(mg)	62.69	
Amino acid content		
Essential amino acids		
Arginine(g)	4.51 ± 0.03	Zia-Ul-Haq et al. (2012)
Histidine(g)	3.87 ± 0.14	
Isoleucine(g)	5.11 ± 0.03	
Leucine(g)	8.21 ± 0.01	
Lysine(g)	2.66 ± 0.09	
Methionine (g)	0.97 ± 0.02	
Phenylalanine(g)	5.65 ± 0.03	
Threonine(g)	6.26 ± 0.39	
Valine(g)	8.04 ± 0.03	
Nonessential amino acids		
Aspartic acid (g)	9.76 ± 0.03	Gokavi et al. (2004) Zia-Ul-Haq et al. (2012)
Alanine(g)	4.83 ± 0.02	
Glutamic acid (g)	19.33 ± 0.19	
Glycine(g)	5.51 ± 0.07	
Proline (g)	2.69 ± 0.09	
Serine(g)	4.96 ± 0.09	
Tyrosine(g)	5.84 ± 0.38	

(continued)

Table 19.1 (continued)

Nutrient	Nutrient amount per 100 gram of garden cress seeds	References
Fatty acid profile		
Arachidic acid(20:0)	2.10 ± 0.57	Zia-Ul-Haq et al. (2012)
Eicosaenoic acid (20:1)	13.40 ± 0.66	
Linoleicacid (18:2)	8.60 ± 0.38	
Linolenic acid(18:3)	32.18 ± 0.59	
Oleic acid (18:1)	30.50 ± 0.16	
Palmitic acid (16:0)	10.30 ± 0.12	
Palmitoleic acid (16:1)	0.70 ± 0.30	
Stearic acid(18:0)	1.90 ± 0.19	
Anti-nutritional factor(raw)		
Phytin phosphorus (mg)	447.2 ± 3.34	Jain et al. (2016)
Oxalates(mg)	134 ± 1.00	

2009). They possess plant substances like estrogen, which promote a mother's milk production and help regulate menstruation (Tukaramrao et al. 2009). These seeds' oil has a balanced proportion of monounsaturated fatty acids (MUFA) (37.6%) and polyunsaturated fatty acids (PUFA) (46.8%). The presence of natural antioxidants, such as eugenol and vitamins A and E, helps shield cells from harm caused by free radicals (Tukaramrao et al. 2009).

19.3 Phytochemical Composition and Therapeutic Properties of Garden Cress

The seeds of garden cress contain a wide variety of phytochemicals, including alpha linolenic acid, methanol, ethyl acetate, imidazole compounds, petroleum ether, aqueous extract, and organic sulfur compounds. Numerous biological activities, including carcinogenic detoxification, apoptosis, antioxidant, antitumor, anti-oral cancer, pulmonary functions, antihypertensive, arterial blood pressure, diuretic, hypoglycemic, antidiabetic, fracture healing, anticarcinogenic, antibacterial and fungal, laxative, and anti-cardiogenic, are responsible for these compounds (Table 19.2).

19.4 Medicinal Properties of Garden Cress

19.4.1 Rich in Vitamins and Minerals

An abundant amount of Vitamin C is contained in garden cress. Almost 32% of the daily requirements of Vitamin C is given to the body by an ounce of cress seeds. The

Table 19.2 Phytochemical composition and therapeutic properties of garden cress

Sr. No.	Chemical constituents	Biological activity	Mechanism	References
1.	Benzyl isothiocyanate, Glucosinolates	Carcinogenic detoxification, cell-cycle arrest, and apoptosis, inhibit tumour cell proliferation (antitumor activity) and free radical scavenging activity	*Benzyl isothiocyanate*: Radiotherapy successfully induces sensitivity to the pancreatic cell lines (PANC-1 and MIAPaCa-2); consequently, apoptosis is increased, which is brought on by ionizing radiation and benzyl isothiocyanate. *Glucosinolates*: This chemical element serves to accelerate the apoptosis of malignant cells by inhibiting angiogenesis, controlling low levels of systemic oxidative stress, and preventing cell cycle progression.	William et al. (2009)
2.	Imidazole compounds	Antihypertensive effect, diuretic activity, antidiabetic activity, bone protective property	*Imidazole*: The inhibiting action on the formation of ergosterol, the fundamental cell terol of fungus, is part of the mechanism. In this process, a fungal cytochrome P-450 enzyme is impacted. The oxidative and peroxidative activity of the enzyme is altered.	Eddouks (2004) Maghrani et al. (2005) Bin Abdullah Juma (2007)
3.	Alphalinolenic acid	Anti-cardiogenic activity, anti-inflammatory activity, and anti-carcinogenic activity	*Alpha linolenic acid*: It helps to prevent the development of carotid plaque. Additionally, a potassium background channel in the neurons is stimulated, which causes the membrane to become hyperpolarized. This, in turn, raises the magnesium block of the calcium channel connected to the NMDA receptor, which plays a crucial part in settling glutamate-mediated excitotoxic neuronal cell death.	Diwakar et al. (2008) Diwakar et al. (2010)
4.	Lepidimoide	Antidiabetic activity	*Lepidimoide*: The development of abscission is heavily inhibited, which results in the automatic prevention of mineral loss that in turn controls the blood glucose level.	Eddouks et al. (2005)
5.	Phenolic acids, flavonoids, anthocyanins	Antioxidant activity, anti-mutagenic activity, anti-carcinogenic activity, ability to modify gene expression	*Polyphenol*: Various anticancer activities have been linked to it by the restriction of apoptosis induction and oxidative stress. *Flavonoids*: They are responsible for the development retardation of mitochondrial apoptotic-dependent gastric cancer stem cells by blocking the P13K-Akt signaling pathway. *Anthocyanin*: This strategy, which includes inhibiting enzymes or compressing trace elements that are further responsible for the creation of free radicals, prevents the generation of reactive species and induces antioxidant activity.	Shen et al. (2016)
6.	Triterpenes	Antioxidant antimicrobial activity	*Tetripenes*: This process involves iso-thiocyanate, which is in charge of the gene activation, involved in inflammation, blocking the signaling of NF-κB.	Bary et al. (2017)

bones, skin, and healthy gums are helped to be maintained by it. The body's needs for Iron are also assured to be covered by it. The body's ability to absorb iron from the foods eaten may be improved by it (Prajapati and Dave 2018).

19.4.2 Anticancer

The seeds of garden cress are a natural remedy for cancer patients because they are rich in vitamins, calcium, iron, manganese, and other beneficial minerals (Prajapati and Dave 2018). The proliferation of MCF-7 breast cancer cells can be inhibited by an aqueous extract of garden seed. Many types of cancer, including brain tumors, prostate cancer, hepatic cancer, lung cancer, and breast cancer, have been shown to be inhibited by kaemferol, which is present in garden cress seeds (Mahassni and Al-Reemi 2013). According to Bhatia and Bhasin (2023), garden cress seeds can shield cells from the harmful effects of free radicals, which can damage DNA and initiate the growth of cancer. Furthermore, a bioactive substance included in garden cress seeds can prevent the action of particular enzymes linked to tumor growth (Ibrahim et al. 2023).

19.4.3 Antimicrobial

In addition to having a strong antifungal effect at concentrations of 2.5% and 10%, the petroleum ether extract of garden cress seed demonstrated dynamic antimicrobial activity against six distinct pathogens at doses of 2.5%, 5%, and 10%. At 2–8% of seed extracts, ethanolic extract of garden cress seed appeared to be very beneficial against the growth of fungi (*Fusarium equiseta, Aspergillus flavus, and Alternaria alternata*) (Sharma et al. 2012).

19.4.4 Antioxidant

The antioxidants that can be found in garden cress seeds are Vitamin E, flavonoids, and other phenolic compounds. They prevent the damaging effects of free radicals, which cause oxidative stress and destroy liver cells. Garden cress seeds aid in preventing liver damage and advancing general liver function by lowering oxidative stress (Hanan et al. 2019).

19.4.5 Anti-inflammatory

Garden cress seeds contain bioactive substances, including flavonoids and alkaloids, which have anti-inflammatory benefits. By modifying pro-inflammatory pathways and reducing the production of inflammatory mediators, flavonoids also help the seeds' anti-inflammatory properties. Alkaloids, on the other hand, have the potential to obstruct inflammatory signaling pathways, helping to reduce inflammation. This interaction of bioactive substances can lessen inflammation-related disorders and enhance general health (Mohamed et al. 2023).

19.4.6 Antidiabetic

Garden cress seeds have excellent power to control blood sugar levels. They have alkaloids that are beneficial against diabetes (Prajapati and Dave 2018). Garden cress aqueous extract has antidiabetic properties and has a hypoglycemic impact without requiring insulin secretion. Seeds also help in the treatment of diabetes mellitus (Eddouks et al. 2002). The enzymes involved in glucose metabolism and insulin signaling pathways may be impacted by the bioactive compounds (Doghmane et al. 2021). Low glycemic index has been seen in seeds because of high fiber content, which slows down the absorption of glucose and prevents abrupt rises in blood sugar after meals (Mulla et al. 2019).

19.4.7 Hepatoprotective

The potential of garden cress seeds to support liver health and shield the liver from harm has drawn attention to their hepatoprotective properties. Many studies have examined the bioactive compounds found in garden cress seeds and how they affect liver function (Jagdale et al. 2021). The seeds' bioactive components promote the growth of new liver cells, aiding in the restoration of damaged liver tissue (Mohamed et al. 2023). The ability of garden cress seeds capacity to control lipid metabolism may help prevent the accumulation of hepatic fat and the onset of linked disorders (Ibrahim et al. 2020).

19.4.8 Antianemic

Garden cress seeds, due to their extraordinary iron and folic acid content, have drawn interest as a potential natural treatment for anemia. Anemia is a condition of a lack of red blood cells or hemoglobin in the blood. These seeds have a robust nutritional composition that makes them an excellent dietary complement for preventing anemia (Chaudhary and Gupta 2021).

19.4.9 Antipyretic

The antipyretic action of these seeds is influenced by the tannins and flavonoids that are present in them. Tannins are well known for being astringent and lowering fever. Through the constriction of blood vessels and a decrease in excessive heat generation, these substances can aid in lowering body temperature. Due to their ability to act as antioxidants, flavonoids can also help lower fever by preventing the oxidative stress brought on by an elevated body temperature (Sharma and Agarwal 2011).

19.4.10 Analgesic

Garden cress seeds have analgesic (pain-relieving) properties because they contain sterols, triterpenes, and alkaloids. By altering neurotransmitter pathways, sterols and triterpenes are known to obstruct the perception and transmission of pain. Saponins are an example of an alkaloid that may bind to pain receptors and alter pain signals. These combined effects provide the seeds with the ability to treat a variety of pains, from minor discomfort to more serious pain disorders (Al-Snafi 2019).

19.5 Health Benefits of Garden Cress

Health benefits of halim seeds are presented in Table 19.3.

19.5.1 Facilitate the Treatment of Anemia

When consumed regularly for about 2 months, garden cress seeds raise the standard of hemoglobin in human blood and offer a significant amount of iron and folic acid. The iron shortage that causes anemia is addressed by this medication. According to studies, Vitamin C supplements improve iron absorption and boost the digestive system's capacity to do so. Without taking any Vitamin C supplements, people can utilize garden cress seeds to cure anemia or iron deficiency because they are high in Vitamin C and iron (Falana et al. 2014).

19.5.2 Treats Menstrual Cycle Disorders

It creates a remedy for ladies who experience irregular or reduced menstrual cycle flow. The hormonal environment is maintained to initiate the menstrual cycle and to sustain the menstrual cycle or menstrual periods on a monthly basis through the consumption of garden cress seeds because of the presence of the hormone estrogen-like photochemical (Falana et al. 2014).

19.5.3 Boosts the Secretion of Breast Milk

Garden cress seeds are widely recognized for their high nutritional content, which is advantageous for nursing mothers. It assists the nursing mother in maintaining the flow and production of breast milk with a high protein and iron content (Prajapati and Dave 2018).

Table 19.3 Health Benefits of Garden Cress

Sr. No.	Health benefits	Components responsible	Key findings	References
1.	Prevents asthma	N,N-dibenzyl thiourea, leporine, N,N-dibenzzyl urea, sinapic acid, Riboflavin etc.	Garden cress seeds containing ethanol extract demonstrated broncho protection against acetylcholine and aid in the prevention of asthma.	Paranjape and Mehta (2006) Mali et al. (2008)
2.	Regulates menstrual cycle	Flavonoids, cardio tonic glycosides, coumarins, glucosinolates, saponins, sterols, Sinapic acid, tannins, triterpene, uric acid, volatile oil, etc.	The existence of estrogenic trails contributes to the maintenance of appropriate menstrual cycle regulation. Emmenagogues, a chemical that causes menstruation by increasing blood flow in the uterus and pelvis.	Pattnaik (2003) Ghante et al. (2011) Singh et al. (2015)
3.	Prevents anemia	Glutamic acid, aspartic acid, potassium, calcium, phosphorus, iron, manganese, zinc, fibers, Linoleic acid, oleic acid, etc.	It is a great source of nonheme iron, or iron-bound hemoglobin, which raises blood hemoglobin levels and guards against anemia. This seed contains a hematic agent.	Paranjape and Mehta (2006)
4.	Increases breast milk secretion	Tocopherols, carotenoids, ascorbic acid, flavonoids, phytosterol, proteins, irons, glutamic acid, Aspartic acid, etc.	It has a high protein level and iron content, which helps a nursing mother produce more milk. It is administered as a postpartum galactogogue to stimulate the production of breast milk.	Pattnaik (2003) Singh et al. (2015)
5.	Diabetes	Alkaloids like lepidine, glucotropaeolin, etc. Antioxidants like tocopherols, carotenoids, ascorbic acid, etc.	Because of its antidiabetic properties, it lowers blood glucose levels; also, the presence of alkaloids helps prevent diabetes. Additionally, diabetes mellitus is treated using it.	Eddouks et al. (2002) Bin Abdullah Juma (2007) Prajapati and Dave (2018)
6.	Protects cancer	Phytochemical & antioxidant components such as tocopherols, carotenoids, eugenol, terpenoids, glucosinolates (glucotropaeolin and 2-phenylethyl glucosinolate), 3′,5′-dimethoxy acetophenone, hexadecanoic acid, methyl ester are present.	Protects against several malignancies because it is a good source of alphalinolenic acid and other phytochemicals, such as phenolic compounds, terpenoids, alkaloids, and organosulfur compounds. Glucosinolates, secondary metabolites that block carcinogens, prevent cancer from developing and spreading.	Kassie et al. (2002) McGee et al. (2002) Kassie et al. (2003) Matthäus and Angelini (2005) Radwan et al. (2007) Khan et al. (2008) Diwakar et al. (2008) Conforti et al. (2008) Barba et al. (2016) and Rajasekaran and Suresh (2021)

19.5.4 Helps to Treat Diabetes

Photochemicals are released from vegetable gum, which is found in garden cress seeds. Glucose levels in diabetics can be managed because of the photochemicals' ability to lower blood sugar (Prajapati and Dave 2018).

19.5.5 Treats Constipation and Stimulates Gastrointestinal Function

The sore throat, asthma, headaches, and coughing are treated with garden cress seeds. People with bronchitis are advised to use it because the characteristics of a bronchodilator are contained in it (Prajapati and Dave 2018).

19.5.6 Maintains Arterial Health and Protects Against Heart Diseases

Garden cress seeds are rich in Vitamin C and Vitamin K (Lahiri and Rani 2020). They protect the arteries from increased level of cholesterol by production of collagen (collagen is a type of protein responsible for elasticity of arterial wall) and helping in prevention of deposition of calcium (calcification of arteries) on the arteries and heart valves (Prajapati and Dave 2018).

Treats High Blood Pressure

It has been established that garden cress seeds can help decrease blood pressure. They have antihypertensive characteristics and diuretic effects, which together help to reduce high blood pressure levels. Seeds have antihypertensive properties, come from bioactive substances that can assist in widening blood vessels, encourage vasodilation, and therefore enhance blood flow. Garden cress seeds' diuretic properties also encourage greater urine production, which helps the body rid itself of extra fluid and sodium. Reducing fluid retention may lessen the stress on blood vessel walls, which would help regulate blood pressure even more (Sharma 2020).

19.5.7 Protects Against Cancer

The scientific investigation has demonstrated that using these seeds to combat cancer is feasible. It contains antioxidants like Vitamin A and Vitamin E, which help shield cells from the harm that free radicals can do. Additionally, there is a bioactive substance that inhibits the synthesis of specific tumor-causing enzymes. Breast cancer cells are killed by the injection (Prajapati and Dave 2018). The presence of omega-3 and omega-6 fatty acids also aids in the prevention of breast cancer (Khan et al. 2008).

19.5.8 Helps to Treat Bone Diseases, Cold, Rheumatism, and Teeth

The health of bones and teeth is promoted by garden cress seeds, as they contain a specific amount of Vitamin K. Calcium levels in bones and joints are increased by Vitamin K. The integrity is maintained, and protection against fracture is provided (Prajapati and Dave 2018).

19.5.9 Treatment of Respiratory Infection

To cure colds, headaches, asthma, sore throats, and coughs, garden cress seeds can be chewed or combined with honey and used as an expectorant (Jabeen et al. 2017).

19.5.10 Maintains Healthy Gums

The growth of destructive bacteria that cause tooth decay is stopped by consuming the garden cress seeds, and at the same time, contributions toward the health of the gums and teeth are helped by them as they are rich sources of vitamins (Singh et al. 2015).

19.5.11 Helps Treat Hemorrhoids

Consumption of garden cress can help to treat hemorrhoids and alleviate their symptoms (Singh et al. 2015).

19.5.12 Beneficial for Skin

Garden cress seeds are rich in fatty acids, iron, calcium, folic acid, Vitamin A, Vitamin C, and Vitamin E and are beneficial in skin disorders related to itching. Also provide treatment for chapped lips, sunburns, skin irritation, and dry skin (Singh et al. 2015).

19.5.13 Benefits of Garden Cress for Hair

Healthy and thick hair is helped to be obtained by garden cress seeds. Damaged hair is treated and hair loss is prevented by boiled seeds because Vitamin C and antioxidant are contained in them (Singh et al. 2015).

19.5.14 Helps to Lose Weight and Slim Down

Weight loss can be achieved naturally by consuming a mixture of garden cress seeds in water before breakfast (Singh et al. 2015).

19.5.15 Memory Booster

The presence of arachidic fatty acid and linoleic acids makes garden cress seeds serve as a potent source of memory booster (Jabeen et al. 2017).

19.6 Uses of Garden Cress

19.6.1 Traditional Use

Garden cress has long been used to treat a variety of illnesses, including cold infusion of seed, high cough, chronic enlargement of the spleen and liver, flatulence, diarrhea, dysentery, indigestion, rheumatic pain, inflammation, viscous humors, tenesmus, secondary syphilis, abortion, anemia, and weakness. Poultices are mostly used for skin ailments, leprosy, sprains, and dysentery. Additionally, it is advised to use the seeds as a diuretic, hemogenic, galactagogue, aphrodisiac, tonic, and depurative. Additionally, the seeds are said to provide benefits such as boosting appetite and preventing hair loss.

Garden cress is known to alleviate the body's allergic reactions to insect stings. It has fiery, bitter, tonic, and aphrodisiac qualities, according to the Ayurvedic medical system (Table 19.4).

19.6.2 Other Use

Garden cress has recently gained attention as a functional food. People take it as a dietary supplement because of its great nutritional value. Noodles, biscuits, health drinks, flakes, cereal drinks, ladoo, snacks, burfi, instant "UPMA" mix, and nutria cookies, among other products made from seeds, are sold in some countries. Because of its acidic flavor and nutritional benefits, garden cress leaves are utilized in salad preparation. Mucilage from seeds is utilized as an excipient in pharmaceuticals. Another usage for garden cress seed cake is as a bed for water purification. Additionally, seeds are utilized as soil purifiers to remove heavy metals from the soil because of their ability to absorb them (Latif and Licek 2004).

Table 19.4 Distinct parts of garden cress are used for different purposes

Plant parts	Used for	References
Plant	Treatment for menstrual disorders, asthma, diarrhea, and bleeding piles Horse and camel feed that lowers blood glucose levels	Mali et al. (2007) Jagdale et al. (2021)
Seeds	Used to treat pleurisy, dropsy, asthma, abdominal colic, sexual debility, throat illness, headaches, diarrhea, and quick bone fracture healing. They are mostly used as poultices for sprains, leprosy, dysentery, and other skin conditions. Renal disorders and hypertension. Roasted seeds in ghee and mixed with sugar are given as a tonic for general weakness in young girls and after childbirth to increase the breast milk	Mali et al. (2007) Dixit Jr. et al. (2020)
Roots	Role in secondary syphilis and tenesmus Syphilis the powder is to be taken with Luke cow's milk.	Mali et al. (2007)
Leaves	Diuretics are taken three times a day after being boiled with water and decocted. Lodhas uses a 2:1 (w/w) mixture of leaf pastes and black pepper paste to relieve boils. Using fresh leaf paste to treat a urinary tract infection.	Falana et al. (2014)
Herb	The plant is crushed and mixed with water to make expectorant, which is taken twice daily. For asthma, the whole herb paste is taken every 4 h for both expectorant and cough.	Falana et al. (2014)

19.7 Application of Garden Cress Seeds in Various Industrial Sectors

19.7.1 Uses in the Food Industry

In order to decrease the amount of oil that fresh-cut potato strips absorb and increase their shelf life, garden cress mucilage may be used as an edible coating (Ali et al. 2021). With or without ascorbic acid (AA), the mucilage of garden cress seed extract *(Lepidium sativum)* was converted into an edible coating. Examining the color, texture, and weight loss of potato strips coated with MSE solutions with or without AA, it was discovered that the coatings effectively delayed browning, decreased weight loss, and preserved texture during cold storage. Seed mucilages are used in place of tragacanth and Arabica gum (Al-Snafi 2019).

The leaves of 74 stimulated garden cress are used as a salad garnish, utilized in the creation of items made from blended garden cress 23 seeds, including soups and faloodas (Shabbir et al. 2018).

19.7.2 Application in Nonfood Industry

19.7.2.1 Uses in Aroma Therapy

Seeds are recommended in rheumatism, inflammation, and muscular pain as they are an excellent source of fatty acids and vitamins (Nasef and Khateib 2021). It is also used by soldiers for warmth and stomach discomfort at night. It also heals skin disorders, sunburns, and amoebic infections. Seed oil is also used to treat dysentery and diarrhea. Gout, rheumatism, glandular swelling, and other conditions are treated with the analgesic medicinal oil made from roasted seeds mixed with sesame oil.

19.7.2.2 Uses in the Cosmetic Industry

Due to their slow accumulation in the brain, liver, heart, and subcutaneous tissue, Surfactants have been shown to have long-term effects on the human body. However, these compounds have a broad impact on the human body, aside from the skin. The situation is further aggravated by the inability of the majority of treatment plants in our country to eliminate surfactants qualitatively, resulting in their gradual deposition in the environment (Tkachuk and Okulovych 2021).

Garden cress seed oil has tocopherol (a natural antioxidant), carotenoid, oleic acid, and alpha-linolenic acid, and it has the ability to neutralize many types of radicals; therefore utilized in the cosmetic industry (El-Saadany et al. 2022). Garden cress is used in bio testing as it is a well-known sensitive test plant that is used to determine the toxicity of various substrates present in cosmetics.

19.7.3 Uses in Pharmaceutical Industry

19.7.3.1 Use as Film Forming Agent

The effects of varying glycerol concentrations (25, 35, and 50 percent w/w) on the synthesis of edible films from cress seed gum were investigated. As the amount of glycerol in the film formulation grew, the edible films' water vapor permeability increased as well, leading to increased flexibility, significantly lower film tensile strength, and higher elongation at break (Fahami and Fathi 2018).

19.7.3.2 Use as an Encapsulating Agent

Water is rapidly absorbed by cress seeds, producing sticky mucilage. These seeds have a mucilaginous composition of 6.5–15%. The hydrocolloids that make up this mucilage have a variety of uses in the food industry. The two sugars that are present in cress seed mucilage are mannose (38.9%) and arabinose (19.4%). Vitamin A's heat stability was significantly improved by using cress seed mucilage as the encapsulating wall material. Additionally, the mucilage was used to encapsulate curcumin via electrostatic interactions with sodium caseinate as a structural material (Moniri et al. 2020; Fahami and Fathi 2018).

19.7.3.3 Utilization as Natural Superdisintegrants

In the pharmaceutical industry, it is commonly used as a disintegrating agent and as herbal medicine. In addition to the dimeric imidazole alkaloids lepidine B, C, D, E, and F, seeds also contain two new monomeric imidazole alkaloids, semilepidinosides A and B, and enhanced mucilage. There are various methods for removing mucilage from seeds (Mali et al. 2022; Nerkar and Gattani 2012).

19.7.3.4 Use as a Gelling Agent

The mucilage of cress seeds is gel-like. This study examined the rheological and mechanical characteristics (adhesiveness, gel strength) of cress seed mucilage at various minimum Carbopol concentrations. It has additive effects as well. Cress seed mucilage and carbopol together resulted in a longer and more efficient buccal route of venlafaxine administration (Mali et al. 2022; Nerkar and Gattani 2012).

19.7.4 Uses in Other Industries

19.7.4.1 Biomonitoring of Pollution

For conducting biomonitoring of pollutants in urban soil on a large scale, garden cress is used as a test plant due to its simplicity, sensitivity, and low cost. Peroxidase is an essential antioxidant enzyme that increases when plants are polluted. It was recently discovered that during garden cress germination, elevated peroxidase activity was strongly associated with high phenolic content and antioxidant activity. In the lignification process, anionic peroxidase (GCP2) is essential for eliminating phenol and p-chlorophenol from contaminated soil and wastewater and fending against the harmful effects of heavy metals (Abdel-Aty et al. 2021).

19.7.4.2 Use of Garden Cress Seed Oil in Improving Productive Traits of Laying Hens

Phenolics, polyphenols, flavonoids, antioxidants, and PUFA are abundant in GC seeds. These active compounds stimulate growth genes, metabolic processes, white blood cell proliferation, cytokine synthesis, and improve final product quality by regulating cholesterol content and boosting PUFA in the bird's body. Supplementation with GC seed oil in the layer diet has a positive influence on the feed conversion ratio, egg mass, and egg production of the birds. Thus, garden cress seeds are essential economic plant seeds that have recently captured global attention due to their nutritional and therapeutic attributes, leading to novel sources of feed additives that can supplement diverse poultry diets (El-Saadany et al. 2022).

19.8 Description of Garden Cress Plant

As a member of the Brassicaceae/cruciferae family, garden cress (Lepidium sativum) is an edible herbaceous plant that grows quickly each year. As such, it shares many characteristics with water cress and mustard. This allogamous plant has

Table 19.5 Description of garden cress plant

Characteristics	
Plant height	47–50 cm
Leaves	Lobed shape with linear segments
Flowers	Small, white to pinkish in color
Pods	Obovate,5 mm length
Seeds	Reddish-brown, Oval Peppery, pungent taste with hot mouth feel

Vaishnavi et al. (2020)

Fig. 19.1 Lobed-shaped piquant leaves of garden cress

Fig. 19.2 Raceme inflorescence of garden cress plant

different levels of tolerance to extended autogamy, including self-compatible and self-incompatible forms. According to Wadhwa et al. (2012), there are diploid forms from 2n = 16 and tetraploid forms from 2n = 32. The tall, glabrous plants that developed reached an average height of 50 cm. The plant's leaves featured linear segments and a lobed form. At the top of the stem were tiny green leaves that resembled feathers, while at the base of the stem were long-shaped leaves (Table 19.5).

The flowers were tiny, white to pinkish, and formed a densely packed inflorescence with numerous branches on top (Figs. 19.1 and 19.2). The siliquae (pod) are glabrous, elliptical elates from the upper half, and 5–6 mm long. The seeds were enclosed in tiny, oblong plant pods (Wadhwa et al. 2012). The seeds had an oval form, were reddish-brown in color, and had a spicy, peppery flavor (Fig. 19.3). As

Fig. 19.3 Dry pod and seeds of garden cress plant

can be seen from their measurements (L-2.60, W-1.20, T-0.94 mm) and thousand seed weight (1.86 g), the seeds were incredibly tiny in both size and shape. The seeds' real density was 1230 kg/m3, and their porosity was 40.67% (Vaishnavi et al. 2020). The seeds are made up of 28% bran and 72% endosperm. Water-soaked seed mucilage (Yadav et al. 2013).

19.9 Varieties of Garden Cress

Varieties available for cultivation by different institutes in India are presented in Table 19.6.

19.10 Soil and Climate

Garden cress may grow in any kind of soil. However, it grows best in well-draining soil that has a pH of 6.0–8.5 (Muralia and Pathak 2007). Because they promote healthy root development and avoid water logging, sandy loam or loamy soils are recommended (Anonymous 2024). Because it can be grown in any climate and has evolved to withstand slightly acidic soil, it doesn't require a lot of agricultural resources (Lahiri and Rani 2020).

As a cool-season crop, garden cress thrives in temperatures that are mild enough for germination and growth. In total, 15–20 °C is the optimal temperature range for germination. The ideal temperature range for growth is between 10 and 20 °C. Garden cress may withstand slightly higher temperatures after it is established, but extended exposure to intense heat can cause bolting, or early flowering, and lower quality. It is prone to freezing. Although it can withstand some shade, garden cress prefers full sun. For tasty leaves to form and flourish robustly, there must be enough sunlight. For best effects, get at least 4–6 h of direct sunlight (Anonymous 2024). It can be grown in areas with 70–125 cm of yearly rainfall (Geda 2009).

It thrives in low-fertility soils, semiarid areas, and temperate climates with 300–2000 mm of annual precipitation (Singh et al. 2022).

Table 19.6 Varieties recommended and release for cultivation

S. No.	Name of variety	Year of release	Centre from Release	Recommended for state	Varietal character
1.	HLS-4	2016	Chaudhary Charan Singh Haryana Agricultural University, Hisar	All state (National)	Early maturing 116 days Resistant to diseases and insect-pests Seed yield: 12–14 q/ha Oil yeild: 306.7 kg/ha
2.	Gujarat Asalio-1 (GA-1)	1998	Anand Agricultural University, Anand, Gujarat	All state	Early maturing variety 117 days Seed yield 15q/ha Excellent for *rabi* season under water logging condition
3.	RVA-1007		RVSKVV, KNS College of Horticulture, Mandsaur, Madhya Pradesh	Madhya Pradesh, Rajasthan, Gujarat, Haryana	Improved Varieties grown by farmers.
4.	Raj Vijay-1007				
5.	RVA-1016				
6.	RVA-1001				
7.	ALS-1	Anand Agricultural University, Anand, Gujarat			
8.	SLS-1	Dr. Yashwant Singh Parmar *University* of Horticulture and Forestry, *Solan, Himachal Pradesh*			

19.11 Field Preparation

The field should be prepared by ploughing two times with cultivator. After ploughing, do the leveling of land.

19.12 Seed Rate, Time, and Method of Sowing

The seed rate is 8–10 kg per hectare, and seeds are treated with 2.5 g thiram per kg of seeds, prior to sowing. Chandrasur is sown as *Rabi* crop. In unirrigated area seeds are sown in month of September-October while in irrigated area, seeds are sown in month of October-November. Seeds are sown either by broadcasting or line sowing with 20–25 cm row distant. The seeds are sown at shallow depth 1–1.5 cm, if seeds are sown at greater depth may affect germination (Muralia and Pathak 2007; Geda 2009). Seeds generally germinate within 5–15 days.

19.13 Manures and Fertilizers

During field preparation, 10–12 tons of FYM per hectare are added to the garden cress crop. 50:50:30 kg of NPK fertilizer is applied each hectare. At the time of seeding, half nitrogen, full phosphorus, and potash are utilized. After 30–35 days of seeding, the remaining half nitrogen is used as a top dressing (Geda 2009; Muralia and Pathak 2007). The recommended fertilizer dose for Rajasthan is 80:60:40 kg/ha of N, P, and K. While the full dose of P and K and the half dose of N were applied as a basal treatment, the remaining N dose was applied with the initial irrigation (Singh et al. 2021).

19.14 Irrigation

In Garden cress crop, 2–3 irrigations are required for better yield. It is very important to have sufficient moisture in the soil at the time of seed setting. If there is sufficient moisture in field at the time of sowing, there is no need of irrigation. If not enough moisture in soil at sowing time, then light irrigation is given after sowing for quick and better germination of seeds. The next at 30 days of sowing and the last during the pod stage when the seeds are milky (Kumar et al. 2024; Muralia and Pathak 2007).

19.15 Intercultural Operations

All medicinal plants should have a weed-free habitat for their young plants. This can be achieved by efficiently managing the first flush of weeds by reducing the frequency of weeding and hoeing cycles (NMPB 2009). In chandrasur, two hands weeding at 20–25 days interval are required initial stage of crop growth. Study suggested that fenoxaprop-p-butyl 100 g/ha post emergence (PoE) and quizalofop-pethyl 50 g/ha PoE applied 30 days after planting could be utilized to control weeds in Chandrashur when labor is limited (Singh et al. 2021). Additionally, he stated that weed-free, twice-hand weeding produced a better seed production and was substantially more cost-effective than the other weed management techniques.

19.16 Plant Protection

The crop is generally less affected by insect-pest and diseases. Sometimes crops are affected by aphid, Alternaria Leaf blight, and powdery mildew.

19.16.1 Aphid

They feed on the sap of the host plants, causing the infected plants to eventually grow deformed and stunted. Plant wilting, yellowing, and stunting are some of the main signs of infestation.

19.16.2 Powdery Mildew

Disease caused by fungi that produce a white powdery growth on leaves, stems, flowers, and pods. For controlling this disease, spraying of sulfur dust @ 0.25% (Kumar et al. 2024).

19.16.3 Alternaria Leaf Blight

Alternaria leaf blight of chandrasur is a destructive disease that damages crops both qualitatively and quantitatively. Tebuconazole + trifloxystrobin (Nativo) and tebuconazole, two systemic fungicides, were shown to be the most effective in controlling this illness. Alternaria leaf blight of chandrasur was successfully managed with neem leaf extract (5.00%) (Meena et al. 2020).

19.17 Harvesting, Threshing, and Storage

In Chandrasur, crops begin to bloom 2 months after they are sown. It is fast-growing crop that becomes mature in about 90 days (Vaishnavi et al. 2020). In total, 90–110 days after sowing, when the leaves become yellow, the pods start drying and the seeds become red, then the crop is ready for harvest. At this time, the crop should be uprooted by hand or cut with sickles. After cutting, it is dried for 2–3 days, and then the seeds are taken out by beating with sticks or the seeds can be separated by rotating the tractor or using a thresher. After extracting the seeds, they should be cleaned and stored in sacks.

19.18 Yield

The average yield of Chadrasur can be obtained 10–12 quintals per hectare in unirrigated areas and 14–16 quintals per hectare in irrigated areas. The market price of its seeds generally ranges from Rs. 7000 to 8000 per kg (Kumar et al. 2024).

19.19 Value-Added Products of Garden Cress Seeds

Garden cress seeds are used to enhance the nutritional and functional qualities of a variety of beverages and foods. The following discusses a few of the products:

19.19.1 Dahiwala Bread

It is a fairly simple breakfast meal to prepare by combining a slice of bread with curd. Following mixing, the bread slice is fried in soybean oil to prevent saturated fat. Garden cress seeds are added to strengthen it. More protein, fat, calcium, iron, and phosphorus are included in this product's formulation. Because of its numerous therapeutic qualities, this substance may be good for nutrition and serve as a therapeutic agent (Agarwal and Sharma 2013).

19.19.2 Brioche Bread

Garden cress seeds were used in the product's development, and the grinder seeds were ground into a powder. Wheat flour was substituted with powdered garden cress seed. The sample that contains the most garden cress seeds has the highest fiber content. The amount of GCS powder used to the brioche's recipe closely correlated with a decrease in carbohydrates. By substituting 1–3% GCS powder for wheat flour in the brioche recipe, the best overall acceptance of the finished brioche bread can be achieved (Dewidar and El-Kherbawy 2019).

19.19.3 Omega-3-Fatty Acid-Rich Biscuits

These are made by incorporating 20 g/100 g of microencapsulated garden cress seed oil with 1.02 g/100 g of alanine (Ala). The biscuits were packaged in 1.05 g/100 g metalized PET film (Umesha et al. 2015).

19.19.4 Health Drinks

Garden Cress seeds can be used to make health beverages with milk since they are high in proteins, carbs, and several important minerals, including calcium, iron, and phosphorus. They also contain 7.6% crude dietary fiber. A milk-based health beverage contains 3% processed Garden Cress seeds, 1% fat, and 5% sugar (w/v) (Singh and Pashwan 2017). It mostly meets the needs of infants, elderly people, growing children, and some convalescent patients. It delivers all the necessary elements that these people require in addition to improving the nutritious qualities of milk. This kind of beverage promotes the growth of lean muscles in people who regularly exercise and helps prevent vitamin deficiencies.

19.19.5 Vegetable Oil Blends with α-Linolenic Acid-Rich Garden Cress Oil

The ratio of n-6 to n-3 PUFA has changed from the advised 10:1 or 2:1 to 50:1 as a result of eating vegetable oils high in n-6 PUFA, such as sunflower, corn, safflower, and soybean oil. Umesha and Naidu (2012) made a blended vegetable oil by mixing vegetable oil with garden cress seed oil and examined how it affected lipid metabolism. They tested with different ratios of garden cress seed oil to rice bran oil, sunflower oil, and sesame oil in order to produce vegetable oils with a PUFA ratio of 2.3–2.6 and look into their modulatory influence on lipid metabolism (Umesha and Naidu 2012; Gerd and Josef 2008).

19.19.6 Fortified Burfi

Burfi is made by roasting Bengal gram flour with ghee over a low heat, then adding sugar syrup and 15% garden cress seeds. The mixture is then spread out and, once it has settled, cut into squares with a knife. It has been shown to be both delicious and incredibly nutritious (Jain et al. 2016).

In 2021, Khomane et al. looked at how to make burfi with garden cress seeds. In order to prepare the burfi, 1% of powdered garden cress seeds and 8% of sugar were added to milk (Khoa). To get the right consistency, the ingredients were well combined and heated gently for 5–8 min. After being removed from the flame and placed on a tray, the product was left to cool and solidify at room temperature under sanitary circumstances until it became just a little bit firm. He discovered that, out of the six versions, the Burfi produced with garden cress seeds (microwave-processed) powder received the highest ratings for color, look, texture, flavor, and general acceptability.

19.19.7 Fortified Chikki

To make syrup, broken jaggery is combined with water and heated to 145 °C. After the heat is extinguished, powdered garden cress seeds and toasted peanuts are added. After the mixture had settled, it was placed onto an oiled surface, cut into squares, and let to cool. Because garden cress seeds are dark brown, chikki fortified with them has a brownish hue, and the crunchiness of the seeds enhances the chikki's texture (Jain et al. 2016).

The nutritional and organoleptic qualities of the value-added fortified chikki with 15% garden cress seed powder (GCSP) are better than those of the fortified chikki without GCSP. Since GCSP is the least expensive source of iron, it is advised that it be used to prepare common dishes in order to enhance their flavor, acceptability, and nutritional value (Katyal et al. 2024).

19.19.8 Garden Cress Seeds Noodles

Noodles were made using powdered garden cress seeds and semolina. The noodle supplemented with the 15% addition of garden cress seeds had somewhat low moisture content. Protein, fiber, phenolic acids, and a host of other nutritious elements can all be found in garden cress seeds. When 15% garden cress was added to noodles, the protein and fiber concentrations increased to 13.06% and 76.80%, respectively, and the noodles' sensory evaluation showed that they were satisfactory (Hanan et al. 2019).

19.19.9 Garden Cress Seeds Cookies

Mix garden cress seeds, semolina, cardamom powder, oat flour, and Bengal gram flour. In addition to having good sensory qualities, cookies made with GCS enrichment had better nutritional profiles in terms of macronutrients and micronutrients like energy, proteins, lipids, carbs, iron, and calcium. The highest concentration of macronutrients and important micronutrients (calcium and iron) was found in the GCS sample with the highest enrichment level (10.5%) among the various cookie samples. In order to improve the nutritional value of children's, adolescents', and malnourished populations' diets without significantly sacrificing their sensory qualities, garden cress seed-enriched cookies, which are high in energy and nutrients, particularly protein, iron, and calcium, can be chosen as a potential health snack (Yadav et al. 2018).

19.19.10 Processed Spread Cheese

Processed cheese products are frequently used as a component in cheese-based dishes such as lasagna, pizza, hamburgers, and sandwiches. In one investigation, processed spread cheese was made with cress seeds instead of salt as an emulsifying agent. Cress seed powder was prepared and added in six different ratios (0.05%, 1.5%, 2.5%, 3.5%, 4.5%, and 5.5%) to processed spread cheese instead of the control (3% commercial emulsifying salt). According to the findings, 3.5% cress seeds significantly improved health, especially when it came to lowering glucose ratios (Punia and Dhull 2019).

19.19.11 Garden Cress Seed Ladoo

A 50 g ladoo with 5 g of garden cress seeds, 20 g of rice flakes, 5 g of bajra, 5 g of roasted Bengal gram dhal, 5 g of samai, and 10 g of jaggery. There were 10 mg of iron in the ladoo. Improvements in the hemoglobin level and clinical signs and symptoms of the chosen moderately anemic teenage females have been observed when garden cress seeds are supplemented with ladoo. This highlights the importance of raising knowledge about the use of wholesome foods that are readily available locally. The supplement can be used to treat anemia since the combination of these chemicals has therapeutic effects that improve hemoglobin levels. Therefore, it may be concluded that these locally accessible, iron-rich foods can help prevent and control anemia (Angel and Devi 2015).

19.19.12 Instant UPMA Mix

The final product was an immediate UPMA blend of garden cress and foxtail millet seeds. This might be the new product development of garden cress and foxtail millet seeds, which have been shown to help people with celiac disease and diabetes. Moisture content ranges from 6.15% to 7.62%, protein ranges from 11.30% to 13.84%, fat ranges from 7.30% to 16.80%, carbohydrate ranges from 55.16% to 64.07%, ash ranges from 2.89% to 4.43%, fiber ranges from 3.90% to 4.31%, and energy values range from 371.06 to 427.50 Kcal/100 g. In comparison to the other four formulations, 10% garden cress semolina received the highest scores for all sensory aspects, including appearance and color (7.8), taste (7.9), texture (7.8), flavor (7.7), and overall acceptability (7.8) (Rodge et al. 2018).

19.19.13 Cereal Beverage

The processing technology of standardization is used for the preparation of cereal milk, which is fortified with garden cress seeds as well as pumpkin seed powder. According to numerous research, adding pumpkin seeds results in the highest fat content (27.83%) while adding barley results in the lowest fat content (2.6%). It has been discovered that cereal grains have a higher carbohydrate content than garden cress and pumpkin seeds. Additionally, it has been reported that LAB starter cultures containing Lactobacillus acidophilus are used to ferment these cereals with lactic acid. The mixture, which contained 70% barley, 30% sorghum, 3% Lactobacillus acidophilus, and a starter culture of the Bifid bacteria bifidum, had the highest overall acceptance score of 8.8 (Shendge and Patharkar 2020).

19.19.14 Millet Garden Cress Seed (MGC) Flakes

Iron-enriched little millet flakes were made by blending 5% garden cress seed with small millet. The shelf life of iron-enriched flakes was 5 months.

19.20 Germplasm Collection and Conservation

A total of 88 germplasm samples have been collected from different states of the country so far. Most of them were collected from Madhya Pradesh (40), Himachal Pradesh (14), and Rajasthan (10). Samples from other States like Delhi, Gujarat, Haryana, Jammu & Kashmir, Karnataka, Manipur, Sikkim, Tripura, Uttar Pradesh, Uttarakhand have been preserved in the National Gene Bank of the National Bureau of Plant Genetic Resources, Indian Council of Agricultural Research, Delhi. Among its other wild species *Lepidium apetalum* (3) and *Lepidium latipfolium* (27) specimens are also preserved in the gene bank (Kumar et al. 2024).

At present, due to the changing climate, it has become necessary to look for plants with more nutritive and medicinal properties so that high-yielding varieties of garden cress crops can be developed, and their use should also be promoted. This is an important resource that needs to be protected. At present, given the medicinal properties of this plant, there is a necessary to collect and conserve its germplasms from various parts of the country in larger quantities so that they can be used in different ways in future.

19.21 Future Potential

Garden Cress is a new, amazing source of a variety of important products. Utilizing these neglected crops for medicinal purposes as well as for the creation of new products can open up new markets and opportunities for the food sector. The seeds of garden cress have a high concentration of phenolic compounds and are an excellent source of fatty acids, minerals, and amino acids. The seeds can be used to make a variety of value-added products because of their high protein, fiber, carbohydrate content, and other minerals. Therefore, the garden cress plant, seeds, and oil present have a great deal of potential for future inventions in terms of their possible preventive benefits toward chronic diseases and in the development of new functional food compositions. Since the seeds are a good source of iron and calcium, patients with iron deficiency anemia, calcium deficiency, and malnutrition, especially women and children, can benefit the most from this product. It should be taken daily to manage nutrition-related issues, and if the government were to make this type of product available to those with lower incomes or those who cannot afford food for their bodies' daily needs, it could be argued that any country would see a significant improvement in population health.

References

Abdel-Aty AM, Salama WH, El-Badry MO, Salah HA, Barakat AZ, Fahmy AS et al (2021) Purification and characterization of peroxidases from garden cress sprouts and their roles in lignification and removal of phenol and p-chlorophenol. J Food Biochem 45(1):e13526

Agarwal N, Sharma S (2013) Garden cress (*Lepidium sativum L.*) a non-conventional traditional plant item for food product. Indian J Tradit Knowl 12(4):699–706

Ali MR, Parmar A, Niedbała G, Wojciechowski T, Abou El-Yazied A, El-Gawad HGA et al (2021) Improved shelf-life and consumer acceptance of fresh-cut and fried potato strips by an edible coating of garden cress seed mucilage. Foods 10(7):1536

Al-Snafi AE (2019) Chemical constituents and pharmacological effects of *Lepidium sativum*. Int J Curr Pharm Res 11(6):1–10

Angel M, Devi KV (2015) Therapeutic impact of garden cress seeds incorporated ladoo among the selected anaemic adolescent girls (12–15 years). J Drug Discov Develop Deliv 3:18–22

Angelini LG, Moscheni E, Colonna G, Belloni P, Bonari E (1997) Variation in agronomic characteristics and seed oil composition of new oil seed crops in central Italy. Ind Crop Prod 6(34):313–323

Anonymous (2024). https://www.agrownet.com/contents/en-us/d340668_Garden-cress_Climate_requirements.html

Balasubramanian (2009) Nutritive value of Indian food. National Institute of Nutrition, ICMR, Hyderabad

Barba FJ, Nikmaram N, Roohinejad S, Khelfa A, Zhu Z, Koubaa M (2016) Bioavailability of glucosinolates and their breakdown products: impact of processing. Front Nutr 16(3):24

Bary EA, Fekri A, Soliman YA, Harmal AN (2017) Novel superabsorbent membranes made of PVA and Ziziphus spina-christi cellulose for agricultural and horticultural applications. New J Chem 41(18):9688–9700

Bhatia K, Bhasin A (2023) Exploring the versatile potential of garden cress seeds: therapeutic applications and industrial utilization: a comprehensive review. Pharma Innov J 12(9):101–108. https://doi.org/10.22271/tpi.2023.v12.i9b.22647

Bin Abdullah Juma AB (2007) The effects of *Lepidium sativum* seeds on fracture-induced healing in rabbits. Medscape Gen Med 9(2):23

Chaudhary P, Gupta R (2021) Impact assessment of dietary and nutrition education intervention on hemoglobin level moderately anemic adolescent girls. Indian J Ext Educ 57(2):145–149

Choudhari S, Keshwa GL, Yadav LR (2010) Effect of sowing dates, row spacing and nitrogen levels on productivity, quality and economics of garden cress (*Lepidium sativum* L.). Indian J Agric Sci 80(8):754–761

Conforti F, Ioele G, Statti GA, Marrelli M, Ragno G, Menichini F (2008) Antiproliferative activity against human tumor cell lines and toxicity test on Mediterranean dietary plants. Food Chem Toxicol 46(10):3325–3332

Dewidar MO, El-Kherbawy MG (2019) Nutritional evaluvation of Brioche bread made from Egyptian wheat and enriched with garden cress seeds (GCS) powder as a functional food. J Food Sci Suez Canal Univ 6(1):27–41. https://doi.org/10.21608/scuj.2019.60153

Diwakar BT, Dutta PK, Lokesh BR, Naidu KA (2008) Bio-availability and metabolism of n-3 fatty acid rich garden cress (*Lepidium sativum*) seed oil in albino rats. Prostaglandins Leukot Essent Fat Acids 78(2):123–125

Diwakar BT, Dutta PK, Lokesh BR, Naidu KA (2010) Physicochemical properties of garden cress (*Lepidium sativum* L.) seed oil. J Am Oil Chem Soc 87(5):539–548

Dixit V Jr, Kumar I, Palandurkar K, Giri R, Giri K (2020) *Lepidium sativum*: bone healer in traditional medicine, an experimental validation study in rats. J Family Med Prim Care 9(2):812–818. https://doi.org/10.4103/jfmpc.jfmpc_761_19

Doghmane A, Aouacheri O, Laouaichia R, Saka S (2021) The investigation of the efficacy ratio of cress seeds supplementation to moderate hyperglycemia and hepatotoxicity in streptozotocin-induced diabetic rats. J Diabetes Metab Disord 20:447–459

Eddouks M (2004) Nutritional, elemental analysis and antioxidant activity of garden cress seeds. Int J Pharm Sci 4:390–395

Eddouks M, Maghrani M, Lemhadri A, Ouahidi ML, Jouad H (2002) Ethno pharmacological survey of medicinal plants used for the treatment of diabetes mellitus, hypertension and cardiac diseases in the south-east region of Morocco (Tafilalet). J Ethnopharmacol 82(2–3):97–103

Eddouks M, Maghrani M, Zeggwagh NA et al (2005) Study of the hypoglycaemic activity of Lepidium sativum L. aqueous extract in normal and diabetic rats. J Ethnopharmacol 97(2):391–395

El-Saadany AS, El-Barbary AM, Shreif EY, Elkomy A, Khalifah AM, El-Sabrout K (2022) Pumpkin and garden cress seed oils as feed additives to improve the physiological and productive traits of laying hens. Ital J Anim Sci 21(1):1047–1057

El-Salam A, Kholoud H, Toliba AO, El-Shourbagy GA, El-Nemr SE (2019) Chemical and functional properties of garden cress (*Lepidium sativum* L.) seeds powder. Zagazig J Agric Res 46(5):1517–1528

Fahami A, Fathi M (2018) Development of cress seed mucilage/PVA nanofibers as a novel carrier for vitamin A delivery. Food Hydrocoll 81:31–38

Falana H, Nofal W, Nakhleh H (2014) A review article *Lepidium sativum* (Garden cress). Pharm-D Program, College of Nursing, Pharmacy and Health Professions, Birzeit University, pp 1–8

Geda AK (2009) Guide to cultivation, processing and marketing of medicinal and aromatic plants. Published by Department of Medicinal and Aromatic Plants, Indira Gandhi Krishi VishwaVidyalaya, Raipur, Chhattisgarh, pp 26–27

Gerd S, Josef E (2008) The opposing effects of n-3 and n-6 fatty acids. Prog Lipid Res 47:147–155

Ghante MH, Badole SL, Bodhankar SL (2011) Health benefits of garden cress (*Lepidium sativum* Linn.) seed extracts. Chapter 62. In: Preedy VR, Watson RR, Vinood B, Patel VB (eds) Nuts and seeds in health and disease prevention. Academic Press. NMPB. 2015. Central sector scheme on conservation, development and sustainable management of medicinal plants. National Medicinal Plants Board (NMBP), Department of AYUSH, Ministry of Health and Family Welfare, Govt. of India, New Delhi, pp 521–525

Gokavi S, Malleshi G, Mingruo G (2004) Chemical composition of garden cress (*Lepidium sativum*) seeds and its fractions and use of bran as a functional ingredient. Plant Foods Hum Nutr 59(3):105–111

Hanan MA, Nahla SZ, Abdelaleem MA (2019) Utilization of garden cress seeds (*Lepidium sativum* L.) as natural source of protein and dietary fiber in noodles. Int J Pharm Res Allied Sci 8(3):17–28

Ibrahim IA, Shalaby AA, Abdallah HML, El-Zohairy NF, Bahr HI (2020) Ameliorative effect of garden cress (*Lepidium sativum* L.) seeds ethanolic extract on high fat diet-prompted non-alcoholic fatty liver disease in the rat model: impact on 3-hydroxy-3-methylglutaryl-coenzyme a reductase and vascular endothelial growth factor. Adv Anim Vet Sci 8(1):1–10

Ibrahim MM, Mounier MM, Bekheet SA (2023) Targeting apoptotic anticancer response with natural glucosinolates from cell suspension culture of *Lepidium sativum*. J Genet Eng Biotechnol 21(1):53

Jabeen A, Rani S, Ibrahim M, Mohammad AS (2017) A review on *Lepidium Sativum*. Indo Am J Pharm Sci 4(8):2223–2227. https://doi.org/10.5281/zenodo.839541

Jagdale YD, Mahale SV, Zohra B, Nayik GA, Dar AH, Khan KA et al (2021) Nutritional profile and potential health benefits of super foods: a review. Sustainability 13(16):9240

Jain T, Grover K, Grewal IS (2016) Development and sensory evaluation of ready to eat supplementary food using garden cress (*Lepidium sativum*) seeds. J Appl Nat Sci 8(3):1501–1506

Kassie F, Rabot S, Uhl M, Huber W, Qin HM, Helma C, Schulte-Hermann R, Knasmüller S (2002) Chemo protective effects of garden cress (*Lepidium sativum*) and its constituents towards 2-amino-3-methyl-imidazo [4, 5-f] quinolone (IQ)-induced genotoxic effects and colonic preneoplastic lesions. Carcinogenesis 23(7):1155–1161

Kassie F, Uhl M, Rabot S, Grasl-Kraupp B, Verkerk R, Kundi M et al (2003) Chemoprevention of 2-amino-3-methylimidazo [4, 5-f] quinolone (IQ)-induced colonic and hepatic preneoplastic lesions in the F344 rat by cruciferous vegetables administered simultaneously with the carcinogen. Carcinogenesis 24(2):255–261

Katyal M, Pandey VK, Tripathi A, Singh K, Rustagi S, Malik S, Kumar R, Pandiselvam R (2024) Development and physico-chemical analysis of value-added Chikki using iron rich garden cress seed and exploring its benefits in treatment of Anaemic patient. Food Chem Adv 5:100763. https://doi.org/10.1016/j.focha.2024.100763

Khan N, Adhami VM, Mukhtar H (2008) Apoptosis by dietary agents for prevention and treatment of cancer. Biochem Pharmacol 76(11):1333–1339

Khomane VB, Kamble DK, Kamble KB, Kubade KB, Shinde SH (2021) Selection of garden cress seeds form for preparation of garden cress seeds *Burfi*. Pharma Innov J 10(1):356–359

Kumar MP, Singh RR, Kumar KP, Anil P, Chand BK (2024) Chandrasur: medicinal plant. Phal-Phool:6–7

Lahiri B, Rani R (2020) Garden cress seeds: chemistry, medicinal properties, application in dairy and food industry: a review. Emergent Life Sci Res 6:1–4

Latif M, Licek E (2004) Toxicity assessment of wastewaters, river waters, and sediments in Austria using cost-effective microbiotests. Environ Toxicol 19(4):302–309

Maghrani M, Zeggwagh NA, Michel JB, Eddouks M (2005) Antihypertensive effect of *Lepidium sativum* L. in spontaneously hypertensive rats. J Ethnopharmacol 100(1–2):193–197

Mahassni SH, Al-Reemi RM (2013) Apoptosis and necrosis of human breast cancer cells by an aqueous extract of garden cress (*Lepidium sativum*) seeds. Saudi J Biol Sci 20(2):131–139

Mali RG, Mahajan SG, Mehta AA (2007) Lepidium Sativum (Garden Cress): a review of contempory literature and medicinal properties. Orient Pharm Exp Med 7(4):331–335

Mali RG, Mahajan SG, Mehta AA (2008) Studies on bronchodilatory effect of *Lepidium sativum* against allergen induced bronchospasm in Guinea pigs. Pharmacogn Mag 4(15):189–192

Mali KD, Patil MA, Vairagde KK, Patil HR, Shevalkar GB (2022) A review on garden cress seeds: mucilage isolation methods, chemical constituents, pharmacological profile, and pharmaceutical application. Pharma Innov J 11(7):639–646

Matthäus B, Angelini LG (2005) Anti-nutritive constituents in oilseed crops from Italy. Ind Crop Prod 21(1):89–99

McGee MM, Hyland E, Campiani G, Ramunno A, Nacci V, Zisterer DM (2002) Caspase-3 is not essential for DNA fragmentation in MCF-7 cells during apoptosis induced by the pyrrolo-1, 5-benzoxazepine, PBOX-6. FEBS Lett 515(1–3):66–70

Meena RP, Saran PL, Kalariya KA, Manivel P (2020) Efficacy of fungicides and plant extracts against Alternaria alternata causing leaf blight of chandrasur (Lepidium sativum). Indian J Agric Sci 90(2):337–340. https://doi.org/10.56093/ijas.v90i2.99017

Mohamed HS, Kholief T, Mohamed RW, Abd El-Rhman A (2023) The modulatory effects of black chia (Salvia hispanica) and garden cress (*Lepidium sativum*) seeds on Nε-CML formation in streptozotocin-injected rats. J Herbmed Pharmacol 12(2):250–261

Moniri H, Farahmandfar R, Motamedzadegan A (2020) Cress seed (*Lepidium sativum*) gum dried by vacuum, freeze, and microwave drying methods: structural, rheological, emulsifying, and foaming properties. J Food Process Eng 43:7. https://doi.org/10.1111/jfpe.13408

Mulla MZ, Ahmed J, Baskaya-Sezer D, Al-Ruwaih N (2019) Effect of high-pressure treatment and cellulase-mediate hydrolysis on functional, rheological and microstructural properties of garden cress seed residual fibre. Int J Food Sci Technol:15464. https://doi.org/10.1111/ijfs.15464

Muralia S, Pathak AK (2007) Medicinal and aromatic plants cultivation and uses. Avishkar Publishers, Distributors, Jaipur, pp 175–176

Nasef AN, Khateib MBR (2021) Study the potential therapeutic effect of garden cress (*Lepidium sativum*) on nephropathy diabetic rats: biological and biochemical studies. Alex Sci Exch J 42(2):263–272

Nerkar PP, Gattani SG (2012) Cress seed mucilage based buccal muco adhesive gel of venlafaxine: in vivo, in vitro evaluation. J Mater Sci Mater Med 23:771–779. https://doi.org/10.1007/s10856-011-4529-7

NMBP (2009) Good agricultural practices for medicinal plants. National Medicinal Plants Board (NMBP), Department of AYUSH, Ministry of Health and Family Welfare, Govt. of India, New Delhi

Paranjape AN, Mehta AA (2006) A study on clinical efficacy of *Lepidium sativum* seeds in treatment of bronchial asthma. Iran J Pharmacol Ther 5(1):55–60

Patil D, Lal A, Nandkule V (2015) Development and quality evaluation of garden cress seed biscuits. Int J Sci Eng Technol 3:770–774

Pattnaik AK (2003) Effect of herbal additives on lactating ruminants with or without subclinical mastitis. Doctoral dissertation, Thesis submitted to Orissa University of Agriculture and Technology, Bhubaneswar, Orissa, India

Prajapati MR, Dave PH (2018) Therapeutic and nutritional importance of Garden cress seeds. J Pharmacogn Phytochem 7(5):140–143

Punia S, Dhull SB (2019) Chia seed (*Salvia hispanica* L.) mucilage (a heteropolysaccharide): functional, thermal, rheological behaviour and its utilization. Int J Biol Macromol 140:1084–1090. https://doi.org/10.1016/j.ijbiomac.2019.08.205

Radwan HM, El-Missiry MM, Al-Said WM, Ismail AS, Shafeek A, Seif-El-Nasr MM (2007) Investigation of the glucosinolates of Lepidium sativum growing in Egypt and their biologicalactivity. Res J Med Med Sci 2:127–132

Rajasekaran R, Suresh PK (2021) Physical and chemical methods of extraction of bioactive molecules from *Lepidium sativum* Linn. and antioxidant activity-based screening and selection of extracts-probable phytochemical, chromatography and mass spectroscopy analysis-based correlates. Res J Pharm Technol 14(6):3082–3092

Rodge SM, Bornare DT, Babar KP (2018) Formulation and quality evaluation of instant UPMA mix of foxtail millet and garden cress seed. Int J Chem Stud 6(3):1854–1857

Shabbir F, Eddouks M, Nadeem F, Azeem MW (2018) A brief review on bioactivities and therapeutic potentials of garden cress (*Lepidium sativum* L.). Int J Chem Biochem Sci 13:36–45

Sharma A (2020) A comprehensive review on pharmacological properties of garden cress (*Lepidium sativum*) seeds. Curr Res Pharm Sci 10(2):13–18. https://doi.org/10.24092/CRPS.2020.100201

Sharma S, Agarwal N (2011) Nourishing and healing prowess of garden cress (*Lepidium sativum* Linn). Indian J Nat Prod Res 2:292–297. http://nopr.niscpr.res.in/handle/123456789/12732

Sharma RK, Vyas K, Manda H (2012) Evaluation of antifungal effect on ethanolic extract of *Lepidium sativum* L. seed. Int J Phytopharmacol 3:117–120

Shen X, Si Y, Wang Z, Wang J, Guo Y, Zhang X (2016) Quercetin inhibits the growth of human gastric cancer stem cells by inducing mitochondrial-dependent apoptosis through the inhibition of PI3K/Akt signaling. Int J Mol Med 38(2):619–626

Shendge SN, Patharkar SR (2020) Standardize the processing technology for preparation of cereal milk fortification with garden cress (*Lepidium sativum*) seed and pumpkin (Cucurbita) seed powder. Pharma Innov J 9(1):423–426

Singh CS, Pashwan VH (2017) The Potential of Garden Cress (*Lepidium sativum* L.) seeds for development of functional foods. In: Jimenez-Lopez JC (ed) Advances in seed biology. IntechOpen. https://doi.org/10.5772/intechopen.70355

Singh CS, Paswan VK, Naik B, Reeta (2015) Exploring the potential of fortification by garden cress (*Lepidium sativum* L) seeds for development of functional foods. Indian J Nat Prod Resour 6:167–175

Singh H, Sethi IB, Kumar S, Jajoria M, Jat LK, Braod NK, Muralia S (2021) Weed management in garden cress (*Lepidium sativum* L.) in Rajasthan, India. Indian J Weed Sci 53(3):281–284. https://doi.org/10.5958/0974-8164.2021.00051.4

Singh B, Lubhani MBM, Sri MB, Arshnoor JK, Singh J (2022) *Lepidium sativum*: its nutritional composition and therapeutic properties. Pharma Innov J 11(7):15–25

Tkachuk N, Okulovych I (2021) Toxicity of aqueous solutions of cosmetics in phytotest with *Lepidium sativum* L. Agrobiodivers Improv Nutr Health Life Qual 5(2):348–354. https://doi.org/10.15414/ainhlq.2021.0034

Tukaramrao DB, Dutta PK, Lokesh B, Naidu A (2009) Physicochemical properties of garden cress (Lepidium sativum L.) seed oil. J Am Oil Chem Soc 87(5):539–548

Umesha SS, Naidu KA (2012) Vegetable oil blends with α-linolenic acid rich garden cress oil Modulate lipid metabolism in experimental rats. Food Chem 135:2845–2851

Umesha SS, Manohar RS, Indiramma AR, Akshitha S, Naidu KA (2015) Enrichment of biscuits with microencapsulated omega-3 fatty acid (alpha-linolenic acid) rich gardencress (*Lepidium sativum*) seed oil: Physical, sensory and storage quality characteristics of biscuits. LWT-Food Sci Technol 60:654–661

Vaishnavi, Gupta R, Choudhary P (2020) Botanical description of garden cress (*Lepidium sativum* L.) plant and physical characteristics of its seeds. J Pharmacogn Phytochem 9(5):2424–2428

Wadhwa S, Panwar M, Agrawal A, Saini N, Patidar L (2012) A review on pharmacognostical study of lepidium sativum. Adv Res Pharm Biol 2(4):216–223

William DJ, Critchley C, Pun S, Chaliha M, O'Hare TJ (2009) Different mechanisms of simple nitrile formation on glucosinolate degradation in *Lepidium sativum* and *Nasturtium officinale* seeds. Phytochemistry 70(11–12):1401–1409. https://doi.org/10.1016/j.phytochem.2009.07.035

Yadav R, Choudhary S, Keshwa GL, Sharma OP (2013) Garden cress (Lepidium sativum) growth, productivity and nutrient uptake under different sowing dates, row spacing and nitrogen levels. Indian J Agron 58(1):114–118

Yadav A, Singh P, Sarma U, Bhatt G, Govila VK (2018) Nutritional and sensory attributes of cookies enriched with garden cress seeds. Int J Recent Sci Res 9(12):30146–30149

Zia-Ul-Haq M, Ahmad S, Calani L, Mazzeo T, Del Rio D, Pellegrini N et al (2012) Compositional study and antioxidant potential of Ipomoea hederacea Jacq. and Lepidium sativum L. seed. Molecules 17(9):10306–10321

MIX
Papier aus verantwortungsvollen Quellen
Paper from responsible sources
FSC® C105338

If you have any concerns about our products,
you can contact us on
ProductSafety@springernature.com

In case Publisher is established outside the EU,
the EU authorized representative is:
Springer Nature Customer Service Center GmbH
Europaplatz 3, 69115 Heidelberg, Germany

Printed by Libri Plureos GmbH
in Hamburg, Germany